PHYSICS RESEARCH AND TECHNOLOGY

DENSITY FUNCTIONAL THEORY

PRINCIPLES, APPLICATIONS AND ANALYSIS

Physics Research and Technology

Additional books in this series can be found on Nova's website under the Series tab.

Additional e-books in this series can be found on Nova's website under the e-book tab.

PHYSICS RESEARCH AND TECHNOLOGY

DENSITY FUNCTIONAL THEORY

PRINCIPLES, APPLICATIONS AND ANALYSIS

JOSEPH MORIN
AND
JEAN MARIE PELLETIER
EDITORS

nova publishers
New York

For permission to use material from this book please contact us:
Telephone 631-231-7269; Fax 631-231-8175
Web Site: http://www.novapublishers.com

LIBRARY OF CONGRESS CATALOGING-IN-PUBLICATION DATA

Density functional theory : principles, applications and analysis / [edited by] Joseph Morin and Jean Marie Pelletier.
pages cm
Includes bibliographical references and index.
ISBN 978-1-62417-954-9 (hardcover)
1. Density functionals. I. Morin, Joseph, 1969- II. Pelletier, Jean Marie, 1972-
QD462.6.D45D467 2013
541'.28--dc23

2012051682

Published by Nova Science Publishers, Inc. † New York

CONTENTS

PREFACE

Density Functional Theory (DFT) is a quantum mechanical modelling method, used in physics and chemistry to investigate the electronic structure (principally the ground state) of many-body systems, in particular atoms, molecules, and the condensed phases. This book provides current research in the study of the principles, applications and analysis of Density Functional Theory (DFT). Topics discussed include density functional treatment of interactions and chemical reactions at interfaces; applications of DFT calculations to lithium carbenoids and magnesium carbenoids; thermoelectric properties of low-dimensional materials by DFT; using DFT computations on the radical scavenging activity studies of natural phenolic compounds; polarizability of C60/C70 fullerene [2+1]- and [1+1]-adducts; DFT application to the calculation of properties of di- and trimethylnaphthalenes; transport calculations of organic materials; the evolution of DFT; the capabilities of DFT for materials design of alloys; and the fundamentals of energy density functionality in nuclear physics. (Imprint: Nova)

Chapter 1 reviews recent applications of density functional theory (DFT) based methods in the study of the interaction of small gaseous molecules with metal nanoparticles, metal surfaces, and porous or biological materials and applications in the study of chemical reactions at catalytic sites of transition metals or enzymes. Focus is given to the interaction of small molecules, *e.g.* H_2O, O_2, CO, CO_2, etc., with the scaffold atoms of metal organic frameworks (MOF) or with zeolites, in the field of gas adsorption, or with the exposed atoms on transition metal surfaces or nanoparticles, in the field of heterogeneous catalysis, and to the interaction of small organic molecules with the capacity to inhibit a catalytic cysteine of the malaria's parasite, in the field of drug design. The roles of under-coordinated atoms on the strength of the interaction and of the type of the exchange-correlation functional considered for the calculations are analyzed. Finally, recent successes of the consideration of DFT based approaches to study, with atomic detail, the reactions of such molecules on these materials are also reviewed.

Metal carbenoids, in which a metal (Li, Mg) and an electronegative element (F, Cl, Br, I) are bound to the same carbon atom, are reactive intermediates that react in a manner similar to that of carbenes. The simultaneous existence of an electropositive metal and an electronegative element on the same carbon atom is responsible for the unexpected reactivity. The elusive reactivity of metal carbenoids is counterintuitive and difficult to understand based on the electronic theory of organic chemistry. Although the experimental direct observation of metal carbenoids may provide insight into their reactivity, this observation is complicated by the thermal instability of metal carbenoids. Recently, quantum chemical calculations, including DFT calculations, have become a versatile tool for investigating the structures and reactivity of unstable chemical species that are not readily studied by experimental means. A number of theoretical studies targeting metal carbenoids have been reported with the development of computational chemistry. In Chapter 2, theoretical studies on the structural elucidation and mechanistic exploration of lithium carbenoids and magnesium carbenoids are overviewed.

Growing concerns regarding energy wastes and environmental changes have stimulated consciousness on the necessity to develop new, more friendly, energetic resources. Among the possibilities offered to find solutions, promoting the advances in materials and technologies for energy conversion appears as a promising route.

In thermoelectric modules the use of the so-called "Seebeck effect" allows to convert heat flow that crosses the module into electricity. The efficiency of a thermoelectric material is given by its figure of merit $ZT=(\alpha 2\sigma/\kappa)T$, where α is the Seebeck coefficient, $\sigma$and $\kappa$are the electrical and thermal conductivities, respectively, and T is the temperature. From this equation we see that improving ZT can be achieved either by increasing the powerfactor $\alpha 2\sigma$ or by reducing the thermal conductivity κ or both. However, for three-dimensional materials these physical properties are interrelated in a counter productive way with regards to our objectives. One way to circumvent this difficulty is to lower the dimensionality of the material down to the nanoscale. In this case, the parameters can be tuned, to some extend, independently from one another. However, the elaboration of nanomaterials, in particular nanowires and quantum dots, and the measurement of their properties are still a challenging task for chemists and physicists. Therefore, theoretical and quantum calculations, and numerical simulations are expected to play an important role for the understanding of the behaviour of these materials and for their developments. The density-functional theory (DFT) is a method of choice for calculating thermoelectric properties of low-dimensional materials. When combined with the semi-classical Boltzmann's transport equation, DFT can predict with reasonable accuracy thermoelectric properties. In addition, with recent developments of the density-functional perturbation theory (DFPT) it has been made easier to calculate properties such as phonon spectra and electron-phonon interactions which are important properties in the field of thermoelectricity.

Chapter 3 aims at delivering an updated picture of the recent advances in the theoretical investigation of thermoelectric properties of low-dimensional materials using DFT and related methods.

Theoretical methods have been recognized nowadays as a useful approach in antioxidant activity studies of phenolic compounds (AH). Toward this direction Density Functional Theory (DFT) calculations employing hybrid functionals such as the Becke's 3 and Lee Yang Parr (B3LYP) are most frequently applied solely or in combination with those obtained using less demanding semi-empirical methods. In Chapter 4, applications of the aforementioned computational methods in the field of food chemistry and life sciences regarding structure–antioxidant activity relationships of various classes of phenolic compounds (simple phenols, phenolic acids and related compounds, coumarins, naphthoquinones, xanthones, stilbenes, chalcones and flavonoids) are presented. Focus is mainly given on the computation of suitable molecular descriptors to characterize the radical scavenging activity of AH via hydrogen atom transfer, electron transfer or by sequential proton loss followed by electron transfer.

The computation of spin distribution on radicals formed by AH, when oxidized during the process, as a useful index of antioxidant activity is also acknowledged; whereas the contribution of other than phenolic hydrogens such as allylic ones to the radical scavenging efficiency is documented.

Except for calculations in the gas phase, simulation of real systems is approximated by the utilization of models that take into account implicitly and/or explicitly solvent effect. The potential of the above approaches, shortcomings and some key points for further studies are commented.

Chapter 5 is devoted to the application of DFT methods to calculation of static polarizabilities of fullerene derivatives. The comparison of the calculated and experimental values has been performed for polarizabilities of C_{60}, C_{70} and some fluoro[60]fullerenes. As it has shown, DFT methods can be effectively used for the prediction of fullerene derivatives polarizabilities, which are necessary for the description of the functioning of fullerene-based nanosystems but have not been measured by the moment.

A model of polarizability for two types of fullerene derivatives $C_{60}X_n$ ([2+1]- and [1+1]-adducts) as a function of the number of added groups has been developed. DFT-calculations show that mean polarizabilities of both classes of fullerene derivatives do not grow up linearly with the increase of the number of addends *n* and are characterized by negative deviations from the additive scheme, *i.e.* depression of polarizability takes place. General formula for calculation of mean polarizability of the functionalized C_{60} fullerene has been derived based on polarizability depression. Its applicability to the related compounds (*e.g.*, C_{70} fullerene derivatives) has been shown.

Predicted by us with DFT methods, this interesting phenomenon may be important in the design of fullerene-containing nanostructures with regulated polar characteristics.

Alkylated naphthalenes are persistent environmental pollutants but there is still little information on their characteristics. In Chapter 6 the authors present the application of Density Functional Theory (DFT) methods for obtaining structural parameters of dimethylnaphthalenes (DMNs) and trimethylnaphthalenes (TMNs) through fully geometry optimization, their vibrational frequencies, IR intensities, Raman activities, and the assignment of vibrational modes in the ground electronic state. Almost all of the investigated molecules are characterized by a planar equilibrium geometry. The knowledge of aromaticity of these molecules based on nucleus-independent chemical shifts (NICS) can lead to a more sophisticated understanding of the reactivity of these molecules. The condensed electrophilic Fukui function calculated at the B3LYP/cc-pVTZ level of theory show the localization of the most positive parts susceptible to electrophilic attack. These regions are particularly important from the point of view of dioxigenation reaction on the aromatic rings of DMNs and TMNs by the naphthalene dioxigenase (NDO) enzymes in the process of biodegradation. The obtained equilibrium geometries and the transition state (TS) geometries enable further investigation of the π^*-σ^* hyperconjugation effects and their influence on the methyl group torsional barriers in the ground electronic state (S_0). Obtained results present basis for the investigation of the conformational flexibility of the aromatic rings of these molecules using *ab initio* techniques which can give important information about geometry relaxation of these molecules in possible intermolecular interactions.

In Chapter 7, the authors review their recent works on transport calculations of organic materials, which include single molecules, molecular wires, and organic semiconductors. First they describe the transport calculation methods based on the nonequilibrium Green's function

(NEGF) and the density-functional theory (DFT) in the localized basis sets and in the plane-wave expansion. Then they apply the method to the transport properties of single molecules and molecular wires. The problems such as conductance and its eigenchannel analysis, bias drop in the molecules, and local heating at the contact to electrodes are explored and presented.

Next, the authors describe the charge transport of organic semiconductors, which are regarded as molecular systems assembling weakly with van der Waals interactions. Recent advances of single-crystal organic transistors with high mobility require us to understand the fundamental transport mechanisms in these mechanically flexible organic materials. Using the time-dependent wave-packet diffusion (TD-WPD) method combined with classical molecular dynamics (MD) with parameters taken from the DFT-D calculations which include the van der Waals energy, they show the temperature dependence of charge transport with strong electron-phonon interaction in polaronic states and treat the crossover from the localized hopping-transport to delocalized band-transport.

Since the 1980's, the Density Functional Theory has evolved to become nowadays the most used tool for the calculation of electron structure of molecules, clusters and solids at a first-principles level of theory. In Chapter 8, the formal evolution of the theory from ancient local models to the modern hybrid meta generalized gradient approximations and double hybrid generalized gradient approximations is discussed. Moreover, the success and limitations of current Density Functional Theory in the prediction of a variety of properties, such as geometries, energies and thermodynamic functions, as well as in the interpretation of spectroscopic data are commented from a critical point of view.

Tailoring and improving material properties, as required in specific application fields, by alloying is a long-known and often used concept. The recent development of computational power has provided the opportunity to perform first principles calculations of systems with larger numbers of atoms ($\approx 10^2$), thus allowing for direct modelling of alloy-related phenomena. In Chapter 9 we focus on applying the concept of *special quasi-random structures* (SQS) to quasi-binary and quasi-ternary nitride alloys, used as protective hard coatings.

In the first part the authors focus on concepts how to construct representative cells with 32-64 atoms. Subsequently, they use those to obtain ground state properties, such as lattice parameters, energies of formation, mixing enthalpies bulk modulus or phase stability, as functions of composition. The particular systems include quasi-binary $Y_{1-x}Al_xN$ and $X_{1-x}Y_xN$ and quasi-ternary $X_{1-x-y}Al_xY_yN$ systems (*X* and *Y* stand for transition-metal elements). It is shown that the lattice parameters do not strictly obey the linear Vegard's-like behaviour for the lattice parameters, but are "bowed out" towards larger values as a consequence of the continuously changing character of the bonds. Comparison of energies of formation for the same material system but various crystallographic variants, allows for discussion of the phase stability. Using the quasiternary models for $Cr_{1-x}Al_xN$ allows for modelling of its paramagnetic state. The authors show that inclusion of magnetism leads to non-negligible alternations in e.g., lattice parameter or bulk modulus, as compared with non-magnetic case. Finally, they discuss the alloying trends in the elastic response of the single crystal and polycrystalline materials.

By comparing the calculated results with extended experimental work the authors demonstrate that the *ab initio* SQS calculations provide reliable trends, and hence can be used for the material selection in the alloy design process.

The fundamentals with regard to energy density functional theory and nuclear physics are studied in Chapter 10. The authors' attention is given to the rigorous mathematical treatment involved in deriving the energy density functional theory. There are specific features that depict this density functional when studying many-nucleon systems, which are quite different from its use in many-electron systems. The intended audience for density functional theory research are physicists, chemists, and mathematicians. In particular, it is also intended for those eager to begin studying the density functional theory.

In: Density Functional Theory
Editors: J. Morin and J.M. Pelletier, pp. 1-57
ISBN: 978-1-62417-954-9

Chapter 1

DENSITY FUNCTIONAL TREATMENT OF INTERACTIONS AND CHEMICAL REACTIONS AT INTERFACES

José R.B. Gomes[1,*], José L.C. Fajín[1,2], M. Natália D.S. Cordeiro[2], Cátia Teixeira[1,3], Paula Gomes[3], Renjith S. Pillai[1], Gerard Novell-Leruth[1], Jordi Toda[1,4] and Miguel Jorge[4]

[1]CICECO – Center for Research in Ceramics and Composite Materials, Departamento de Química, Universidade de Aveiro, Aveiro, Portugal
[2]REQUIMTE – Chemistry and Technolology Network,Departamento de Química e Bioquímica, Faculdade de Ciências, Universidade do Porto, Porto, Portugal
[3]CIQUP – Centro de Investigação em Química da Universidade do Porto, Departamento de Química e Bioquímica, Faculdade de Ciências, Universidade do Porto, Porto, Portugal
[4]LSRE – Laboratory of Separation and Reaction Engineering, Faculdade de Engenharia, Universidade do Porto, Porto, Portugal

Abstract

This chapter reviews recent applications of density functional theory (DFT) based methods in the study of the interaction of small gaseous molecules with metal nanoparticles, metal surfaces, and porous or biological materials and applications in the study of chemical reactions at catalytic sites of transition metals or enzymes. Focus is given to the interaction of small molecules, *e.g.* H_2O, O_2, CO, CO_2, etc., with the scaffold atoms of metal organic frameworks (MOF) or with zeolites, in the field of gas adsorption, or with the exposed atoms on transition metal surfaces or nanoparticles, in the field of heterogeneous catalysis, and to the interaction of small organic molecules with the capacity to inhibit a catalytic cysteine of the malaria's parasite, in the field of drug design. The roles of under-coordinated atoms on the strength of the interaction and of the type of the exchange-correlation functional considered for the calculations are analyzed. Finally, recent successes of the consideration of DFT based approaches to study, with atomic detail, the reactions of such molecules on these materials are also reviewed.

* E-mail address: jrgomes@ua.pt (preferred contact)

Keywords: Computational chemistry; density functional theory; heterogeneous catalysis; enzymatic catalysis; adsorption; porous materials

1. Introduction

In the last two decades, the field of density functional theory (DFT) treatment of interaction and chemical reactions at surfaces and interfaces has seen rapid expansion and developments. This is due to important implementations that made DFT approaches very accurate and competitive with the most advanced many electron wavefunction methods but with concomitant lower computational costs. The combination of DFT approaches with modern computers enable to perform accurate calculations in a reasonable time for molecular systems that are more realistic and, therefore, are more interesting for researchers in areas as diverse as chemistry, biochemistry, physics, chemical engineering, catalysis, materials science, etc. Within these scientific areas, gas adsorption, gas separation and heterogeneous or enzymatic catalysis are fields that are benefiting from such developments in a way that calculated interaction sites and sorption, diffusion, activation or reaction energies are in remarkable agreement with the best experimental data available. Therefore, DFT methods are not seen anymore as just interesting tools to interpret at the atomic level observations taken from research performed in the experimental laboratory, but are now considered as approaches with predictive power and able to compete with state-of-the-art experimental techniques. In fact, DFT data may also be used to develop classical force fields or microkinetic models for gathering additional relevant information and/or to understand the effects of the pressure, temperature, etc., on the molecular systems under study.

The present chapter is organized as follows: In section 2, a brief history of the most important developments in the field of DFT will be presented. Also in this section we will introduce the different models that are commonly used to treat a variety of different molecular systems, ranging from small aggregates of atoms, which can be fully modeled with DFT approaches, up to very large systems, which often require the combination of accurate and less accurate approaches, respectively, for the most relevant part of the system (inner region) and for the region embedding the latter (external region).

In section 3, different examples of application of DFT approaches in our groups for the study of the energetics of the interaction with different substrates will be presented, namely, i) the interaction of small gaseous molecules with bulk porous materials, ii) the interaction and reaction of small gaseous molecules on catalyst surfaces based on transition metals, and iii) the interaction of an organic compound with the active site of an enzyme. In the case of the porous materials, we will analyze i) the adequacy of different exchange-correlation functionals to adequately describe the energy and local geometry for the interaction of water, ethane, ethylene and acetylene with the open metal sites in the HKUST-1 metal organic framework (MOF), and ii) how DFT approaches can be used to interpret unusual experimental adsorption/desorption by identification of the most probable sites for water adsorption into $AlPO_4$-5 and $AlPO_4$-11 aluminophosphate molecular sieves or for nitric oxide adsorption into the channels of the Engelhard titanosilicate NaETS-4. The adequacy of DFT approaches for studies in the field of heterogeneous catalysis will also be reviewed in this chapter. In particular, by comparison with available experimental data, it is possible to analyze the quality of the calculated adsorption energies and activation energy barriers for the reaction of water dissociation on the most stable planar copper surface, Cu(111). The most interesting DFT approach in terms of quality and speed of the calculations was further employed in the study of other surface reactions catalyzed by pure or doped gold surfaces. In the latter, the effects of low-coordinated atoms and doping will be analyzed for the first O-H bond dissociation in the water molecule, a crucial step for the occurrence of the reaction of water gas shift. It will be concluded that the presence of low-coordinated atoms on the catalytic surface increases the adsorption energies of the reactants and decreases the activation energy barrier for the dissociation reaction, with concomitant lower overall reaction rates. Also, the effect of doping gold surfaces with other transition metal atoms will be shown to be desirable but only under specific conditions, *i.e.*, concentration and dispersion of the dopant species. In the field of enzymatic catalysis, this chapter will review recent advances in the understanding of how ligands bind to proteases of the malaria parasite disabling its capacity to degrade host haemoglobin to provide free amino acids for parasite protein synthesis. Particular attention will be given to recent examples where DFT-based approaches are being used to analyze the interaction of non-covalent and covalent inhibitors, mostly peptide-based compounds, with the catalytic center of falcipains, cysteine proteases from the malaria parasite *Plasmodium falciparum* that play a key role in haemoglobin degradation. Preliminary results from molecular dynamics simulations on the binding of a peptidyl vinyl sulfone inhibitor to the falcipain active site are reported.

Finally, section 4 summarizes our main results obtained by the use of DFT-based methods in studies in the fields of materials science, heterogeneous catalysis and drug design.

2. Computational Methods and Structural Models

Density functional theory (DFT) is presently one of the most successful and also most popular quantum mechanical approaches to describe matter. Its applicability ranges from atoms to molecules and solids, including nuclei and quantum and classical fluids. Indeed, DFT is routinely used nowadays for calculating a great variety of properties in chemistry (*e.g.* molecular structures, reaction paths, etc.) and in physics (*e.g.* band structures of solids).

The ultimate goal of most approaches in solid state physics and quantum chemistry is the solution of the time-independent, non-relativistic Schrödinger equation [1], toward gathering full information of a system. But as it is well known, an exact solution of such equation is not possible for non-hydrogenoid species, and several approximations must be made. The most universally accepted approximation is the so-called Born-Oppenheimer approximation [2], based on the fact that nuclei are several thousand times heavier than electrons, thus allowing the wavefunction of a system to be broken into its electronic and nuclear components. Then, for a many-electron system and given nuclear potential, the variational principle defines a procedure to determine its approximate ground-state energy and wavefunction, along with other properties of interest.

Conventional approaches developed so far for coping with this many-body problem can be divided into wavefunction-based methods and DFT methods. The former include the often called *ab initio* methods like the Hartree-Fock (HF), Møller-Plesset perturbation theory (MP), configuration interaction (CI), and coupled cluster (CC) methods [3]. To be noticed here is that the wavefunction for a system of N electrons depends on $3N$ spatial variables, and so its complexity increases with the number of electrons. In contrast DFT methods [4], as the name implies, use functionals of the electron density, which is a function of only 3 spatial variables and thus independent of the system size.

The basic principle of the density functional theory emerged in the nineteen-twenties with the work on the uniform electron gas of Thomas and Fermi [5, 6], who came up with the idea that the energy of a system is given completely in terms of its electron density. However, only in 1964, with the publication of the Hohenberg and Kohn (H-K) theorems [7], and in 1965, with the derivation of the set of monoelectronic equations with which one can obtain the ground-state density (Kohn-Sham equations) [8], further developments and practical applications of DFT methods have arisen. Essentially, the first H-K theorem demonstrates that the electron density uniquely determines the Hamiltonian operator and thus all the properties of the system, while the second one states that the functional pertaining to the ground-state energy of the system delivers the lowest energy if and only if the input density is the true ground state density (*i.e.* nothing but the variational principle). At present, the accuracy and efficiency of DFT-based methods depend on the basis-set for the expansion of the Kohn-Sham orbitals, but particularly upon the quality of the applied exchange-correlation (XC) functional [8].

The first approach initially proposed for modeling XC functionals is the local density approximation (LDA) [9]. At the centre of this approach is the idea of a uniform electron gas, so these functionals depend only on the local electron density. The local-spin density approximation (LSDA) proposed by Slater [10] constitutes a straightforward generalization of LDA to include electron spin into the functionals. Indeed, one of the earliest LSDA-XC functionals is the well-known SVWN functional, with its exchange part developed by Slater [11] and its correlation part by Vosko, Wilk and Nusair [12]. But even though LSDA gives bond lengths of molecules and solids typically with an astonishing accuracy (of ~2%), experience has shown that it leads to overbinding energies and underestimation of barrier heights. The moderate accuracy that LSDA provides is thus insufficient for most applications in chemistry.

The first logical step to go beyond LSDA, taking into account the non-homogeneity of the true electron density is to add also the gradient of the local density. This is the so-called generalized gradient approximation (GGA). Thanks to much thoughtful work, important progresses have been made in deriving successful GGA functionals during the last years. Among the most commonly used XC functionals of this type are the Perdew 86 (P86), Becke (B86, B88), Perdew-Wang 91 (PW91), Laming-Termath-Handy (CAM), Perdew-Burke-Ernzerhof (PBE), revised Perdew-Burke-Ernzerhof (RPBE), Perdew-Burke-Ernzerhof revised for solids (PBEsol), Becke exchange and Lee-Yang-Parr correlation (BLYP), and Armiento-Mattsson 2005 (AM05) [13 - 27]. For example, the PW91 is the most applied functional to model metallic surfaces and the PBE and RPBE to metaloxides in heterogeneous catalysis studies [28]. In general, GGA functionals have been shown to provide better predictions for total energies, atomization energies, and structural properties in comparison to LSDA functionals. However, they still give too low barrier heights and as a rule fail in describing van der Waals interactions.

A new class of more flexible XC functionals, labelled meta-GGAs [29], include additionally the Kohn-Sham kinetic energy for the occupied orbitals. A typical example of meta-GGAs is the TPSS functional [30], which has been shown to give, for instance, more accurate equilibrium properties of densely packed solids and their surfaces [31]. The great disadvantage of this type of functionals is that they are numerically more subject to convergence problems.

In addition, XC functionals known as hybrid (or meta-hybrid) functionals are combinations of GGA (or meta-GGA) functionals with nonlocal HF exchange [21]. Even though hybrid functionals are much more expensive than pure GGAs, they are progressively approaching the desired accuracy in many molecular properties, and in several cases they can even deliver results comparable with highly sophisticated post-HF methods. Examples of this type of functionals currently applied in computational chemistry include the hybrid B3LYP [15, 21] and the meta-hybrid M06 [32]. For instance, the M06 functional has been shown to afford good accuracy for transition metal thermochemistry, while the so popular B3LYP functional is now known to properly describe many properties of covalently bound compounds of main-group elements [32,33]. However, for periodic systems, B3LYP is a wrong option because the correlation energy part is incorrect in the homogeneous electron gas limit [34]. Other hybrid functionals, such as the PBE0 [35] and HSE03 [36], have been developed to obtain a better description in such cases. Indeed, these hybrid functionals overall improve predictions for lattice parameters of most solids, and the band gaps in semiconductors and insulators, as well as offering an excellent description of insulating antiferromagnetic rare-earth and transition-metal oxides where GGAs failed. However, the results obtained with hybrid functionals for solids are sometimes ambiguous, apart from requiring a very high computational cost. But recently, it has been found that graphical processing units (GPU) could strongly reduce the computational time required for hybrid functionals [37, 38]. Yet other approaches can be also less computationally demanding and reduce failures, such as too small band gaps, wrong dissociation energies for molecules, and problems with systems having localized *f* electrons. That is, semi-empirical terms can be inserted into a typical GGA functional to correct the electronic description. This is the case of the so-called GGA+U approach where U, the fitted Hubbard parameter, can be fitted to reproduce experimental band-gaps, geometries, etc. The GGA+U approach has been extensively used for modeling NiO, Ce_2O_3 or the vacancy systems of CeO_2 and TiO_2 [39, 40].

Improvements of hybrid functionals have additionally been sought by including also an *exact partial correlation*, such as the functional recently proposed by Grimme [41], thereafter termed double-hybrid. In essence, double-hybrid functionals are based on combining standard GGA XC functionals with HF exchange and a perturbative second-order correlation part (MP2) that is obtained from the Kohn-Sham (GGA) orbitals and eigenvalues. Such type of functionals are not yet widely used but nevertheless, for equilibrium thermochemistry, they seem to offer a performance clearly superior to conventional DFT functionals and to approach that of composite *ab initio* methods (*e.g.* G1 and G2 [42]), at a small fraction of the computational cost of composite methods.

Last, but not least, the several tests performed so far for evaluating the performance of current density functionals clearly illustrate their limitations in correctly describing van der Waals interactions (vdW). However, several approaches have appeared lately to cope with that aspect. For example, in the DFT-D approach [43], a semi-empirical dispersion potential is added to the conventional Kohn-Sham DFT energy to mimic such interactions. A simpler dispersion correction is offered by the semi-empirical force fields of Grimme *et al.* (DFT-D2) [44], where the dispersion contributions are calculated by pairwise interactions from the London formula [45].

Another similar approach (DFT-DF) has been proposed by Lundqvist and co-workers [46, 47], *i.e.* the authors suggested a non-local correlation functional that approximately accounts for the dispersion interactions. Yet the results achieved with a nonlocal vdW-DFT correlation functional certainly depend on the choice of the particular variant of the nonlocal vdW-correlation and on the exchange functional to be combined with it. The DFT-DF approach can resort to exchange functionals like the RPBE or the optimized PBE (optPBE) [48], B88 (optB88) [48], or to the rPW86 exchange functional (DFT-DF2 approach) [47,49], but the optimal functional seems to be the optB86b exchange functional [50].

In the past years, a different approach has been proposed, known as the self-consistent-charge density functional tight binding (SCC-DFTB, hereafter abbreviated with DFTB) method [51]. The DFTB method is based on a second-order expansion of the DFT total energy expression, and constitutes an alternative to the conventional semi-empirical methods in quantum chemistry like the popular MNDO, AM1 and PM3 schemes, which are derived from HF theory. However, DFTB is not a semi-empirical method in a strict sense, since its parameterization procedure is completely based on DFT calculations, and no fit to empirical data has to be performed.

In contrast to most semi-empirical methods, DFTB is a non-orthogonal method, *i.e.* it is based on a non-orthogonal basis set, which has been emphasized to be a key factor for transferability [52]. In terms of computational speed, DFTB is comparable with semi-empirical methods, the main cost is the solution of the generalized eigenvalue problem in a minimal basis as no integrals have to be evaluated. DFTB has been applied mostly to solid state and clusters but it appears to be quite reliable also in predicting structures and dynamic behavior of many biomolecules [53, 54]. For the latter, its interest relies especially on using it as a fast method to apply in dual-level calculation schemes.

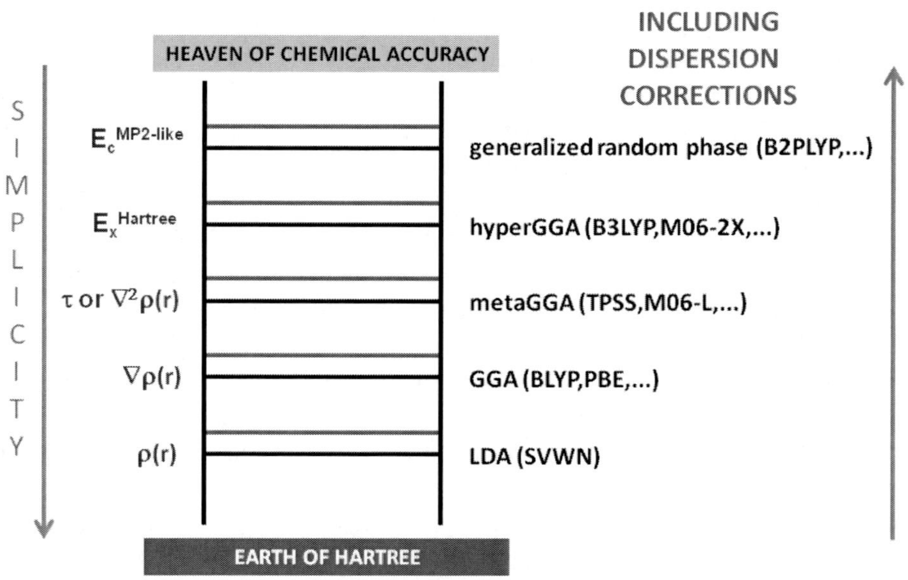

Figure 1. Jacob's ladder approach for the systematic improvement of DFT functionals according to the metaphor of Perdew and Schmidt [55], showing some of the most common functionals within each rung.

In summary, the several functionals applied in this work contain elements of all the five rungs according to Perdew and Schmidt's "Jacob's ladder" approach [55] (see Figure 1). As can be seen in Figure 1, in Perdew and Schmidt's metaphor, ground level would be Hartree theory and Heaven the exact density functional, and successive rungs of the ladder correspond to additional pieces of information that enter the functional. Rung one employs just the density (ρ) and corresponds to the local spin density approximation (*e.g.* SVWN). Rung two introduces the density gradient (∇ρ) and is occupied by the various GGAs, such as the XC BLYP and PBE functionals. Rung three additionally introduces the kinetic energy density (τ) or the Laplacian ($\nabla^2\rho$) and is occupied by the various meta-GGAs like TPSS. Hybrid GGAs and meta-GGAs that include the HF exchange constitute the fourth rung (*e.g.* B3LYP and M06). Rung five would include exact exchange and exact partial correlation terms, and double-hybrid functionals constitute one special case thereof. DFT users might climb the ladder to gain greater accuracy (at greater cost), but also tend to descend the ladder for simplicity depending upon their needs.

The correct inclusion of dispersion corrections to any of the DFT classes shown in Figure 1 would bring the approaches closer to Heaven, *c.f.* illustrated by the red rungs in Figure 1.

The structural models usually considered to model large systems such as bulk materials, surfaces or biomolecules have to be reduced somehow for being tractable with DFT approaches. For very accurate results, the DFT calculations should consider models of up to a few tens of atoms while standard DFT calculations will be possible for systems composed by a few hundreds of atoms.

Systems composed by a much larger number of atoms will have to consider several approximations, *e.g.* DFTB, or will have to mix DFT and more approximate approaches, *e.g.* ONIOM calculations mixing different approaches for treating different layers of atoms [56].

In the case of models for extended surfaces, usually considered for studies in the field of heterogeneous catalysis, there are two approaches that can be used, namely, the so-called cluster model and the periodic slab model. The description of the electrons of the atoms in the former method employs basis sets of atomic orbitals centered on atoms while the periodic approaches typically consider plane-wave functions. Furthermore, in both approaches, computer time may be saved if some of the inner electrons are described with the pseudopotential approach [57].

In the first approach, the cluster comprises a finite number of atoms cut from the bulk material following a specific direction in order to simulate a portion of the surface and a specific Miller index orientation. Thus, in the case of the cluster model approach, spurious effects arise related to the limited cluster size and to the presence of boundary effects. Cluster models have been mainly used in the earliest DFT calculations in surface science. The results obtained with this approach critically depend on the choice of the cluster size, stoichiometry and shape. Nevertheless, the cluster model can provide useful information if the cluster is properly chosen and the influence of its size is investigated *a priori*.

Some authors developed strategies that can be used to circumvent problems related with the finite size of the clusters, i.e., the so-called embedding schemes [58]. In the embedding schemes, the original cluster constitutes the inner region, which is treated with high-level approaches, and is surrounded by an external region that is treated differently from one class of embedding to another. The external region can be treated with a low-level approach (similar to ONIOM type calculations [56]), or it can be an array of point charges, a crystal field or even a homogeneous dielectric medium [58]. Clusters embedded in large arrays of point charges were proven to be quite useful to study metal oxide surfaces such as those of MgO [59, 60] or α-Al_2O_3 [61 - 63].

In Figure 2 is shown a schematic representation of the inner and external regions used to model the α-Al_2O_3(0001) surface with the cluster model approach. The inner region is composed of 26 aluminum and 15 oxygen atoms, i.e., a cluster with $Al_{26}O_{15}$ stoichiometry, which includes 18 total ion potentials (outermost Al atoms) that were used to avoid spurious polarization of the outer cluster oxygen atoms. The external region contains 2354 point charges with values of +3 and -2 for cations and anions, respectively, which were used to provide an adequate embedding for a satisfactory representation of the Madelung potential at the center of the cluster model [61 - 63]. In the periodic model, a simple portion of atoms — unit cell — is expanded periodically, usually in the three spatial dimensions (XYZ).

This periodicity mimics the properties of infinite systems, *i.e.* bulk materials or surfaces, using just a small portion of the infinite material, and eliminates the spurious effects affecting the cluster models. The computational cost is thus directly linked with the number (size) and the symmetry of the elements in the unit cell, *i.e.*, small crystalline systems such as bulk metals are treated at very low computational cost since the infinite system can be obtained by periodical repetition of just a few atomic positions (Figure 3a). The study of alloys, contaminants, or vacancies is possible by expanding the model and replacing or eliminating one or more atoms in the periodic system (Figure 3b).

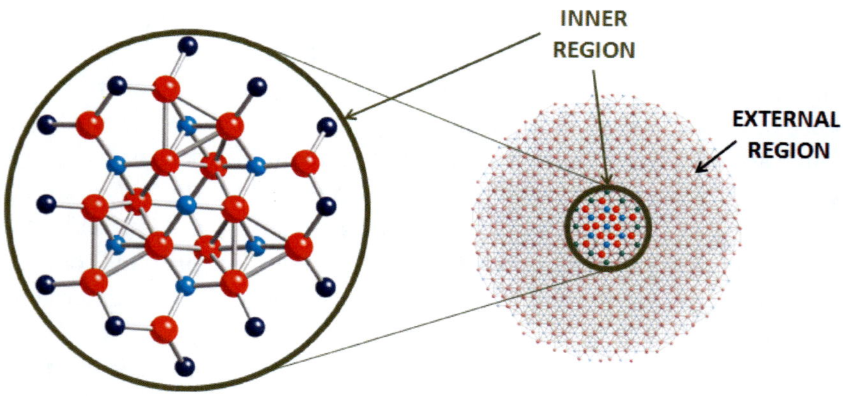

Figure 2. Cluster model used to represent the α-Al_2O_3(0001) surface for studying the adsorption of palladium atoms and small palladium clusters on this oxide. In the inner region, oxygen atoms appear in red while aluminum atoms appear in cyan (atoms treated with an all-electron basis set) or in blue (TIPs, *i.e.*, atoms treated with just a pseudopotential mimicking an Al^{3+} cation, *i.e.*, without basis functions) [61 - 63].

In the approaches repeating periodically the unit cell in the three spatial dimensions, which constitutes the most common kind of repetition in computer codes, the surfaces of the materials are obtained by using a unit cell that is composed by several layers of the material (slab) and a vacuum region in the direction normal to the surface (Figure 3c).

Usually, metal surfaces are well represented by four or five metal layers while, for instance, the surfaces of ionic oxides, *e.g.* alumina or titania [64, 65], have to include several metal and oxygen layers for a good representation. Thus, a study of the dependence of the

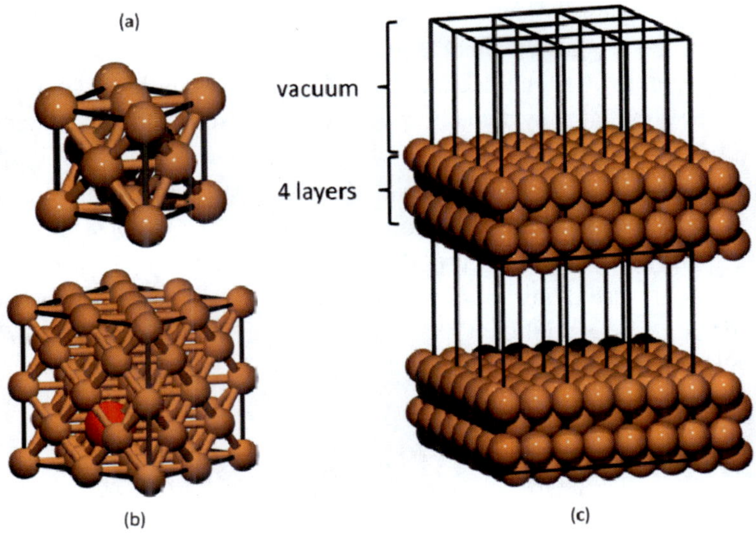

Figure 3. Examples of periodic unit cells that can be used to model a bulk metal (a), a defect (impurity, vacancy, etc. shown as a red sphere) on a bulk metal (b) and to model a metal surface (c). The model illustrated for the representation of a surface was obtained by repeating 18 times a 2×2 unit cell containing 4 atomic layers and a vacuum region.

calculated properties with the number of layers used in the slab has to be performed *a priori* until convergence is achieved. When studying adsorption or catalysis, the sizes of the slabs in the X and Y directions, presuming that the vacuum region is in the Z direction, are dependent on the volumes of the adsorbates and on the surface coverage. It is also important to check the effect of adsorbate-adsorbate lateral interactions, *i.e.*, interaction of an adsorbate with the replicas in neighboring cells. In order to make the calculations feasible, the vacuum region has to be as small as possible; vacuum regions of approximately 10 Å are enough for the study of small adsorbates but larger regions very probably will be necessary for large adsorbates, especially those weakly bound to the surface.

It is important to notice that the periodicity of the slab model approach reduces computational costs because the periodicity of the unit cell allows for a reduction of the number of wavefunctions needed to represent the infinite system to a finite number of orbitals that are needed to represent the nuclei in the unit cell (Bloch's theorem) [4].

The approaches above are not useful to model large but non-periodic systems such as biomolecules. The study of interactions involving these systems where bonds are cleaved or are formed, e.g. enzymatic catalysis, requires a different strategy. The strategy used resembles the embedded cluster approach, i.e., the region where important contacts are made or where bonds are cleaved and formed is treated with a quantum mechanics (QM) approach while the surrounding region, which usually also includes the solvent molecules, is treated with an inexpensive computational approach, e.g. molecular mechanics (MM) at the atomistic scale. Calculations employing the so-called QM/MM methodology may consider the full optimization of the systems and, therefore, the treatment of the QM/MM interface requires special care (Figure 4). Considering the QM part in the center, a second region (weight zone) is a region neighboring the QM center where the forces are interpolated between MM and QM forces.

A third region or layer (buffer zone) is a part of the total system where the forces are taken from the MM and the last zone is the MM layer. The low computational costs required for the MM part together with the accuracy of the methods considered for the QM region enables the study of quite large systems in reasonable time and with an accuracy comparable to that obtained by treating the entire system with a QM approach (*e.g.* DFT). A speed-up in the calculations can also be achieved if, instead of an atomistic approximation, coarse-grained models are used for handling the MM region. At present, the QM/MM methodology is not implemented for periodic systems but preliminary trials have been performed by Zhao *et al.* [66, 67], and tested for studying mechanical deformation of metallic systems and other similar applications. For example, in this case, QM methods are used for coping with impurities and other interesting defects.

Figure 4. Schematic representation of the QM/MM method. See text for details.

3. DFT Applications

In a quick survey of the scientific literature it is possible to notice a significant number of research articles describing results obtained within DFT just in the past decade. In the next three sub-sections, recent works using DFT applications in quite impacting areas of science will be reviewed.

3.1. Interaction of Gaseous Molecules with Porous Materials

In the past years, DFT approaches have been used to explore the catalytic and sorption properties of highly porous materials, such as aluminosilicates [68, 69], aluminophosphates [70 - 77], and titanosilicates [78 - 81]. These methods were also successfully implemented to develop suitable force field for classical simulations, especially adsorption of sorbates in zeolites. For example, DFT approaches were used for developing force fields for CO_2 adsorption in zeolites [82] or to study the unusual water sorption mechanism in aluminophosphates [77]. Still, one of the most recent applications of DFT in research is in the understanding of the properties of metal-organic frameworks (MOF) [83, 84]. These materials, first reported in 1994 by Yaghi *et al.* [85], are constituted by two major components, i) a metal ion or cluster of metal ions and ii) an organic molecule. The latter bridges the metal ions and, for this reason, receives the name of linker. The linker builds the MOF architecture but it is the metal's coordination that influences the dimension of the structure by dictating how many ligands can bind to the metal and in which orientation, Figure 5.

The linkers are usually mono-, bi-, tri-, or tetravalent ligands, depending on the MOF. Examples of monovalent ligands used in the synthesis of MOFs are NH_3 or Cl, examples of bivalent ligands are bipirydine or ethylenediamine and examples of trivalent or tetravalent ligands are terpyridine or triethylenetetramine, respectively.

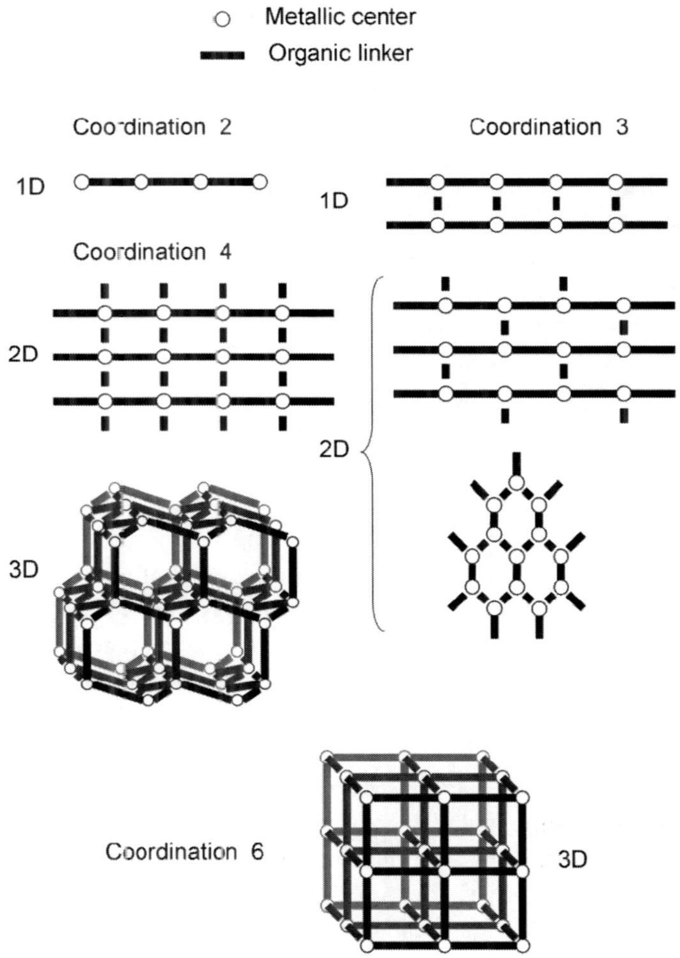

Figure 5. The effect of the coordination of the metal vs. the dimension of the MOF. Adapted from [86].

The 3D structures of MOFs assumed recently great importance due to their porous structures with variable channels or cavity sizes through which some molecules can be stored, may diffuse or can meet others to chemically react. The size of the pores and the surface areas of MOFs were found to be larger than those determined for zeolites (microporous aluminosilicate minerals commonly used as commercial adsorbents). Furthermore, the framework flexibility of MOFs and the presence of unsaturated metal sites (under-coordinated open sites) has been suggested to have a vital role in their interaction with some molecules [87, 88]. These are the main reasons for the growing scientific interest in MOFs since the first example of this class of porous material was reported. MOFs are considered as very promising materials in several areas of research and significant efforts are being undertaken in order to synthesize and to evaluate their use in practice [89]. Examples of potential applications of MOFs are in gas/energy storage, *e.g.* storage of hydrogen [90] or methane [91], in carbon dioxide sequestration [92], in gas separation [93], in catalysis [94], in the development of highly selective and sensitive sensors [95], etc. It is precisely due to the great demand on the synthesis and on the characterization of novel materials for specific applications that computational studies are being thought as cheap and fast alternatives for understanding and predicting the properties of MOFs.

In the last decade, the number of publications focused on the synthesis, characterization and application of MOFs increased dramatically, which is somehow related with the consideration and use of computational techniques to handle such systems [96, 97]. Computational studies in the field started only quite recently, very probably due to the huge size of the crystallographic cells for these hybrid materials, which require the use of more approximated but reliable approaches, *e.g.* DFTB methodology that became recently available, or the use of large computational facilities.

In fact, as it will be shown below, there are other structural specificities of MOFs that make their study difficult by using only computational approaches. Examples of research topics involving MOFs that were studied during the recent years were the adsorption energies and adsorption sites of gases into the pores [98], separation of mixtures of gases by specific interactions [99], doping effects on the adsorption of gases [100] and mechanical properties of these materials [101].

The topic of gas adsorption is that with the highest number of publications in the literature due to the large size of the pores and due to the possibility to tune adsorption by just modifying the physical and chemical properties of the organic linkers. A critical review on the ligand design for functional MOFs was published in the literature very recently [102].

Several different computational methods have been used to investigate the adsorption of gases by MOFs, namely, grand-canonical Monte Carlo (GCMC) [103] or molecular dynamics (MD) simulations [104], density functional theory [105, 106], MP2 [106, 107], coupled cluster [107], etc. Historically, several gases have been studied such as molecular hydrogen, CO_2, CH_4, CO and hydrocarbons, but the two former species, H_2 and CO_2, were the most studied ones due to commercial and environmental applications.

There is another adsorbate that has been also studied but due to different reasons: H_2O. Water is present in MOFs as a guest molecule, originated by the synthetic procedure and coordinated to the metal centers to satisfy its coordination needs, such water molecules can be removed by heating under vacuum which activates the material. It was shown in previous studies that guest water molecules play a crucial role in MOF structures [108, 109].

The MOFs with guest water molecules differ from those where the metal atoms are coordinatively saturated by the framework and are quite appealing for gas adsorption (they were found to improve H_2 uptake) and gas separation (preferential adsorption of certain gas components in a mixture *e.g.* high selectivity for CO_2 over CH_4).

As stated above, MOFs are not easy to be handled by just computational approaches due to several reasons. First, the models considered for MOFs are based on their perfect crystalline structures but the real materials frequently contain impurities and/or defects that are responsible for dead volumes in these porous materials. Usually, discrepancies between calculated and experimental adsorption data are attributed to this fact. Another difficulty in the computational modeling of MOFs is caused by the flexibility of the material, and computational models very often rely on a rigid structure. The third problem is related with the inaccuracies of several computational approaches in the correct handling of the adsorbate-MOF interaction by deficiencies in the treatment of dispersion forces (long-range weak, van der Waals, interactions) between the adsorbate and the organic linkers or in the description of the specific interactions between the unsaturated metal centers and certain adsorbates. Nevertheless, several different methods have been used to investigate the adsorption over different MOFs and important information was obtained.

Zhou and co-workers [110] considered GCMC and the OPLS-AA force field to investigate the adsorption isotherms and adsorption sites of H_2 in zeolitic imidazolate frameworks (ZIFs), which are a new type of MOFs that are topologically isomorphic with zeolites. These authors were also interested in the understanding of the differences between H_2 adsorption in ZIFs and in MOFs, which are suggested to have origin in the different electronegativities of the atoms and in the steric hindrance close to the adsorption site. In the classical MOF structures, *e.g.* MOF-5, the main adsorption site is surrounded by oxygen atoms, grouped in COO- units, while in ZIFs, *e.g.* ZIF-8, the preferred adsorption position is close to a C=C bond of the imidazolate ligand. Since the electronegativity of the O atom is greater than that of the double bond, the H_2 adsorption is stronger in MOFs than in ZIFs porous materials. Furthermore, Zhou *et al.* found that in ZIFs, the free space available for the adsorption of H_2 molecules close to the preferred adsorption site, *i.e.*, near the C=C bonds, is smaller than the space available in regions close to the oxygen atoms in MOFs. In principle, the information obtained from their GCMC simulations about the differences in the preferred adsorption sites in ZIFs and MOFs will be very useful to design new materials with improved adsorption capacities.

The preferred sites for H_2 adsorption in another Zn-based MOF, *i.e.*, MOF-177 where 1,3,5-benzenetribenzoate (BTB) is the linker, were also analyzed by Dangi *et al.* but using DFT. [111] MOF-177 has pores larger than those in MOF-5 and was reported as having the highest hydrogen uptake capacity beyond the U.S. Department of Energy (DOE) target, so that there has been very keen interest to understand the interaction of H_2 molecule in MOF-177. These authors considered the PW91 exchange-correlation functional and double numerical plus polarization (DNP) basis sets to compute the binding energy of H_2 molecule(s) to the inorganic cluster or to the organic linker (BTB) of MOF-177. It was found that the interaction of hydrogen was stronger with the inorganic cluster than with the organic linker. The PW91-calculated interaction energies were 2.96-4.50 $kJ{\cdot}mol^{-1}$ or 2.6-3.8 $kJ{\cdot}mol^{-1}$ at the inorganic cluster or at the organic linker, respectively. These energies are far from the thermodynamic requirement for an adsorbent capable of storing hydrogen at ambient temperature, *i.e.*, the H_2-adsorbent interaction energy should be higher than 15 $kJ{\cdot}mol^{-1}$.[112] From the energy values calculated by Dangi *et al.* [111] it can be concluded that both the organic linker and the inorganic cluster play an important role in the adsorption of hydrogen. Furthermore, it was suggested that the H_2 binding energy at the inorganic cluster was affected by the organic linker. The authors considered also in their work the adsorption of multiple H_2

molecules in MOF-177. Figure 6 shows a view of the interaction of hydrogen molecules with the cluster model used to study the adsorption of H_2 in MOF-177. Interestingly, the calculated data for multiple H_2 adsorptions suggest that already adsorbed hydrogen molecules contribute to the binding of additional H_2 molecules. However, this cooperative effect was found to be insufficient to turn MOF-177 into a suitable material for room temperature storage of hydrogen. Watanabe *et al.* [104] conducted MD and GCMC simulations complemented with DFT/PW91 plane-wave calculations to examine the adsorption (GCMC) and diffusion (MD) of CO_2 and CH_4 in the MOF Cu(hfipbb)(H_2hfipbb)$_{0.5}$, either as single components or as mixtures, and tested the selectivity of this MOF for their separation. The DFT calculations were used to compute adsorption energies and diffusion activation energies for CO_2 and CH_4 in the pores of Cu(hfipbb)(H_2hfipbb)$_{0.5}$. They anticipated that the calculated DFT energies strongly underestimate the attractive dispersion forces with the pore walls and this constitutes the reason why some authors used MP2 or coupled cluster approaches instead of DFT [106, 107], to study the adsorption of gases in MOFs. Nevertheless, the results by Watanabe *et al.* [104] show that the adsorption of CO_2 is stronger than that of CH_4, which is common in most MOFs. However, more important, is the fact that the diffusivity of CO_2 species is several orders of magnitude larger than that of CH_4, which is crucial for gas separation purposes. Watanabe *et al.* also calculated the diffusion activation energy of C_2H_6 and it resulted to be identical to that of CH_4.

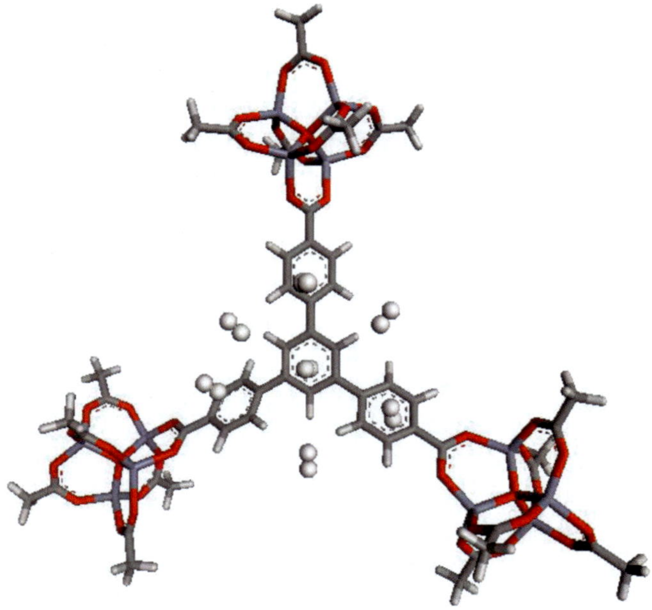

Figure 6. MOF-177 cluster model centered in the organic linker with H_2 molecules at various adsorbed position used for DFT studies. The atoms are colored as follows: grey for carbon, red for oxygen, and white for hydrogen.

This implies that the activation energy for propane and butane to diffuse in Cu(hfipbb)(H_2hfipbb)$_{0.5}$ is likely to be similar to that of CH_4. Watanabe and Scholl [113] also studied the chemisorption of several molecules, *e.g.*, H_2O, CO, NO, pyridine, and NH_3, on open Cu sites in the HKUST-1 MOF by DFT and GCMC approaches. They concluded that CO molecules bind to the Cu^{2+} metal centers and that they can occupy a fraction of Cu sites even in the presence of some water (water is usually coordinated to Cu sites in non activated samples of HKUST-1 [114]). Similar predictions were made for NO adsorption, *i.e.*, a small fraction of NO was able to adsorb on the metal sites even in the presence of water, while without the presence of the water, NO was able to cover only half of all the adsorption sites. Ammonia and pyridine showed the highest binding energies. Finally, these authors alerted that with the DFT approach considered, i.e., the PW91 functional, it is not possible to accurately model the interaction of molecules with MOFs through van der Waals forces.

In the last years, new methodologies based on DFT have emerged to address the issue of adsorption in MOFs. For instance, Vilhelmsen *et al.* [115] combined DFT calculations (PBE exchange-correlation functional) with a genetic algorithm (GA) to study the structure and the mobility of Au, Pd and AuPd clusters in MOF-74. The GA was used to generate and to select several nanocluster structures in the pores of the MOF from an initial random population and to discard unreasonable structures. The GA kept a population of the 15 lowest energy and structurally different candidates which were used to propose new structures. At the end, a small population of genetically more stable structures was selected to perform DFT calculations and to determine the most stable adsorbed structure in the pores of the MOF. Their results showed that Pd clusters were bound more strongly than Au clusters to the MOF structure, i.e., the latter are more mobile than the former in the MOF-74 framework. Interestingly, they found no energy gain from alloying the two metals and found also that nanoparticle adsorption is equally favorable on the organic parts or on the open metal sites of the MOF.

The density-functional based tight-binding method (DFTB) with self consistent charge (SCC) is another approach that has been proposed recently by Lukose and co-workers [116] to describe materials with large crystallographic unit cells such as MOFs. This method is based on a second-order expansion of the Kohn-Sham total energy in DFT with respect to charge density fluctuations.

The zeroth order approach is equivalent to a common standard non-self-consistent (TB) scheme, while at second order a transparent, parameter-free, and readily calculable expression for generalized Hamiltonian matrix elements can be derived. These are modified by a self-consistent redistribution of Mulliken charges (SCC). Besides the usual "band structure" and short-range repulsive terms, the final approximate Kohn-Sham energy additionally includes a Coulomb interaction term between charge fluctuations. At large distances, it accounts for long-range electrostatic forces between two point charges and approximately includes self-interaction contributions of a given atom if the charges are located at one and the same atom. Due to the SCC extension, DFTB can be successfully applied to problems where deficiencies within the non-SCC standard TB approach become obvious. Lukose *et al.* have also validated the DFTB with SCC and London dispersion corrections for various types of MOFs. The method provides excellent geometrical parameters when compared to experimental data and, for small model systems, also compares well with DFT calculations. Moreover, it seems to be also useful for catalytically active MOFs containing Cu atoms and whose electronic description is difficult.

Very recently, Grajciar *et al.* [117, 118] used a combined DFT/CC method, which includes a correction function to account for the discrepancy between DFT and CCSD(T) energies, to treat the adsorption of water and carbon dioxide in HKUST-1. An important conclusion from their work was to find that the dispersion component accounted only for a fraction of the discrepancy between the PBE and CCSD(T) results, suggesting that DFT approaches based on the GGA are not able to handle this type of systems, and that many of the results reported in the literature for the interaction of different molecules with open metal sites of MOFs have to be treated with caution. Their results showed also a high dependence between the coverage and the preferential adsorption sites for CO_2. At low coverage, the CO_2 molecules adsorb onto the Cu sites while at high coverage, up to CO_2/Cu ratio of 20:12, the CO_2 species preferentially occupy sites in the windows of small cages. Moreover, the interaction of CO_2 molecules with other molecules adsorbed on Cu sites results in increased stability of CO_2 in cage windows sites.

Peralta *et al.* [119] considered a different approach to study the adsorption mechanism on MOFs containing open metal sites. These authors used dispersion corrected DFT (PBE-D2) to rationalize the experimental data for the adsorption of aromatic, alkene and alkane model compounds, *e.g.*, o- and p-xylene, 1-octene and n-octane, respectively, in MOFs and in zeolites. Apart from concluding that MOFs are better adsorbents than zeolites, due to their large pore volumes, they conclude also that apolar MOFs behave similarly to apolar zeolites, MOFs with extraframework cations behave similarly to zeolites with extraframework cations, but MOFs with coordinatively unsaturated sites, *e.g.* HKUST-1, show a peculiar behavior. The latter MOFs behave like zeolites with extraframework cations, which are polar adsorbents, with the exception that alkenes with flexible side chains adsorb stronger than aromatics. Usually, the π-electrons of the aromatic ring interact strongly with extraframework cations and are more strongly adsorbed than alkenes, but in the case of MOFs containing coordinatively unsaturated sites, the approach of the aromatic ring to the metal site is sterically hindered by the surrounding ligands, which considerably weakens the interaction. The steric hindrance felt by compounds containing aromatic rings does not occur for linear alkenes.

Dispersion corrected PBE calculations were also employed by Sillar *et al.* [106] to study the adsorption of H_2 in an isoreticular metal-organic framework (IRMOF) called MOF-5, which has oxide-centered Zn_4O nodes linked by organic molecules. These authors compared the results from periodic DFT calculations with those calculated with MP2 and finite-sized models, $OZn_4(CO_2H)_6$ with formate anions as ligands and $OZn_4(CO_2Ph)_6$ with benzoate anions as the organic linkers, that represent individual structural elements of the whole MOF framework. The results obtained for the *ab initio* calculations showed a good description of the H_2-MOF interaction energies but also an important dependency on the size of the basis set used and on the BSSE. In the case of DFT calculations, a semi-empirical $1/r^6$ term was added to the DFT energies [43] to get a better description of the dispersion forces that are present in this type of interaction. The comparison of DFT with DFT+D and with MP2 indicates that dispersion is the major contribution to H_2 binding. Moreover, agreement between DFT+D results for the benzoate model and the periodic structure shows that it is mainly the local environment around the adsorption site that determines the adsorption properties of MOF-5. Consequently, the use of DFT with an improper description of dispersion, or the use of too small models with methods that include dispersion, does not result in converged adsorption energies.

Wu and co-workers [120] also combining GCMC and DFT approaches studied the storage of methane in three MOFs (HKUST-1, PCN-11 and PCN-14) having high methane storage capacities. The DFT calculations were based on the LDA with the Perdew-Zunger exchange-correlation since they argue that the GGA (PBE functional) severely underestimates the CH_4-MOF binding, particularly between the open metal site and the methane molecule.

However, these authors did not provide any comparison of LDA results with benchmark data and decided to rely just on the LDA approach because it is known to give overestimated binding for weak interactions [121]. These authors concluded that methane adsorption takes place through two major binding mechanisms, i) enhanced Coulomb interaction with the coordinatively unsaturated metals, and ii) enhanced vdW interaction at potential pocket sites.

From what has been exposed above, it seems that a good description of the MOF-adsorbate interaction by an affordable computational approach, *e.g.* an approach based on DFT, is still needed, especially for MOFs containing open metal sites. Due to that, it is not surprising that recent theoretical studies [118, 122] in the area of MOFs were focused on the interaction of small molecules with probably the most paradigmatic example of a MOF containing open metal sites, *i.e.*, HKUST-1 [123, 124] or copper benzene tricarboxylate ($Cu_3(BTC)_2$ also called CuBTC), which is illustrated in Figure 7.

Its crystalline structure consists of copper paddlewheels, where each copper atom is coordinated to four oxygen atoms from the BTC linkers and to one water molecule (solvent).

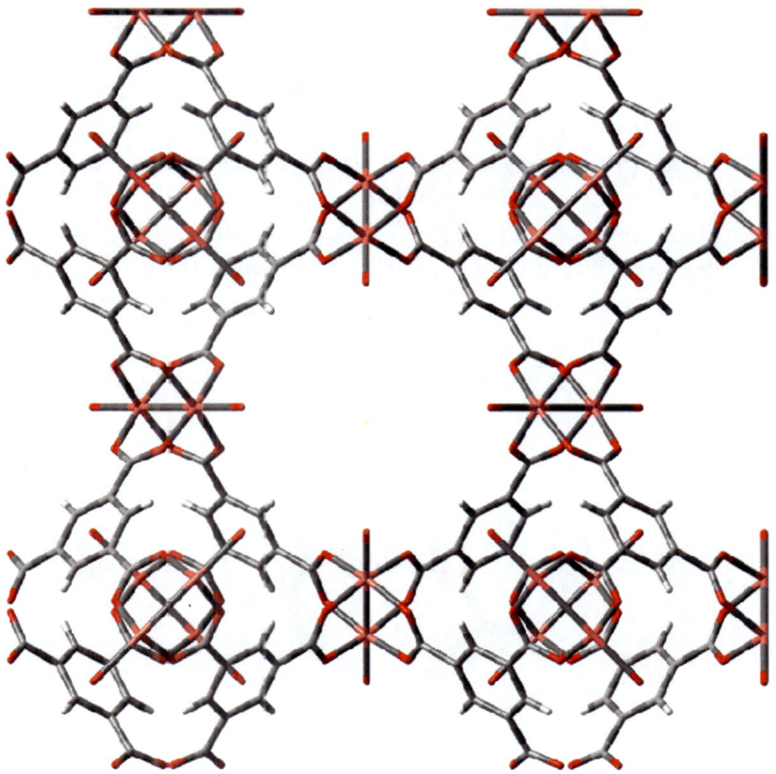

Figure 7. Tube representation of the CuBTC crystalline framework [123]. Color code: pink is copper, red is oxygen, grey is carbon, and white is hydrogen.

The water molecules interact directly with the copper atoms [125, 126] and when removed, for instance by heating under vacuum, these copper sites become available for adsorption of other molecules, which constitutes the method for the activation of CuBTC. This MOF is hard to be described by computational approaches due to the large size of its unit cell and to the open-shell nature of the copper sites [125, 126]. Furthermore, dispersion interactions were found to have an important role in adsorbate-Cu interaction [118, 127]. In order to perform very accurate calculations on the mechanisms of adsorption on CuBTC, some authors [118] choose a copper formate paddlewheel (Figure 8) as a reduced model of the extended CuBTC framework to study adsorption on the open metal sites.

Despite the quite drastic reduction in the size of the model when compared with the number of atoms in the HKUST-1 unit cell, it was found recently that the copper formate paddlewheel describes the local environment of copper atoms in the CuBTC structure as it is expressed by much larger cluster models [127, 128].

The use of a much smaller model enables the consideration of more complete basis sets, *e.g.* basis sets augmented with several polarization and diffuse functions but, unfortunately, the problems concerning the description of the dispersion interactions and the open shell nature of the copper atoms by DFT approaches do not have such an easy solution. That is why some authors decided to introduce a correction function (DFT-CC) in order to overcome this problem. However, this requires the *a priori* calculation of this correction function for each type of open metal site and each open metal site neighborhood in a MOF, which is far from being straightforward. It would be interesting to find a DFT exchange-correlation functional that is able to correctly treat the interaction of adsorbates with open metal sites of MOFs. This seems to be possible since there are many different exchange-correlation functionals in the literature, some including dispersion corrections, and very probably there is one that is able to deal with these systems. In fact, very difficult cases were successfully treated with DFT, such as the adsorption of CO on MgO(001) [59] and the adsorption of CO and NO on Ni-doped MgO(001) [60], which is very encouraging.

In order to check if DFT approaches can in fact be used to model adsorption on the open metal sites of MOFs, we performed a very complete study on the interaction of water with CuBTC [129], modeled by the copper formate paddlewheel, with several different DFT based approaches spanning the different rungs in the Jacob's ladder proposed by Perdew and Schmidt [55].

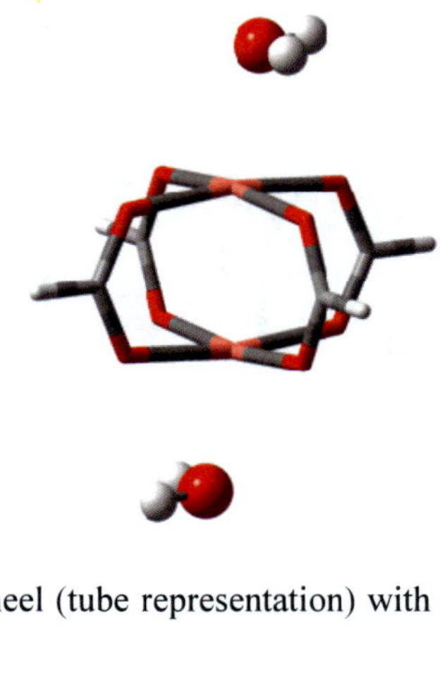

Figure 8. Copper formate paddlewheel (tube representation) with two adsorbed water molecules (ball-and-stick representation).

The DFT energies and optimized geometries were compared directly with the benchmark CCSD(T)/CBS result from Grajciar *et al.* [118] and with our own MP2/CBS calculations [129]. The H_2O-CuBTC geometry used in most of the cluster model calculations performed by Grajciar *et al.* [118], i.e., copper formate paddlewheel with two water molecules each one interacting with a copper site, was optimized at the PBE/aug-cc-pVTZ(-PP) level of theory and high-symmetry was considered. In our calculations, symmetry was reduced significantly in order to allow the two water molecules bound to the copper formate paddlewheel to have different orientations (Figure 8). The H_2O-Cu and Cu-Cu distances determined experimentally for the crystalline material are 2.17 Å and 2.63 Å, respectively [123]. The CCSD(T)/CBS enthalpy of interaction, at $T = 0$K, reported by Grajciar *et al.* is -51.2 kJ·mol^{-1} [118], while our MP2/CBS result, based on the MP2/cc-pVDZ(-PP) fully optimized geometry and thermal corrections [130, 131], as included in the Gaussian 09 code [132], is -55.4 kJ·mol^{-1} [129]. In our calculations, the all-electron basis set cc-pVDZ was used for H, C, and O atoms while the cc-pVDZ-PP basis set, including a pseudopotential for the inner electrons, was considered for copper atoms.

The DFT calculations were carried out with several different exchange-correlation functionals, namely, the hybrid M06 [133], M05-2X [134], M06-2X [133] and B3LYP approaches [15,21], the double-hybrid B2PLYP [41] and mPW2PLYP functionals [135], the pure M06-L method [136], the dispersion-corrected B97D [43] and wB97XD functionals [137], and the long-range corrected LC-BP86 [14,16], LC-PBE [138,19] and LC-wPBE functionals [139 - 141]. The cc-pVTZ(-PP) basis sets were used in these calculations [129].

As expected, the computed results confirm that most of the DFT approaches are unable to provide a proper geometrical and energetic description of the H_2O-CuBTC interaction. The best values obtained for the H_2O-Cu distance were calculated with the LC-BP86 or the LC-PBE functionals, with values of 2.20 Å and 2.22 Å, respectively, while the worst results were calculated with the B97D approach (2.37 Å).

In the case of the Cu-Cu distance, the best results were calculated with the B97D, B3LYP and M06-2X approaches, with values of 2.62 Å, 2.62 Å and 2.65 Å, respectively. The calculated Cu-Cu distances farthest from the experimental value were computed with the LC-PBE and LC-BP86 approaches (2.47 Å with both approaches). The analysis of the calculated interaction enthalpies shows the LC-BP86 as the best approach, enthalpy is -51.1 kJ·mol^{-1}, while the M06 family provides also very good results. The results calculated with the M06, M06L and M06-2X approaches are -45.9 kJ·mol^{-1}, -49.2 kJ·mol^{-1}, and -53.8 kJ·mol^{-1}, respectively. The B3LYP and the LC-wPBE enthalpies of interaction are those farthest from the CCSD(T)/CBS and MP2/CBS benchmark enthalpies with values of -27.9 kJ·mol^{-1} and -35.8 kJ·mol^{-1}, respectively.

From the results above it seems that even some dispersion-corrected functionals such as the B97D approach have problems in the correct description of this kind of systems, which constitute indeed hard tests for DFT approaches and that may be used in the future development of novel density functionals to test their accuracies. Interestingly, the GGA functionals combined with the LC corrections of Hirao and co-workers [138], yield the best description for the H_2O-Cu distance but on the other hand give too short Cu-Cu distances. Still, the interaction enthalpy calculated with the LC-BP86 approach is in very good agreement with the benchmark results and may be considered in future calculations involving MOFs with open metal sites. The M06 family of functionals presents also a very good compromise in terms of calculated geometrical and energetic parameters, especially, the M06-2X approach. Finally, the comparison of calculations performed with cc-pVDZ(-PP) and cc-pVTZ(-PP) showed a great dependency of interaction enthalpies with the basis set considered in the calculations.

Preliminary results for the interaction of ethane, ethylene and acetylene with CuBTC, also modeled with the copper formate paddlewheel, suggest that the LC-BP86 approach is a very good alternative to the very expensive post-Hartree-Fock methods for studying adsorption in MOFs with open metal sites. The LC-BP86 approach was found to handle well the interaction of the C-C, C=C and C≡C with the open metal sites in CuBTC.

In spite of the problems registered in the literature, DFT approaches are extensively used for structure prediction in the field of material science since, in many cases, the results of DFT calculations of solid state systems agree quite satisfactorily with the experimental findings. Furthermore, the computational costs are low as compared to traditional methods which are based on using the many electron wave functions like the Hartree-Fock theory [142]. As highlighted above, theoretical calculations have so far been performed on porous materials like zeolites and MOFs to predict essentially their acidic nature, interaction energy and adsorption phenomena for guest molecules inside the cavities. [68 - 72] Porous zeolite-type materials are widely used in oil refinery cracking processes and as adsorbents for gas separations [143].

A zeolite is a crystalline aluminosilicate with a three dimensional framework structure, which forms uniformly sized pores of molecular dimension. Several aluminosilicates with various structural specificities are available today, including zeo-type materials where silicon or aluminum atoms are replaced by phosphorous or titanium atoms for generating aluminophosphates or titanosilicates, respectively [143 - 145].

The understanding of the properties of each existing porous material or the prediction of the possible properties of a new material for a particular use by experimental methods is quite challenging, *i.e.*, it is economically very expensive and time consuming. For those reasons, computational approaches such as those based on the density functional theory are assuming an important role in the material science domain and are becoming general practice for providing atomic details about the interaction and reaction of gases in porous materials.

Adsorption studies of CO_2 in zeolites have largely utilized DFT to interpret the sorption phenomena but also to generate reliable force field parameters for performing molecular simulations required to predict adsorption isotherms and diffusion data. For example, Plant *et al.* [82] performed calculations at the PW91/DNP level of theory to obtain the Coulombic charges of the sorbate and of the adsorbent required to study the interaction of CO_2 with NaX and NaY faujasites.

They considered also the DFT optimized geometry of the CO_2-zeolite cluster as the starting configuration for generating a potential energy curve. They performed a series of single point energy calculations with a constrained geometry in order to maintain the previously optimized equilibrium angle and where the distance between the carbon dioxide molecule and the cations in the adsorbent was varied between 1.0 Å to 8.0 Å with increments of 0.1 Å (Figure 9).

At each increment a single point energy calculation was done using the B3LYP hybrid functional with the triple-zeta valence basis set (TZV) augmented with a double set of polarization d functions on the C and O atoms of CO_2 and with a double set of diffuse p functions on Na^+ [18, 21, 146, 147]. They performed also a single point energy calculation with the CO_2 molecule at a distance of 100 Å along the same CO_2-cation vector.

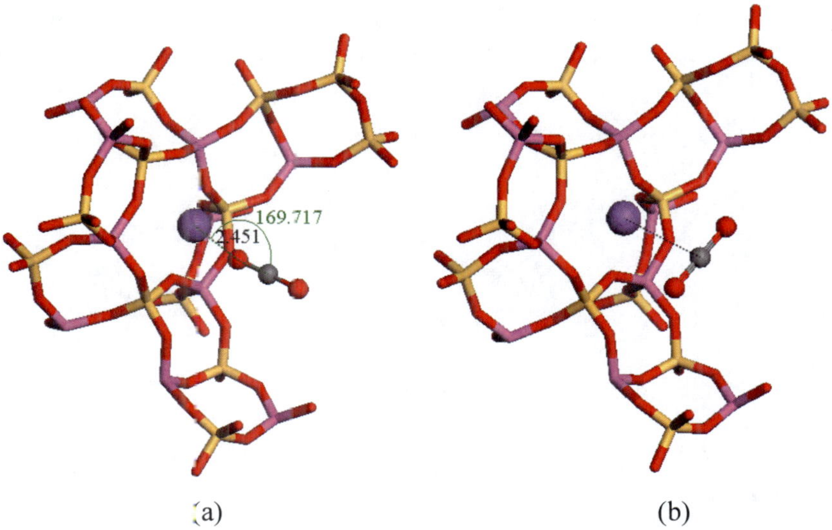

Figure 9. Schematic representation of the CO_2–Na^+/zeolite geometry considered for the calculation of the potential energy curve corresponding to the Na^+–$O(CO_2)$ (a) and Na^+–$C(CO_2)$ (b) interactions. The atoms are colored as follows: yellow for silicon, pink for aluminum, red for oxygen, violet for sodium and grey for carbon.

The calculated energy value at this large distance was set as the zero point reference and used to calculate interaction energies for each Na^+–$O(CO_2)$ or Na^+–$C(CO_2)$ increment used to construct the energy profile. The potential energy curve was then fitted by the combination of a Coulombic and a Buckingham potential. The methodology and the force field derived by Plant *et al.* [82] were validated against experimental microcalorimetry data for both NaX and NaY, and were further used to understand the microscopic adsorption mechanisms for CO_2 in these materials.

Using DFT and canonical Monte Carlo (CMC) approaches, Pillai *et al.* [77] conducted research work on the understanding of the unusual water sorption phenomena observed in aluminophosphate molecular sieves $AlPO_4$-5 and $AlPO_4$-11. The DFT calculations, performed with the CASTEP code, ultrasoft pseudopotentials and the PBE exchange-correlation functional according to the method described by White and Bird, [148 - 152] were considered for the optimization of the structure (Broyden-Fletcher-Goldfarb-Shanno, BFGS, method) of these materials with and without water molecules inside the aluminophosphate crystal. The effect of the concentration of water in the cavities was analyzed by performing the DFT calculations with a number of adsorbed water molecules ranging from 1 to 12. The most favorable positions were located after CMC simulation runs. Selected optimized structures for $AlPO_4$-5 without and with adsorbed water molecules are shown in Figure 10. Based on their calculations, they found that the optimized cell parameters are increased slightly due to the water sorption and that with the increase in water loading in the main cavity of the aluminophosphate, the distances between framework oxygen atoms and the hydrogen atoms of water decrease. The DFT calculations have demonstrated hydrogen bonding between adsorbed water molecules inside the main 12-membered ring of aluminophosphate. In addition to this, the comparison of the optimized structures for the dehydrated $AlPO_4$-5 material and for the material with just a single water molecule loaded into the unit cell showed a decrease of 2.2° in the water molecule facing Al-O-P angle.

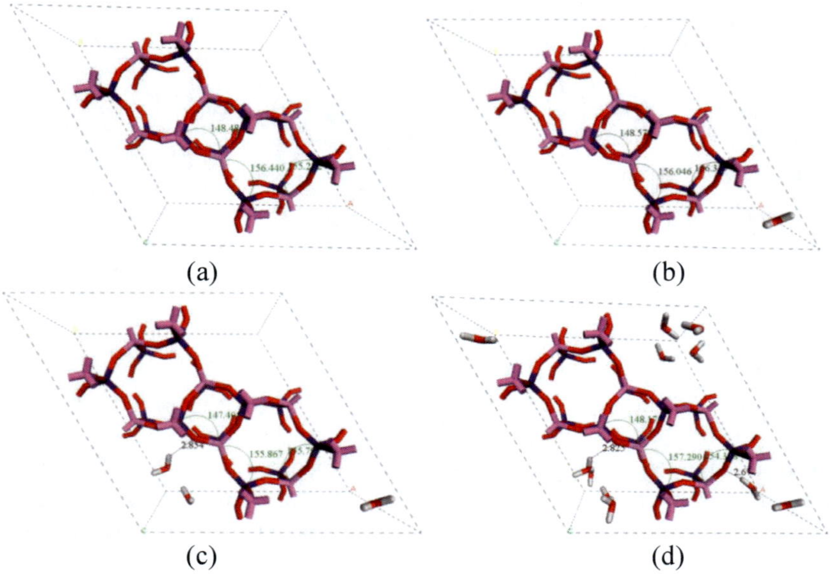

Figure 10. The selected Al-O-P angle and the selected shortest distance between hydrogen and framework oxygens in the energy minimized unit cell of (a) $AlPO_4$-5, (b) $AlPO_4$-5($1H_2O$), (c) $AlPO_4$-5 ($3H_2O$) and (d) $AlPO_4$-5 ($12H_2O$). The atoms are colored as follows: pink for aluminum, violet for phosphorus, red for oxygen, and white for hydrogen. Adapted from [77].

The decrease in the value of the Al-O-P angle is directly linked with a decrease of the acidity of $AlPO_4$-5 [153]. Similar Al-O-P angle values to those calculated for the structure with just a single water molecule in the unit cell were found for the optimized geometries of $AlPO_4$-5 having 3 and 12 adsorbed water molecules per unit cell. This is due to the fact that upon water adsorption, the additional water molecules coordinate directly with already adsorbed water molecules and new direct contacts with the aluminophosphate framework do not take place.

It is suggested that the initial loading of water in the large cavity is due to the mild hydrophilicity in the framework but the isobaric increase in loading is due to the chain of multiple hydrogen bonds between adsorbed water molecules. Finally, the DFT results have suggested that the hydrophilicity in aluminophosphates plays a major role in adsorption.

Kuznicki [154] reported in 1989 the synthesis of new microporous zeolite-type titanium silicates, namely Engelhard titanosilicates (ETS). Built of perfect or defected TiO_6-octahedra and SiO_4-tetrahedra, microporous titanosilicates have been the focus of researchers' attention due to their adsorption and catalytic properties. Due to the one dimensional −Ti−O−Ti− chain isolated in a siliceous matrix [155], these titanosilicates show good optical and catalytic properties. Engelhard titanosilicates have been applied as potential heterogeneous catalysts [156], photo catalysts [157, 158] and have been also utilized in the stabilization of radical cations [159].

For those reasons, computational chemists paid great attention to understand the electronic structure of titanosilicates, especially in ETS. For example, Sinclair and Catlow

[78] used various cluster models for finding the mechanism of alkene oxidation over titatanosilicate catalyst using DFT, BP86 functional, and a DZVP basis. Zimmerman *et al.* [160] reported that the optical properties of ETS could be calculated via the ONIOM method [56] using a 3Ti, 5Ti finite cluster model (Figure 11).

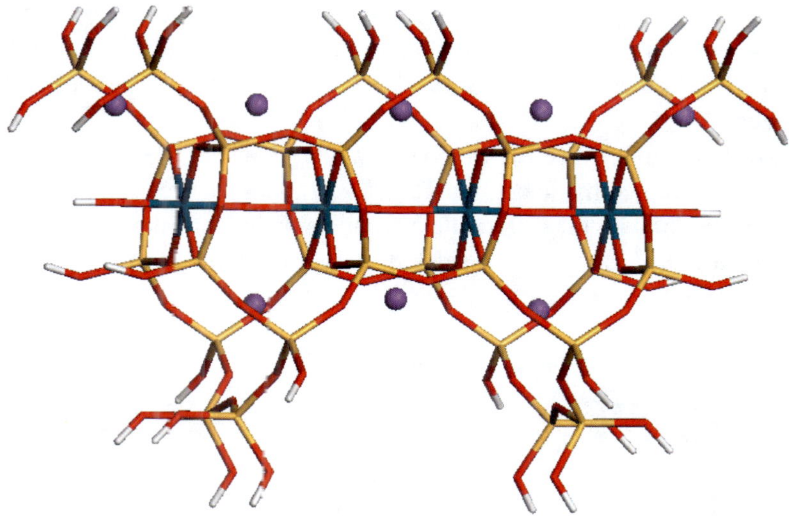

Figure 11. The cluster model of NaETS-10 used to determine its electronic structure by DFT. The atoms are colored as follows: yellow for silicon, green for titanium, red for oxygen, violet for sodium and white for hydrogen. Adapted from [56].

The DFT part in their ONIOM calculations considered the PBE functional and the CEP-121G* basis set for Na, Si and O atoms or the CEP-121G basis set for Ti and H atoms, while the MM part was modeled with the universal force field (UFF) approach.

The electronic and geometric properties of ETS optimized models showed that any asymmetry within the –O–Ti–O– chain does not affect the electronic structure of ETS-10. Recently, Shough *et al.* [79] used also the ONIOM approach and the PBE functional to describe the DFT region, with different embedded clusters to understand how the framework transition metal (V^{IV}, V^{V}, Nb^{V}, Mo^{V}, Cr^{III}, Fe^{III}, and Cu^{II}) substitution into the one-dimensional –O–M–O– chain structure affected the valence and the conduction bands and midgap states of the different TM substituted ETS-10. Linear combinations of the key factors affecting the electronic structure trends, *i.e.*, energies of the valence and conduction bands and midgap states, were used to elaborate predictive models for investigating the band structure of all possible three-metal combinations. The transition metal combinations with the lowest predicted ligand to metal charge transfer (LMCT) transition energies were suggested to enhance the photocatalytic activity of ETS-10 [79, 81]. In conclusion, it is possible to derive predictive models from electronic structure calculations of materials by using cluster models and DFT or DFT/MM approaches, which may be used to exhaustively search all possible metal combinations in a time span shorter than what it would take to investigate each system individually with DFT.

Cluster model approaches together with a DFT method can be used to obtain atomic charges and potential energy curves for the development of force fields to model this type of systems but also to calculate other data, *e.g.* vibrational frequencies, which can be further used to interpret results from experimental work. Recently, we have studied the interaction of NO with another Engelhard titanosilicate, NaETS-4, this one presenting unsaturated titanium atoms, which has been modeled by both periodic and cluster model approaches [161].

The calculations considered two different DFT approaches, namely, the GGA/PBE and the B3LYP hybrid exchange-correlation functionals. The periodic calculations considered plane-wave basis sets and were employed to understand the preferred sites for NO adsorption. The role of water molecules in the pores of NaETS-4 on the mechanism of NO adsorption was also analyzed.

It was shown that the presence of water in the pores of ETS-4 promotes the NO adsorption at the unsaturated (pentacoordinated) Ti^{4+} framework ions (Figure 12), which was found to be in agreement with the different experimental behavior found for NO adsorption in the ETS and in the other materials considered, *e.g.*, the zeolites mordenite and CaA, and natural and pillared clays.

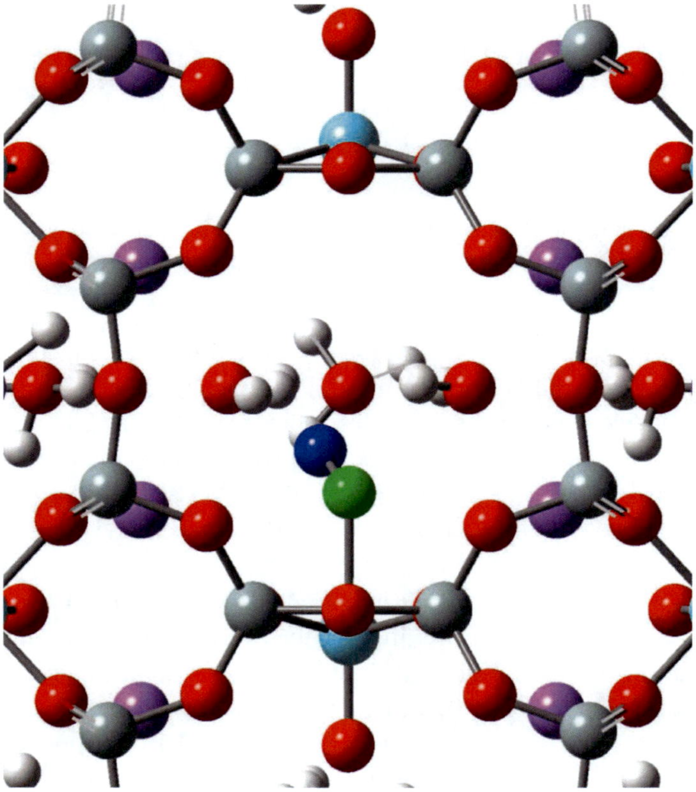

Figure 12. Optimized configuration for NO (O-down) interacting with the five-coordinated Ti atom in NaETS-4 modeled with the periodic approach. The atoms are colored as follows: green and blue for oxygen and nitrogen atoms of NO, grey for silicon, red for framework and water oxygen atoms, white for hydrogen, cyan for titanium, white for hydrogen and violet for sodium. Adapted from [161].

As it happens with dehydrated zeolites, NO was suggested to bind to extra-framework cations when some water molecules (three per unit cell, 1×2×1, used in the calculations) were removed from the pores of NaETS-4. A cluster model, $(TiO_5)Si_4O_{10}H_8$, centered on the unsaturated Ti^{4+} framework ions (Figure 13) was used for the calculation of the vibrational frequencies of several different N_xO_y species, namely, NO, N_2O, NO_2 and N_2O_2, adsorbed directly on or close to Ti^{4+} framework ions. The calculations were performed for aiding the experimental assignments of the infrared bands determined by diffuse reflectance infrared Fourier transform spectroscopy (DRIFT).

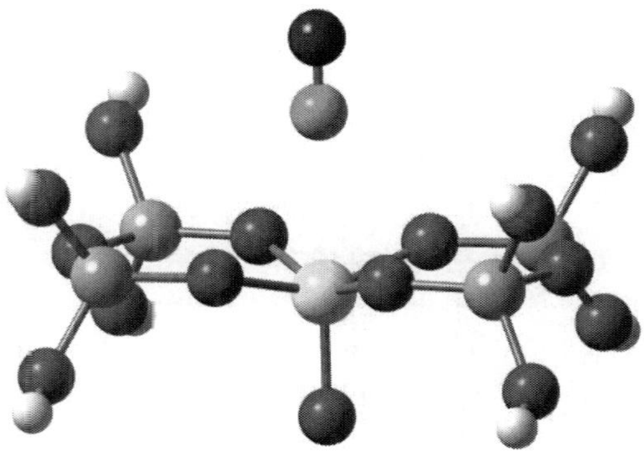

Figure 13. Optimized configuration for NO (O-down) interacting with the five-coordinated Ti atom in NaETS-4 modeled with the cluster approach. Adapted from [161].

The calculated vibrational frequencies were used to support the experimental evidences that NO molecules can also react inside the cavities of the different materials studied to yield other nitrogenated species [161].

3.2. Chemical Reactions Catalyzed by Metal Surfaces and Metal Nanoparticles

Theoretical methods are being widely used in heterogeneous catalysis in order to understand the catalytic processes at micro and also at macro levels, being of special relevance those based on the density functional theory due to their relatively high accuracy at moderate computational cost.

The combination of theoretical and experimental methods leads to a better understanding of the catalytic properties of extended surfaces and nanoparticles, allowing the determination of important mechanistic details that are difficult to obtain only from experimental results and that are very useful to understand the catalytic processes [162 - 168 A few examples are given below that illustrate how DFT can be used to provide very useful information at the atomic level, which is not easy to be obtained with experimental approaches, even for quite simple reactions and for very well controlled conditions. For instance, Michaelides *et al.* [163] conclude from their DFT calculations carried out at the PW91 level that water acts as an autocatalyst in its formation on platinum (111) from chemisorbed oxygen and hydrogen adatoms. Honkala *et al.* [165] used DFT to obtain the rate constants, within the harmonic transition state theory (TST) formalism, for the ammonia production on ruthenium nanoparticles supported on a magnesium aluminum spinel and using as unique experimental input the particle size. The theoretical data obtained are in good agreement with rate measurements performed experimentally over a wide range of synthesis conditions. Furthermore, Rodríguez *et al.* [167] concluded from DFT calculations that the high activity detected experimentally for the water gas shift (WGS) reaction on TiO_2 and CeO_2 nanoparticles supported on the Au(111) surface is due to the occurrence of the WGS reaction in the interface between the oxide nanoparticle and the gold surface, where the reaction has lower activation energy. DFT calculations were also used for the elucidation of mechanisms as those occurring in the metal-oxide interface of platinum-ceria catalysts, which are responsible for the enhancement of the performance of supported catalysts [168]. Vayssilov *et al.* identified two different oxidative interaction types between the platinum and the ceria support, *i.e.*, interaction based on a pure electronic effect involving electron transfer from the metal to the oxide or a mechanism with transport of activated oxygen from the oxide to the metal [168].

In the computational study of chemical reactions catalyzed by metal surfaces and metal nanoparticles, for instance gas-solid catalyzed reactions, the catalytic systems and the reacting species (adsorbates) have to be represented properly. Great care has to be taken in the choice of the size of systems under study, *i.e.*, they have to be representative of the real systems but have to be tractable computationally. Nowadays, in this type of computational studies, the periodic model approach is usually the most suitable one since it represents more accurately the size and the type of exposed surfaces which are encountered in a real catalyst. The most important problem affecting the periodic model approach is the difficulty in the study of low-coverage regimes, which can only be studied with this approach if very large unit cells are considered, with obvious implications in the computational cost of the calculations. This problem can be overcome by considering the cluster model approach, *i.e.*, a model that consists of a finite portion of the catalyst (a cluster or a nanoparticle).

The problem with this approach is connected with the different physicochemical nature of the border atoms which can be in a certain manner reduced by consideration of an embedding scheme. Both simulation techniques are useful since quite often catalysts are made of nanoparticles, clusters, rows or rods dropped on an infinite surface and, therefore, a realistic representation of the catalyst surfaces will require the consideration of high and low-coordinated atoms [167, 169].

The reaction of CO oxidation on gold-based catalysts is one of the most studied reactions by experimental and computational approaches since the discovery by Haruta *et al.* [170] that gold nanoparticles highly dispersed on an oxide support are able to catalyze the CO oxidation at temperatures as low as -70 ºC. Most of the computational works devoted to the study of this reaction on gold were focused on explaining why the dispersion of the gold increases the reactivity toward the CO oxidation [171, 172]. The main conclusion retrieved is that the presence of low-coordinated atoms in the catalyst surface, which is directly related with the high dispersion of the nanoparticles, increases the adsorption energy of the CO and decreases the activation energy barrier for the oxidation reaction. On gold surfaces, it was proposed that the reaction follows the elementary paths below:

$$CO^* + O_2^* \rightarrow OCOO^* + * \quad (1)$$

$$OCOO^* \rightarrow CO_2\uparrow + O^* \quad (2)$$

$$CO^* + O^* \rightarrow CO_2\uparrow + 2^* \quad (3)$$

where the symbol * denotes an adsorption site available on the catalyst surface (Figure 14). Thus, the reaction of CO oxidation by molecular oxygen proceeds via the formation of the

OCOO intermediate, which is decomposed to gaseous carbon dioxide and an oxygen adatom. This implies that, on gold surfaces, the O-O bond cleavage in the OCOO intermediate, resulting from the direct reaction with molecular oxygen, is more favorable than the reaction of O_2 dissociation [171 - 174].

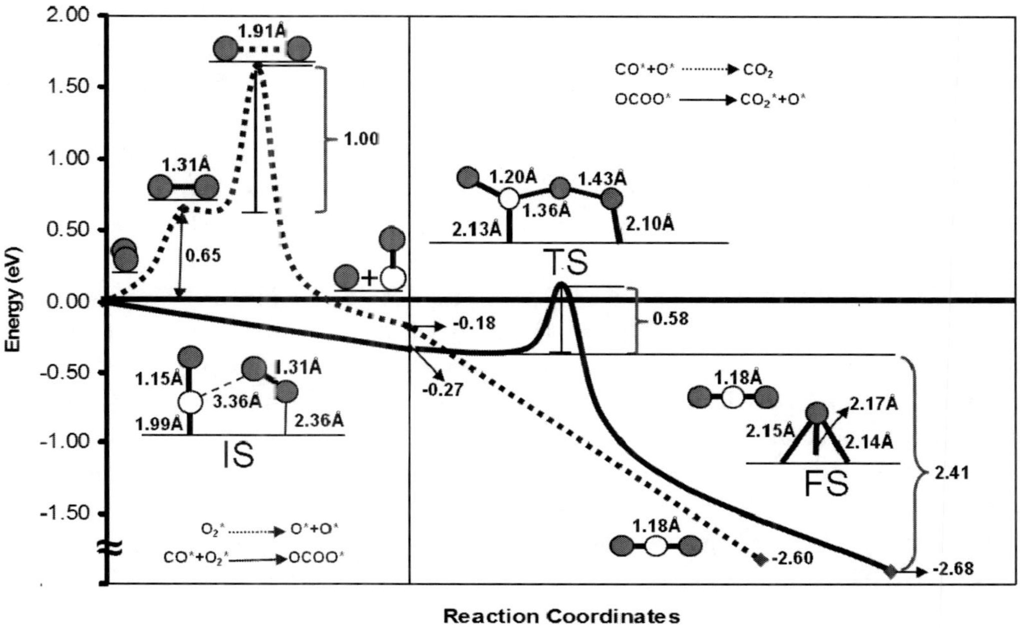

Figure 14. Reaction profile for the CO oxidation by direct reaction with molecular oxygen on the Au(321) surface. Dark and light spheres are used for oxygen and carbon atoms, respectively. Adapted from [174].

The energies calculated in these computational studies confirm that the reaction on pure and defect-free gold surfaces will follow the mechanism described by the reactions shown in equations 1-3. This is the case for both the planar surfaces of gold, including the most stable Miller index, *i.e.*, Au(111), and also for the stepped surfaces, the latter having some low-coordinated atoms on the exposed surface.

However, the global picture can change if the gold active species are different from those usually found on extended pure metal surfaces. For instance, the supporting of gold nanoparticles on a nanostructured dual-oxide support was found to introduce significant mechanistic changes since the reaction of O_2 dissociation becomes energetically accessible [169]. In this type of catalysts, the oxidation reaction occurs directly between CO* and O* surface species.

Very recently, the remarkable oxidative catalytic activity of nanoporous gold was suggested to have its origins in silver impurities available at the outermost layers of the catalyst, which resulted from the process by which the nanoporous catalysts are obtained [175]. In fact, the structure of nanoporous gold is obtained by leaching a less noble metal, such as silver, out of the corresponding alloy [176]. Nanoporous gold was claimed to be a pure and unsupported gold catalyst but the analysis of its morphology, surface composition, and catalytic properties reveal that silver, even if removed almost completely from the bulk, segregates to the surface. Therefore, nanoporous gold should be considered as a bimetallic catalyst rather than as a pure Au catalyst. The segregation of silver to the surface of gold is supported by very recent Monte Carlo simulations [177]. In the simulations of Deng *et al.*, the segregation of Ag to the surface in several different sized AgAu bimetallic particles as observed [177]. The understanding of the role of the silver species in aiding the dissociation of the oxygen molecule on gold-based catalysts was obtained from periodic DFT calculations with the PW91 exchange-correlation functional. [164] Fajín *et al.* considered several different pure and Ag-doped gold surfaces as models of nanoporous gold catalysts where O_2 was suggested to be activated. [164] Since the structure of nanoporous gold is not known, Fajín *et al.* considered both planar and stepped gold surfaces, namely, Au(111), Au(110) and Au(321), where silver atoms were deposited on these surfaces (atomic deposition by addition of atoms) or were used to substitute gold atoms (atomic substitution by replacing atoms in the original surface) in these surfaces (Figure 15).

The DFT calculations showed that residual Ag species on gold-based catalysts are able to promote adsorption and to favorably dissociate O_2 with high rate constants. Based on the calculated energies and reaction rates, O_2 activation on nanoporous gold catalysts most probably occurs on Ag particles or rows deposited on Au or on corrugated Au surfaces with a few incrusted Ag atoms. In such cases, the reaction of O_2 dissociation was found to be possible after surmounting moderate energy barriers. Previous works showed that negatively charged gold clusters with even number of atoms were found to chemisorb O_2. Fajín *et al.* [164] performed additional calculations considering negatively charged slabs and found that charged catalysts were in fact able to stabilize more efficiently the adsorbates with respect to gaseous oxygen than the uncharged slabs, with a slightly more decisive role for the initial state, *i.e.*, adsorption of O_2 on the surface (O_2*), which enabled the catalysis of the reaction of dissociation of molecular oxygen.

The reaction of water gas shift (WGS) constitutes another example of a catalytic reaction with an enormous importance for several different industrial applications, *e.g.*, the methanol synthesis, the methanol steam reforming and the cleaning of hydrogen streams for fuel cells, [178 - 180] which has been heavily studied by means of several different experimental and computational techniques. In industry, the WGS reaction is catalyzed by copper-based catalysts, being the active Cu species dispersed on a oxide support. [181, 182]. Recently, catalysts using gold as the active species were suggested as promising alternatives for the WGS reaction [183, 184].

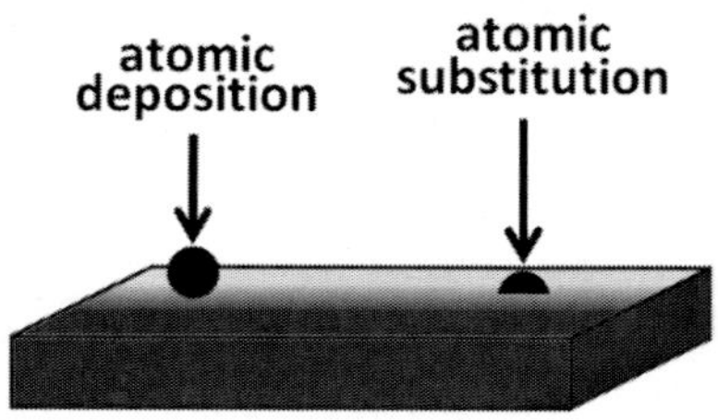

Figure 15. Incorporation of silver atoms in gold surfaces by atomic deposition or atomic substitution.

In the catalysts used for the WGS reaction, the metal species are accepted as the active ones [185 - 187] but, the nature of the support, [188 - 192], the existence of oxygen vacancies [193, 194] or the catalyst preparation process [195] may significantly affect the performance of the catalysts. Doping [196, 197] or formation of alloys [198, 199] can also affect the catalytic activity.

The DFT works devoted to the understanding of the WGS reaction on extended surfaces evidenced the importance of low-coordinated atoms in the catalysis [200, 201]. Still, the mechanism is suggested to be the same on flat and stepped copper surfaces [200, 201]. In most surfaces studied so far, the reaction of water dissociation into OH and H surface species is considered to be the rate limiting step, which is the case for copper surfaces [200, 201].

Most of the calculations that we have developed so far in our group are based on the PW91 functional since this DFT approach was found to provide adsorption energies for the water molecule on Cu(111), and for the energy barrier for its dissociation on the same surface, which were in good agreement with available experimental data [202]. Furthermore, most of the calculations in the condensed phase, i.e., for adsorbates interacting with the surface of metallic catalysts, were performed with a restricted formalism since the calculations run much faster than those performed with an unrestricted formalism and the differences between the results calculated with both approaches were found to be quite small [203], even when dealing with adsorbates that in the gas phase have one or more unpaired electrons.

The first reaction step of the WGS reaction, *i.e.*, the reaction of water dissociation, on several metallic flat and stepped surfaces was used to obtain Brønsted-Evans-Polanyi (BEP) relationships [204, 205], linking the activation energy barrier of the reaction of water dissociation with the energy of reaction, or with the adsorption energy of the reaction products, or with the adsorption energy of an oxygen adatom (see Figure 16) [206]. These three BEP relationships were found to be approximately linear functions [206]. Similar relationships were obtained recently for the reaction of water dissociation on platinum nanoparticles and on bimetallic surfaces [207, 208]. These relationships are of great interest for screening purposes since the determination of the activation energy barrier in other systems not included in the derivation of the BEPs is understandably easier and with concomitant computational costs that are lower than those required to fully study the WGS reaction, especially the calculation of the activation energy barriers for the reaction on several different catalyst surface models. In fact, it is of particular interest the BEP relationship relating the activation energy barrier with the adsorption energy of an oxygen atom due to the easy calculation of this quantity. However, the BEP relationships obtained in these works are quite simple and do not take into account the influences of several parameters that usually play an important role catalysis, such as pressure, temperature and concentration of the reacting species. Usually, these deficiencies are overcome through consideration of microkinetic models based on DFT calculated adsorption energies and activation energy barriers [200]. The WGS reaction on much more complex systems such as CeO_x and TiO_x nanoparticles grown on the Au(111) surface was also studied by means of DFT calculations [167], for obtaining a clearer picture of the causes for the high activity of these catalytic systems towards the WGS reaction. The activity was related with the direct participation of the oxide in the catalytic process, *i.e.*, the rate-limiting elementary path for the WGS reaction, namely, the water dissociation, takes place on oxygen vacancies at the oxide nanoparticles. Then, the reaction of CO with the products of water dissociation occurs at the metal-metal oxide interface. DFT methods were also successfully used in the interpretation of other catalytic reactions, such as the NO oxidation [209 - 212], the NO dissociation [213], the hydrogen activation [214, 215], the oxidation of alcohols [216], the ethylene hydrogenation

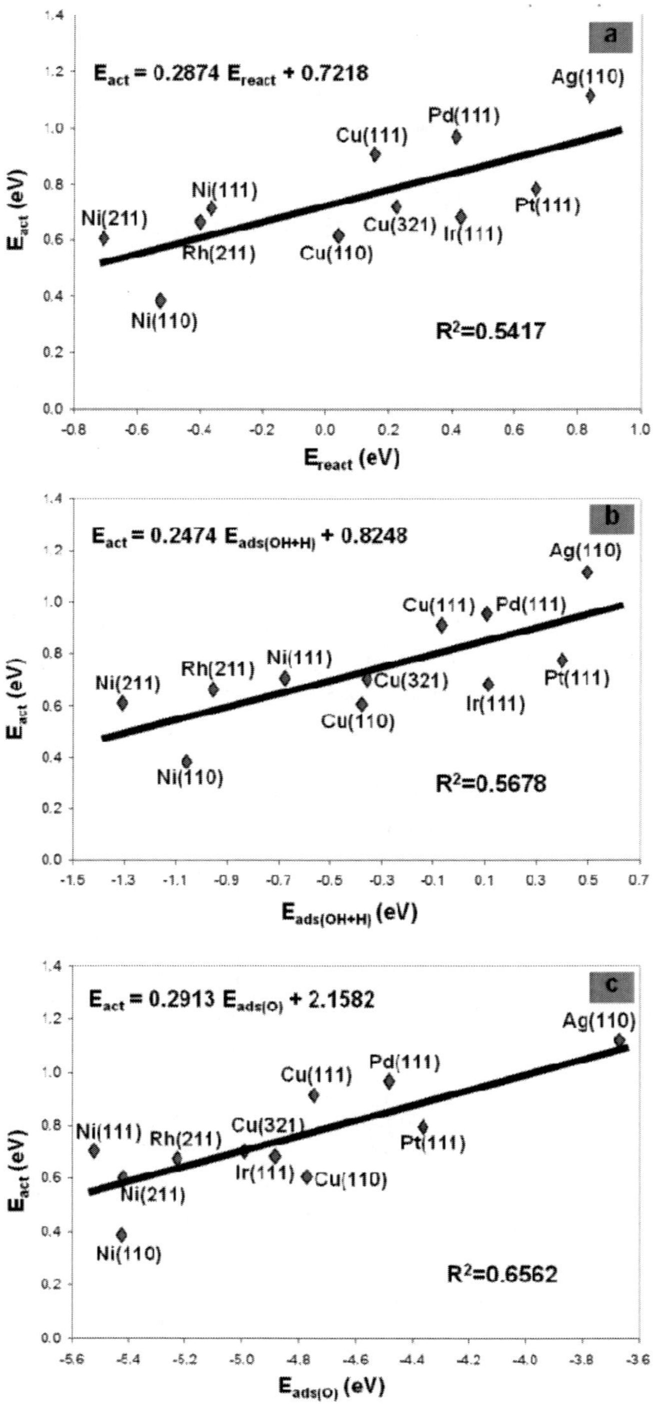

Figure 16. BEP$_s$ relationships between activation energy barriers (E_{act}) and: (a) reaction energy (E_{react}), (b) co-adsorption energy of HO and H species; (c) adsorption energy of atomic oxygen. Note that (a) corresponds to the standard BEP relationship while (c) provides a convenient descriptor for the reaction of water dissociation. Figure adapted from [206].

and dehydrogenation [217], occurring on surfaces, on nanoparticles or on supported nanoparticles.

In the case of reaction involving the NO molecule, a very complete study considering gold as the catalyst active species was conducted recently. [209, 210, 213, 218] The reaction of NO oxidation was studied using periodic models of the Au(111) surface either in the presence of atomic or of molecular oxygen. The DFT calculations show that the reaction with atomic oxygen is exothermic. Both in the case of the reaction with atomic or with molecular oxygen, the desorption of gaseous NO_2 constitutes the bottleneck of the reactions, since the energy required to desorb NO_2 from the surface is 0.78 eV (75 $kJ \cdot mol^{-1}$), which constitutes the highest of the calculated barriers for the reaction of NO oxidation [209].

The reaction of NO reduction on gold surfaces, either in the presence or absence of CO and/or H_2 on the catalyst, was also studied by DFT, more precisely with the PW91 functional and with plane-wave basis sets [218].

The calculations considered a periodic model of a stepped gold surface presenting low-coordinated atoms, i.e., the Au(321) surface. The study of the NO reduction with CO on gold surfaces allowed the determination of a complex reaction scheme with several competing routes, which lead to several different reaction products such as ammonia (NH_3), molecular nitrogen (N_2), nitric dioxide (NO_2) or nitrous oxide (N_2O). The extensive analysis of the several elementary steps enables to suggest that under oxygen-rich conditions, either in the presence or in the absence of surface hydrogen species, NO will be preferentially oxidized to NO_2, which will stay on the catalyst surface. Under oxygen-poor conditions, the most favorable mechanism leading to the formation of gaseous N_2 and CO_2 in the absence of hydrogen species proceeds via $(NO)_2$ dimers, which eventually decompose into N_2O and N_2. The oxygen adatoms will combine with CO to form CO_2. This reaction is described by the following equation, $2NO+2CO \rightarrow ON_2O+2CO \rightarrow N_2O+O+2CO \rightarrow N_2+2O+2CO \rightarrow N_2+2CO_2$. In the presence of hydrogen, the most favorable mechanism leading to the formation of gaseous N_2 and CO_2 follows through the NOH and N_2O intermediates and water vapor is one of the products of the reaction.

In this case, the global reaction is described by the following equation, $2NO+CO+H_2 \rightarrow 2NO+CO+2H \rightarrow NO+NOH+CO+H \rightarrow NO+N+CO+H_2O \rightarrow N_2O+CO+H_2O \rightarrow N_2+O+CO+H_2O \rightarrow N_2+CO_2+H_2O$. The activation energies for the bottleneck steps in these two reactions schemes are moderate and have approximately the same value, *i.e.*, ~0.9 eV (87 $kJ \cdot mol^{-1}$). However, under oxygen-poor conditions but in the presence of hydrogen, much more favorable reaction routes leading to NH_3 as the final nitrogenated species (instead of N_2) are possible.

In such cases, successive addition of H atoms to N or NOH surface species yields NH, NH_2 and, finally, NH_3 with a maximum energy barrier of ~0.7 eV (67 $kJ \cdot mol^{-1}$), or yields NHOH, NH_2OH and, finally, NH_3 with a maximum energy barrier of ~0.2 eV (14 $kJ \cdot mol^{-1}$). In the latter situation, the global reaction is described by the following equation, $2NO+CO+4H_2 \rightarrow 2NO+CO+8H \rightarrow 2NOH+CO+6H \rightarrow 2NHOH+CO+4H \rightarrow 2NH_2OH+CO+2H \rightarrow 2NH_3+2OH+CO \rightarrow 2NH_3+OH+OCOH \rightarrow 2NH_3+CO_2+H_2O$. When the reaction occurs in the presence of hydrogen, the bottleneck step is the formation of the surface NOH species, which has an activation energy of ~0.5 eV (48 $kJ \cdot mol^{-1}$). The activation energy barrier in the presence of hydrogen is ~40% smaller than in the other two cases above. The possible formation of N_2 and NH_3 agrees with available experimental results on gold catalysts but also on other transition metal ones. In fact, despite the simplicity of the model, DFT approaches are found to be quite useful to accurately study heterogeneous catalytic reactions. This gives enhanced confidence to the calculated activation energies and paves the way for the optimization of the catalytic process, for instance, via microkinetic modeling in which the effects of the pressure, of the temperature and of different combinations of the admission gases can be studied in detail.

3.3. Interaction of Drugs with Enzymes (Enzymatic Inhibition)

Malaria is well known as an infectious lethal disease since ancient times, and remains the most widespread (via mosquito) and severe tropical disease, with ~1 million deaths/year [219]. Five parasite species cause human malaria: *Plasmodium* (*P.*) *vivax*, *P. malariae*, *P. ovale*, *P. falciparum* (*Pf*, the deadliest) and *P. knowlesi* (recently found to infect humans) [220]. Several organized efforts to control the dissemination of malaria, and achieve its eradication have been made through history [221]. Despite initial success, with regional elimination in Southern Europe and in some countries in North Africa and in the Middle East, there has been complete failure to eradicate malaria in many other countries due to a number of factors [222]. Since then, we have been witnessing an increase in occurrences globally, though the number of fatal cases has been decreasing over the past few years [219].

Major reasons for failure in malaria eradication are (i) parasite's complex life cycle (asexual and sexual cycles in the human and mosquito hosts, respectively) and (ii) fast-emerging resistance to many available drugs, as antifolates or chloroquine, which were classical antimalarial drugs for half a century [222]. Recently, drug resistance against the state of the art antimalarial drug, artemisinin, is also emerging [223, 224]. Therefore, the development of new classes of effective and affordable antimalarials, especially compounds that act against novel biochemical targets, is required.

Since the unveiling of the *Pf* genome in 2002 [225], a number of potential targets for drug intervention have emerged [226]. Among them, falcipains (FP), cysteine proteases from *Pf* that play a key role in haemoglobin degradation, have been referred as highly promising targets [227]. Haemoglobin is the most abundant protein in erythrocytes, and becomes completely degraded after parasite entry. During the erythrocytic stage, the parasite relies on human haemoglobin hydrolysis to supply amino acids for protein synthesis and to maintain osmotic stability [228, 229]. Thus, it is understandable that falcipains have generated substantial interest [230]. Inhibitors of these enzymes have been shown to block the hydrolysis of haemoglobin [231] and to inhibit the rupture of erythrocytes [232], thus suggesting that they are valuable candidate therapeutic targets in the development of new antimalarials. The best characterized *Pf* cysteine proteases are the falcipains and among the four *P. falciparum* cysteine proteases, falcipain-2 (FP2) and falcipain-3 (FP3) appear to be the principal food vacuolar haemoglobinases [222]. Studies have demonstrated that undegraded haemoglobin accumulates in the food vacuole when the FP2 gene is disrupted, confirming that this enzyme participates in haemoglobin hydrolysis [233]. Although disruption of FP3 could not be achieved, when the gene was replaced with a tagged functional copy, evidences were found indicating that FP3 is essential for erythrocytic parasites [234]. FP2 and FP3 share 67% sequence identity and contribute more or less equally to the digestion of haemoglobin in the food vacuole. Thus, it is commonly accepted that a potent inhibitor against both FP2 and

FP3 will most probably impair haemoglobin degradation to levels that are lethal for the parasite [222].

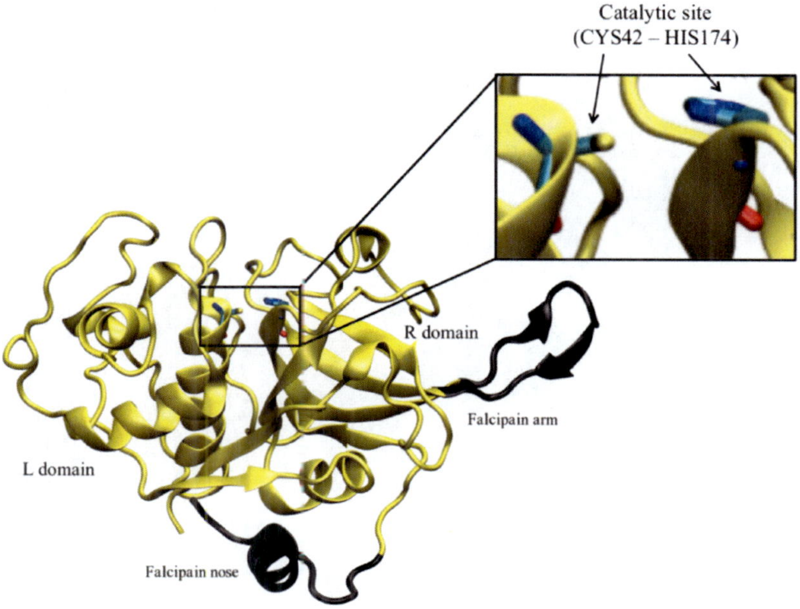

Figure 17. Representation in new cartoon of the structure of falcipain-2 (PDB code: 3BPF) with the L domain (Left) and the R (Right) domain colored in light grey. The N-terminal extension (FPnose) at the lower end and the C- terminal insertion (FParm) at the rightmost end are colored in dark grey. Adapted from [235].

The falcipains are fairly typical papain-family cysteine proteases adopting the classical papain-like fold in which the protease is divided into L (left) and R (right) domains (Figure 17), though they have some rather uncommon features, *e.g.* FP2 and FP3 structures contain two unique features assigned as $FP2/3_{nose}$ and $FP2/3_{arm}$ (Figure 17) [235].

The $FP2/3_{nose}$ motif, situated at the N-terminus, is required for folding, while the $FP2/3_{arm}$ contains a C-terminal insertion and mediates interaction between the falcipains and haemoglobin independent of the enzyme's active site [235]. The active site of *Pf* cysteine proteases is composed of the catalytic triad CYS42, HIS174 and ASN204 for FP2 and CYS44, HIS176 and ASN206 for FP3, which are located at the junction between domains L and R. The catalytic site is formed by a cysteine and a histidine, whose side chains form a thiolate/imidazolium ion pair, and also by an asparagine, which has a crucial role in the appropriate orientation of the ion pair [236].

A schematic representation of the mechanism of action is outlined in Figure 18. One of the key events in the catalytic hydrolysis of haemoglobin is the nucleophilic attack of the thiolate anion to the appropriate electron deficient carbonyl group of the substrate leading to a negatively charged tetrahedral intermediate that was found to be stabilized by the "oxyanion hole" formed by the side chains of GLN36 and TRP206, in FP2, and by GLN38 and TRP208 in FP3 [222].

A wide range of compounds has been identified as inhibitors of falcipains, which are able to block the enzyme's activity by forming a reversible or irreversible covalent bond with the active site cysteine. Most of the falcipain inhibitors identified so far are peptide-based inhibitors (Figure 19) [237].

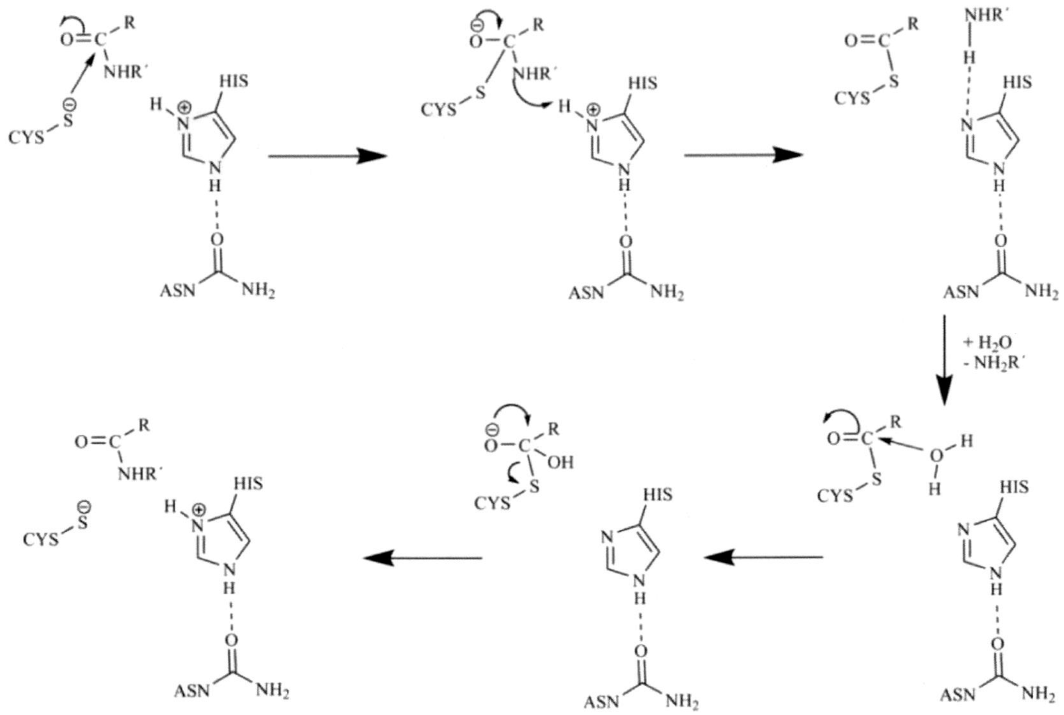

Figure 18. Schematic reaction mechanism of the cysteine protease mediated cleavage of a peptide bond.

This category includes peptidyl fluoromethyl ketones [238], peptidyl vinyl sulfones [239], cinnamic/4-aminoquinoline peptidyl conjugates [240], peptidyl α-ketoamide derivatives [241], epoxysuccinyl derivatives [242] and peptidyl aziridines [242]. Historically, novel enzyme inhibitors are discovered either by serendipity or by screening of natural and synthetic compounds. Over the past few years, through a better understanding of biochemical processes and the emergence of powerful computational tools, rational drug design has emerged complementarily to these techniques. The use of density functional theory modeling towards the development of novel falcipain inhibitors is reviewed in this section.

Despite the explosive growth of computer power over the past decades that led to the development of large-scale simulation techniques, a direct application of first principle approaches, which treat explicitly all electrons and determine interactions between atoms solely from their electronic configurations and positions, to the study of biomolecules is still limited due to the size of the model systems. While highly accurate, these methods are very expensive and can only realistically be implemented for small sized models, representative of the biological system, and very short time scales when applied to molecular dynamics. A successful strategy is the coupling of classical molecular dynamics (MD) simulations [243], based on empirical force fields, to quantum mechanics resulting in the so-called quantum mechanics/molecular mechanics (QM/MM) MD method [244].

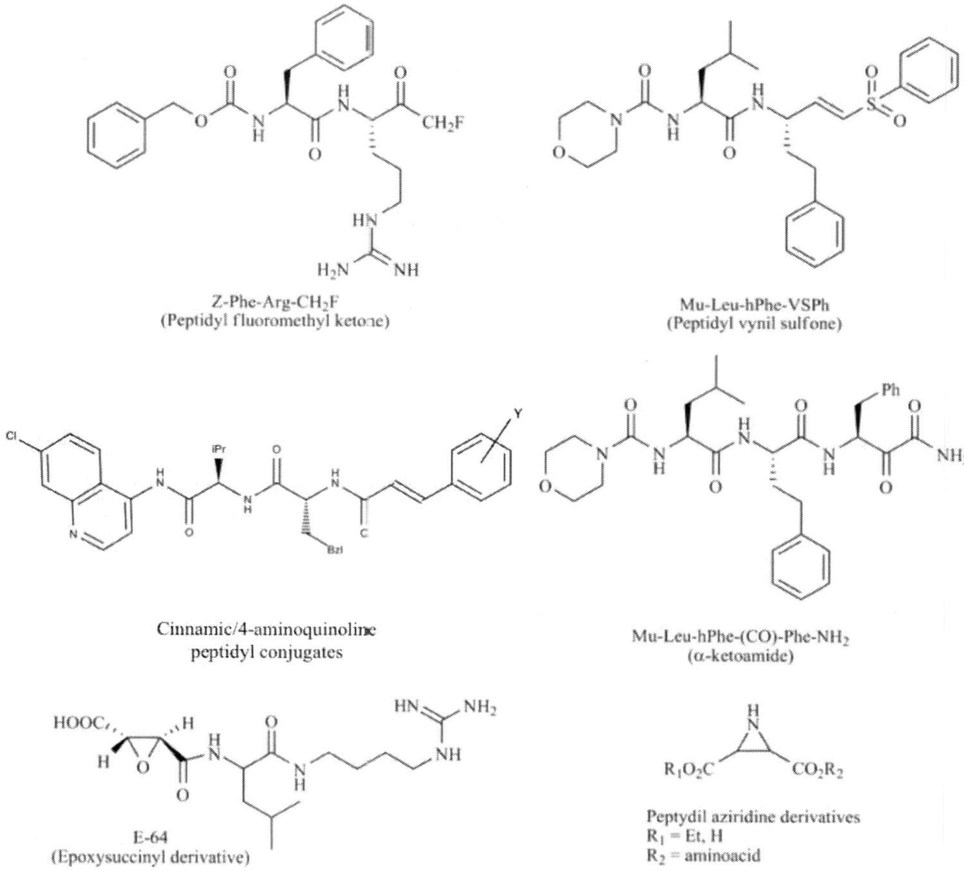

Figure 19. Molecular structures of some peptide-based falcipain inhibitors.

Although QM and QM/MM methods have given important insights into the catalytic mechanism of falcipains and into the falcipain inhibitors reactivity, the studies performed so far are mainly focused on epoxides and aziridines. Kinetics, thermodynamics and regioselectivity of the ring-opening reaction of epoxide- and aziridine-based agents have been investigated by QM calculations [245, 246].

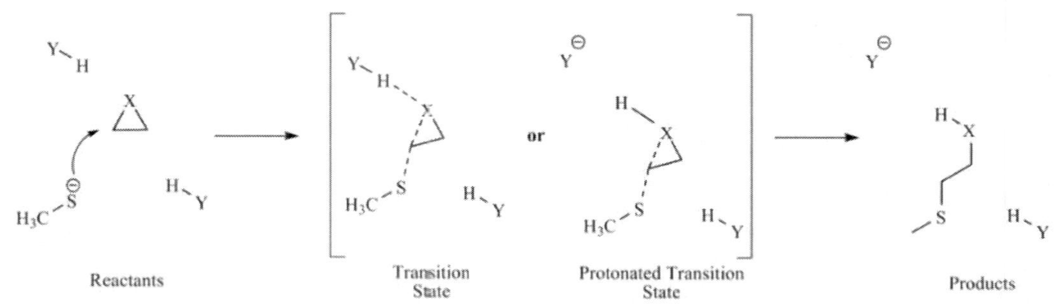

Figure 20. Schematic representation of the ring opening mechanism of the alkylation of a methyl thiolate by an N-substituted aziridine ring. X = O, N-R. Adapted from [247].

Generally, the mode of action of these molecules is described as a two-step mechanism. Firstly, there is a formation of a non-covalent enzyme-inhibitor complex assured by a favorable arrangement of the molecule into the binding pocket. Then, ring opening takes place (Figure 20), when one of the carbon atoms of the ring is attacked by the thiolate group of the catalytic CYS, which leads to irreversible alkylation of this amino acid.

The quantum chemical calculations performed by Helten and co-workers [245,246], evaluated the influence of environment at the ring carbon atoms on the kinetics and thermodynamics. These calculations indicated that the aziridine ring opening is influenced by acidic media to a greater extent than what was observed for epoxide.

They observed that protonation of the nitrogen center of aziridine-based inhibitors occurs at the beginning of the reaction course, more precisely, prior to the transition state for the ring opening described above. Importantly, it was also found that, without such previous *N*-protonation step, the aziridine is not active. Therefore, it is suggested that the protonation stabilizes both the transition state and the product of ring opening, significantly increasing the reaction rate. Hence, electron-withdrawing substituents at the nitrogen atom should lead to lower reaction barriers and, consequently, to higher inhibition potencies [246]. For epoxide-based inhibitors, the protonation of the oxygen center takes place only after occurrence of the transition state; therefore, since the products are strongly stabilized, only the thermodynamics of this reaction is favored by the *O*-protonation, while the kinetics remains unchanged, in contrast with the behavior of aziridines [246]. The authors defined a simplified model as representative of the enzyme inhibition process, where the attacking cysteine was mimicked by a methyl thiolate ($H_3C\text{-}S^-$) while the inhibitors were modeled by the three membered ring systems $(H_2C)_2X$ with X = O, N-R. To capture the effect of a decreasing pH value on the reaction profile, solvent molecules with increasing proton donor ability were placed in the vicinity of the heteroatom of the three membered rings and in the vicinity of the methyl thiolate. Water molecules were employed to mimic environments with weak proton donor ability (p*K*a = 15.74), while NH_4^+ (p*K*a ≈ 9.3) and HCO_2H (p*K*a ≈ 3.8) molecules were used to simulate environments with higher proton donor abilities [245]. The geometry optimizations of the relevant stationary points, checked by frequency calculations, were computed with the BLYP [14,18] functional while the energies were obtained from B3LYP [21] single point calculations. Both functionals were combined with a TZVP basis set [248]. The accuracy of the theoretical approach was assessed and the authors reported that the reaction profile obtained with B3LYP showed an excellent agreement with CCSD(T) results, whereas BLYP considerably underestimated the barrier heights [245].

Further work was done by Vicik and co-workers to study the effect of substitution at the aziridine nitrogen atom in order to modulate the properties of the aziridine ring related to the nucleophilic ring-opening reaction and, consequently, to provide a rational background for the design of aziridine-based inhibitors with improved inhibition potency [249]. Quantum chemical calculations, employing the same theoretical approach and more simplified models when compared with those in the studies of Helten *et al.* [245,246], predicted that the attachment of a CHO- group at the N-position of the aziridine ring should lower the reaction barrier for nucleophilic attack by the methylthiolate species by about 14 $kcal \cdot mol^{-1}$ (59 $kJ \cdot mol^{-1}$).

This indicated that N-formylated derivatives should possess increased potencies since the first-order rate constant of inhibition, k_i, is enhanced. Reactions with N-formyl aziridine-2,3-dicarboxylate (**b1**, Figure 21), which were performed to verify the predictions, showed a

5000-fold enhanced inhibition potency relative to the N-unsubstituted analogue (**b2**, Figure 21), against the *Pf* cysteine protease FP2 [250].

This inhibition increase was mainly a result of very low dissociation constants K_i and not a consequence of an increased k_i value. Additional computations were performed for clarifying these unexpected findings [249]. It was shown that the attack of the cysteine moiety at the carbonyl group of the CHO-substituent possesses a much lower barrier than the corresponding attack at the ring- carbon. However, this reaction is endothermic, *i.e.*, the formed complex will decompose on a fast time scale resulting in a reversible attack at the carbonyl carbon of the formyl moiety.

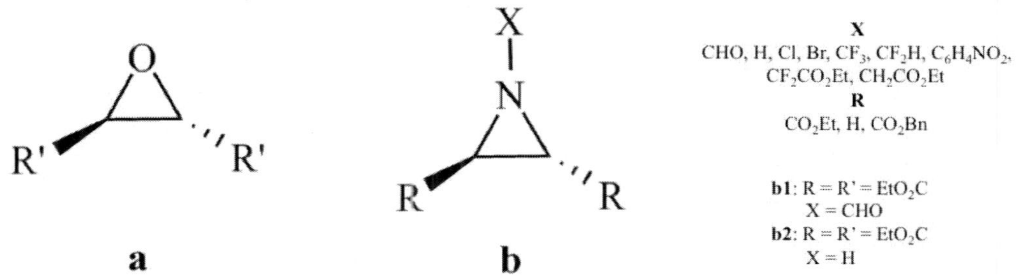

Figure 21. General structure of (a) epoxide- and (b) aziridine-based inhibitors. The structure of N-formyl aziridine-2,3-dicarboxylate and its unsubstituted analogue are represented by b1 and b2, respectively.

Based on the previous findings that N-formylated substituents at the aziridine nitrogen strongly accelerate the alkylation step of the enzymatic inhibition process, Buback and co-workers focused on the *N*-substitution pattern by varying the electron-withdrawing power of substituents like Cl, Br, CF_3, CF_2H and $C_6H_4NO_2$ [247].

Insights into the kinetics and thermodynamics of ring opening were obtained for the different *N*-substituted aziridines (Figure 21), through the achievement of two-dimensional potential energy surfaces by employing the same theoretical approach and the most simplified model reported so far, and compared with the corresponding non-substituted analogues. The computed reaction barriers for the substituted aziridines ranged from 13-16 kcal·mol^{-1} (54-67 kJ·mol^{-1}), representing a decrease in the activation barriers of up to 15 kcal·mol^{-1} (63 kJ·mol^{-1}), when compared with the barriers calculated for the unsubstituted compounds. Furthermore, the new substituents possessed the additional advantage that they do not bring up reaction pathways other than the nucleophilic ring opening, as observed for the N-formylated substituents. To verify these hypotheses, selected compounds were synthesized and tested, confirming the predictions and supporting the correlation between compound activity and the presence of such groups attached to the N atom.

With the models reported so far it was possible to explain the influence of the pH value of the surroundings and of the substituents on the potency of aziridine- and epoxide-based inhibitors. The computations were even able to contribute to the design of improved cysteine protease inhibitors [249]. Despite their success, these models cannot account for the explicit molecular structure of the protein environment, which may also be important for the inhibition mechanism as the role of the HIS residue of the active site of cysteine proteases. In this context, Mladenovic and co-workers performed a study to elucidate to what extent and how the HIS unit takes part in the inhibition mechanism of epoxide- and aziridine-based inhibitors [251]. For this purpose, they computed the reaction profiles of the corresponding ring-opening reactions. The model system used in the QM/MM computations consisted of the enzyme, the warhead of the inhibitors, and water as the solvent. The part treated quantum mechanically corresponded to the electrophilic warhead of the inhibitors and the side chains of the catalytic residues CYS and HIS in their zwitterionic form. To test if water can mediate the proton transfer, they performed computations with and without the mediating water molecule in the QM part. During geometry optimizations and scans of the potential energy surface, QM calculations were done at the BLYP/TZVP level of theory while single-point computations employing the B3LYP/TZVP level were performed to determine more precisely the activation and reaction energies. The CHARMM force field was used for the MM part [252]. The computed potential energy surface captured the energy changes during the inhibition reaction, which involved the ring opening of the inhibitor and the creation of the S(CYS)-C bond. The results excluded a direct proton shift from HIS to the organic ligand, showing instead that a single water molecule is sufficient to establish a very efficient relay system, which facilitates the proton transfer from HIS to the inhibitor. This finding strongly suggests that substituents susceptible to block the proton transfer might diminish the activity of the inhibitors.

An essential step in studying an enzyme inhibition process is to establish its mechanism. Modeling can analyze transition states directly, which are central in enzyme inhibition and cannot be studied experimentally because they are extremely short lived. Specific interactions that stabilize transition states or intermediates might not be apparent from experimental structures. Therefore, their identification using molecular modeling techniques can potentially offer enhanced affinity if they are exploited in designed ligands, as many enzymes show exceptionally high affinities for transition states and intermediates. However, the above reported investigations on falcipain inhibition were performed on very simplified models, due to the high level of QM methods used, and hence such models are not expected to be qualitatively accurate as many enzyme residues were not included. Equally important in modeling a reaction is the need for extensive reaction-pathway calculations and conformational sampling, which constitute significant challenges for large molecules. Because of the complexity of protein internal motions, many conformational states can exist and a single structure might not be truly representative [253]. However, the reaction profiles described earlier for falcipain inhibition by epoxides and aziridines are based on computed energies rather than free energies mainly because standard free energy calculations at the applied high level QM/MM approaches are extremely demanding in terms of computational efforts.

The speed-up of the calculations can be even greater if a parameterized quantum mechanics technique such as the DFTB approach is considered or if the latter is combined with molecular mechanics. This allows QM/MM MD simulations in the ns regime to be easily feasible, which is an essential ingredient for a reliable determination of the free energies for biomolecular reactions.

Given that earlier investigations related to enzyme inhibition have mainly focused on epoxides and aziridines ring-opening upon the thiolate attack from the catalytic cysteine, a detailed examination of falcipain inhibition, for example, will determine free energy barriers and will elucidate any inhibitor specificity. Reactivity insights into other typical FP inhibitors such as that of vinyl sulfones may be obtained, which are of great interest. In this context, our group has recently performed a hybrid QM/MM MD simulation to elucidate the irreversible

FP3 inhibition by a peptidyl vinyl sulfone. To reduce computational cost, only residues within 8 Å of the ligand were kept active in the QM/MM MD calculations. The system was solvated in an octahedral box of explicit water under periodic boundary conditions. Once the initial system was minimized and equilibrated, we switched from a pure MM potential to a hybrid QM/MM Hamiltonian, using SCC-DFTB level as implemented in AMBER [254]. The QM part (44 atoms) consisted of part of the ligand, CYS and HIS side-chains (in their zwitterionic form), while the rest of the system (remaining ligand atoms, protein and water) was described using the GAFF [255, 256], ff03 [257] and TIP3P [258] force fields, respectively (Figure 22). Finally, 1 ns QM/MM MD was accomplished at NpT (1 bar, 300K) conditions.

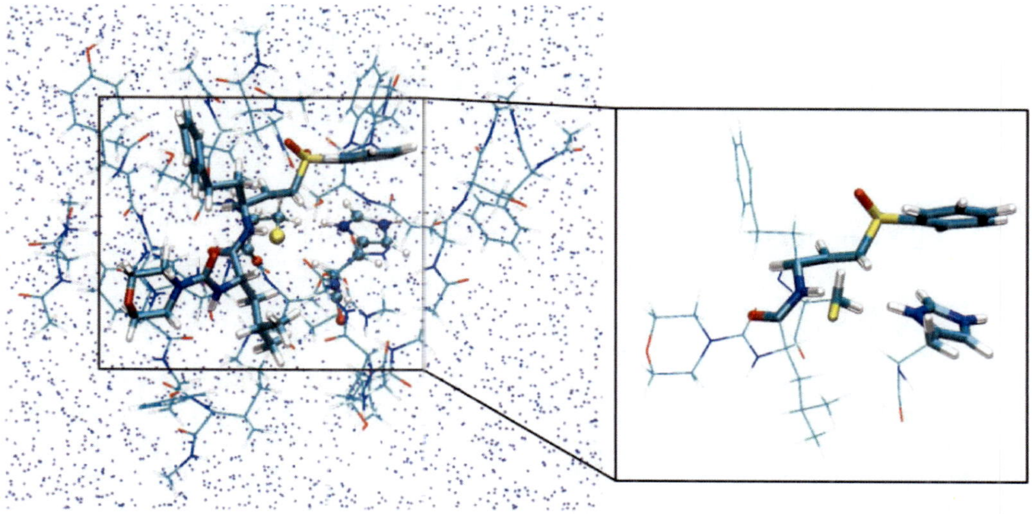

Figure 22. The overall structure of the model system used in the QM/MM MD simulations (left) consists of all FP3 residues within 8 Å of the ligand, the inhibitor, and water (blue dots) as solvent. The atoms in the QM region are represented in licorice in the right panel.

The QM/MM approach allows bonds to be formed and broken in this region, making it possible to study the alkylation of CYS thiolate by the peptidyl vinyl sulfone under the simulation conditions, as observed when monitoring the distances between i) the thiolate and the C atom of the ligand susceptible to suffer the attack, ii) the histidine nitrogen and the H atom that will be transferred to the ligand and iii) the C atom of the ligand and the H atom that will be transferred to it from the histidine residue (Figure 23).

The preliminary results suggest that the alkylation is a concerted process, *i.e.*, the nucleophilic attack from CYS thiolate to the putative site of attack of the vinyl bond is concerted with the proton-transfer reaction from the side chain of HIS to the other C atom of the double bond. Although these preliminary results are highly encouraging, further work has to be done in order to obtain the free energy profile of the inhibition process. We hereby briefly describe the procedure we intend to implement for that purpose. The QM/MM complex has first to be optimized with an iterative minimization procedure while following a specific reaction coordinate. For each minimized complex determined along the path, the MM subsystem has to be equilibrated with a classical short MD simulation with the QM subsystem frozen. The resulting snapshots are then ready to be used as the starting structure for QM/MM MD calculation of the free energy profile (potential of mean force) using the

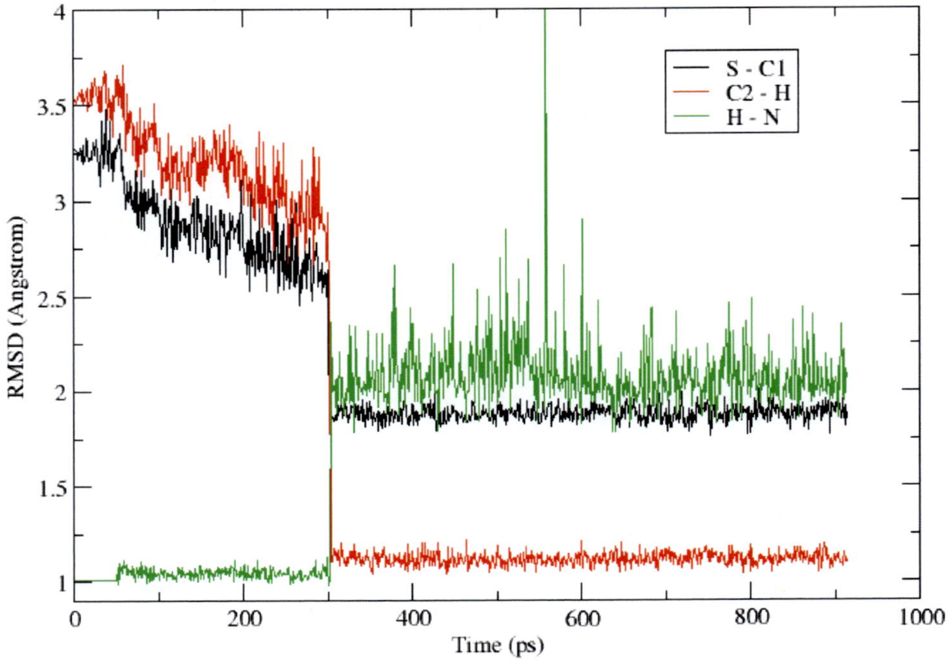

Figure 23. Interatomic distances during QM/MM MD of the model system, which consisted of all FP3 residues within 8 Å of the ligand and a vinyl sulfone The distances between i) the thiolate and the C atom of the ligand susceptible to suffer the attack, ii) the histidine nitrogen and the H atom that will be transferred to the ligand and iii) the C atom of the ligand and the H atom that will be transferred to it from the histidine, are represented in black, green and red colors, respectively.

umbrella sampling [259]. The free energy surfaces could then be obtained by calculating the (normalized) probability P of finding our complex in a conformation at a particular region in space from the QM/MM MD trajectories, then converting this number to free energies by $G = -RT \ln(P)$, where G is the Gibbs free energy, R is the general gas constant, and T is the temperature. Hopefully, applying this procedure to falcipain inhibition will allow the study of the effect of different inhibitors on the energy profile of the reaction and, consequently, to understand how the reactivity and reversibility of the inhibitors could be modulated in order to design safe and effective drugs against malaria.

If the ideas above are successful, the methodologies considered can be used to study other enzymatic reactions which can be very relevant for the design and for guiding the experimental synthesis of novel drugs.

Conclusion

In this chapter, recent examples of the application of density functional theory (DFT) based approaches in three very relevant areas in the fields of (bio)chemistry, materials science, heterogeneous catalysis and chemical engineering were reviewed.

It was shown that DFT approaches combined with suitable models of the systems under study are important to interpret and to complement the results from experiments, and yield very relevant information at the atomic level that can be used to improve the systems for specific applications in those areas.

In the fields of gas storage and gas separation, it was shown that DFT studies are very useful for the development of classical force fields and are very promising for the understanding of the most favorable positions for the interaction of different types of gaseous species with the frameworks of materials, even for very difficult cases such as materials possessing under-coordinated, open-shell, metal sites. In fact, preliminary results from our group suggest that some DFT approaches are able to provide adsorption geometries and energies for the interaction of gases with different materials that are in quite good agreement with experimental or with benchmark results computed with very accurate, but also computationally very demanding, post-Hartree-Fock approaches. In the field of heterogeneous catalysis, it was shown that DFT methods have high applicability in the study of reactions catalyzed by solid catalysts, allowing the determination of important mechanistic details not accessible by purely experimental work. The importance of the low-coordinated atoms in the catalysis of several reactions was demonstrated as well as the applicability of Brønsted-Evans-Polanyi relationships for the fast determination of activation energy barriers from well-chosen descriptors. These results are very encouraging for the future design of novel materials and catalysts based only on computational studies. Finally, DFT approaches are also emerging as a most valuable tool toward a better understanding of enzymatic catalysis at the electronic level, using realistic models.

This is particularly useful in the design of irreversible enzyme inhibitors, tunable to covalently bind the enzyme's catalytic amino acids.

Acknowledgments

This work is supported by projects PTDC/EQU-EQU/099423/2008, PTDC/QUI-QUI/109914/2009, PTDC/QUI-QUI/116864/2010, PTDC/QUI-QUI/117439/2010, PEst-C/QUI/UI0081/2011, PEst-C/EQB/LA0020/2011, Pest-C/EQB/LA0006/2011 and PEst-C/CTM/LA0011/2011, financed by FEDER through COMPETE - Programa Operacional Factores de Competitividade and by FCT - Fundação para a Ciência e a Tecnologia. Thanks are also due to FCT for Programa Ciência 2007 (funding for JRBG and MJ) and for the post-doctoral fellowships with references SFRH/BPD/64566/2009, SFRH/BPD/62967/2009 and SFRH/BPD/70283/2010 awarded to JLCF, CT and RSP, respectively.

References

[1] Schrödinger, E. An Undulatory Theory of the Mechanics of Atoms and Molecules. *Phys. Rev.*, 1926, 28, 1049-1070.

[2] Born, M.; Oppenheimer, R.; Quantum Theory of Molecules. *Ann. Phys.*, 1927, 84, 0457-0484.

[3] Hehre, W.J.; Radom, L.; Schleyer, P.V.R.; Pople, J.A. *Ab initio Molecular Orbital Theory*. New York: John Wiley, 1986.

[4] Jensen, F. *Introduction to Computational Chemistry*, 1st ed.; John Wiley and Sons: Chichester, England, 2004.

[5] Thomas, L.H. The effect of the orbital velocity of the electrons in heavy atoms on their stopping of alpha-particles. *P. Camb. Philos. Soc.*, 1927, 23, 713-716.

[6] Fermi, E. Un Metodo Statistico per la Determinazione di alcune Priopriet̀à dell'Atomo. *Rend. Accad. Naz. Lincei.*, 1927, 6, 602-607.

[7] Hohenberg, P.; Kohn, W. Inhomogeneous Electron Gas. *Phys. Rev. B*, 1964, 136, 864-871.

[8] Kohn, W.; Sham, L.J. Self-Consistent Equations Including Exchange and Correlation Effects. *Phys. Rev. A*, 1965, 140, 1133-1138.

[9] Dirac, P.A.M. Note on Exchange Phenomena in the Thomas Atom. *P. Camb. Philos. Soc.*, 1930, 26, 376-385.

[10] Slater, J.C. A Simplification of the Hartree-Fock Method. *Phys. Rev.*, 1951, 81, 385-390.

[11] Slater, J.C. The *Self-Consistent Field for Molecules and Solids*; McGraw-Hill: New York, 1974.

[12] Vosko, S.H.; Wilk, L.; Nusair, M. Accurate Spin-Dependent Electron Liquid Correlation Energies for Local Spin-Density Calculations - A Critical Analysis. *Can. J. Phys.*, 1980, 58, 1200-1211.

[13] Becke, A.D. Density Functional Calculations of Molecular-Bond Energies. *J. Chem. Phys.*, 1986, 84, 4524-4529.

[14] Becke, A.D. Density-Functional Exchange-Energy approximation with correct asymptotic-behavior, *Phys. Rev. A*, 1988, 38, 3098-3100.

[15] Perdew, J.P.; Wang, Y. Accurate and simple Density Functional for the electronic Exchange Energy – Generalized Gradient Approximation. *Phys. Rev. B*, 1986, 33, 8800-8802.

[16] Perdew, J.P. Density-Functional approximation for the correlation-energy of the inhomogeneous electron-gas. *Phys. Rev. B*, 1986, 33, 8822-8824.

[17] Perdew, J.P. Correction. *Phys. Rev. B*, 1986, 34, 7406-7406.

[18] Lee, C.T.; Yang, W.T.; Parr, R.G. Development of Colle-Salvetti Correlation-Energy formula into a Functional of the Electron-Density. *Phys. Rev. B*, 1988, 37, 785-789.

[19] Perdew, J.P.; Burke, K.; Ernzerhof, M. Generalized gradient approximation made simple. *Phys. Rev. Lett.*, 1996, 77, 3865-3868.

[20] Hammer, B.; Hansen, L.B.; Nørskov, J.K. Improved adsorption energetics within density-functional theory using revised Perdew-Burke-Ernzerhof functionals. *Phys. Rev. B*, 1999, 59, 7413-7421.

[21] Becke, A.D. Density-Functional Thermochemistry. III. The Role of Exact Exchange. *J. Chem. Phys.*, 1993, 98, 5648-5652.

[22] Sauer, J. Molecular-Models in *ab initio* Studies of Solids and Surfaces - From Ionic-Crystals and Semiconductors to Catalysts. *Chem. Rev.*, 1989, 89,199-255.

[23] Ashcroft, N.W.; Mermin, N.D. Solid State Physics. Philadelphia, Holt-Saunders, 1976.

[24] Hamann, D.R.; Schluter, M.; Chiang, C. Norm-Conserving Pseudopotentials. *Phys. Rev. Lett.*, 1979, 43, 1494-1797.

[25] Perdew, J.P.; Chevary, J.A.; Vosko, S.H.; Jackson, K.A.; Pederson, M.R.; Singh, D.J.; Fiolhais, C. Atoms, Molecules, Solids, and Surfaces - Applications of the Generalized Gradient Approximation for Exchange and Correlation. *Phys. Rev. B*, 1992, 46, 6671-6687.

[26] Hammer, B.; Hansen, L.B.; Nørskov, J.K. Improved adsorption energetics within density-functional theory using revised Perdew-Burke-Ernzerhof functionals. *Phys. Rev. B,* 1999, 59, 7413-7421.

[27] Armiento, R.; Mattsson, A.E. Functional designed to include surface effects in self-consistent density functional theory. *Phys. Rev. B*, 2005, 72, 085108-1:13.

[28] Krieger, J. B.; Chen, J.; Iafrate, G.J.; Savin, A. *In Electron Correlations and Materials Properties;* Gonis, A.; Kioussis, N., Eds.; Plenum: New York, 1999; p. 463.

[29] Sun, G.Y.; Kurti, J.; Rajczy, P.; Kertesz, M.; Hafner, J.; Kresse, G. Performance of the Vienna ab initio simulation package (VASP) in chemical applications, *J. Mol. Struct. (Theochem)*, 2003, 624, 37-45.

[30] Tao, J.M.; Perdew, J.P.; Staroverov, V.N.; Scuseria, G.E. Climbing the Density Functional Ladder: Nonempirical Meta-Generalized Gradient Approximation Designed for Molecules and Solids. *Phys. Rev. Lett.*, 2003, 91, 146401-1:4.

[31] Perdew, J.P.; Ruzsinszky, A.; Gábor, I.; Vydrov, O.A.; Scuseria, G.E.; Constantin, L.A., Zhou, X.; Burke, K. Restoring the Density-Gradient Expansion for Exchange in Solids and Surfaces. *Phys. Rev. Lett.*, 2008, 100, 136406-1:4.

[32] Zhao, Y.; Truhlar, D.G. Density Functionals with Broad Applicability in Chemistry. *Acc. Chem. Res.,* 2008, 41, 157-167.

[33] Sousa, S.F.; Fernandes, P.A.; Ramos, M.J. General Performance of Density Functionals. *J. Phys. Chem. A*, 2007, 111, 10439-10452.

[34] Paier, J.; Hirschl, R; Marsman, M.; Kresse, G. The Perdew-Burke-Ernzerhof Exchange-Correlation Functional Applied to the G2-1 Test Set Using a Plane-Wave Basis Set. *J. Chem. Phys.,* 2005, 122, 234102-1:13.

[35] Adamo, C.; Barone, V. Toward Reliable Density Functional Methods without Adjustable Parameters: The PBE0 Model, *J. Chem. Phys.*, 1999, 110, 6158-6170.

[36] Heyd, J.; Scuseria, G.E.; Ernzerhof, M. Hybrid Functionals Based on a Screened Coulomb Potential. *J. Chem. Phys.*, 2003, 118, 8207-8215.

[37] Maintz, S.; Eck, B.; Dronskowski, R. Speeding up plane-wave electronic-structure calculations using graphics-processing units. *Comput. Phys. Commun.,* 2011, 182, 1421-1427.

[38] Hutchinson, M.; Widom, M. VASP on a GPU: Application to exact-exchange calculations of the stability of elemental boron. *Comput. Phys. Commun.*, 2012, 183, 1422-1426.

[39] da Silva, J.L.F.; Ganduglia-Pirovano, M.V.; Sauer, J. Formation of the Cerium Orthovanadate $CeVO_4$: DFT+U Study. *Phys. Rev. B,* 2007, 76, 045121-1:8.

[40] Ganduglia-Pirovano, M.V.; Hofmann, A.; Sauer, J. Oxygen vacancies in transition metal and rare earth oxides: Current state of understanding and remaining challenges. *Surf. Sci. Rep.*, 2007, 62, 219-270.

[41] Grimme, S. Semiempirical Hybrid Density Functional with Perturbative Second-Order Correlation. *J. Chem. Phys.,* 2006, 124, 034108-1:16.

[42] Raghavachari, K.; Curtiss, L.A. In *Quantum-Mechanical Prediction of Thermochemical Data.* Cioslowski, J., Ed.; Kluwer Academic Publishers: Dordrecht, The Netherlands, 2001; pp 67-98.

[43] Grimme, S. Semiempirical GGA-type density functional constructed with a long-range dispersion correction. *J. Comput. Chem.*, 2006, 27, 1787-1799.

[44] Grimme, S.; Antony, J.; Ehrlich, S.; Krieg, H. A. A conbsistent and accurate ab initio parametrization of density functional dispersion correction (DFT-D) for the 94 elements H-Pu. *J. Chem. Phys.,* 2010, 132, 154104-1:19.

[45] London, F. The General Theory of Molecular Forces. *Trans. Faraday Soc*., 1937, 33, 8-26.

[46] Dion, M.; Rydberg, H.; Schroder, E.; Langreth, D.C.; Lundqvist, B.I. Van der Waals Density Functional for General Geometries. *Phys. Rev. Lett.,* 2004, 92, 246401-1:4.

[47] Lee, K.; Murray, E.D.; Kong, L. Z.; Lundqvist, B.I.; Langreth, D.C. Higher-Accuracy van der Waals Density Functional. *Phys. Rev. B*, 2010, 82, 081101-1:4.

[48] Klimeš, J.; Bowler, D.R.; Michaelides, A. Chemical accuracy for the van der Waals density functional. *J. Phys. Condens. Matter*, 2010, 22, 022201-1:5.

[49] Murray, E.D.; Lee, K.; Langreth, D. C. Investigation of Exchange Energy Density Functional Accuracy for Interacting Molecules. *J. Chem. Theory Comput*., 2009, 5, 2754-2762.

[50] Klimeš, J.; Bowler, D.R.; Michaelides, A. Van der Waals density functionals applied to solids. *Phys. Rev. B*, 2011, 83, 195131-1:13.

[51] Elstner, M.; Porezag, D.; Jungnickel, G.; Elsner, J.; Haugk, M.; Frauenheim, T.; Suhai, S.; Seifert, G., Self-Consistent-Charge Density-Functional Tight-Binding Method for Simulations of Complex Materials Properties. *Phys. Rev. B,* 1998, 58, 7260-7268.

[52] Goringe, C.M.; Bowler, D.R.; Hernández, E. Tight-Binding Modelling of Materials. *Rep. Prog. Phys*., 1997, 60, 1447-1512.

[53] Elstner, M. The SCC-DFTB Method and its Application to Biological Systems. *Theor. Chem. Acc.,* 2006, 116, 316-325.

[54] Riccardi, D.; Schaefer, P.; Yang, Y.; Yu, H.; Ghosh, N.; Prat-Resina, X.; Konig, P.; Li, G.; Xu, D.; Guo, H.; Elstner, M.; Cui, Q., Development of effective quantum mechanical/molecular mechanical (QM/MM) methods for complex biological processes. *J. Phys. Chem. B,* 2006, 110, 6458-6469.

[55] Perdew, J.P.; Schmidt, K. Jacob's Ladder of Density Functional Approximations for the Exchange-Correlation Energy. *AIP Conf. Proc.,* 2001, 577, 1-20.

[56] Dapprich, S.; Komáromi, I.; Byun, K.S.; Morokuma, K.; Frisch, M.J.A New ONIOM Implementation in Gaussian 98. Part 1. The Calculation of Energies, Gradients and Vibrational Frequencies and Electric Field Derivatives. *J. Mol. Struct. (Theochem),* 1999, 461-462, 1-21.

[57] Hellmann, H. A New Approximation Method in the Problem of Many Electrons. J. Chem. Phys., 1935, 3, 61-61.

[58] Gutdeutsch, U.; Birkenheuer, U.; Krüger, S.; Rösch, N. On cluster embedding schemes based on orbital space partitioning. *J. Chem. Phys*., 1997, 106, 6020-6030.

[59] Valero, R.; Gomes, J.R.B.; Truhlar, D.G.; Illas, F. Good performance of the M06 family of hybrid meta generalized gradient approximation density functionals on a difficult case: CO adsorption on MgO(001). *J. Chem. Phys*., 2008, 129, 124710-1:7.

[60] Valero, R.; Gomes, J.R.B.; Truhlar, D.G.; Illas, F. Density functional study of CO and NO adsorption on Ni-doped MgO(100). *J. Chem. Phys.,* 2010, 132, 104701-1:13.

[61] Gomes, J.R.B.; Illas, F.; Cruz-Hernández, N.; Márquez, A.; Sanz, J.F. Interaction of Pd with α-Al_2O_3(0001): A case study of modeling the metal-oxide interface on complex substrates. *Phys. Rev. B*, 2002, 65, 125414-1:9.

[62] Gomes, J.R.B.; Illas, F.; Cruz-Hernández, N.; Sanz, J.F.; Wander, A.; Harrison, N. M. Surface model and exchange-correlation functional effects on the description of Pd/α-Al_2O_3(0001). *J. Chem. Phys.*, 2002, 116, 1684-1691.

[63] Gomes, J.R.B.; Lodziana, Z.; Illas, F. Adsorption of small palladium clusters on the relaxed α-Al_2O_3(0001) surface. *J. Phys. Chem. B*, 2003, 107, 6411-6424.

[64] Gomes, J.R.B.; Prates Ramalho, J.P. Adsorption of Ar atoms on the relaxed defect-free TiO_2(110) surface. *Phys. Rev. B*, 2005, 71, 235421-1:8.

[65] Gomes, J.R.B.; Prates Ramalho, J.P.; Illas, F. Adsorption of Xe atoms on the TiO_2(110) surface: A density functional study. *Surf. Sci.*, 2010, 604, 428-434.

[66] Zhao, Y.; Wang, C.Y.; Lu, G. Magnetism-Driven Dislocation Dissociation, Cross Slip, and Mobility in NiAl. *Phys. Rev. B*, 2010, 82, 060404-1:8.

[67] Zhao, Y.; Wang, C.Y.; Peng, Q.; Lu, G. Error Analysis and Applications of a general QM/MM Approach. *Comput. Mater. Sci.*, 2010, 50, 714-719.

[68] Catlow, C.R.A.; Smit, B.; Van Santen R.A, *Computer Modelling of Microporous Materials*. Ed.1, London: Academic Press, 2004.

[69] Catlow, C.R.A. *Modeling of structure and Reactivity in zeolites*. London: Academic Press, 1997.

[70] Corà, F.; Catlow, C.R.A.; D'Ercole, A. Acid and redox properties of co-substituted aluminium phosphates. *J. Mol. Catal. A-Chem.*, 2001, 166, 87-99.

[71] Saadoune, I.; Corà, F.; Alfredsson, M.; Catlow, C.R.A. Computational study of the structural and electronic properties of dopant ions in microporous AlPOs. 2. Redox catalytic activity of trivalent transition metal ions. *J. Phys. Chem. B*, 2003, 107, 3012-3018.

[72] Elanany, M.; Vercauteren, D.P.; Kubob, M.; Miyamoto, A. The acidic properties of H-MeAlPO-5 (Me = Si, Ti, or Zr): A periodic density functional study. *J. Mol. Catal. A-Chem.*, 2006, 248, 181-184.

[73] Hartmann, M.; Kevan, L. Transition-metal ions in aluminophosphate and silicoaluminophosphate molecular sieves: Location, interaction with adsorbates and catalytic properties. *Chem. Rev.*, 1999, 99, 635-664.

[74] Saadoune, I.; Corà, F.; Catlow, C.R.A. Computational study of the structural and electronic properties of dopant ions in microporous AlPOs. 1. Acid catalytic activity of divalent metal ions. *J. Phys. Chem. B*, 2003, 107, 3003-3011.

[75] Elanany, M.; Koyama, M.; Kubo, M.; Selvam, P.; Miyamoto, A. Periodic density functional investigation of Bronsted acidity in isomorphously substituted chabazite and AlPO-34 molecular sieves. *Micropor. Mesopor. Mater.*, 2004, 71, 51-56.

[76] Fois, E.; Gamba, A.; Tilocca, A. Structure and dynamics of the flexible triple helix of water inside VPI-5 molecular sieves. *J. Phys. Chem. B*, 2002, 106, 4806-4812.

[77] Pillai, R.S.; Jasra, R.V. Computational Study for Water Sorption in $AlPO_4$-5 and $AlPO_4$-11 Molecular Sieves. *Langmuir*, 2010, 26, 1755-1764.

[78] Sinclair, P.E.; Catlow, C.R.A. Quantum Chemical Study of the Mechanism of Partial Oxidation Reactivity in Titanosilicate Catalysts: Active Site Formation, Oxygen Transfer, and Catalyst Deactivation. *J. Phys. Chem. B*, 1999, 103, 1084-1095.

[79] Shough, A.M. Doren, D.J., Ogunnaike, B. Transition Metal Substitution in ETS-10: DFT Calculations and a Simple Model for Electronic Structure Prediction. *Chem. Mater.*, 2009, 21, 1232-1241.

[80] Damin, A.; Xamena, F.X.L.; Lamberti, C.; Civalleri, B.; Zicovich-Wilson, C.M.; Zecchina, A. Structural, Electronic, and Vibrational Properties of the Ti-O-Ti Quantum Wires in the Titanosilicate ETS-10. *J. Phys. Chem. B,* 2004, 108, 1328-1336.

[81] Guo, M.; Pidko, E.A.; Fan, F.; Feng, Z.; Hofmann, J.P.; Weckhuysen, B.M.; Hensen, E.J.M.; Li, C. Structure and Basicity of Microporous Titanosilicate ETS-10 and Vanadium-Containing ETS-10. *J. Phys. Chem. C,* 2012, 116, 17124-17133.

[82] Plant, D.F.; Maurin, G.; Deroche, I.; Llewellyn, P.L. Investigation of CO_2 adsorption in Faujasite systems: Grand canonical Monte Carlo and molecule dynamics simulations based on a new derived Na^+-CO_2. *Micropor. Mesopor. Mater.,* 2007, 99, 70-78.

[83] Getman, R.B.; Bae, Y-S.; Wilmer, C.E.; Snurr, R.Q. Review and Analysis of Molecular Simulations of Methane, Hydrogen, and Acetylene Storage in Metal-Organic Frameworks. *Chem. Rev*., 2012, 112, 703-723.

[84] Sagara, T.; Klassen, J.; Ganz, E. Computational study of hydrogen binding by metal-organic framework-5. *J. Chem. Phys*., 2004, 121, 12543-12547.

[85] Yaghi, O.M.; Sun, Z.; Richardson, D.A.; Groy, T. L. Directed Transformation of Molecules to Solids: Synthesis of a Microporous Sulfide from Molecular Germanium Sulfide Cages. *J. Am. Chem. Soc*., 1994, 116, 807-808.

[86] Ribas Gispert, J. *Coordination Chemistry.* Barcelona, WILEY-VCH, 2008.

[87] Bourrelly, S.; Llewellyn, P.L.; Serre, C.; Millange, F.; Loiseau, T.; Ferey, G. Different Adsorption Behaviors of Methane and Carbon Dioxide in the Isotypic Nanoporous Metal Terephthalates MIL-53 and MIL-47. *J. Am. Chem. Soc*., 2005, 127, 13519-13521.

[88] Wu, H.; Zhou, W.; Yildirim, T. High-Capacity Methane Storage in Metal–Organic Frameworks M_2(dhtp): The Important Role of Open Metal Sites. *J. Am. Chem. Soc*., 2009, 131, 4995-5000.

[89] Kuppler, R.J.; Timmons, D.J.; Fang, Q.-R.; Li, J.-R.; Makal, T.A.; Young, M.D.; Yuan, D.; Zhao, D.; Zhuang, W.; Zhou, H.-C. Potential applications of metal-organic frameworks. *Coordin. Chem. Rev*., 253, 2009, 3042-3066.

[90] Chen, B.; Contreras, D.S.; Ockwig, N.; Yaghi, O.M. High H_2 Adsorption in a Microporous Metal-Organic Framework with Open-Metal Sites. *Angew. Chem. Int. Ed.,* 2005, 44, 4745-4749.

[91] Duren, T.; Sarkisov, L.; Yaghi, O.M.; Snurr, R.Q. Design of New Materials for Methane Storage. *Langmuir,* 2004, 20, 2683-2689.

[92] Millward, A.; Yaghi, O. M. Metal-organic frameworks with exceptionally high capacity for storage of carbon dioxide at room temperature. *J. Am. Chem. Soc.,* 2006, 127, 17998-17999.

[93] Chen, B.; Liang, C.; Yang, J.; Yaghi, O.M. A microporous metal-organic framework for gas-chomatographic separation of alkanes. *Angew. Chem. Int. Ed*., 2006, 118, 1390-1393.

[94] Luz, I.; Llabrés i Xamena, F.X.; Corma, A. MOFs as catalyst: Activity, reusability and shape-selectivity of a Pd-containing MOF. *J. Catal*., 2010, 276, 134-140.

[95] Chen, B.; Yang, Y.; Zapata, F.; Lin, G.; Qian, G.; Lobkovsky, E.B. Luminescent Open Metal Sites within a Metal–Organic Framework for Sensing Small Molecules. *Adv. Mater.,* 2007, 19, 1693-1696.

[96] Yang, Q.; Zhong, C. Molecular simulation of carbon dioxide/methane/hydrogen mixture adsorption in metal-organic frameworks. *J. Phys. Chem. B,* 2006, 110, 17776-17783.

[97] Ferey, G.; Mellot-Draznieks, C.; Serre, C.; Millange, F. Crystallized frameworks with giant pores: Are there limits to the possible?. *Acc. Chem. Res.,* 2005, 38, 217-225.

[98] Valenzano, L.; Civalleri, B.; Chavan, S.; Palomino, G.T.; Areán, C.O.; Bordiga, S. Computational and experimental studies on the adsorption of CO, N_2, and CO_2 on Mg-MOF-74. *J. Phys. Chem. C,* 2010, 114, 11185-11191.

[99] Yang, Q.; Xue, C.; Zhong, C.; Chen, J.-F. Molecular simulation of separation of CO_2 from flue gases in Cu-BTC meta -organic Framework. *AIChE J.,* 2007, 53, 2832-2840.

[100] Mavrandonakis, A.; Tylianakis, E.; Stubos, A.K.; Froudakis, G.E. Why Li doping in MOFs enhances H_2 storage capacity? A multi-scale theoretical study. *J. Phys. Chem. C,* 2008 112, 7290-7294.

[101] Tan, J.C.; Cheetham, A.K. Mechanical properties of hybrid inorganic-organic framework materials: Establishing fundamental structure-property relationships. *Chem. Soc. Rev.,* 2011, 40, 1059-1080.

[102] Almeida Paz, F.A.; Klinowski, J.; Vilela, S.M.F.; Tomé, J.P.C.; Cavaleiro, J.A.S.; Rocha, J. Ligand design for functional metal–organic frameworks. *Chem. Soc. Rev.,* 2012, 41, 1088-1110.

[103] Zhou, M.; Wang, Q.; Zhang, L.; Liu, Y.C.; Kang, Y. Adsorption sites of hydrogen in zeolitic imidazolate frameworks. *J. Phys. Chem. B,* 2009, 113, 11049-11053.

[104] Watanabe, T.; Keskin, S.; Nair, S.; Sholl, D.S. Computational identification of a metal organic framework for high selectivity membrane-based CO_2/CH_4 separations: Cu(hfipbb)(H_2hfipbb)$_{0.5}$. *Phys. Chem. Chem. Phys.,* 2009, 11, 11389-11394.

[105] Keskin, S.; Liu, J.; Rankin, R.B.; Johnson, J.K.; Sholl, D.S. Progress, opportunities, and challenges for applying atomically detailed modeling to molecular adsorption and transport in metal - Organic framework materials. *Ind. Eng. Chem. Res.,* 2009, 48, 2355-2371.

[106] Sillar, K.; Hofmann, A.; Sauer, J. Ab initio study of hydrogen adsorption in MOF-5. *J. Am. Chem. Soc.,* 2009, 131, 4143-4150.

[107] Sagara, T.; Klassen, J.; Ortony, J.; Ganz, E. Binding energies of hydrogen molecules to isoreticular metal-organic framework materials. *J. Chem. Phys.,* 2005, 123, 014701-1:4.

[108] Greathouse, J.A.; Allendorf, M.D. The interaction of water with MOF-5 simulated by molecular dynamics. *J. Am. Chem. Soc.,* 2006, 128, 10678-10679.

[109] Cheng, Y.; Kondo, A.i; Noguchi, H.; Kajiro, H.; Urita, K.; Ohba, T.; Kaneko, K.; Kanoh, H. Reversible Structural Change of Cu-MOF on Exposure to Water and Its CO_2 Adsorptivity. *Langmuir,* 2009, 25, 4510-4513.

[110] Zhou, M; Wang, Q; Zhang, L; Liu, Y.C; Kang, Y. Adsorption sites of hydrogen in zeolitic imidazolate frameworks. *J. Phys. Chem. B,* 2009, 113, 11049-11053.

[111] Dangi, G.P.; Pillai, R.S.; Somani, R.S.; Bajaj, H.C.; Jasra, R.V. A density functional theory study on the interaction of hydrogen molecule with MOF-177. *Mol. Simul.,* 2010, 36, 373-381.

[112] Bhatia, S.K.; Myers, A.L. Optimum conditions for adsorptive storage. *Langmuir,* 2006, 22, 1688-1700.

[113] Watanabe, T.; Sholl, D.S. Molecular chemisorption on open metal sites in Cu_3(benzenetricarboxylate)$_2$: A spatially periodic density functional theory study. *J. Chem. Phys.*, 2010, 133, 094509-1:12.

[114] Lamia, N.; Jorge, M.; Granato, M.A.; Almeida Paz, F.A.; Chevreau, H.; Rodrigues, A. E. Adsorption of propane, propylene and isobutane on a metal–organic framework: Molecular simulation and experiment. *Chem. Eng. Sci.*, 2009, 64, 3246-3259.

[115] Vilhelmsen, L.B.; Walton, K.S.; Sholl, D.S. Structure and Mobility of Metal Clusters in MOFs: Au, Pd, and AuPd Clusters in MOF-74. *J. Am. Chem. Soc.*, 2012, 134, 12807-12816.

[116] Lukose, B; Supronowicz, B; Petkov, P; Frenzel, J; Kuc, A.B; Seifert, G; Vayssilov, G. N; Heine, T. Structural properties of metal-organic frameworks within the density-functional based tight-binding method. *Phys. Status Solidi B,* 2012, 249, 335-342.

[117] Grajciar, L; Wiersum, A.D; Llewellyn, P.L; Chang, J.-S; Nachtigall, P. Understanding CO_2 Adsorption in CuBTC MOF: Comparing Combined DFT-ab Initio Calculations with Microcalorimetry Experiment. *J. Phys. Chem. C*, 2011, 115, 17925-17933.

[118] Grajciar, L.; Bludsky, O.; Nachtigall, P. Water Adsorption on Coordinatively Unsaturated Sites in CuBTC MOF. *J. Phys. Chem. Lett.* 2010, 1, 3354-3359.

[119] Peralta, D; Chaplais, G; Simon-Masseron, A; Barthelet, K; Chizallet, C; Quoineaud, A.-A; Pirngruber, G.D. Comparison of the behavior of metal-organic frameworks and zeolites for hydrocarbon separations. *J. Am. Chem. Soc.*, 2012, 134, 8115-8126.

[120] Wu, H; Simmons, J.M; Liu, Y; Brown, C.M; Wang, X.-S; Shengqian, M; Peterson, V.K; Zhou, W. Metal-organic frameworks with exceptionally high methane uptake: Where and how is methane stored?. *Chem.-Eur. J.*, 2010, 16, 5205-5214.

[121] Wu, H.; Zhou, W.; Yildirim, T. High-Capacity Methane Storage in Metal−Organic Frameworks M_2(dhtp): The Important Role of Open Metal Sites. *J. Am. Chem. Soc.*, 2009, 131, 4995-5000.

[122] Castillo, J.M.; Vlugt, T.J.H.; Calero, S. Understanding Water Adsorption in Cu-BTC Metal-Organic Frameworks. *J. Phys. Chem. C,* 2008, 112, 15934-15939.

[123] Chui, S.S.Y.; Lo, S.M.F.; Charmant, J.P.H.; Orpen, A.G.; Williams, I.D. A Chemically functionalizable nanoporous material $[Cu_3(TMA)_2(H_2O)_3]_n$. *Science,* 1999, 283, 1148-1150.

[124] Farrusseng, D.; Daniel, C.; Gaudillère, C.; Ravon, U.; Schuurman, Y.; Mirodatos, C.; Dubbeldam, D.; Frost, H.; Snurr, R.Q. Heats of adsorption for seven gases in three metal - Organic frameworks: Systematic comparison of experiment and simulation. *Langmuir*, 2009, 25, 7383-7388.

[125] Poater, J.; Solà, M.; Rimola, A.; Rodríguez-Santiago, L.; Sodupe, M.. Ground and Low-Lying States of Cu^{2+}-H_2O. A Difficult Case for Density Functional Methods. *J. Phys. Chem.* A, 2004, 108, 6072-6078.

[126] Ducéré, J.M.; Goursot, A.; Berthomieu, D. Comparative density functional theory study of the binding of ligands to Cu^+ and Cu^{2+}: Influence of the coordination and oxidation state. *J. Phys. Chem.* A, 2005, 109, 400-408.

[127] Fischer, M.; Gomes, J.R.B.; Fröba, M.; Jorge, M. Modeling Adsorption in Metal-Organic Frameworks with Open Metal Sites: Propane/Propylene Separations. *Langmuir*, 2012, 28, 8537-8549.

[128] Liang, Z.; Marshall, M.; Caffee, A.L. CO_2 adsorption-based separation by metal organic framework (Cu-BTC) versus zeolite (13X). *Energ. Fuel.*, 2009, 23, 2785-2789.

[129] Toda, J.; Fischer, M.; Jorge, M.; Gomes, J.R.B. *Water adsorption on a copper formate paddlewheel model: A comparative MP2 and DFT study*. Submitted for publication.

[130] Møller, C.; Plesset, M.S. Note on an Approximation Treatment for Many-Electron Systems. *Phys. Rev.*, 1934, 46, 618-622.

[131] Peterson, K.A.; Puzzarini, C. Systematically convergent basis sets for transition metals. II. Pseudopotential-based correlation consistent basis sets for the group 11 (Cu, Ag, Au) and 12 (Zn, Cd, Hg) elements. *Theor. Chem. Acc.*, 2005, 114, 283-296.

[132] Frisch, M.J.; Trucks, G.W.; Schlegel, H.B.; Scuseria, G.E.; Robb, M.A.; Cheeseman, J.R.; Scalmani, G.; Barone, V.; Mennucci, B.; Petersson, G. A.; Nakatsuji, H.; Caricato, M.; Li, X.; Hratchian, H. P.; Izmaylov, A. F.; Bloino, J.; Zheng, G.; Sonnenberg, J.L.; Hada, M.; Ehara, M.; Toyota, K.; Fukuda, R.; Hasegawa, J.; Ishida, M.; Nakajima, T.; Honda, Y.; Kitao, O.; Nakai, H.; Vreven, T.; Montgomery, Jr., J.A.; Peralta, J. E.; Ogliaro, F.; Bearpark, M.; Heyd, J.J.; Brothers, E.; Kudin, K.N.; Staroverov, V.N.; Kobayashi, R.; Normand, J.; Raghavachari, K.; Rendell, A.; Burant, J.C.; Iyengar, S. S.; Tomasi, J.; Cossi, M.; Rega, N.; Millam, J.M.; Klene, M.; Knox, J.E.; Cross, J.B.; Bakken, V.; Adamo, C.; Jaramillo, J.; Gomperts, R.; Stratmann, R.E.; Yazyev, O.; Austin, A. J.; Cammi, R.; Pomelli, C.; Ochterski, J.W.; Martin, R.L.; Morokuma, K.; Zakrzewski, V.G.; Voth, G.A.; Salvador, P.; Dannenberg, J.J.; Dapprich, S.; Daniels, A.D.; Farkas, Ö.; Foresman, J.B.; Ortiz, J.V.; Cioslowski, J.; Fox, D.J. Gaussian 09, Revision B.01, Gaussian, Inc., Wallingford CT, 2009.

[133] Zhao, Y.; Truhlar, D.G. The M06 suite of density functionals for main group thermochemistry, thermochemical kinetics, noncovalent interactions, excited states, and transition elements: Two new functionals and systematic testing of four M06-class functionals and 12 other functionals. *Theor. Chem. Acc.*, 2008, 120, 215-241.

[134] Zhao, Y.; Schultz, N.E.; Truhlar, D.G. Design of density functionals by combining the method of constraint satisfaction with parametrization for thermochemistry, thermochemical kinetics, and noncovalent interactions. *J. Chem. Theory Comput.*, 2006, 2, 364-382.

[135] Schwabe, T.; Grimme, S. Towards Chemicals acuracy for the thermodynamics of large molecules: new hybrid density functionals including non-local correlation effects. *Phys. Chem. Chem. Phys.*, 2006, 8, 4398-4401.

[136] Zhao, Y.; Truhlar, D.G. A new local density functional for main-group thermochemistry, transition metal bonding, thermochemical kinetics, and noncovalent interactions. *J. Chem. Phys.*, 2006, 125, 194101-1:18.

[137] Chai, J.-D.; Head-Gordon, M. Long-range corrected hybrid density functionals with Damned atom-atom dispersion corrections. *Phys. Chem. Chem. Phys.*, 2008, 10, 6615-6620.

[138] Tawada, Y.; Tsuneda, T.; Yanagisawa, S.; Yanai, T.; Hirao, K. A long-range-corrected time-dependent density functional theory. *J. Chem. Phys.*, 2004, 120, 8425-8433.

[139] Vydrov, O.A.; Scuseria, G.E. Assessment of a long-range corrected hybrid functionals. *J. Chem. Phys.*, 2006, 125, 234109-1:9.

[140] Vydrov, O.A.; Heyd, J.; Krukau, A.; Scuseria, G.E. Importance of short-range versus long-range Hartree-Fock exchange for the performance of hybrid density functionals. *J. Chem. Phys.*, 2006, 125, 074106-1:9.

[141] Vydrov, O.A.; Scuseria, G.E.; Perdew, J.P. Tests of functionals for systems with fractional electron Lumber. *J. Chem. Phys.*, 2007, 126, 154109-1:6.

[142] Koch, W.; Holthausen, M.C. *A chemist's guide to Density Functional theory.* 2nd Ed. New York: Wiley-VCH, 2001.
[143] Breck,W. Zeolite *Molecular Sieves: Structure, Chemistry and Use*. New York: Wiley-Interscience, 1974.
[144] Barrer, R.M. *Zeolites and clay minerals as sorbents and molecular sieves.* London: Academic Press, 1978.
[145] Baerlocher, Ch.; Meier, W.M.; Olson, D.H. Atlas of Zeolite Framework Types. *Structure Commission of the International Zeolite Association*, Amsterdam: Elsevier, 2001.
[146] Schafer, A.; Huber, C.; Ahlrichs, R. Fully optimized contracted Gaussian basis sets of triple zeta valence quality for atoms Li to Kr. *J. Chem. Phys.*, 1994, 100, 5829-5834.
[147] Ugliengo, P.; Garrone, E.; Ferrari, A.M.; Zecchina, A.; Otero Arean, C. Quantum Chemical Calculations and Experimental Evidence for O-Bonding of Carbon Monoxide to Alkali Metal Cations in Zeolites. *J. Phys. Chem. B,* 1999, 103, 4839-4846.
[148] Accelrys, *Materials Studio Release Notes*, Release 4.1; Accelrys Software, Inc.: San Diego, CA, 2006.
[149] Marzari, N.; Vanderbilt, D.; Payne, M.C. Ensemble Density-Functional Theory for Ab Initio Molecular Dynamics of Metals and Finite-Temperature Insulators. *Phys. Rev. Lett.*, 1997, 79, 1337-1340.
[150] Poulet, G.; Sautet, P.; Tuel, A. Structure of Hydrated Microporous Aluminophosphates: Static and Molecular Dynamics Approaches of AlPO4-34 from First Principles Calculations. *J. Phys. Chem. B*, 2002, 106, 8599-8608.
[151] Vanderbilt, D. Soft self-consistent pseudopotentials in a generalized eigenvalue formalism. *Phys. Rev. B,* 1990, 41, 7892-7895.
[152] White, J. A.; Bird, D.M. Implementation of gradient-corrected exchange-correlation potentials in Car-Parrinello total-energy calculations. *Phys. Rev. B*, 1994, 50, 4954-4957.
[153] Nur, H.; H. Hamdan, H. The ionic size of metal atoms in correlation with acidity by the conversion of cyclohexanol over MeAPO-5. *Mater. Res. Bull., 2001*, 36, 315-322.
[154] Kuznicki, S.M. US *Patent* 4453, 202, 1989.
[155] Kuznicki, S.M. Bell V.A., Petrovic, I. Desai, B.T. US *Patent* 6068, 682, 2000.
[156] Philippou, A.; Rocha, J.; Anderson, M.W., The strong basicity of the microporous titanosilicate ETS-10. *Catal. Lett.,* 1999, 57, 151-153.
[157] Howe, R.F.; Krishnanadi, Y.K. Photoreactivity of ETS-10. *Chem. Commun.*, 2001, 17, 1588-1589.
[158] Surolia, P.K.; Tayade, R.J.; Jasra, R.V. Photocatalytic Degradation of Nitrobenzene in an Aqueous System by Transition-Metal-Exchanged ETS-10 Zeolites. *Ind. Eng. Chem. Res.*, 2010, 49, 3961-3966.
[159] Krishna, R.M.; Prakash, A.M.; Kurshev, V.; Kevan, L. Photoinduced charge separation of methylphenothiazine in microporous titanosilicate M-ETS-10 (M = Na^++K^+, H^+, Li^+, Na^+, K^+, Ni^{2+}, Cu^{2+}, Co^{2+}) and Na^+, K^+-ETS-4 molecular sieves at room temperature. *Phys. Chem. Chem. Phys.* 1999, 1, 4119-4124.
[160] Zimmerman, A.M.; Doren, D.J.; Lobo, R.F. Electronic and Geometric Properties of ETS-10: QM/MM Studies of Cluster Models. *J. Phys. Chem. B*, 2006, 110, 8959-8964.
[161] Pinto, M.L.; Rocha, J.; Gomes, J.R.B.; Pires, J. Slow Release of NO by Microporous Titanosilicate ETS-4. *J. Am. Chem. Soc.*, 2011, 133, 6396-6402.

[162] Greeley, J.; Nørskov, J.K.; Mavrikakis, M. Electronic structure and catalysis on metal surfaces. *Annu. Rev. Phys. Chem.* 2002, 53, 319-348.

[163] Michaelides, A.; Hu, P. Catalytic water formation on platinum: a first principal study. *J. Am. Chem. Soc.*, 2001 123, 4235-4242

[164] Fajín, J.L.C.; Cordeiro, M.N.D.S.; Gomes, J.R.B. On the theoretical understanding of the unexpected O2 activation by nanoporous gold. *Chem. Comm.*, 2011, 47, 8403-8405.

[165] Honkala, K.; Hellman, A.; Remediakis, I.N.; Logadottir, A.; Carlsson, A.; Dahl, S.; Christensen, C.H.; Nørskov, J.K. Ammonia synthesis from first-principles calculations. *Science*, 2005, 307, 555-558.

[166] Bond, G. Mechanisms of the gold-catalyzed water-gas shift. *Gold Bull.*, 2009, 42, 337-342.

[167] Rodríguez, J.A.; Ma, S.; Liu, P.; Hrbek, J.; Evans, J.; Pérez, M. Activity of CeO_x and TiO_x Nanoparticles Grown on Au(111) in the Water-Gas Shift Reaction. *Science,* 2007, 318, 1757-1760.

[168] Vayssilov, G.N.; Lykhach, Y.; Migani, A.; Staudt, T.; Petrova, G. P.; Tsud, N.; Skála, T.; Bruix, A.; Illas, F.; Prince, K.C.; Matolín, V.; Neyman, K.M.; Libuda, J. Support nanostructure boosts oxygen transfer to catalytically active platinum nanoparticles. *Nature Mater.*, 2011, 10, 310-315.

[169] Liu, Z.-P.; Jenkins, S.J.; King, D.A. Role of nanostructured dual-oxide supports in enhanced catalytic activity: theory of CO oxidation over $Au/IrO_2/TiO_2$. *Phys. Rev. Lett.*, 2004, 93, 156102-1:4.

[170] Haruta, M.; Yamada, N.; Kobayashi, T.; Iijima. S. Gold catalysts prepared by coprecipitation for low-temperature oxidation of hydrogen and of carbon monoxide. *J. Catal.,* 1989, 115, 301-309.

[171] Mavrikakis, M.; Stoltze, P.; Nørskov, J.K. Making gold less noble. *Catal. Lett.,* 2000, 64, 101-106.

[172] Liu, Z.–P.; Hu, P.; Alavi, A. Catalytic role of gold in gold-based catalysts: a density functional theory study on the CO oxidation on gold. *J. Am. Chem. Soc.*, 2002, 124, 14770-14779.

[173] Fajín, J.L.C.; Cordeiro, M.N.D.S.; Gomes, J.R.B. Adsorption of atomic and molecular oxygen on the Au(321) surface: DFT study. *J. Phys. Chem. C.*, 2007, 111, 17311-17321.

[174] Fajín, J.L.C.; Cordeiro, M.N.D.S.; Gomes, J.R.B. DFT study of the CO oxidation on the Au(321) surface. *J. Phys. Chem. C* 2008, 112, 17291-17302.

[175] Wittstock, A.A.; Zielasek, V.; Biener, J.; Friend, C.M.; Bäumer, M. Nanoporous Gold catalysts for selective gas-phase oxidative coupling of methanol at low temperature. *Science*, 2010, 327, 319-322.

[176] Wittstock, A.; Neumann, B.; Schaefer, A.; Dumbuya, K.; Kübel, C.; Biener, M.M.; Zielasek, V.; Steinrück, H.-P.; Gottfried, J.M.; Biener, J.; Hamza, A.; Bäumer, M. Nanoporous Au: An Unsupported Pure Gold Catalyst?. *J. Phys. Chem. C,* 2009, 113, 5593-5600.

[177] Deng, L.; Hu, W.; Deng, H.; Xiao,S.; Tang, J. Au–Ag Bimetallic Nanoparticles: Surface Segregation and Atomic-Scale Structure. *J. Phys. Chem. C,* 2011, 115, 11355-11363.

[178] Rozovskii, A. .; Lin, G.I. Fundamentals of methanol synthesis and decomposition. *Top. Catal.*, 2003, 22, 137-150.

[179] Larrubia-Vargas, M.A.; Busca, G.; Costantino, U.; Marmottini, F.; Montanari, T.; Patrono, P.; Pinzari, F.; Ramis, G. An IR study of methanol steam reforming over ex-hydrotalcite Cu–Zn–Al catalysts. *J. Mol. Catal. A: Chem.*, 2007, 266, 188-197.
[180] Lee, S.H.D.; Applegate, D.V.; Ahmed, S.; Calderone, S.G.; Harvey, T.L. Hydrogen from natural gas: part I—autothermal reforming in an integrated fuel processor. *Int. J. Hydrogen Ener.*, 2005, 30, 829-842.
[181] Ovesen, C.V.; Clausen, B.S.; Hammershøi, B.S.; Steffensen, G.; Askgaard, T.; Chorkendorff, I.; Nørskov, J.K.; Rasmussen, P.B.; Stoltze, P.; Taylor, P. J. A Microkinetic Analysis of the Water–Gas Shift Reaction under Industrial Conditions. *J. Catal.*, 1996, 158, 170-180.
[182] Schumacher, N.; Boisen, A.; Dahl, S.; Gokhale, A.A.; Kandoi, S.; Grabow, L.C.; Dumesic, J.A.; Mavrikakis, M.; Chorkendorff, I. Trends in low-temperature water–gas shift reactivity on transition metals. *J. Catal.*, 2005, 229, 265-275.
[183] Fu, Q.; Saltsburg, H.; Flytzani-Stephanopoulos, M. Active Nonmetallic Au and Pt Species on Ceria-Based Water-Gas Shift Catalysts. *Science,* 2003, 301, 935-938.
[184] Rodríguez, J.A.; Evans, J.; Graciani, J.; Park, J.-B.; Liu, P.; Hrbek, J.; Sanz, J.F. High Water−Gas Shift Activity in TiO_2(110) Supported Cu and Au Nanoparticles: Role of the Oxide and Metal Particle Size. *J. Phys. Chem. C*, 2009, 113, 7364-7370.
[185] Li, L.; Zhan, Y.; Zheng, Q.; Zheng, Y.; Lin, X.; Li, D.; Zhu, J. Water–Gas Shift Reaction Over Aluminum Promoted Cu/CeO_2Nanocatalysts Characterized by XRD, BET, TPR and Cyclic Voltammetry (CV). *Catal. Lett.*, 2007, 118, 91-97.
[186] Boccuzzi, F.; Chiorino, A.; Manzoli, M.; Andreeva, D.; Tabakova, T. FTIR Study of the Low-Temperature Water–Gas Shift Reaction on Au/Fe_2O_3 and Au/TiO_2 Catalysts. *J. Catal.*, 1999, 188, 176-185.
[187] Liu, Z.-P.; Jenkins, S.J.; King, D.A. Origin and Activity of Oxidized Gold in Water-Gas-Shift Catalysis. *Phys. Rev. Lett.*, 2005, 94, 196102-1:4.
[188] Burch, R. Gold catalysts for pure hydrogen production in the water–gas shift reaction: activity, structure and reaction mechanism. *Phys. Chem. Chem. Phys.*, 2006, 8, 5483-5500.
[189] Yahiro, H.; Murawaki, K.; Saiki, K.; Yamamoto, T.; Yamaura, H. Study on the supported Cu-based catalysts for the low-temperature water–gas shift reaction. *Catal. Today*, 2007, 126, 436-440.
[190] Rodríguez, J.A.; Liu, P.; Hrbek, J.; Pérez, M.; Evans, J. Water–gas shift activity of Au and Cu nanoparticles supported on molybdenum oxides. *J. Mol. Catal. A: Chem.*, 2008, 281, 59-65.
[191] Kumar, P.; Idem, R.A Comparative Study of Copper-Promoted Water−Gas-Shift (WGS) Catalysts. *Energ. Fuel.*, 2007, 21, 522-529.
[192] Rodriguez, J.A.; Liu, P.; Wang, X.; Wen, W.; Hanson, J.; Hrbek, J.; Pérez, M.; Evans, J. Water-gas shift activity of Cu surfaces and Cu nanoparticles supported on metal oxides. *Catal. Today.* 2009, 143, 45-50.
[193] Wang, X.; Rodriguez, J.A.; Hanson, J.C.; Gamarra, D.; Martínez-Arias, A.; Fernández-García, M. In Situ Studies of the Active Sites for the Water Gas Shift Reaction over Cu−CeO_2 Catalysts: Complex Interaction between Metallic Copper and Oxygen Vacancies of Ceria. *J. Phys. Chem. B.*, 2006, 110, 428-434.

[194] Chen, Y.; Cheng, J.; Hua, P.; Wang, H. Examining the redox and formate mechanisms for water–gas shift reaction on Au/CeO_2 using density functional theory. *Surf. Sci.* 2008, 602, 2828-2834.
[195] Zhang, L.; Wang, X.; Millet, J.-M.M.; Matter, P.H.; Ozkan, U. S. Investigation of highly active Fe-Al-Cu catalysts for water-gas shift reaction. *Appl. Catal. A: Gen.*, 2008, 351, 1-8.
[196] Du, X.; Yuana, Z.; Cao, L.; Zhanga, C.; Wanga, S. Water gas shift reaction over Cu–Mn mixed oxides catalysts: Effects of the third metal. *Fuel Proces. Technol.*, 2008, 89, 131-138.
[197] Nishida, K.; Atake, I.; Li, D.; Shishido, T.; Oumi, Y.; Sano, T.; Takehira, K. Effects of noble metal-doping on $Cu/ZnO/Al_2O_3$ catalysts for water–gas shift reaction: Catalyst preparation by adopting “memory effect” of hydrotalcite. *Appl. Catal. A: Gen.*, 2008, 337, 48-57.
[198] Knudsen, J.; Nilekar, A.U.; Vang, R.T.; Schnadt, J.; Kunkes, E.L.; Dumesic, J.A.; Mavrikakis, M. Besenbacher, F. A Cu/Pt Near-Surface Alloy for Water−Gas Shift Catalysis. *J. Am. Chem. Soc.*, 2007, 129, 6485-6490.
[199] Zhao, X.; Ma, S.; Hrbek, J.; Rodriguez, J.A. Reaction of water with Ce–Au(111) and CeO_x/Au(111) surfaces: Photoemission and STM studies. *Surf. Sci.*, 2007, 601, 2445-2452.
[200] Gokhale, A.; Dumesic, J. A.; Mavrikakis, M. On the mechanism of low-temperature water gas shift reaction on copper. *J. Am. Chem. Soc.*, 2008, 130, 1402-1414.
[201] Fajín, J.L.C.; Cordeiro, M.N.D.S.; Illas, F.; Gomes, J.R.B. Influence of step sites in the molecular mechanism of the water gas shift reaction catalyzed by copper. *J. Catal.*, 2009, 268, 131-141.
[202] Fajín, J.L.C.; Cordeiro, M.N.D.S.; Illas, F.; Gomes, J.R.B. Effect of the exchange-correlation potential and of surface relaxation on the description of the H_2O dissociation on Cu(111). *J. Chem. Phys.*, 2009, 130, 224702-1:8.
[203] Fajín, J.L.C.; Cordeiro, M.N.D.S.; Gomes, J.R.B.; Illas, F. On the Need for Spin Polarization in Heterogeneously Catalyzed Reactions on Nonmagnetic Metallic Surfaces. *J. Chem. Theory Comput.*, 2012, 8, 1737-1743.
[204] Brønsted, J.N. Acid and basic catalysis. *Chem. Rev.*, 1928, 5, 231-338.
[205] Evans, M.G.; Polanyi, M. Inertia and driving force of chemical reactions. *Trans. Faraday Soc.*, 1938, 34, 11-24.
[206] Fajín, J.L.C.; Cordeiro, M.N.D.S.; Illas, F.; Gomes, J.R.B. Descriptors controlling the catalytic activity of metallic surfaces towards water splitting. *J. Catal.*, 2010, 276, 92-100.
[207] Fajín, J.L.C.; Bruix, A.; Cordeiro, M.N.D.S.; Gomes, J.R.B.; Illas, F. Density functional theory model study of size and structure effects on water dissociation by platinum nanoparticles. *J. Chem. Phys.*, 2012, 137, 034701-1:10.
[208] Fajín, J.L.C.; Cordeiro, M.N.D.S.; Gomes, J.R.B. Water dissociation on bimetallic surfaces: general trends. *J. Phys. Chem. C.*, 2012, 111, 10120-10128.
[209] Fajín, J.L.C.; Cordeiro, M.N.D.S.; Gomes, J.R.B. DFT study on the NO oxidation on a flat gold surface model. *Chem. Phys. Lett.*, 2011, 503, 129-133.
[210] Fajín, J.L.C.; Cordeiro, M.N.D.S.; Gomes, J.R.B. DFT study on the reaction of NO oxidation on a stepped gold surface. *App. Catal. A: Gen.*, 2010, 379, 111-120.

[211] Torres, D.; González, S.; Neyman, K.M.; Illas, F. Adsorption and oxidation of NO on Au(111) surface: Density functional studies *Chem. Phys. Lett.,* 2006, 422, 412-416.

[212] Zhang, W.; Li, Z.; Luo, Y.; Yang, J.A first-principles study of NO adsorption and oxidation on Au(111) surface. *J. Chem. Phys*., 2008, 129, 134708-1:5.

[213] Fajín, J.L.C.; Cordeiro, M.N.D.S.; Gomes, J.R.B. The role of preadsorbed atomic hydrogen in the NO dissociation on a zigzag stepped gold surface: A DFT study. *J. Phys. Chem. C*, 2009, 113, 8864-8877.

[214] Gómez, T.; Flórez, E.; Rodríguez, J.A.; Illas, F. Reactivity of transition metals (Pd, Pt, Cu, Ag, Au) toward molecular hydrogen dissociation: extended surfaces versus particles supported on TiC(001) or small is not always better and large is not always bad. *J. Phys. Chem. C*, 2011, 115, 11666-11672.

[215] Corma, A.; Boronat, M.; González, S.; Illas, F. On the activation of molecular hydrogen by gold: a theoretical approximation to the nature of potential active sites. *Chem. Commun*., 2007, 3371-3373.

[216] Boronat, M.; Corma, A.; Illas, F.; Radilla, J.; Ródenas, T.; Sabater, M.J. Mechanism of selective alcohol oxidation to aldehydes on gold catalysts: Influence of surface roughness on reactivity. *J. Catal.,* 2011, 278, 50-58.

[217] Pallassana, V.; Neurock, M. Electronic Factors Governing Ethylene Hydrogenation and Dehydrogenation Activity of Pseudomorphic Pd_{ML}/Re(0001), Pd_{ML}/Ru(0001), Pd(111), and Pd_{ML}/Au(111) Surfaces. *J. Catal*., 2000, 191, 301-317.

[218] Fajín, J.L.C.; Cordeiro, M.N.D.S.; Gomes, J.R.B. Unraveling the mechanism of the NO reduction by CO on gold based catalysts. *J. Catal*., 2012, 289, 11-20.

[219] World Health Organization (WHO), World Malaria Report 2011, 2011 [October 15, 2012]. URL: http://www.who.int/malaria/world_malaria_report_2011/WMR2011_noprofiles_lowres.pdf.

[220] Ta, T.T.; Salas, A.; Ali-Tammam, M.; Martinez Mdel, C.; Lanza, M.; Arroyo, E.; Rubio, J.M., First case of detection of Plasmodium knowlesi in Spain by Real Time PCR in a traveller from Southeast Asia. *Malar. J.*, 2010, 9, 219-224.

[221] Alilio, M.S.; Bygbjerg, I.C.; Breman, J.G., Are multilateral malaria research and control programs the most successful? Lessons from the past 100 years in Africa. *Am. J. Trop. Med. Hyg.,* 2004, 71, 268-278.

[222] Teixeira, C.; Gomes, J.R.B.; Gomes, P., Falcipains, Plasmodium falciparum cysteine proteases as key drug targets against malaria. *Curr. Med. Chem*., 2011, 18, 1555-1572.

[223] Dondorp, A.M.; Nosten, F.; Yi, P.; Das, D.; Phyo, A.P.; Tarning, J.; Lwin, K.M.; Ariey, F.; Hanpithakpong, W.; Lee, S.J.; Ringwald, P.; Silamut, K.; Imwong, M.; Chotivanich, K.; Lim, P.; Herdman, T.; An, S.S.; Yeung, S.; Singhasivanon, P.; Day, N.P.; Lindegardh, N.; Socheat, D.; White, N.J., Artemisinin resistance in Plasmodium falciparum malaria. *N. Engl. J. Med.,* 2009, 361, 455-467.

[224] Wongsrichanalai, C.; Meshnick, S.R., Declining artesunate-mefloquine efficacy against falciparum malaria on the Cambodia-Thailand border. *Emerg. Infect. Dis.,* 2008, 14, 716-719.

[225] Gardner, M.J.; Hall, N.; Fung, E.; White, O.; Berriman, M.; Hyman, R.W.; Carlton, J.M.; Pain, A.; Nelson, K.E.; Bowman, S.; Paulsen, I.T.; James, K.; Eisen, J.A.; Rutherford, K.; Salzberg, S.L.; Craig, A.; Kyes, S.; Chan, M.S.; Nene, V.; Shallom, S.J.; Suh, B.; Peterson, J.; Angiuoli, S.; Pertea, M.; Allen, J.; Selengut, J.; Haft, D.; Mather, M.W.; Vaidya, A.B.; Martin, D.M.; Fairlamb, A.H.; Fraunholz, M.J.; Roos,

D.S.; Ralph, S.A.; McFadden, G.I.; Cummings, L.M.; Subramanian, G.M.; Mungall, C.; Venter, J.C.; Carucci, D.J.; Hoffman, S.L.; Newbold, C.; Davis, R.W.; Fraser, C.M.; Barrell, B., Genome sequence of the human malaria parasite Plasmodium falciparum. *Nature*, 2002, 419, 498-511.

[226] Padmanaban, G., Drug targets in malaria parasites. *Adv. Biochem. Eng. Biotechnol.*, 2003, 84, 123-141.

[227] Rosenthal, P.J., Cysteine proteases of malaria parasites. *Int. J. Parasitol.*, 2004, 34, 1489-1499.

[228] Goldberg, D.E., Haemoglobin degradation. *Curr. Top. Microbiol. Immunol.*, 2005, 295, 275-291.

[229] Lew, V.L.; Tiffert, T.; Ginsburg, H., Excess haemoglobin digestion and the osmotic stability of Plasmodium falciparum-infected red blood cells. *Blood*, 2003, 101, 4189-4194.

[230] Rosenthal, P.J.; Sijwali, P.S.; Singh, A.; Shenai, B.R., Cysteine proteases of malaria parasites: targets for chemotherapy. *Curr. Pharm. Des.,* 2002, 8, 1659-1672.

[231] Rosenthal, P.J.; Wollish, W.S.; Palmer, J.T.; Rasnick, D., Antimalarial effects of peptide inhibitors of a Plasmodium falciparum cysteine proteinase. *J. Clin. Invest.,* 1991, 88, 1467-1472.

[232] Salmon, B.L.; Oksman, A.; Goldberg, D.E., Malaria parasite exit from the host erythrocyte: a two-step process requiring extraerythrocytic proteolysis. *Proc. Natl. Acad. Sci. U.S.A.*, 2001, 98, 271-276.

[233] Sijwali, P.S.; Rosenthal, P.J., Gene disruption confirms a critical role for the cysteine protease falcipain-2 in haemoglobin hydrolysis by Plasmodium falciparum. *Proc. Natl. Acad. Sci. U.S.A.*, 2004, 101, 4384-4389.

[234] Sijwali, P.S.; Koo, J.; Singh, N.; Rosenthal, P.J., Gene disruptions demonstrate independent roles for the four falcipain cysteine proteases of Plasmodium falciparum. *Mol. Biochem. Parasitol.,* 2006, 150, 96-106.

[235] Wang, S.X.; Pandey, K.C.; Somoza, J.R.; Sijwali, P.S.; Kortemme, T.; Brinen, L.S.; Fletterick, R.J.; Rosenthal, P.J.; McKerrow, J.H., Structural basis for unique mechanisms of folding and haemoglobin binding by a malarial protease. *Proc. Natl. Acad. Sci. U.S.A.*, 2006, 103, 11503-11508.

[236] Leung, D.; Abbenante, G.; Fairlie, D.P., Protease inhibitors: current status and future prospects. *J. Med. Chem.*, 2000, 43, 305-341.

[237] Ettari, R.; Bova, F.; Zappala, M.; Grasso, S.; Micale, N., Falcipain-2 inhibitors. *Med. Res. Rev.*, 2010, 30, 136-167.

[238] Powers, J.C.; Asgian, J.L.; Ekici, O.D.; James, K.E., Irreversible inhibitors of serine, cysteine, and threonine proteases. *Chem. Rev.,* 2002, 102, 4639-4750.

[239] Shenai, B.R.; Lee, B.J.; Alvarez-Hernandez, A.; Chong, P.Y.; Emal, C.D.; Neitz, R.J.; Roush, W.R.; Rosenthal, P.J., Structure-activity relationships for inhibition of cysteine protease activity and development of Plasmodium falciparum by peptidyl vinyl sulfones. *Antimicrob. Agents Chemother.*, 2003, 47, 154-160.

[240] Pérez, B.C.; Teixeira, C.; Figueiras, M.; Gut, J.; Rosenthal, P.J.; Gomes, J.R.B.; Gomes, P., Novel cinnamic acid/4-aminoquinoline conjugates bearing non-proteinogenic amino acids: towards the development of potential dual action antimalarials. *Eur. J. Med. Chem.*, 2012, 54, 887-899.

[241] Lee, B.J.; Singh, A.; Chiang, P.; Kemp, S.J.; Goldman, E.A.; Weinhouse, M.I.; Vlasuk, G.P.; Rosenthal, P.J., Antimalarial activities of novel synthetic cysteine protease inhibitors. *Antimicrob. Agents Chemother.*, 2003, 47, 3810-3814.

[242] Schulz, F.; Gelhaus, C.; Degel, B.; Vicik, R.; Heppner, S.; Breuning, A.; Leippe, M.; Gut, J.; Rosenthal, P.J.; Schirmeister, T., Screening of protease inhibitors as antiplasmodial agents. Part I: Aziridines and epoxides. *Chem. Med. Chem.*, 2007, 2, 1214-1224.

[243] Karplus, M.; McCammon, J.A., Molecular dynamics simulations of biomolecules. *Nat. Struct. Biol.*, 2002, 9, 646-652.

[244] Friesner, R.A.; Guallar, V., Ab initio quantum chemical and mixed quantum mechanics/molecular mechanics (QM/MM) methods for studying enzymatic catalysis. *Annu. Rev. Phys. Chem.*, 2005, 56, 389-427.

[245] Helten, H.; Schirmeister, T.; Engels, B., Model Calculations about the Influence of Protic Environments on the Alkylation Step of Epoxide, Aziridine, and Thiirane Based Cysteine Protease Inhibitors. *J. Phys. Chem. A*, 2004, 108, 7691-7701.

[246] Helten, H.; Schirmeister, T.; Engels, B., Theoretical studies about the influence of different ring substituents on the nucleophilic ring opening of three-membered heterocycles and possible implications for the mechanisms of cysteine protease inhibitors. *J. Org. Chem.*, 2005, 70, 233-237.

[247] Buback, V.; Mladenovic, M.; Engels, B.; Schirmeister, T., Rational design of improved aziridine-based inhibitors of cysteine proteases. *J. Phys. Chem. B*, 2009, 113, 5282-5289.

[248] Schafer, A.; Huber, C.; Ahlrichs, R., Fully optimized contracted gaussian basis sets of triple zeta valence quality for atoms Li to Kr. *J. Chem. Phys.*, 1994, 100, 5829-5835.

[249] Vicik, R.; Helten, H.; Schirmeister, T.; Engels, B., Rational design of aziridine-containing cysteine protease inhibitors with improved potency: studies on inhibition mechanism. *Chem. Med. Chem.*, 2006, 1, 1021-1028.

[250] Sijwali, P.S.; Brinen, L.S.; Rosenthal, P.J., Systematic optimization of expression and refolding of the Plasmodium falciparum cysteine protease falcipain-2. *Protein Expr. Purif.*, 2001, 22, 128-134.

[251] Mladenovic, M.; Schirmeister, T.; Thiel, S.; Thiel, W.; Engels, B., The importance of the active site histidine for the activity of epoxide- or aziridine-based inhibitors of cysteine proteases. *Chem. Med. Chem*, 2007, 2, 120-128.

[252] MacKerell, A.D.; Bashford, D.; Bellott; Dunbrack, R.L.; Evanseck, J.D.; Field, M.J.; Fischer, S.; Gao, J.; Guo, H.; Ha, S.; Joseph-McCarthy, D.; Kuchnir, L.; Kuczera, K.; Lau, F.T.K.; Mattos, C.; Michnick, S.; Ngo, T.; Nguyen, D.T.; Prodhom, B.; Reiher, W.E.; Roux, B.; Schlenkrich, M.; Smith, J.C.; Stote, R.; Straub, J.; Watanabe, M.; Wiórkiewicz-Kuczera, J.; Yin, D.; Karplus, M., All-Atom Empirical Potential for Molecular Modeling and Dynamics Studies of Proteins. *J. Phys. Chem. B*, 1998, 102, 3586-3616.

[253] Zhang, Y.; Kua, J.; McCammon, J.A., Influence of Structural Fluctuation on Enzyme Reaction Energy Barriers in Combined Quantum Mechanical/Molecular Mechanical Studies. *J. Phys. Chem. B*, 2003, 107, 4459-4463.

[254] Seabra, G.d.M.; Walker, R.C.; Elstner, M.; Case, D.A.; Roitberg, A.E., Implementation of the SCC-DFTB Method for Hybrid QM/MM Simulations within the Amber Molecular Dynamics Package. *J. Phys. Chem. A*, 2007, 111, 5655-5664.

[255] Wang, J.; Wang, W.; Kollman, P.A.; Case, D.A., Automatic atom type and bond type perception in molecular mechanical calculations. *J. Mol. Graph. Model.,* 2006, 25, 247-260.

[256] Wang, J.; Wolf, R.M.; Caldwell, J.W.; Kollman, P.A.; Case, D.A., Development and testing of a general amber force field. *J. Comput. Chem.*, 2004, 25, 1157-1174.

[257] Duan, Y.; Wu, C.; Chowdhury, S.; Lee, M.C.; Xiong, G.; Zhang, W.; Yang, R.; Cieplak, P.; Luo, R.; Lee, T.; Caldwell, J.; Wang, J.; Kollman, P., A point-charge force field for molecular mechanics simulations of proteins based on condensed-phase quantum mechanical calculations. *J. Comput. Chem.*, 2003, 24, 1999-2012.

[258] Jorgensen, W.L.; Chandrasekhar, J.; Madura, J.D.; Impey, R.W.; Klein, M.L., Comparison of simple potential functions for simulating liquid water. *J. Chem. Phys.*, 1983, 79, 926-935.

[259] Babin, V.; Karpusenka, V.; Moradi, M.; Roland, C.; Sagui, C., Adaptively biased molecular dynamics: An umbrella sampling method with a time-dependent potential. *Int. J. Quantum Chem.*, 2009, 109, 3666-3678.

In: Density Functional Theory
Editors: J. Morin and J.M. Pelletier, pp. 59-93
ISBN: 978-1-62417-954-9

Chapter 2

APPLICATIONS OF DENSITY FUNCTIONAL THEORY CALCULATIONS TO LITHIUM CARBENOIDS AND MAGNESIUM CARBENOIDS

Tsutomu Kimura*
Department of Chemistry, Faculty of Science Division II,
Tokyo University of Science, Tokyo, Japan

Abstract

Metal carbenoids, in which a metal (Li, Mg) and an electronegative element (F, Cl, Br, I) are bound to the same carbon atom, are reactive intermediates that react in a manner similar to that of carbenes. The simultaneous existence of an electropositive metal and an electronegative element on the same carbon atom is responsible for the unexpected reactivity. The elusive reactivity of metal carbenoids is counterintuitive and difficult to understand based on the electronic theory of organic chemistry. Although the experimental direct observation of metal carbenoids may provide insight into their reactivity, this observation is complicated by the thermal instability of metal carbenoids. Recently, quantum chemical calculations, including DFT calculations, have become a versatile tool for investigating the structures and reactivity of unstable chemical species that are not readily studied by experimental means. A number of theoretical studies targeting metal carbenoids have been reported with the development of computational chemistry. In this chapter, theoretical studies on the structural elucidation and mechanistic exploration of lithium carbenoids and magnesium carbenoids are overviewed.

1. Introduction

Metal carbenoids are a class of organometallic compounds in which an electronegative heteroatom such as a halogen or oxygen is found at the α-position. Various types of metal carbenoids, including lithium carbenoids (I) and magnesium carbenoids (II), have been developed and studied in great depth (Figure 1). The chemical properties of metal carbenoids

*E-mail address: kimtwo@rs.tus.ac.jp

are dependent on both the covalent/ionic character of the carbon–metal bond and the leaving-group ability of the heteroatom at the α-position. The simultaneous existence of an electropositive metal and an electronegative element on the same carbon atom brings about their unexpected and intriguing reactivity. Typical carbenoid reactions include [2+1] cycloaddition with alkenes and insertion into unreactive bonds such as C–H and C–C bonds (Figure 1, Eqs. 1–3). Another characteristic feature of metal carbenoids is their ambiphilic reactivity, that is, they can react with both electrophiles and nucleophiles (Figure 1, Eqs. 4 and 5). Furthermore, substitution reactions can occur at the vinylic carbon atom of metal alkylidene carbenoids (Figure 1, Eq. 6).

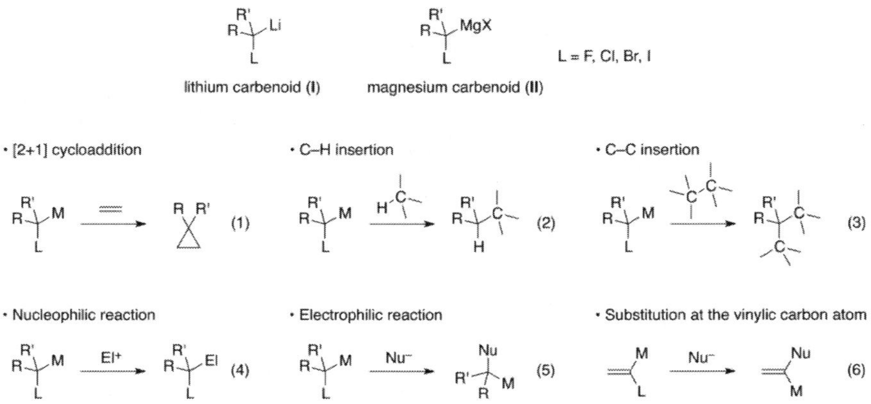

Figure 1. Chemical structures and typical reactions of metal carbenoids I and II.

Since the pioneering work of Köbrich [1,2], metal carbenoids have become a versatile synthetic tool and play an important role in modern synthetic organic chemistry. However, the elusive reactivity of metal carbenoids is difficult to understand based on the electronic theory of organic chemistry. Although the experimental direct observation of metal carbenoids may provide insights into their reactivity, metal carbenoids are generally thermally labile reactive intermediates, and the isolation of metal carbenoids remains a formidable task. Recently, quantum chemical calculations, including DFT calculations, have become a versatile tool for investigating the structures and reactivity of unstable chemical species that cannot be readily studied by experimental means. A number of theoretical studies targeting metal carbenoids have been reported with the development of computational chemistry. As several excellent reviews on lithium carbenoids and magnesium carbenoids are available [1–11], this chapter will mainly focus on theoretical studies of the structural elucidation and mechanistic exploration of α-halo-substituted organolithiums and organomagnesiums, namely, lithium carbenoids and magnesium carbenoids.

2. Molecular Structures of Metal Carbenoids

Metal carbenoids can be depicted by four canonical structures A–D (Figure 2). In canonical structure A, the halogen atom is connected to the carbon atom, whereas the metal cation is dissociated from the carbon atom [12]. By contrast, in canonical structure B, the halogen atom departs from the carbon atom as a leaving group, and metal is covalently bound to the carbon atom [13]. These structures reflect the nucleophilic and electrophilic properties of metal carbenoids, respectively. In canonical structure C, both the metal and halogen atoms form covalent bonds with the central carbon atom; carbene D corresponds to the α-elimination product of the tetrahedral metal carbenoid C [14]. The identity of the structure that primarily contributes to the chemical properties of metal carbenoids is a subject of great interest.

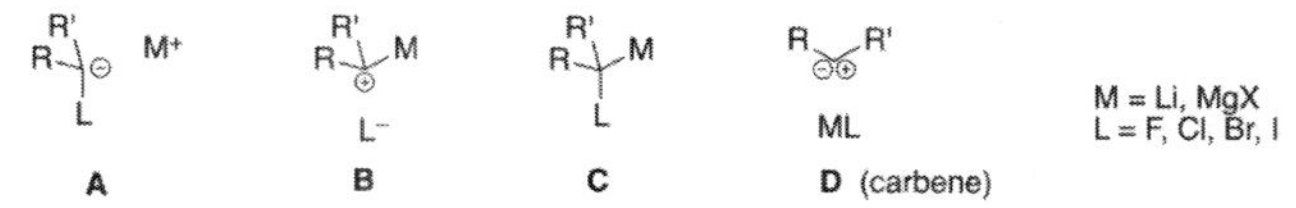

Figure 2. Canonical structures of metal carbenoids.

Andrews and co-workers observed several vibrational frequencies of lithium carbenoids via matrix isolation infrared spectra [15,16]. The NMR spectroscopic properties of various types of lithium carbenoids were investigated by Seebach and co-workers in the 1980s [17,18]. However, mainly due to their instability, further progress in the structural elucidation of metal carbenoids by experimental means did not occur until Boche and colleagues determined the crystal structures of metal carbenoids by X-ray molecular structure analysis [19–21]. Accordingly, much effort has been devoted to elucidating the structures of metal carbenoids by quantum chemical calculations. In this section, theoretical studies on the structures of lithium carbenoids and magnesium carbenoids are highlighted.

2.1. Molecular Structures of α-Halo-Substituted Metal Carbenoids

Schleyer, Pople, and colleagues investigated the structures and properties of a series of fundamental organic molecules, including lithium carbenoids, by ab initio calculation as early as the 1970s [22,23]. In 1979, Clark and Schleyer reported the structures of mono-, di-, and tri-halo-substituted methyllithiums in the gas phase (Figure 3) [24–26]. Geometry optimization of possible structures of (fluoromethyl)lithium, CH_2FLi (1), using SCF theory with the 4-31G basis set (5-21G for lithium), provided three local minima 1a–c [24]. The

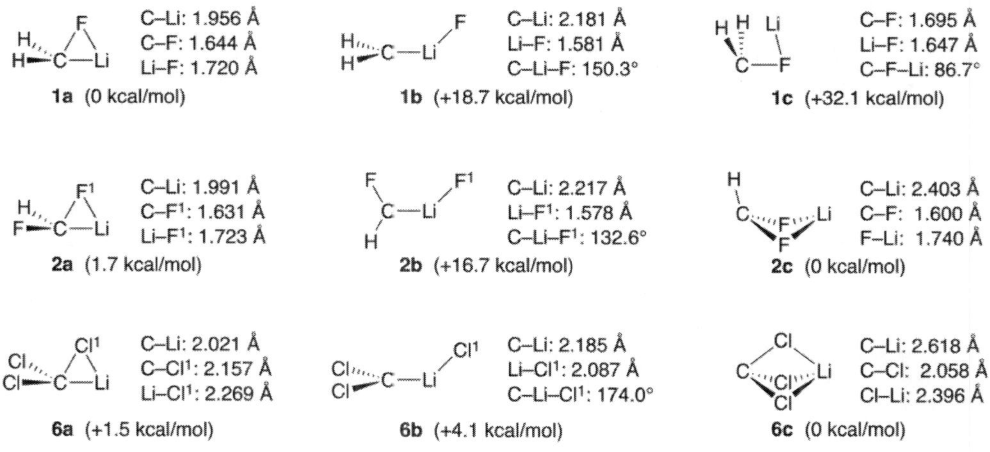

Figure 3. Structures of the lithium carbenoids CH_2FLi, CHF_2Li, and CCl_3Li.

lowest energy structure 1a adopted a peculiar "umbrella-shaped" geometry in which the carbon, fluorine, and lithium atoms formed a three-membered ring and all four valences around the carbon atom extended into a single hemisphere. The C–F bond (1.644 Å) was significantly elongated in comparison with those of typical organic fluorides. The local minima 1b and 1c could be regarded as complexes of singlet methylene with lithium fluoride (CH_2–LiF and CH_2–FLi), in which methylene acted as an electron donor and acceptor, respectively. The barriers for the conversion of isomers 1b and 1c into the lowest energy structure 1a were estimated to be only 1.4 and 2.5 kcal/mol, respectively, indicating that isomers 1b and 1c were metastable species. One year later, Abronin and co-workers reported calculations of the potential hypersurface of CH_2FLi (1) in the singlet and triplet states using the SCF MO LCAO method in the semi-empirical INDO approximation and non-empirical STO-3G basis set [27]. Their computational study revealed that one local minimum was present in both states.

The molecule was predicted to adopt a C_S structure and have a short C–F bond length (1.355 Å) and a large F–C–Li bond angle (174.21°). Pross and Radom reported the calculated C–Li and C–F bond lengths of CH_2FLi (1) (C–Li: 2.027 Å, C–F: 1.454 Å) using the 4-31G basis set in the course of their theoretical study on a series of disubstituted methanes [28].

(Difluoromethyl)lithium, CHF_2Li (2), also existed in three isomeric forms 2a–c [25]. In the most stable structure 2c, the lithium atom was associated with the backside of the pyramidal CHF_2 anion, with two fluorine atoms interacting with the lithium atom. Clark and Schleyer proposed that the equilibria between the structures of 2a and 2c might be important in determining the reactivity of carbenoids and that the relative energies of isomers 2a–c in the gas phase might significantly differ from those in the donor solvent. (Trichloromethyl)lithium, CCl_3Li (6), was found to have the triply bridged geometry 6c, in which the lithium atom is situated on the "wrong side" of the molecule [26]. Tetrahedral lithium carbenoids such as the canonical structure C in Figure 2 were not local minima in these studies.

Vincent and Schaefer revisited the structural study of lithium carbenoid 1 to validate Clark–Schleyer's predictions by using a double-zeta plus polarization (DZ+P) basis set [29]. Three structural isomers similar to 1a–c were found as local minima (1a: 0 kcal/mol; 1b: +17.1 kcal/mol; 1c: +26.1 kcal/mol), whereas the C–Li–F bond angle in structure 1b increased to 180°. The barrier for the rearrangement of structure 1b to the most stable structure 1a was estimated to be only 1.0 kcal/mol, suggesting that isomer 1b readily converted to structure 1a and indicating that the two structures 1a and 1c would exist.

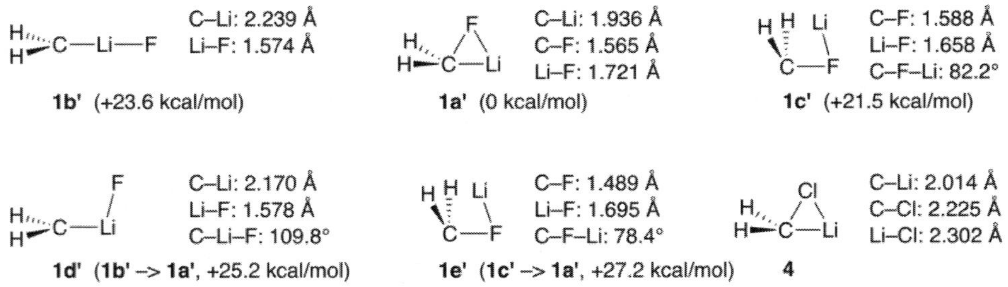

Figure 4. Structures of the lithium carbenoids CH_2FLi and CH_2ClLi.

Because Vincent and Schaefer did not examine the rearrangement pathway from 1c to 1a, Clark and Schleyer reexamined the transition structures of lithium carbenoid 1 using an ab initio method including electron correction and zero-point energy correction (Figure 4) [30]. Single-point energy calculations were performed at the MP4SDTQ/6-31G* level of theory for the geometries 1a'–e', which were optimized at the RHF/6-31G* level. Structure 1a' was more stable than structures 1b' and 1c' by 23.6 and 21.5 kcal/mol, respectively. The barriers for the conversions of structures 1b' and 1c' into the ground minimum 1a' via transition states 1d' and 1e' were 1.6 and 5.7 kcal/mol, respectively. Based on these results, Clark and Schleyer predicted that the experimental observation of both isomers 1b' and 1c' was difficult and that only the most stable structure 1a' was detectable.

Clark and Schleyer theoretically investigated the structures and stabilities of a series of organolithium and organosodium compounds having a substituent such as an amino, hydroxy, or fluoro group at the α-position as well as the corresponding anionic species [31]. In the course of their systematic study, chloromethyllithium, CH_2ClLi (4), was found to adopt a lithium-bridged structure similar to that of CH_2FLi (1a) (Figure 4). The authors concluded that the chlorine atom stabilized lithium carbenoids more efficiently than the electronegative fluorine atom. Qiu and Deng followed Clark–Schleyer's study on the structures and isomerization of CH_2FLi (1) shown in Figure 4 with CH_2ClLi (4) [32]. The chlorine analog 4 had three equilibrium structures and two isomeric transition states, which were similar to those of CH_2FLi 1a'–e'.

To elucidate the influence of the substituent at the α-position on the structures of lithium carbenoids, Boche and co-workers investigated the C–L bond length and hybridization of the carbenoid carbon atom of a series of lithium carbenoids CH_2LLi (1: L = F, 4: L = Cl, 7: L = Br, 10: L = I) at the MP2/6-311++G(d,p) level of theory [33]. Compared to the C–L bond lengths in methyl halides CH_3L, the C–L bond lengths in lithium carbenoids were elongated by 5.5–12.3% (1: 12.3%, 4: 7.0%, 7: 6.7%, 10: 5.5%). Based on these results, they speculated that CH_2FLi (1) had the strongest carbenoid character. The hybridizations of carbenoid carbon atoms in the C–Li bond of lithium carbenoids 1, 4, 7, and 10 were in the range of $sp^{1.7}$–$sp^{1.9}$, indicating that the carbenoid carbon atom in the C–Li bond had a high *s*-character. By contrast, those in the C–L bonds of lithium carbenoids were $sp^{8.5}$–$sp^{11.8}$, indicating that the carbenoid carbon atoms bore a strong *p*-character.

Subsequently, Boche and co-workers reported the extensive investigation of the substituent effects on the carbenoid character of the mono-, di-, and tri-halo-substituted lithium carbenoids CH_2LLi (1: L = F, 4: L = Cl, 7: L = Br, 10: L = I), CHL_2Li (2: L = F, 5: L = Cl, 8: L = Br, 11: L = I), and CL_3Li (3: L = F, 6: L = Cl, 9: L = Br, 12: L = I) [6]. Calculations at the MP2/6-311++G(d,p) level of theory revealed a strong preference for the lithium-bridged geometries in almost all cases. In the series of mono-halo-substituted lithium carbenoids 1, 4, 7, and 10, the singlet methylene-lithium halide complexes (CH_2–LiL and CH_2–LLi) like 1b and 1c in Figure 3 were less stable than the lithium-bridged structures by 27.1–30.8 and 21.5–24.5 kcal/mol, respectively. In the case of di-halo-substituted lithium carbenoids 2, 5, 8, and 11, mono-bridged (like 2a in Figure 3) or di-bridged geometries (like 2c in Figure 3) were the most stable structures, and the differences in the relative energies between them were small (2: 2.7 kcal/mol, 5: –0.4 kcal/mol, 8: –0.7 kcal/mol). Five local minima were identified for the tri-halo-substituted lithium carbenoids 3, 6, 9, and 12. The mono-lithium-bridged structures had the lowest energies, and the differences in the relative energies of the five isomers were small.

Kvíčala and co-workers investigated the equilibrium structures of a series of fluorine-containing mono-, di-, and tri-halo-substituted methyllithiums, CH_2FLi (1), CHF_2Li (2), CF_3Li (3), CHClFLi (13), $CClF_2Li$ (14), CCl_2FLi (15), CHBrFLi (16), $CBrF_2Li$ (17), CBr_2FLi (18), and CBrClFLi (19) at the RHF/6-31+G(d,p) and MP2/6-31+G(d,p) levels of theory [34]. Five distinct minimum structures were recognized: tetrahedral structure (T), trigonal bidentate with coordination of carbon and halogen to lithium (BCL), tetrahedral tridentate with coordination of carbon and two halogens to lithium (TCL_2), quadrilateral bidentate with coordination of two halogens to lithium (BL_2), and bipyramidal tridentate with coordination of three halogens to lithium (TL_3) (Figure 5). In almost all cases, several minima were located, and the BCL structure was the most stable geometry. For example, the BCF structure had minimal energies in the case of mono-, di-, and tri-fluoro-substituted lithium carbenoids 1–3.

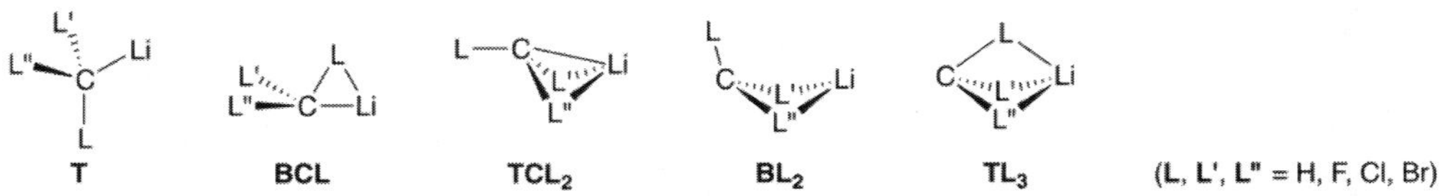

Figure 5. Equilibrium structures of lithium carbenoids.

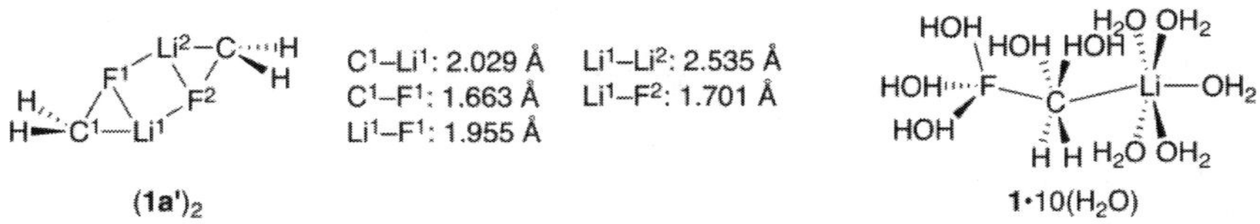

Figure 6. Structures of dimeric lithium carbenoid and decahydrated lithium carbenoid.

Similarly, the chlorine-containing lithium carbenoids 13–15 adopted the BCCl geometry as the most stable forms. In the case of the bromine-containing lithium carbenoids 16–19, the BCBr structure was the most stable geometry with the exception of the dibromo-substituted lithium carbenoid 18. Lithium carbenoid 18 had $TCBr_2$ geometry in the lowest energy structure.

The aggregation state and solvation effect are important considerations when reproducing the true properties of metal carbenoids under experimental conditions. Clark and Schleyer investigated the effect of dimerization on the structure and dissociation energy of CH_2FLi (1) (Figure 6) [35]. The dimerization energy of CH_2FLi (1) was –56.2 kcal/mol, which was intermediate between that of lithium fluoride (–87.3 kcal/mol) and methyllithium (–46.3 kcal/mol). The geometry of the monomeric units in the dimer $(1a')_2$ was essentially same as that in the monomer 1a'. Dimerization of lithium carbenoid 1 was thought to facilitate the dissociation of singlet methylene.

In 1985, Simonetta and co-workers reported the effect of solvent on the dissociation of CH_2FLi (1) by semi-empirical MO-SCF theory at the CNDO/2 level of approximation [36]. The lithium atom in the decahydrated lithium carbenoid $1{\bullet}10(H_2O)$ was coordinated with five oxygen atoms of water molecules, and the fluorine atom was surrounded with three water molecules (Figure 6). Two water molecules formed hydrogen bonds with the carbenoid carbon atom. The dissociation of the fluoride anion from the hydrated structure 1•10(H_2O) leading to the formation of the CH_2Li^+ cation was predicted to be energetically favored with respect to the dissociation of the lithium cation leading to the formation of the CHF_2^- anion and the dissociation of the lithium fluoride leading to the formation of the methylene CH_2.

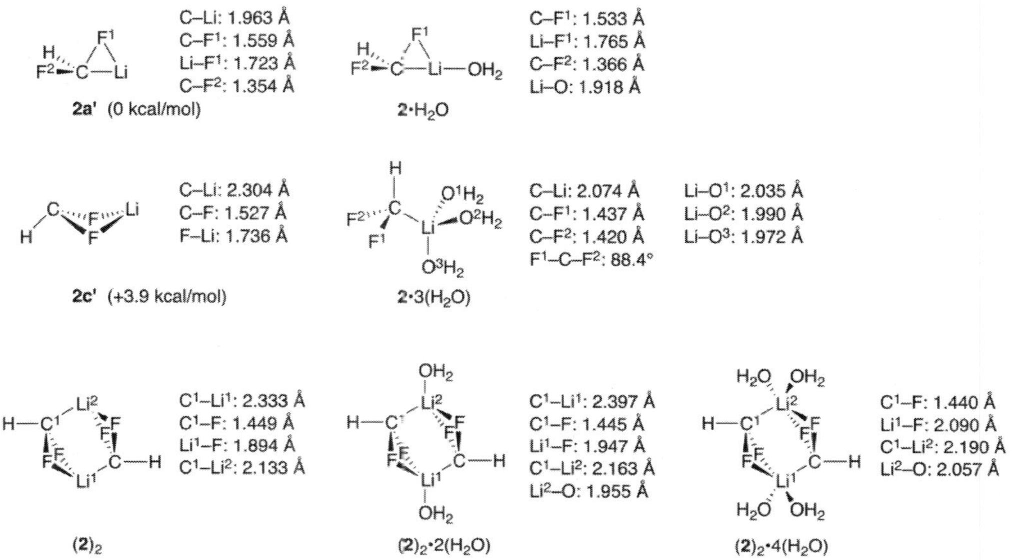

Figure 7. Structures of non-hydrated and hydrated lithium carbenoid CHF_2Li in monomers and dimers.

Tonachini and colleagues reported the monomeric and dimeric structures of CHF_2Li (2) as well as their hydrated structures in the course of their theoretical studies on the structures and reactivity of *gem*-di-halo-substituted allyl anions (Figure 7) [37]. Two stable structures, 2a' and 2c', which correspond to structures 2a and 2c in Figure 3, were observed for CHF_2Li (2) at the RHF/6-31G(d) level of theory. The free energy of the dimerization of 2 to $(2)_2$ at 183K was –41.9 kcal/mol, indicating that monomer 2 had an inclination to dimerize. They also performed calculations based on the mono- and tri-hydrated lithium carbenoids 2•H_2O and 2•3(H_2O). In the case of tri-hydrated lithium carbenoid 2•3(H_2O), the lithium atom was coordinated with three oxygen atoms of water molecules and a carbenoid carbon atom, and there were no significant interactions between the lithium atom and the two fluorine atoms. The dimerization energies of trihydrated lithium carbenoid 2•3(H_2O) to the dimers surrounded by two water molecules $(2)_2$•2(H_2O) and four water molecules $(2)_2$•4(H_2O) were calculated to be –13.6 and –17.5 kcal/mol, respectively. These results demonstrated that solvation significantly reduced the preference for dimerization.

Tonachini and colleagues also theoretically studied the monomer-dimer equilibrium in the chlorine analog $CHCl_2Li$ (5) with and without water molecules (Figure 8) [38]. Only one minimal structure was observed for $CHCl_2Li$ (5), in which one chlorine atom interacted with the lithium atom to form a lithium-bridged triangular geometry.

The dimerization energy of 5 to $(5)_2$ was –24.3 kcal/mol, which was smaller than that of fluorine analog 2. In the trihydrated lithium carbenoid bearing tetra-coordinated lithium atom 5•3(H_2O), the interactions between the lithium atom and chlorine atom were lost. In contrast to the results of fluorine analog 2, the dimerization energies of the monomer 5•3(H_2O) to the dihydrated dimer $(5)_2 \cdot 2(H_2O)$ and tetrahydrated dimer $(5)_2 \cdot 4(H_2O)$ were 14.8 and 17.0 kcal/mol, respectively, which were unfavorable to dimerization.

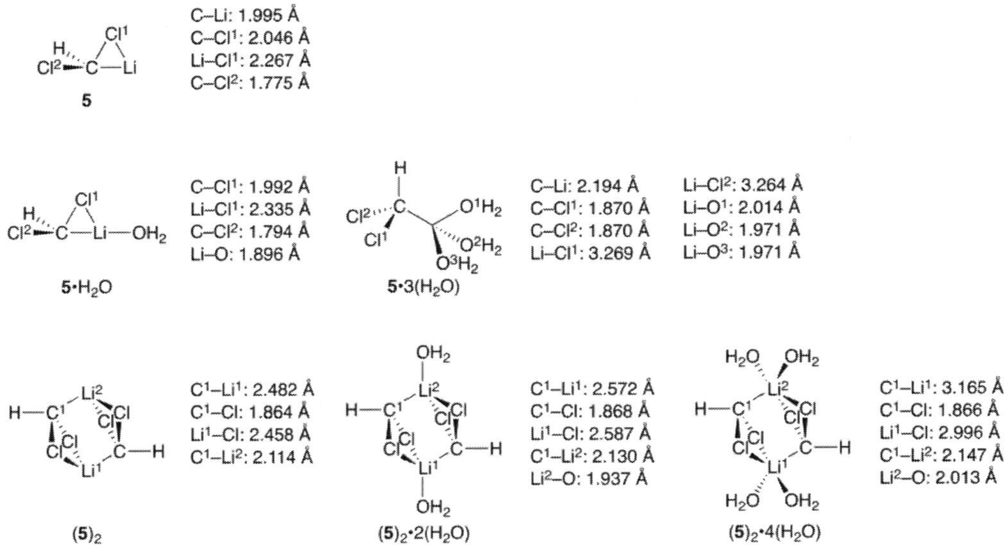

Figure 8. Structures of non-hydrated and hydrated lithium carbenoid $CHCl_2Li$ in monomers and dimers.

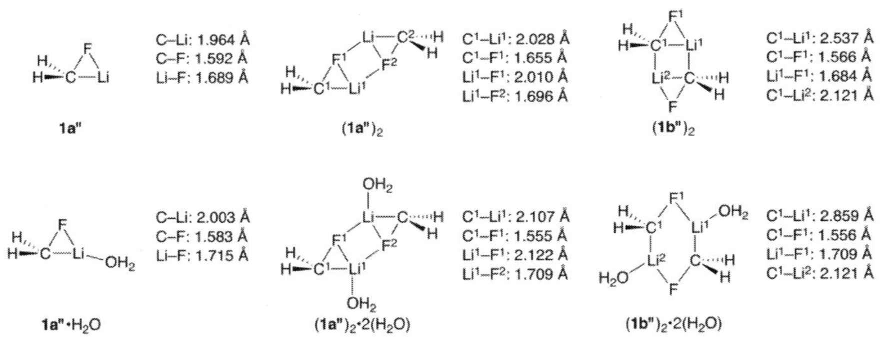

Figure 9. Structures of non-hydrated and hydrated lithium carbenoid CH_2FLi in monomers and dimers.

In 2001, Bohce and co-workers described a theoretical study on the aggregation and solvation of lithium carbenoids using the MP2/6-31G(d)//3-21G level of theory (Figure 9) [6]. In addition to the dimer $(1a")_2$, which was more stable than the two monomers by –40 kcal/mol, a much more stable dimer $(1b")_2$ (–50 kcal/mol) was identified. A similar trend of dimerization energies was also observed in the hydrated systems [$(1a")_2 \cdot 2(H_2O)$: –25.3 kcal/mol, $(1b")_2 \cdot 2(H_2O)$: –39.8 kcal/mol]. Solvation of one water molecule to each lithium atom in monomer 1a" and dimers $(1a")_2$ and $(1b")_2$ led to a slight lengthening of the carbon–lithium bond and shortening of the carbon–fluorine bond. Based on these results, the authors proposed that solvation should stabilize lithium carbenoids.

Bohce and co-workers also calculated the structures of a series of trihydrated lithium carbenoids CH_2FLi [$1a" \cdot 3(H_2O)$], CH_2ClLi [$4' \cdot 3(H_2O)$], $CHCl_2Li$ [$5' \cdot 3(H_2O)$], and CCl_3Li [$6' \cdot 3(H_2O)$] (Figure 10). The solvation of three water molecules to the lithium atom weakened the halogen–lithium bonds and shortened the carbon–halogen bonds. Lithium carbenoids bearing three solvated water molecules adopted tetrahedral structures.

In 2002, Pratt and co-workers reported DFT calculations of monomeric and dimeric structures of CH_2FLi (1) and CH_2ClLi (4) in the gas phase and in dimethyl ether solvent

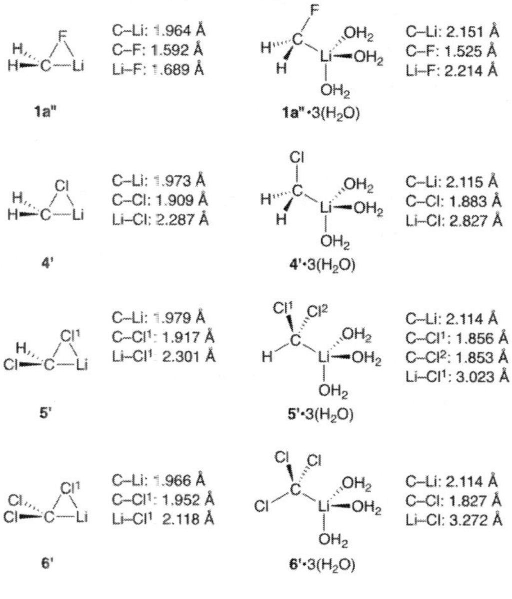

Figure 10. Structures of non-solvated and trihydrated lithium carbenoids.

(Figure 11) [39]. In the gas phase, both monomers and dimers of CH_2FLi (1) and CH_2ClLi (4) were predicted to existed as single isomers 1a''', 4'', (1a''')$_2$, and (4'')$_2$ on the basis of the results of calculations performed at the *m*PW1PW91/MG3S level of theory. The carbon–halogen bonds of monomers 1a''' and 4'' were longer than those observed for standard carbon–halogen bonds by approximately 0.2 Å. Upon dimerization, the carbon–halogen bonds shortened and the carbon–lithium bonds lengthened in both cases. The free energies for

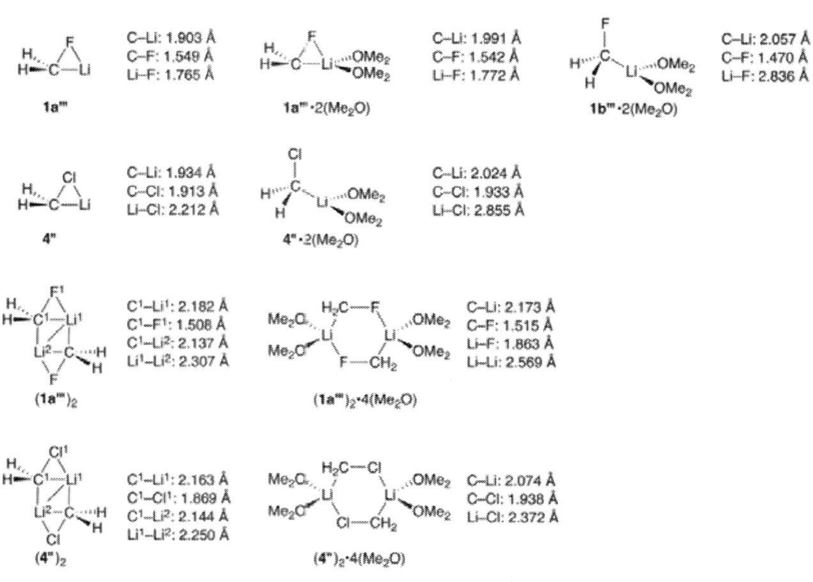

Figure 11. Monomeric and dimeric structures of lithium carbenoids in the gas phase and dimethyl ether solvent.

dimerization of lithium carbenoids 1 and 4 were exothermic (1: –38.6 kcal/mol, 4: –33.4 kcal/mol). A combination of explicit microsolvation and a continuum reaction field (bulk solvation) were used to incorporate the solvation effects. When two dimethyl ether molecules were coordinated to each lithium atom, two structures, 1a'''•2(Me_2O) and 1b'''•2(Me_2O), were found in lithium carbenoid 1, and one structure, 4''•2(Me_2O), was obtained for lithium carbenoid 4. The distances between the lithium and halogen atoms of the solvated structures [1b'''•2(Me_2O): 2.836 Å, 4''•2(Me_2O): 2.855 Å] were much greater than those of the gas phase structures (1a''': 1.765 Å, 4'': 2.212 Å), indicating the absence of interactions between the lithium and halogen atoms. In the case of solvated dimers, three distinct structures including the structures $(1a''')_2$•4(Me_2O) or $(4'')_2$•4(Me_2O) were found for each lithium carbenoid 1 and 4. The dimerization free energies of lithium carbenoids 1 and 4 in dimethyl ether at 173 K were estimated to be –0.9 kcal/mol and 3.7 kcal/mol, respectively. The difference in the dimerization energy of lithium carbenoids 1 and 4 appeared to reflect the greater affinity of the lithium atom for the fluorine atom than the chlorine atom. The authors anticipated that the chloro-substituted lithium carbenoid 4 existed as the monomer and the fluorine analog 1 existed preferentially as the monomer at concentrations below 0.2 M.

Organolithium compounds are known to form mixed aggregates. The formation of mixed aggregates may result in changes in reactivity relative to the monomeric state. Pratt and co-workers used DFT calculations to investigate mixed aggregates of lithium carbenoid with lithium dimethylamide, which is often used as a strong base in organic synthesis (Figure 12) [40]. The gas-phase free energies of mixed aggregates consisting of lithium carbenoid 4 and lithium dimethylamide calculated at the B3LYP/6-31G+G(d) level indicated that the mixed trimers and mixed tetramers were major species. By contrast, the calculated free energies of the THF-solvated mixed aggregates demonstrated that the aggregation states were in favor of mixed dimer and trimers.

The halogen-metal exchange reaction to generate metal carbenoids gives lithium halides as co-products.

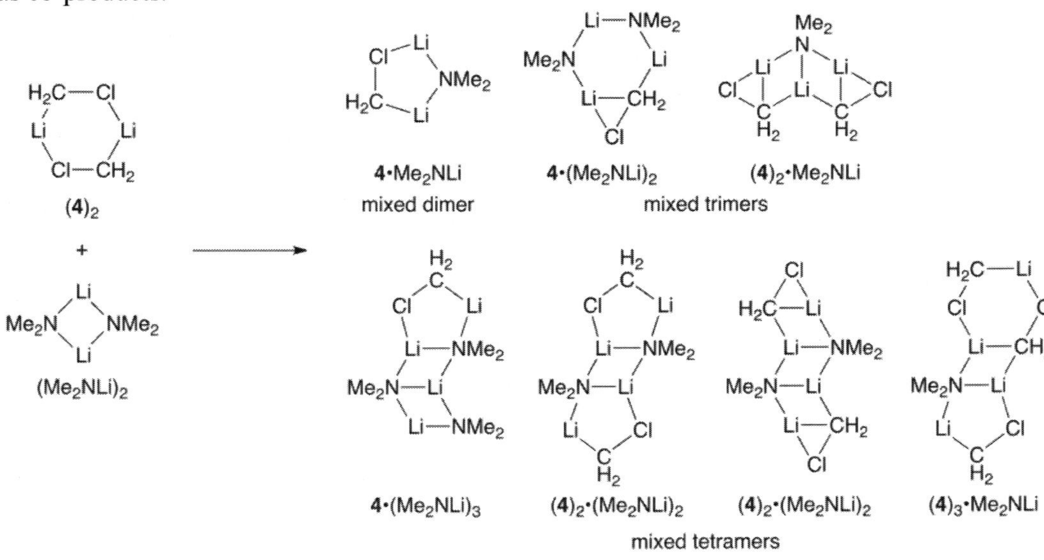

Figure 12. Mixed aggregates of lithium carbenoid with lithium dimethylamide.

Exposure of organometallic reagent to air results in the formation of metal alkoxides. Taking into account the aggregation of lithium carbenoids with these species, Pratt and co-workers used DFT calculations to investigate the structures of mixed aggregates of lithium carbenoids 4 and 7 with lithium methoxide (MeOLi) and lithium halides (LiCl, LiBr) [41]. The calculations for the monomer (CH_2LLi); a mixed dimer (CH_2LLi•MeOLi); two mixed trimers [CH_2LLi•2(MeOLi), $(CH_2LLi)_2$•MeOLi]; and four mixed tetramers [CH_2LLi•3(MeOLi), two isomers of $(CH_2LLi)_2$•2(MeOLi), $(CH_2LLi)_3$•MeOLi] performed at the B3LYP/6-31+G(d) level of theory revealed that mixed aggregates of lithium carbenoids with lithium methoxide would be present almost exclusively as mixed tetramers $(CH_2LLi)_2$•2(MeOLi) and $(CH_2LLi)_3$•MeOLi. By contrast, mixed trimers and tetramers seemed to exist in the case of lithium halides. In several cases, the THF-solvation of the lithium atom in the mixed aggregates reduced the tendency to form mixed aggregates.

To determine which levels of theory are able to provide accurate geometries and dimerization free energies for lithium carbenoids, systematic investigations of the basis set and electron correlation effects were performed by Pratt and co-workers [42]. Specifically, geometry optimization and dimerization free energy calculations of three lithium carbenoids, 1, 4, and 7, and three lithium alkylidene carbenoids, H_2C=CLLi (20: L = F, 21: L = Cl, 22: L = Br), in the gas phase were performed at the B3LYP and MP2 levels with several different basis sets (MIDIX, 6-31G(d), 6-31+G(d), 6-31++G(d,p), 6-311++G(d,p), aug-cc-pvdz) and at the CCSD(T)/aug-cc-pvdz level (Figure 13). The optimized geometries of monomeric and dimeric halomethyllithiums 1, 4, and 7 were not very sensitive toward the basis sets and methods with the exception of the MIDIX basis set, whereas those of lithium alkylidene carbenoids 20, 21, and 22 performed using the B3LYP method were not always consistent with the results obtained with the MP2 and CCSD methods. The B3LYP method predicted the planar dimeric structure (dimer P) to be a local minimum, but this structure was not a minimum potential energy surface at the MP2 level and failed to find a chair conformation (dimer C) for H_2C=CFLi (20). The B3LYP method appeared to generate erroneous results in the case of the flat potential surface. The dimerization free energies tended to be overestimated by small basis sets compared to the basis sets containing polarization and diffuse functions. The thermal corrections to the dimerization free energies were relatively independent of the basis sets larger than 6-31+G(d).

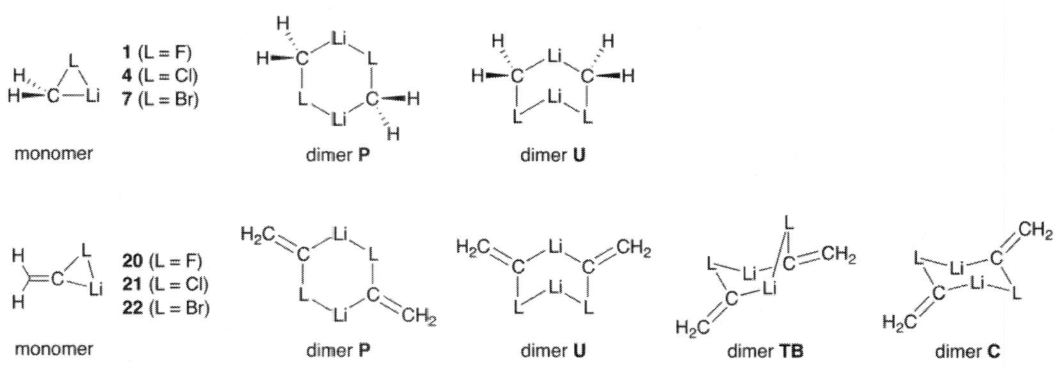

Figure 13. Structures of monomeric and dimeric lithium carbenoids and lithium alkylidene carbenoids.

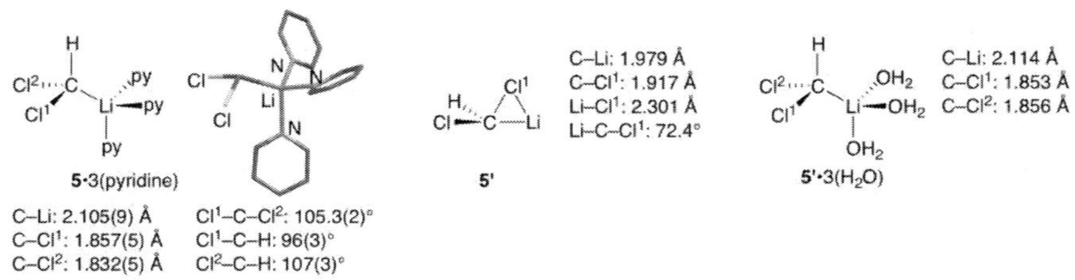

Figure 14. X-ray molecular structure of lithium carbenoid [5•3(pyridine)] and the calculated structures of the non-solvated and trihydrated lithium carbenoids.

In 1996, Boche and co-workers revealed the molecular structure of the lithium carbenoid $CHCl_2Li$•3(pyridine) [5•3(pyridine)] by X-ray crystallography (Figure 14) [21]. The nitrogen atoms of the three pyridine molecules were coordinated to the lithium atom, and the carbenoid carbon atom adopted a slightly distorted tetrahedral structure. The distances between the lithium atom and the two chlorine atoms were 3.226 Å and 3.301 Å, respectively, and there were no significant interactions between the lithium atom and chlorine atoms. Both of the C–Cl bond lengths (1.857(5) Å and 1.832(5) Å) were longer than those of typical organic chlorides by approximately 0.1 Å. The sum of the bond angles Cl^1–C–Cl^2, Cl^1–C–H, and Cl^2–C–H was 308°, which was smaller than that of the tetrahedral compound (328°). These results indicated the high *p*-character of the C–Cl bonds. Boche and co-workers also studied the structures of the non-solvated lithium carbenoid 5' and trihydrated lithium carbenoid 5'•3(H_2O) at the MP2/6-31G(d) level. Non-solvated lithium carbenoid 5' had a lithium-bridged structure, whereas the trihydrated lithium carbenoid 5'•3(H_2O) had a tetrahedral structure similar to that indicated by X-ray crystallography. The chemical shifts in the ^{13}C NMR spectra for model compounds 5' and 5'•3(H_2O) were estimated to be 62.8 and 49.5 ppm by IGLO calculations, respectively, and the experimental chemical shift measured in THF was 50.0 ppm.

In contrast to the large number of theoretical studies on lithium carbenoids, reports on magnesium carbenoids are limited. Hoffmann and co-workers reported an experimental and theoretical study of magnesium carbenoids in 1998 [43]. Based on the chemical shifts of the carbenoid carbon atom in the ^{13}C NMR spectra, they predicted that the carbenoid character of metal carbenoids should decrease in the order lithium to magnesium and zinc. In addition, theoretical calculations on (bromomethyl)magnesium chloride•dihydrate (23) at the MP2/6-31G(d) (MWB-31G(d) for Cl and Br) level indicated that magnesium carbenoid 23 had a longer carbon–bromine bond (2.03 Å) and a smaller Mg–C–Br bond angle (100°) in comparison with those of typical organic compounds (Figure 15). From these results, they speculated that the carbenoid carbon atom exhibited partial sp^2-hybridization character and the carbon–bromine bond had a high *p*-character. In fact, Kimura and Satoh found that the (chloromethyl)magnesium chloride•bis(dimethyl ether) complex 24 optimized at B3LYP/6-311++G(d,p) level had a high *p*-character in the C–Cl bond and a high *s*-character in the C–Mg bond [44]. The structure of magnesium carbenoid 25 in the gas phase was reported by Satoh and Ando [45]. The optimized geometry of magnesium carbenoid 25 at B3LYP/6-31+G* level adopted a magnesium-bridged three-membered structure and had structural features similar to those of lithium carbenoids in the gas phase.

As described above, the bond lengths and angles around the carbenoid carbon atoms in solvation-free metal carbenoids significantly deviated from those of conventional organic molecules, and the unusual structures were believed to contribute to their unconventional reactivity. These unusual structures were proposed to originate from the high affinity of metal toward halogens. However, the metal in the solvation-free state does not satisfy the octet rule and unavoidably accepts a lone pair from the neighboring heteroatom, apparently resulting in the artificial metal-bridged triangular structures.

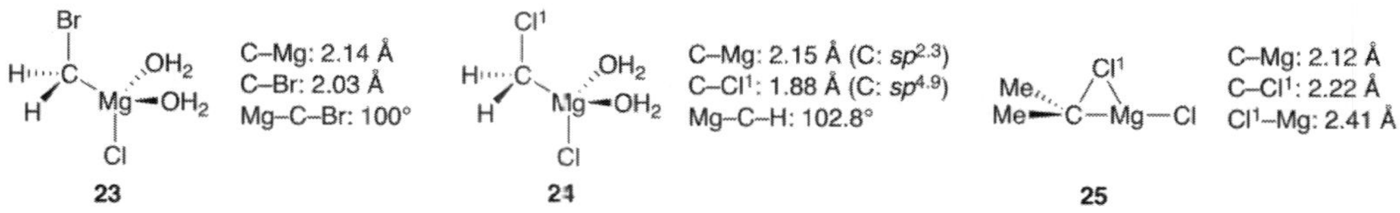

Figure 15. Optimized geometries and hybridization of magnesium carbenoids.

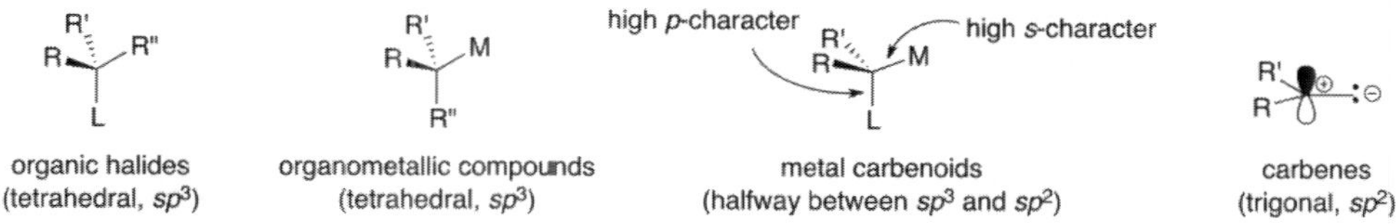

Figure 16. Hybridization of metal carbenoids and related compounds.

Therefore, it is inadequate to discuss the structures and reactivity of metal carbenoids in solution based on the results of solvent-free model compounds. However, it is noteworthy that even the properly solvated metal carbenoids adopted curiously bent structures without interactions between the metal and halogen atom. The simultaneous existence of the electropositive metal and the electronegative halogen atom on the same carbon atom leads to the significant change in the hybridization of the carbon atom (Figure 16). That is, the *p*-character in the carbon–halogen bond increases, and accordingly, the *p*-character of residual bonds around the carbenoid carbon atom decreases. Therefore, the true nature of the carbenoid carbon atom is halfway between a sp^3-hybridized carbon atom such as those of organic halides and conventional organometallic compounds and a sp^2-hybridized carbon atom such as those of carbenes.

2.2. Molecular Structures of Cyclopropylmetal Carbenoids

(1-Halocyclopropyl)lithiums and (1-halocyclopropyl)magnesiums, also referred to as cyclopropylmetal carbenoids, are another type of metal carbenoid that contain a highly strained cyclopropane ring. Cyclopropylmetal carbenoids are good precursors for the synthesis of allenes via the Skattebøl rearrangement. Wang and Deng reported the structure of (1-fluorocyclopropyl)lithium (26) at the RHF/6-31G*//RHF/3-21G level in 1989 (Figure 17) [46]. In contrast to the existence of several equilibrium structures of lithium carbenoids, (1-fluorocyclopropyl)lithium had only one minimum, 26a. Similar to the structures of lithium carbenoids, the molecule adopted a lithium-bridged triangular geometry. The energy barrier for the isomerization of 26a to 26a' via the transition state 26b was 6.7 kcal/mol, suggesting that structure 26a could readily isomerize to 26a'. Boche and co-workers also theoretically investigated the structures of the (1-halocyclopropyl)lithiums c-C_3H_4LLi (26: L = F, 27: L =

Cl) at the HF/6-311++G(d,p) level [6]. The carbene–LiF complex like 26b was a local minimum in their study.

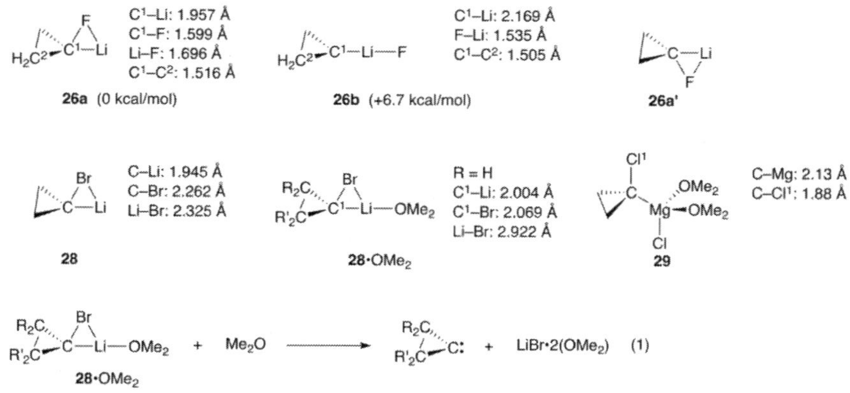

Figure 17. Structures of cyclopropylmetal carbenoids.

The energy difference between the lithium-bridged structure 26a and the carbene–LiF complex 26b was small (2.2 kcal/mol), and only a carbene–LiCl complex was located for *c*-C_3H_4ClLi (27).

Azizoglu and colleagues reported the structures of (1-bromocyclopropyl)lithiums (Figure 17) [47]. Non-substituted (1-bromocyclopropyl)lithium (28) optimized at the B3LYP/6-31G(d) level had a long C–Br bond, and the lithium atom interacted with both the carbon and bromine atoms, similar to the fluorine analog 26a. Kaszynski and co-workers studied the dissociation energies of the mono(dimethyl ether)-solvated cyclopropyllithium carbenoids 28•OMe_2 to carbenes and LiBr•2(OMe_2) at the B3LYP/6-31+G(d,p) level of theory (Figure 17, Eq. 1) [48]. The dissociation process was endothermic by 15.5–22.5 kcal/mol. For cyclopropylmagnesium carbenoid, Kimura and Satoh determined the structure of *c*-$C_3H_4ClMgCl$•2(OMe_2) (29) [44]. Bis(dimethyl ether)-solvated cyclopropylmagnesium carbenoid 29 had a long C–Cl bond and expanded bond angles around the carbenoid carbon atom. There were no significant interactions between the chlorine atom on the carbenoid carbon atom and the magnesium atom.

2.3. Molecular Structures of Metal Alkylidene Carbenoids

Metal alkylidene carbenoids are another type of metal carbenoid in which both the metal and halogen atoms are connected to the terminal alkenyl carbon atom. Boche and co-workers successfully determined the molecular structures of lithium alkylidene carbenoid 30 and magnesium alkylidene carbenoid 31 by single crystal X-ray diffraction in 1993 and 1994 (Figure 18) [19,20]. Two nitrogen atoms of *N,N,N',N'*-tetramethylethylenediamine, an oxygen atom of THF molecule, and a carbon atom of the vinyl group were connected to the lithium atom in the structure of lithium alkylidene carbenoid 30. An extended carbon–chlorine bond length (1.855(7) Å) and expanded C^2–C^1–Li bond angle (137.1(6)°) were significant structural features of lithium alkylidene carbenoid 30. Similar features were also observed in magnesium alkylidene carbenoid 31.

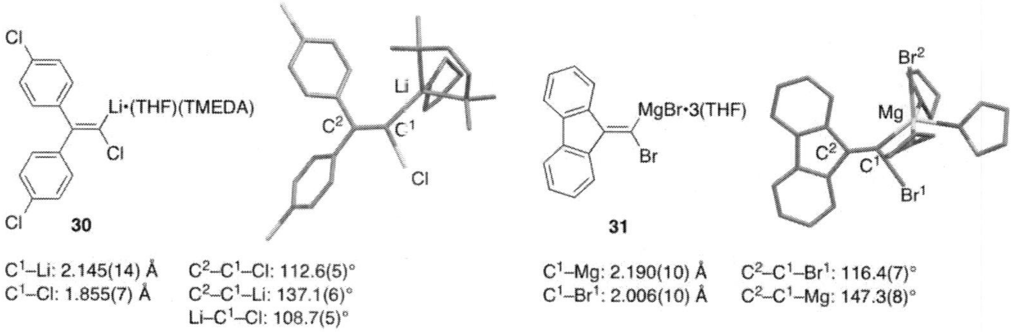

Figure 18. Crystal structures of [1-chloro-2,2-bis(4-chlorophenyl)vinyl]lithium•TMEDA•2(THF) 30 and [bromo(9*H*-fluoren-9-ylidene)methyl]magnesium bromide•4(THF) 31. One THF molecule was omitted for clarity in both drawings.

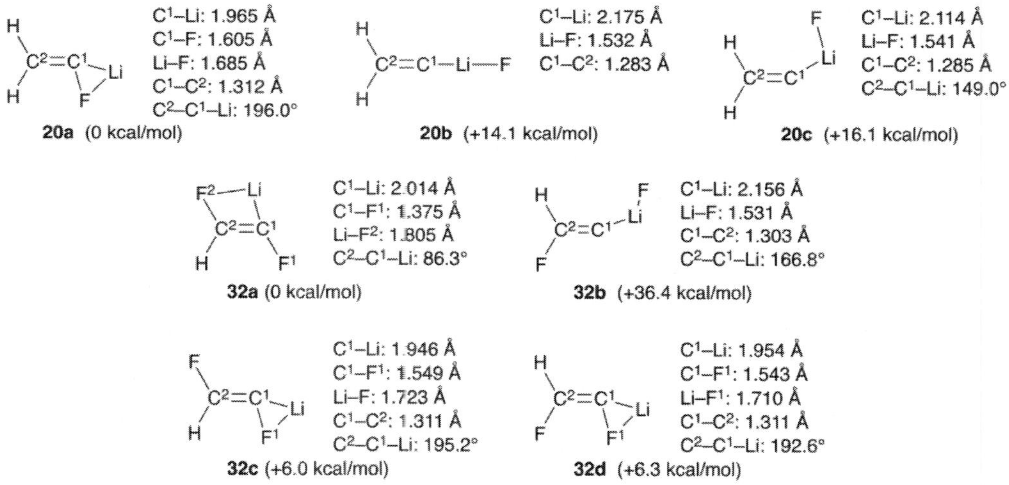

Figure 19. Structures of lithium alkylidene carbenoids.

Before the first X-ray molecular structure analysis of lithium alkylidene carbenoid 30 was attained, Wang and Deng reported a theoretical study on the structures and decomposition pathways of (1-fluorovinyl)lithium, H_2C=CFLi (20), and (1,2-difluorovinyl)lithium, HFC=CFLi (32), at the RHF/3-21 level (Figure 19) [49,50]. Two local minima 20a and 20b were located for H_2C=CFLi (20), and only isomer 20a was predicted to be detectable because of the low isomerization barrier (2.0 kcal/mol) through the transition state 20c. The ground minimum 20a had a lithium-bridged three-membered ring geometry that was similar to that of lithium carbenoid 1a in Figure 3. Zheng and co-workers also reported that the lithium-bridged structure 20a had the lowest energy in the equilibrium structures [51]. The structure 20b could be regarded as the singlet vinylidene–LiF complex. Four equilibrium structures were found for HFC=CFLi (32). The distance between the lithium atom and fluorine atom at the β-position in the quadrilateral structure 32a was considerably short (1.805 Å), and two carbon atoms of the vinylidene unit, the lithium atom, and the fluorine atom formed a four-membered ring. The structure 32b corresponded to the vinylidene–LiF complex, and two local minima

32c and 32d were the geometrical isomers of the lithium-bridged three-membered ring structure.

In 2001, Boche and colleagues theoretically investigated the equilibrium structures of a series of lithium alkylidene carbenoids $H_2C{=}CFLi$ (20), $H_2C{=}CClLi$ (21), and $H_2C{=}CBrLi$ (22) in the gas phase at the HF/6-311++G(d,p) level [6]. The fluorine derivative 20 had two local minima 20a' and 20b' corresponding to the geometries 20a and 20b, and the difference in energy between these minima 20a' and 20b' was only 1.3 kcal/mol. By contrast, the chlorine and bromine derivatives 21 and 22 had only one local minimum corresponding to the vinylidene–lithium halide complex.

As described in the section on lithium carbenoids, solvation and aggregation significantly affect the structures of lithium carbenoids. To determine the effect of solvation on the structures of lithium alkylidene carbenoids, Schoeller performed ab initio calculations of (1-chlorovinyl)lithium, $H_2C{=}CClLi$ (21), and (1-chloro-2-methylprop-1-en-1-yl)lithium, $Me_2C{=}CClLi$ (33), in the gas phase and their trihydrated complexes, $H_2C{=}CLiCl{\bullet}3(H_2O)$ [21•3(H_2O)] and $Me_2C{=}CClLi{\bullet}3(H_2O)$, at the RHF/6-31+G* and MP2/6-31+G* levels (Figure 20) [52]. The structural parameters of 21a and 21a•3(H_2O) calculated at the RHF level and non-solvated structure 21b calculated at the MP2 level did not agree with the experimental values shown in Figure 18.

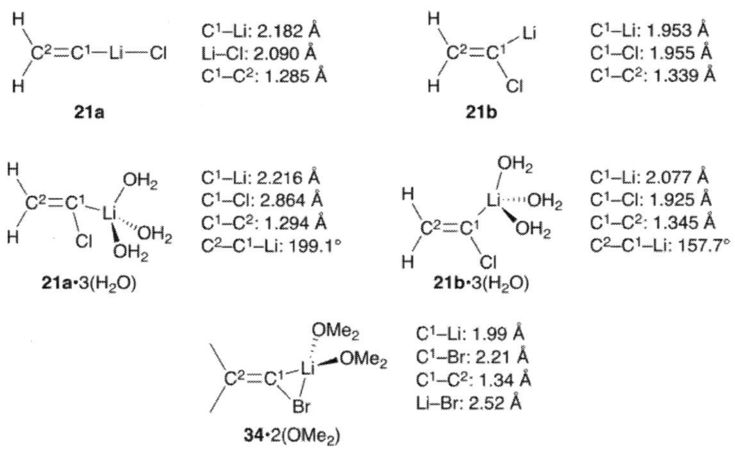

Figure 20. Structures of non-hydrated, bis(dimethyl ether)-solvated, and trihydrated lithium alkylidene carbenoids.

However, the structural features observed in experimental structure 30 were well reproduced to give the structure 21b•3(H_2O) when both solvation and electron correlation effects were incorporated into the geometry optimization. In the case of the lithium alkylidene carbenoid solvated with two dimethyl ether molecules 34•2(OMe_2), the distance between the lithium and bromine atoms was 2.52 Å [53].

The importance of the solvation effect was also shown by Rodríguez and co-workers in their theoretical study on the structures of the (*E*)- and (*Z*)-(1-iodo-2-methoxy-2-phenylvinyl)lithiums *E*-35 and *Z*-35 (Figure 21) [54]. The interactions between the lithium and oxygen atoms or the lithium atom and the phenyl group were present in the optimized structures of *E*-35 and *Z*-35 at the HF/3-21G level when solvent-free lithium alkylidene carbenoids were used as model compounds. However, these interactions were absent in the

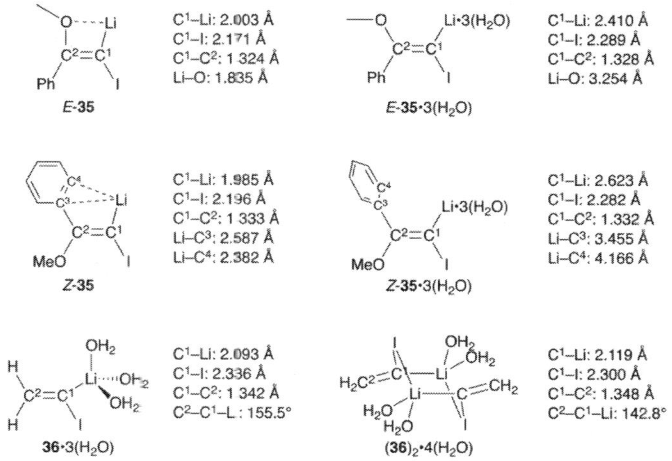

Figure 21. Structures of non-solvated and trihydrated lithium alkylidene carbenoids and dimeric tetrahydrated lithium alkylidene carbenoid.

trihydrated lithium alkylidene carbenoids *E*-35•3(H_2O) and *Z*-35•3(H_2O). Rodríguez and co-workers also investigated the equilibrium in the hydrated monomer 36•3(H_2O) and dimer $(36)_2$•4(H_2O) at the MP2/DZV(d) level of theory [55]. The tetrahydrated dimer $(36)_2$•4(H_2O) adopted a chair conformation. The monomeric structure 36•3(H_2O) was found to be more stable than the dimeric structure $(36)_2$•4(H_2O) by 7.0 kcal/mol.

Nelson and co-workers disclosed the isomerization pathways of (1-chloro-2-methylprop-1-en-1-yl)lithium (33) in the gas phase using the semi-empirical MNDO method [56]. The activation energy of the direct isomerization pathway from 33a to 33a' through the transition state 33b was estimated to be 42.7 kcal/mol (Figure 22, Eq. 1). Metal-assisted isomerization is an alternative pathway to geometrical isomerization. The activation energy for the migration of a chlorine atom from the carbon atom to the lithium atom through transition structure 33c was 26.8 kcal/mol (Figure 22, Eq. 2). When one or two NH_3 molecules were coordinated to the lithium atom of lithium alkylidene carbenoid 33, the situation was slightly different (Figure 22, Eq. 3) [57]. In the solvated systems, intermediate 33c•n(NH_3), which was similar to the transition structure 33c, retained the stereochemistry, and the activation energies for the conversion of ground minimum 33a•n(NH_3) to 33c•n(NH_3) through the transition state 33e•n(NH_3) were 27.5 (n = 1) and 26.7 kcal/mol (n = 2), respectively. The rotation barriers for the C–Li bond from 33c•n(NH_3) to 33c'•n(NH_3) were small, suggesting a facile geometrical isomerization of the ionized structures 33c•n(NH_3) and 33c'•n(NH_3). Rodríguez and co-workers speculated that the substitution reaction of nucleophiles with lithium alkylidene carbenoids occurs via nucleophilic attack not on 33a•n(NH_3) but on the ionized intermediate 33c•n(NH_3).

Kvíčala and co-workers studied the ^{19}F NMR chemical shifts of (*E*)-(1,2-difluorovinyl)lithium (*E*-32) and (1,2,2-trifluorovinyl)lithium (37) and their tris(dimethyl ether)-solvated structures *E*-32•3(OMe_2) and 37•3(OMe_2) [58]. Specifically, the chemical shifts of the ^{19}F NMR spectra of the structures *E*-32, 37, *E*-32•3(OMe_2), and 37•3(OMe_2) were calculated by the SOS-DFPT-IGLO method. The lithium alkylidene carbenoids *E*-32 and 37 had two or three local minima, including quadrilateral geometry like 32a (Figure 19) and lithium-bridged triangular geometry like 32c (Figure 19), whereas the solvation of

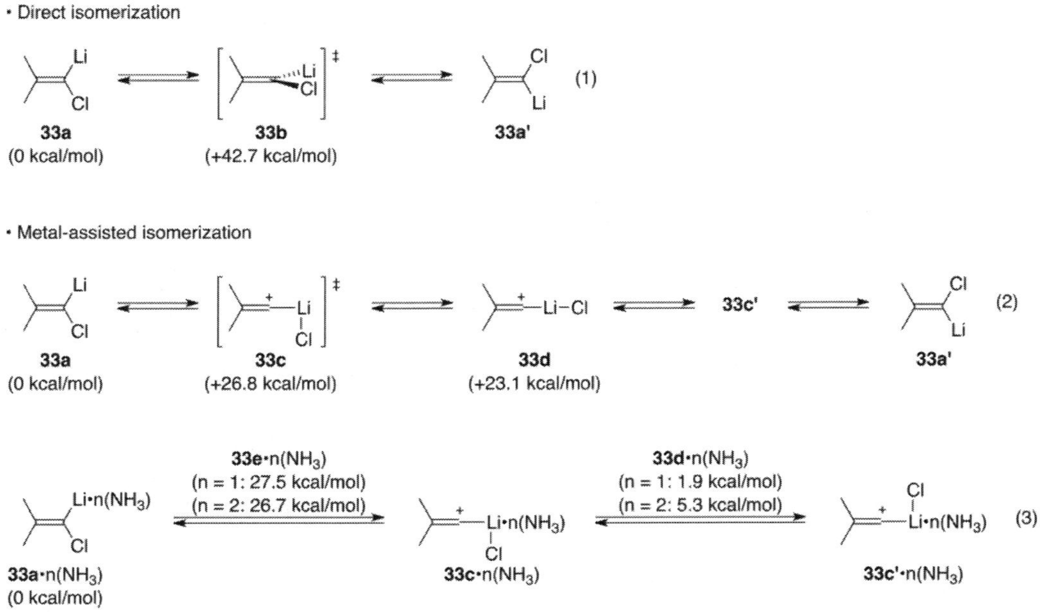

Figure 22. Geometrical isomerization pathways of lithium alkylidene carbenoid.

lithium alkylidene carbenoids *E*-32 and 37 with dimethyl ether resulted in only one minimum with a geometry similar to 21b•3(H_2O) (Figure 20) and 35•3(H_2O) (Figure 21). The relative chemical shifts of the triangular structures and the solvated structures were consistent with the experimental chemical shifts.

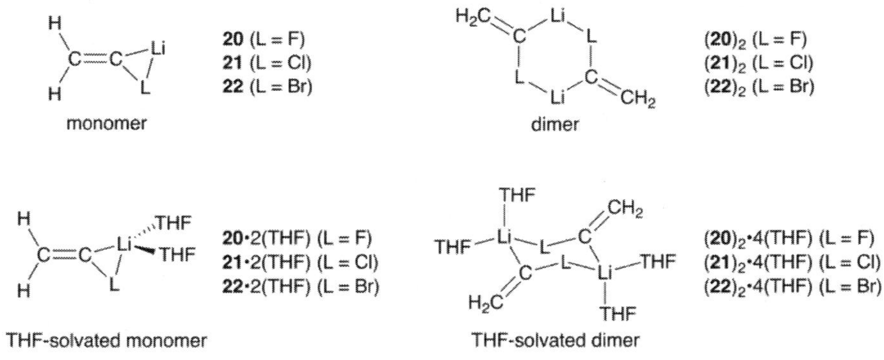

Figure 23. Monomeric and dimeric structures of non-solvated and THF-solvated lithium alkylidene carbenoids.

Pratt and colleagues investigated the structures of a series of lithium alkylidene carbenoids 20–22 in the gas phase and in THF solution (Figure 23) [59]. The monomeric lithium alkylidene carbenoids 20–22 had lithium-bridged three-membered geometries in the gas phase. The dimers $(20)_2$, $(21)_2$, and $(22)_2$ in the gas phase had planar geometries in which the lithium atoms, halogen atoms, and vinyl moieties were on the same plane. The dimerization free energies of lithium alkylidene carbenoids 20, 21, and 22 to $(20)_2$, $(21)_2$, and $(22)_2$ at 200 K calculated at the B3LYP/6-311+G(2df,2dp) level were –34.5, –21.2, and –21.4 kcal/mol, respectively, indicating that each of the lithium alkylidene carbenoids would exist exclusively as a dimer. By contrast, the dimerization free energies of the bis(THF)-solvated lithium alkylidene carbenoids 20•2(THF), 21•2(THF), and 22•2(THF) to tetrakis(THF)-solvated chair-like dimers $(20)_2$•4(THF), $(21)_2$•4(THF), and $(22)_2$•4(THF) were –11.2, 0.4, and –0.1 kcal/mol, respectively. That is, the dimeric structure was favored for the fluorine derivative 20, and the chlorine and bromine derivatives 21 and 22 were predicted to be mixtures of monomer and dimer in THF solution at 200 K.

Satoh and Ando reported the structures of the magnesium alkylidene carbenoids 38, *E*-39, and *Z*-39 optimized at the B3LYP/6-31+G* level in the gas phase (Figure 24) [45]. Similar to the structures of lithium alkylidene carbenoids in the gas phase, the magnesium atom and chlorine atom were very close to each other, and these molecules had magnesium-bridged triangular geometries.

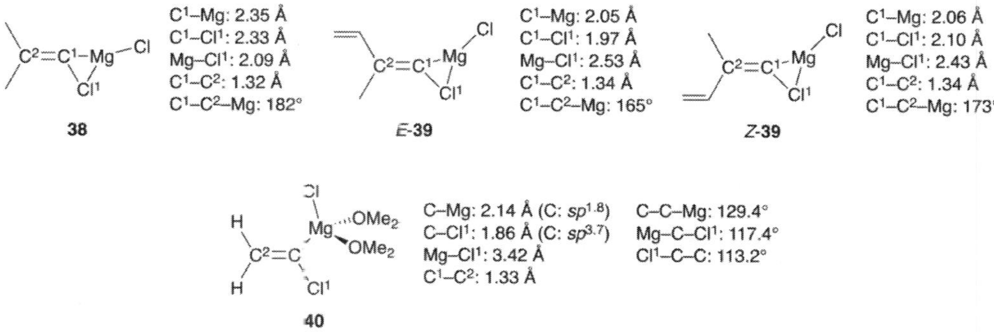

Figure 24. Structures of magnesium alkylidene carbenoids.

Kimura and Satoh studied the structure of the bis(dimethyl-ether)-solvated magnesium alkylidene carbenoid 40 at the B3LYP/6-311++G(d,p) level [44]. In contrast to the structures in the gas phase, there were no significant interactions between the magnesium and chlorine atoms on the carbenoid carbon atom. The carbenoid carbon atom of the magnesium alkylidene carbenoid 40 had a high *p*-character in the C–Cl bond.

3. Reactions

Providing insight into the reactivity of metal carbenoids is another subject of theoretical investigations. Considerable effort has been invested in the elucidation of the reaction pathways and transition structures of the [2+1] cycloaddition of metal carbenoids, Skattebøl rearrangement of cyclopropylmetal carbenoids, and Fritsch-Buttenberg-Wiechell rearrangement and nucleophilic substitution at the vinylic carbon atom of metal alkylidene carbenoids. These studies provided new insight into the reactivity of metal carbenoids. In this section, computational studies of the reactions of metal carbenoids are highlighted.

3.1. [2+1] Cycloaddition of Metal Carbenoids with Alkenes

[2+1] Cycloaddition, also referred to as cyclopropanation, of metal carbenoids with alkenes is representative of the carbenoid reaction and is often used to synthesize cyclopropanes (Figure 1, Eq. 1). Two reaction pathways are involved in the cyclopropanation reaction (Figure 25). One of the possible reaction pathways is a methylene transfer pathway in which metal carbenoids react with alkenes in a concerted manner via [2+1] cycloaddition (Figure 25, Eq. 1). Another pathway is the carbometalation of alkenes. In this pathway, [2+2] addition of metal carbenoids with alkenes occurs to give organometallic compounds bearing a leaving group at the γ-position, and a subsequent intramolecular substitution reaction provides cyclopropanes (Figure 25, Eq. 2).

In 1983, Clark and Schleyer reported a computational investigation of the concerted reaction pathway of the cyclopropanation of lithium carbenoid with ethylene (Figure 26) [60]. The calculations were performed using ab initio gradient techniques and the 3-21G basis set. Initially, ethylene and CH_2FLi (1) formed a reactant complex 41.

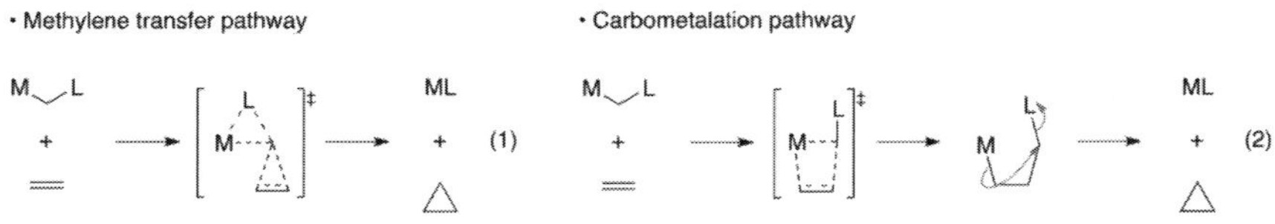

Figure 25. Possible reaction pathways for cyclopropanation of metal carbenoids with alkenes.

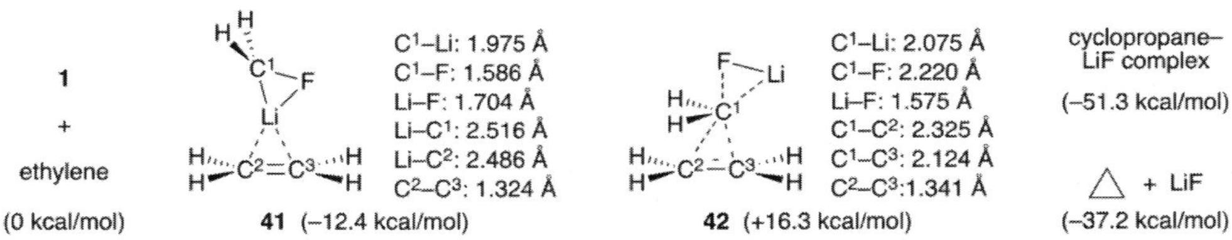

Figure 26. Structures of the CH_2FLi–ethylene complex and the transition structure of cyclopropanation.

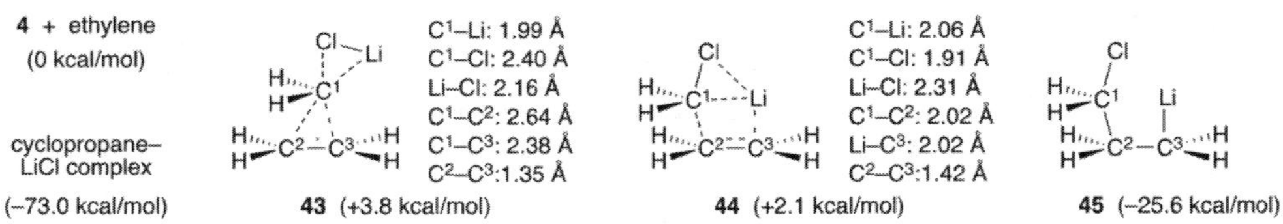

Figure 27. Transition state structures for the reaction of lithium carbenoid with ethylene.

In the transition structure 42, the methylene plane was nearly parallel to the ethylene plane, indicating that the HOMO of ethylene interacted with the LUMO of lithium carbenoid 1 (π–σ* interaction). In addition, the C–F and C–Li bonds were stretched by 0.576 Å and 0.119 Å, respectively. These results demonstrated that lithium fluoride was substantially decomplexed. That is, transition structure 42 could be regarded as a lithium carbenoid with strongly decomplexed lithium fluoride, approaching ethylene. Around the same time, Clark and Schleyer estimated the energy gain for the cyclopropanation of lithium carbenoid 1 with ethylene at the MP4SDQ/6-31G* level to be –59.0 kcal/mol [30].

Nakamura and co-workers used DFT calculations to investigate the feasibility of the methylene transfer pathway and carbometalation pathway for the cyclopropanation reaction of CH_2ClLi (4) with ethylene (Figure 27) [61,62]. The methylene transfer pathway involved the butterfly-type transition state 43, and the carbometalation pathway proceeded through a four-centered transition structure 44 to give the [2+2] adduct 45. The activation energy for the methylene transfer pathway calculated at the B3LYP/6-31G* level was +3.8 kcal/mol, and that for the carbometalation pathway was +2.1 kcal/mol. These results demonstrated that lithium carbenoids could react with ethylene either in a concerted manner or a stepwise manner, with competition between the two pathways.

To elucidate the influence of the leaving group on the reactivity of lithium carbenoids, Boche and co-workers investigated the cyclopropanation of a series of lithium carbenoids CH_2LLi (1: L = F, 4: L = Cl, 7: L = Br, 10: L = I) with ethylene [33]. The theoretical study revealed that lithium carbenoids 1, 4, 7, and 10 had almost equal activation energies for the cyclopropanation reaction through the concerted pathway (1: 7.4 kcal/mol, 4: 6.9 kcal/mol, 7: 6.5 kcal/mol, 10: 6.1 kcal/mol). These results suggested that the carbenoid character of all four lithium carbenoids was essentially the same in the cyclopropanation reaction. However, the bond dissociation energy of the C–F bond in lithium carbenoid 1 (194.8 kcal/mol) was much larger than that of the C–I bond in lithium carbenoid 10 (156.3 kcal/mol). On the basis of the bond dissociation energies of the lithium halides, Boche and co-workers concluded that the formation of the lithium–halogen bonds compensated for the cleavage of carbon–halogen bonds.

Zhou and Cao also investigated substituent effects on the cyclopropanation reaction of a series of lithium carbenoids 1, 4, 7, and 10 through the methylene transfer pathway and the carbometalation pathway at the CCSD(T)/6-311G**//B3LYP/6-311G** level of theory [63]. The activation energy for the methylene transfer cyclopropanation of CH_2FLi (1) (9.8 kcal/mol) was larger than that of CH_2ClLi (4) (7.6 kcal/mol), CH_2BrLi (7) (7.4 kcal/mol), and CH_2ILi (10) (7.5 kcal/mol). By contrast, the activation energies for the cyclopropanation via the carbometalation pathway increased in the order CH_2FLi (1) (6.1 kcal/mol), CH_2ClLi (4) (7.1 kcal/mol), CH_2BrLi (7) (8.2 kcal/mol), and CH_2ILi (10) (8.5 kcal/mol). In the case of CH_2FLi (1), the carbometalation pathway was more favored than the methylene transfer pathway, whereas the methylene transfer pathway was favored in the reaction with CH_2BrLi (7) and CH_2ILi (10). Zhou and Cao took note of the σ orbital of the C–Li bond (HOMO) and the σ^* orbital of the C–L bond (LUMO–1) in lithium carbenoids and the π and π^* orbitals in ethylene. The interaction between the σ^* orbital of the C–L bond in lithium carbenoids and the π orbital in ethylene mainly contributed to the formation of a transition structure for the methylene transfer pathway, while that between the σ orbital of the C–Li bond in lithium carbenoids and the π^* orbital in ethylene formed a transition structure for the carbometalation pathway. The difference in substituents at the α-position of lithium carbenoids led to a change in the hybridization of the carbenoid carbon atom. A higher *p*-character of the carbenoid carbon atom in the C–L bonds resulted in a σ^* orbital of C–L bond with lower energy, and a lower *s*-character of the carbenoid carbon atom in the C–Li bond led to a σ orbital of the C–Li bond with higher energy. These factors appeared to affect the activation energies for the cyclopropanation reaction in both pathways.

Phillips and colleagues performed a DFT investigation for the cyclopropanation reaction of CH_2ILi (10) with ethylene, *trans*-2-butene, and styrene (Figure 28) [64]. There were small changes between the structure of the monomeric lithium carbenoid and those of the CH_2ILi moiety in the transition states 46, 47, and 48. The barrier heights of the reaction were estimated to be 6.6, 6.9, and 5.3 kcal/mol, respectively, at the B3LYP/6-311G** level of theory. The formation of the cycloadducts and lithium iodide was exothermic by 46.5, 43.6, and 43.2 kcal/mol. By contrast, the activation energy for the carbometalation pathway with ethylene via transition state 49 was 7.7 kcal/mol [65]. These results demonstrated that the reaction could proceed by competition between a methylene transfer mechanism and a carbometalation mechanism in the gas phase.

To investigate the conflict between the two reaction pathways, Phillips and colleagues also studied the cyclopropanation reaction of lithium carbenoids with ethylene using more complex models that reflected aggregation and solvation states [66]. The reaction barriers for the methylene transfer pathway of the lithium carbenoid dimer $(1)_2$ and tetramer $(1)_4$ were 10.1 and 8.0 kcal/mol, respectively, at the B3LYP/6-311G** level of theory.

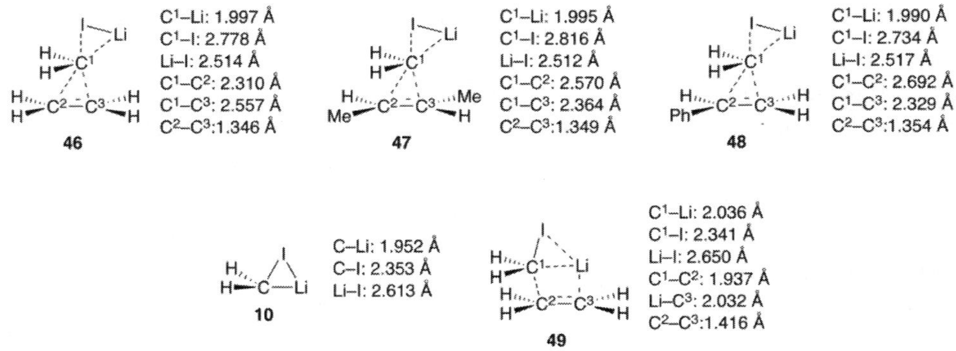

Figure 28. Transition state structures of the cyclopropanation of lithium carbenoid with alkenes.

These values were significantly different from that for the monomeric lithium carbenoid 1 (16.0 kcal/mol). By contrast, the reaction barriers for the carbometalation pathway of the lithium carbenoid dimer $(1)_2$ and tetramer $(1)_4$ were 26.8 and 33.9 kcal/mol, respectively, which were considerably larger than that for the monomeric lithium carbenoid 1 (12.5 kcal/mol) and those for the methylene transfer pathway of the dimer $(1)_2$ and tetramer $(1)_4$. In the transition structures of dimeric lithium carbenoid 50 via the methylene transfer pathway, an exiting fluorine atom interacted with two lithium atoms, facilitating the elimination of lithium fluoride, whereas the fluorine atom interacted with one lithium atom in transition structure 51 via the carbometalation pathway (Figure 29). The difference in these interactions led to a lower transition state energy for the methylene transfer pathway. To investigate the solvation effect, dimethyl ether molecules were employed as explicit solvent molecules. In the case of the solvated dimers $(1)_2{\bullet}2(OMe_2)$ and $(1)_2{\bullet}4(OMe_2)$, in which each lithium atom was coordinated with one or two dimethyl ether molecules, the activation energies for the methylene transfer pathway further decreased [$(1)_2{\bullet}2(OMe_2)$: 7.2 kcal/mol, $(1)_2{\bullet}4(OMe_2)$: 7.8 kcal/mol]. The solvation led to the destabilization of the reactant complex like 41 in Figure 26 because the coordination of dimethyl ether molecules to the lithium atom weakened the interaction of the lithium carbenoid with ethylene. These results suggested that the methylene transfer pathway was highly favored over the carbometalation pathway in the reaction system that included aggregation and solvation effects.

In addition to the above intermolecular reaction, Phillips and colleagues investigated the feasibility of the methylene transfer pathway and carbometalation pathway in the intramolecular cyclopropanation reaction of γ-siloxy-substituted lithium carbenoids (Figure 30) [67].

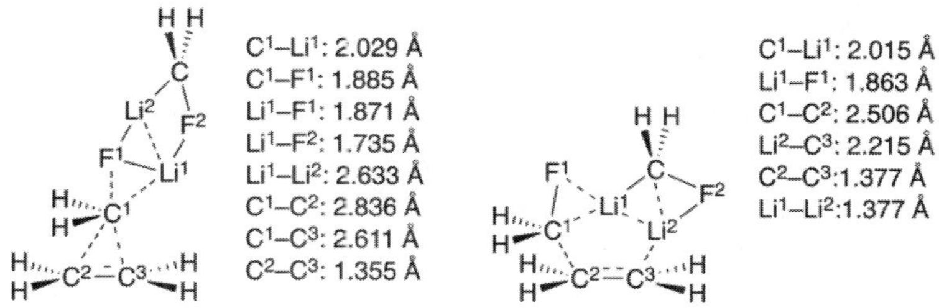

50 **51**

Figure 29. Transition state structures for the methylene transfer pathway and carbometalation pathway of the reaction of a dimeric lithium carbenoid with ethylene.

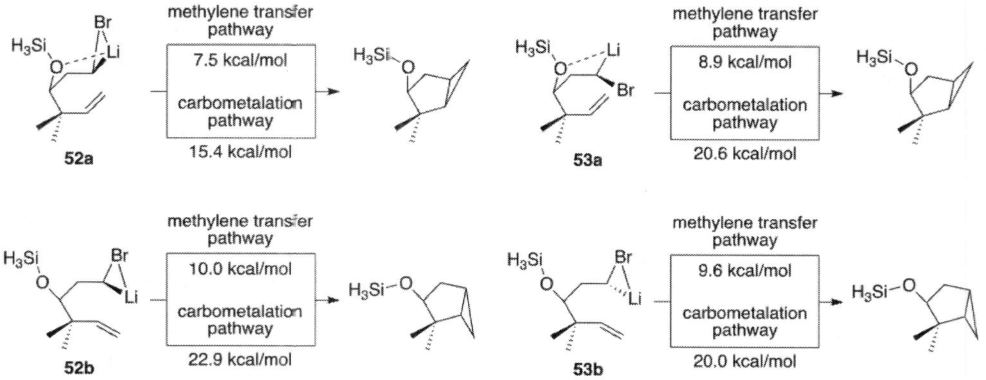

Figure 30. Reaction pathways and energy barriers for the intramolecular cyclopropanation of diastereomers in a Lewis base-assisted manner and non-Lewis base-assisted manner.

The reaction barriers were calculated for diastereomer 52 and 53 in a Lewis base-assisted (siloxy-coordinated) manner (52a, 53a) and non-Lewis base-assisted manner (52b, 53b) at the B3LYP/6-311G** level of theory with the application of a polarized continuum model for tetrahydrofuran. In each case, the reaction barriers for the methylene transfer pathway were smaller than those for the carbometalation pathway. The carbometalation mechanism did not appear to compete with the methylene transfer mechanism. For isomer 52, the Lewis base-assisted methylene transfer pathway had the lowest activation energy. By contrast, there was a small energy difference between the Lewis base-assisted methylene transfer pathway and the non-Lewis base-assisted methylene transfer pathway for the other isomer 53. These results were in good agreement with the experimental observations [68].

Pratt and co-workers investigated the activation free energies for the cyclopropanation reaction of ethylene with the monomeric and dimeric lithium carbenoids 4 and $(4)_2$ and mixed aggregates 4•LiCl consisting of lithium carbenoid 4 and lithium chloride [41]. The gas phase activation free energies were calculated at the HF, B3LYP, and MP2 levels with the 6-31+G(d) basis set. There were no significant differences between the activation free energies for 4, $(4)_2$, and 4•LiCl calculated at the B3LYP (14.2–15.4 kcal/mol) and MP2 levels (16.6–17.6 kcal/mol). However, the activation free energy of the mixed aggregate 4•LiCl (9.5 kcal/mol) was considerably smaller than those of the monomer 4 (14.2 kcal/mol) and dimer $(4)_2$ (11.1 kcal/mol) for calculations at the HF level.

Pratt and co-workers also investigated the reaction of monomeric and dimeric lithium carbenoids with ethylene and 2,3-dimethyl-2-butene in detail to examine the viability of the methylene transfer and carbometalation pathways [69]. The activation free energies for the cyclopropanation of the monomeric lithium carbenoids 1, 4, and 7 with ethylene via the methylene transfer pathway (1: 11.4 kcal/mol, 4: 10.5 kcal/mol, 7: 10.5 kcal/mol) calculated at the B3LYP level with the 6-31+G(d) basis set were comparable to those via the carbometalation pathway (1: 9.4 kcal/mol, 4: 12.1 kcal/mol, 7: 11.7 kcal/mol) in the gas phase. Reactant complexes like 41 in Figure 26 were not observed in their study. In the case of the carbometalation pathway, the second step was an intramolecular substitution reaction. There are two types of possible reaction pathways. One is a S_N2-type reaction via the *anti*-transition state 54, and the other is a *syn*-elimination of lithium halides via the *syn*-transition state 55 (Figure 31). The activation free energies for the *syn*-elimination pathway (1: 5.5 kcal/mol, 4: 10.2 kcal/mol, 7: 10.7 kcal/mol) were considerably lower than those for the S_N2-type reaction pathway (1: 24.9 kcal/mol, 4: 15.9 kcal/mol, 7: 15.4 kcal/mol). The activation free energies for the cyclopropanation of the two distinct dimeric lithium carbenoids P and U (Figure 13) via the methylene transfer pathway (P: 11.2–13.7 kcal/mol, U: 9.8–13.9 kcal/mol) were significantly lower than those for the carbometalation pathway (P: 24.0–29.2 kcal/mol, U: 21.7–23.8 kcal/mol). In the case of monomeric bis(THF)-solvated systems, the methylene transfer pathway (10.3–13.3 kcal/mol) was preferred over the carbometalation pathway (18.6–23.1 kcal/mol). By contrast, the activation free energies for both the methylene transfer pathway (P: 13.9–19.0 kcal/mol, U: 16.6–21.3 kcal/mol) and the carbometalation pathway (P: 29.6–36.6 kcal/mol, U: 30.4–31.7 kcal/mol) of dimeric THF–solvated lithium carbenoids calculated at the MP2/6-31+G(d) level increased significantly in comparison with non-solvated systems. The preference for the methylene transfer mechanism was also observed in the reaction of lithium carbenoids with 2,3-dimethyl-2-butene. The overall results were consistent with those reported by Ke, Zhao, and Phillips [66], and these studies demonstrated that the consideration of aggregation and solvation in computational modeling is of great importance.

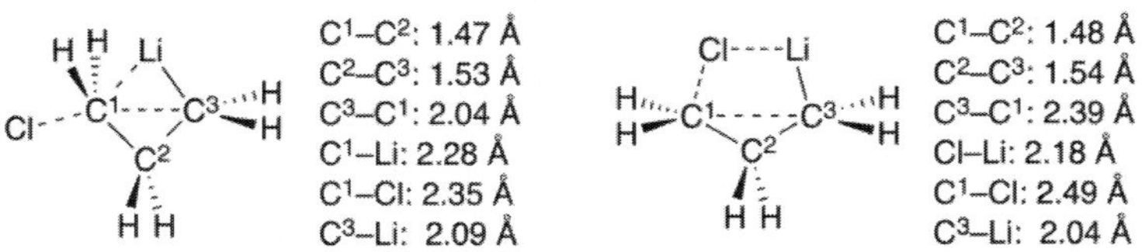

54 **55**

Figure 31. Transition state structures of the intramolecular substitution reaction.

Pratt and colleagues examined the performance of 13 modern DFT functionals, GGA (PW91PW91, mPWPW91, PBEPBE), meta-GGA (TPSSTPSS, M06-L), hybrid GGAs (B3LYP, B3PW91, mPW1LYP, mPW1PW91, PBE1PBE), and hybrid meta-GGAs (BMK, M06, M06-2X), with the 6-31+G(d) basis set to determine which was able to reproduce the results of the MP2/6-31+G(d) level of theory [70]. MP2/6-31+G(d) single-point energy calculations were performed at DFT/6-31+G(d)-optimized geometries, and the energy differences between MP2 and DFT levels were investigated. Specifically, 84 molecules (51 gas phase molecules including 24 transition state structures and 33 THF-solvated molecules including 12 transition structures) and 78 reactions (45 gas phase reactions and 33 reactions involving THF-solvated molecules) concerning cyclopropanation of ethylene with the lithium carbenoids 1, 4, and 7 were calculated with the above 13 functionals. The hybrid meta generalized gradient approximation function M06-2X was determined to be the best of the tested functionals, and the most widely used B3LYP functional was a rather poor choice for these systems.

Zhang and co-workers investigated the cyclopropanation reaction of the lithium carbenoids CH_2LLi (4: L = Cl, 7: L = Br, 10: L = I) with the ketene $CH_2{=}C{=}O$ leading to the formation of cyclopropanone and lithium halides in the gas phase and in THF solvent using the B3LYP/6-311++G** (LANL2DZ basis set for halogens) level of theory (Figure 32) [71]. Three-centered transition states 56, 57, and 58 were found for the reaction that were similar to each other. In the transition structures, the C–L bonds were significantly elongated, and the halogen atoms were attracted to the lithium atom. Zhang and co-workers used a polarized continuum model to take into account the solvent effects (THF) on the reaction. The barriers for the cyclopropanation reaction increased in the order CH_2ClLi (6.06 kcal/mol), CH_2BrLi (6.47 kcal/mol), and CH_2ILi (7.34 kcal/mol).

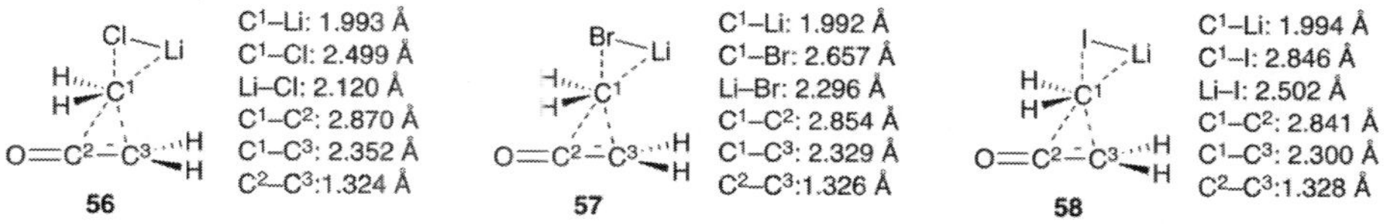

Figure 32. Transition state structures of the reaction of lithium carbenoids with ketene.

3.2. Skattebøl Rearrangement

The Skattebøl rearrangement is a ring-opening reaction of cyclopropylmetal carbenoids leading to the formation of allenes (Figure 33, Eq. 1). The reaction mechanism of the Skattebøl rearrangement was studied by Wang and Deng at the RHF/6-31G*//RHF/3-21G level in 1989 [46]. The energy barrier for the rearrangement of (1-fluorocyclopropyl)lithium 26a to propa-1,2-diene via the transition state 26b was calculated to be 16.4 kcal/mol (Figure 33). The process from 26a to 26b occurred with disrotation of two methylene units. When the C_2–C_1–C_3 bond angle reached 105°, the process changed to conrotation. The product complex (allene-LiF complex) 26c, in which lithium fluoride interacted with allene at both ends, was more stable than the reactant 26a by 41.3 kcal/mol.

Azizoglu and co-workers studied substituent effects in the mechanism of the Skattebøl rearrangement at the B3LYP/6-31G(d) level in detail (Figure 34) [47]. The ring-opening reaction of mono-substituted (1-bromocyclopropyl)lithiums 59a to allenes can proceed according to two possible pathways, that is, a concerted pathway and a stepwise pathway. Therefore, a concerted pathway from 59a to allene and lithium bromide via the transition state 59b (Figure 34, Eq. 1) and a stepwise pathway including initial carbon–bromine bond cleavage in the lithium carbenoid 59a leading to the formation of intermediate 59d via the transition state 59c and the following step from the intermediate 59d to the allene through the transition state 59e (Figure 34, Eq. 2) were examined. Both the concerted pathway and the stepwise pathway were found to be operative, and electron-donating groups facilitated the conversion of lithium carbenoids to allenes. Azizoglu and co-workers also studied the ring-opening mechanism of mono-, di-, and tri-silacyclopropyllithium carbenoids [72,73].

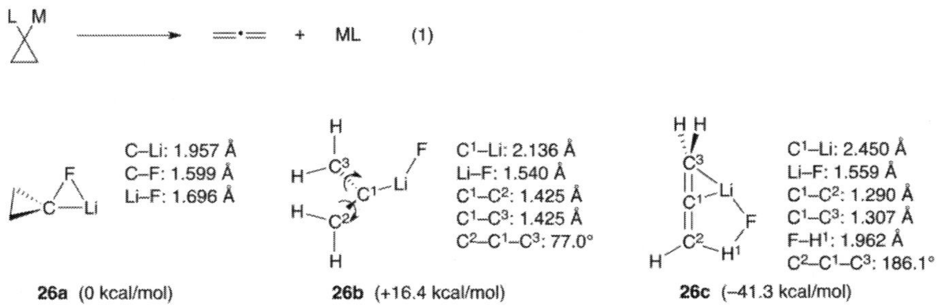

Figure 33. Skattebøl rearrangement of cyclopropyllithium carbenoids and the transition state structure of the reaction.

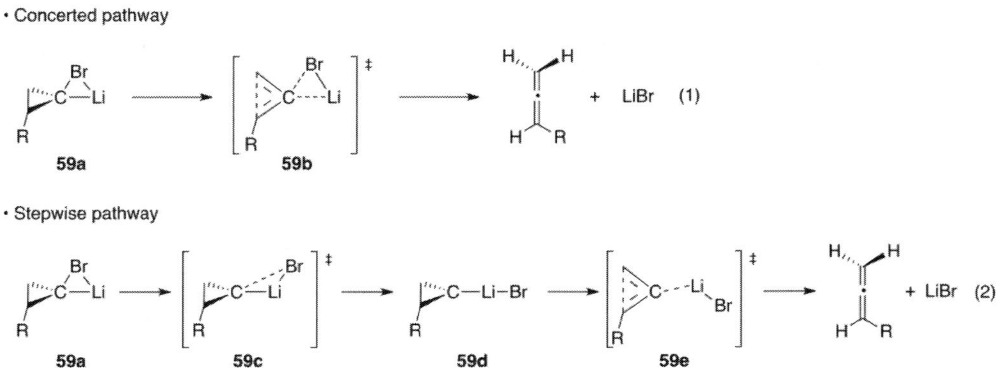

Figure 34. Skattebøl rearrangement of cyclopropyllithium carbenoids bearing a substituent.

3.3. Fritsch-Buttenberg-Wiechell Rearrangement

1,2-Migration of metal alkylidene carbenoids results in the formation of alkynes (Figure 35, Eq. 1) [74,75]. This reaction is known as the Fritsch-Buttenberg-Wiechell rearrangement. As a decomposition pathway of lithium alkylidene carbenoids, Wang and Deng investigated the Fritsch-Buttenberg-Wiechell rearrangement of H_2C=CFLi (20) and HFC=CFLi (32) at the RHF/3-21G level (Figure 35) [49]. The rearrangement of H_2C=CFLi (20) provides acetylene and lithium fluoride. The initial cleavage of the C–F bond resulted in a vacant *p*-orbital at the carbenoid carbon atom, and the subsequent hydrogen shift to the *p*-orbital via the transition state 20d led to the formation of product complex 20e. Wang and Deng speculated that the hydrogen atom located at the *cis*-position to the fluorine atom might preferentially migrated. However, this speculation conflicts with the experimental observations, in which the substituent located at the *trans*-position to the leaving group preferentially migrates. This preference seems to be due to the position of the substituent on the same side of the back lobe of the C–L antibonding orbital with respect to the C=C double bond. The energy difference between the ground minimum 20a and the transition state 20d was 37.2 kcal/mol. In the case of HFC=CFLi (32), the rearrangement from 32d to fluoroacetylene occurred through the transition state 32e, and the energy difference was 50.9 kcal/mol.

Rodríguez and colleagues reported a theoretical study on three mechanisms for the hydrogen migration of the trihydrated lithium alkylidene carbenoid 36•3(H_2O)

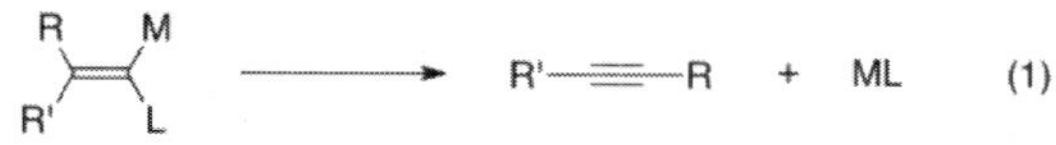

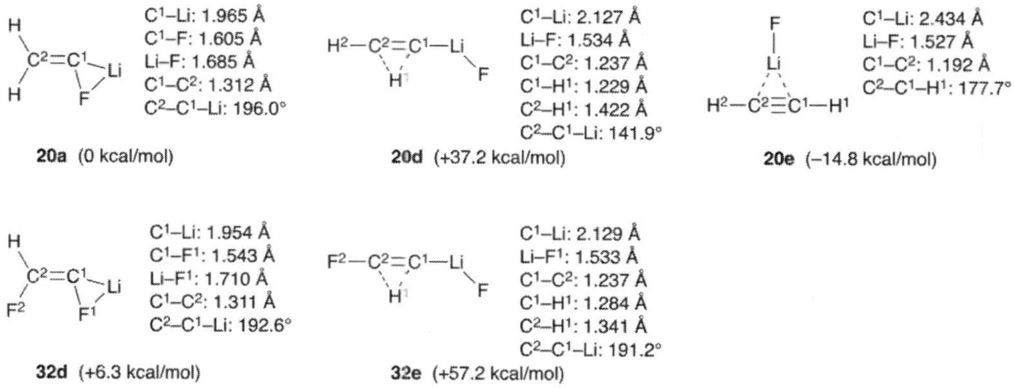

Figure 35. Transition state structures of lithium alkylidene carbenoids in the Fritsch-Buttenberg-Wiechell rearrangement.

[H_2C=CILi•3(H_2O)] (Figure 36) [55]. Two mechanisms were involved in the concerted processes in which the *cis*- or *trans*-hydrogen atom in the lithium alkylidene carbenoid 36•3(H_2O) migrated to the carbenoid carbon atom. The other mechanism was the stepwise pathway, in which α-elimination of lithium iodide from 36•3(H_2O) occurred to give carbene 36d and the subsequent migration of the hydrogen atom via the transition state 36e led to the formation of acetylene. Rodríguez and colleagues used the STQN method to search for the transition structures. The transition state structures 36a•3(H_2O), 36b•3(H_2O), and 36c•3(H_2O) were less stable than 36•3(H_2O) by 13.3, 14.0, and 16.0 kcal/mol at the MP4/TZV(2df+,2p) + SCRF level, indicating the preference of the concerted hydrogen migration rather than the stepwise mechanism.

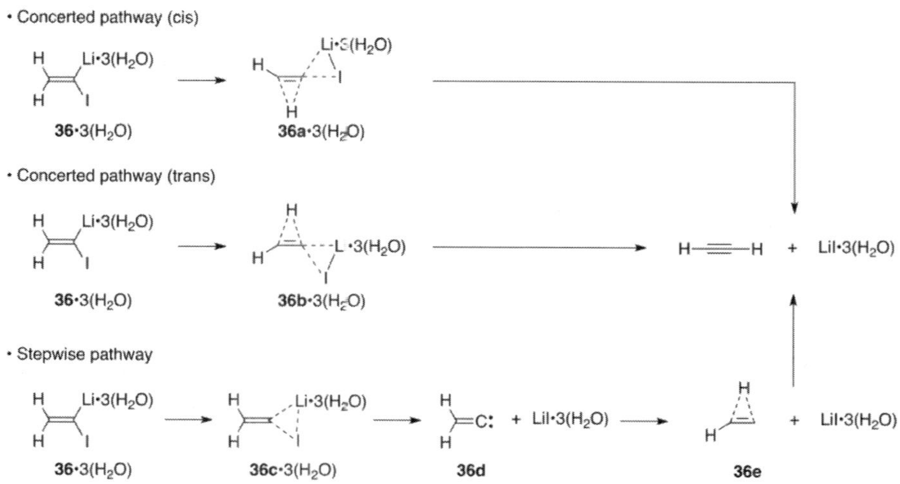

Figure 36. Three possible reaction pathways for the Fritsch-Buttenberg-Wiechell rearrangement of lithium alkylidene carbenoid.

Poisson and co-workers investigated the α- and β-elimination pathways of (*E*)-{1,2-dichloro-2-[(2-phenylcyclohexyl)oxy]vinyl}lithium (60) leading to the formation of chloroynol ether 61 (Figure 37) [76]. The α-elimination process corresponded to the Fritsch-Buttenberg-Wiechell rearrangement, and both the α- and β-elimination pathways afforded the same product, 61. Metadynamics ab initio calculations of mono-THF-solvated vinyllithium revealed that the activation free energies for the α- and β-elimination pathways were 6 and 11 kcal/mol, respectively.

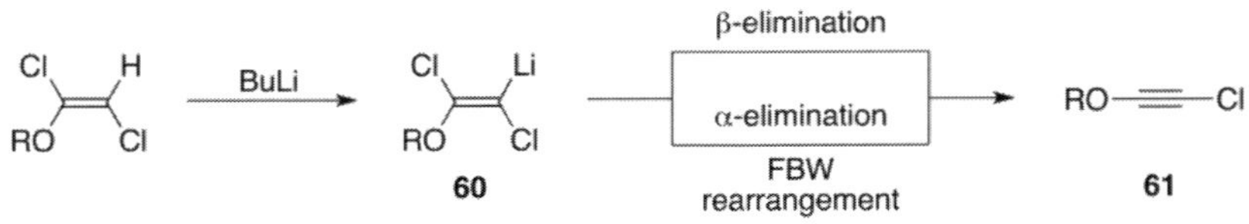

Figure 37. Two possible pathways for the formation of chloroynol ether.

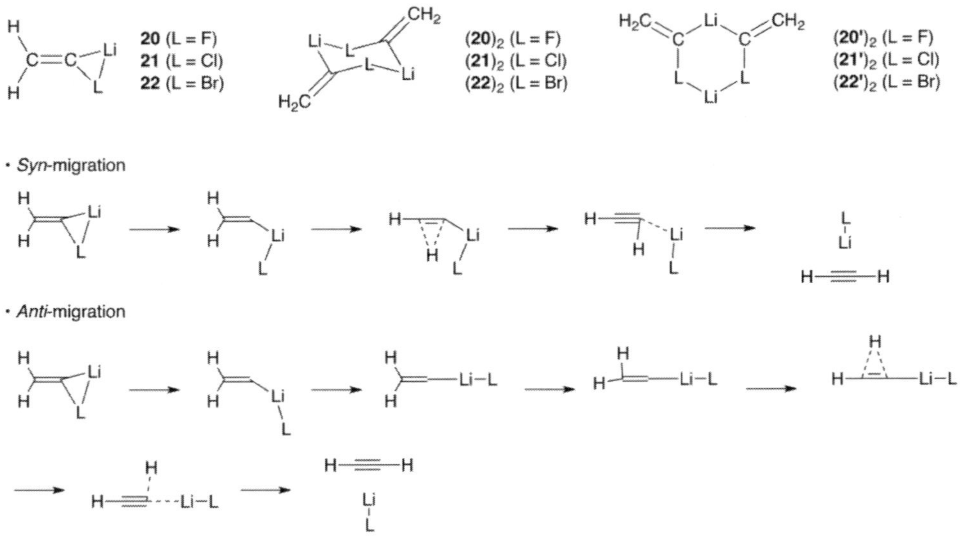

Figure 38. Fritsch-Buttenberg-Wiechell rearrangement of monomeric and dimeric lithium alkylidene carbenoids.

Pratt and co-workers investigated the *syn*- and *anti*-migration pathways of the Fritsch-Buttenberg-Wiechell rearrangement of lithium alkylidene carbenoids in the monomers 20, 21, and 22 and the two isomeric dimers $(20)_2$, $(21)_2$, $(22)_2$, $(20')_2$, $(21')_2$, and $(22')_2$ at the MP2/6-31+G(d) level (Figure 38) [77]. The activation energies along the reaction pathways of *syn*- and *anti*-hydrogen migrations for the monomers 20, 21, and 22 were 7.1–10.6 kcal/mol (*syn*-migration) and 7.5–10.6 kcal/mol (*anti*-migration), respectively. By contrast, the activation energies for the dimers $(20)_2$, $(21)_2$, and $(22)_2$ (1.7–9.0 kcal/mol) and $(20')_2$, $(21')_2$, and $(22')_2$ (0.2–7.0 kcal/mol) were considerably smaller than those of monomers 20, 21, and 22. The authors concluded that the most facile pathways occur in aggregated species.

3.4. Substitution of Metal Alkylidene Carbenoids with Nucleophiles

It is well known that the substitution reaction of primary or secondary alkyl halides with nucleophiles proceeds according to a S_N2 mechanism. By contrast, the substitution reaction of alkenyl halides with nucleophiles is an unfavorable process (Figure 39, Eq. 1). For example, Glukhovtsev and co-workers estimated the activation energy of the nucleophilic substitution of chloroethene with a chloride anion at the vinylic position via a concerted in-plane S_N2 pathway to be 32.6 kcal/mol [78]. The difficulty in substitution at the vinylic carbon atom appears to be mainly due to the following factors: (i) an approach of nucleophiles to alkenyl halides from the backside of the C–L bond is difficult due to a steric effect arising from the substituents around the alkene unit; (ii) delocalization of a lone pair on the halogen atom to the *p*-orbital of the alkene unit strengthens the C–L bond; and (iii) the decrease in *p*-character in the C_{sp2}–L bond compared with that in the C_{sp3}–L bond results in a small C–L antibonding orbital. By contrast, a substitution reaction occurs at the vinylic carbon atom of metal alkylidene carbenoids (Figure 39, Eq. 2).

Lucchini and co-workers investigated the S_N2 and Ad_N-E mechanisms in bimolecular nucleophilic substitutions at the vinylic carbon atom.

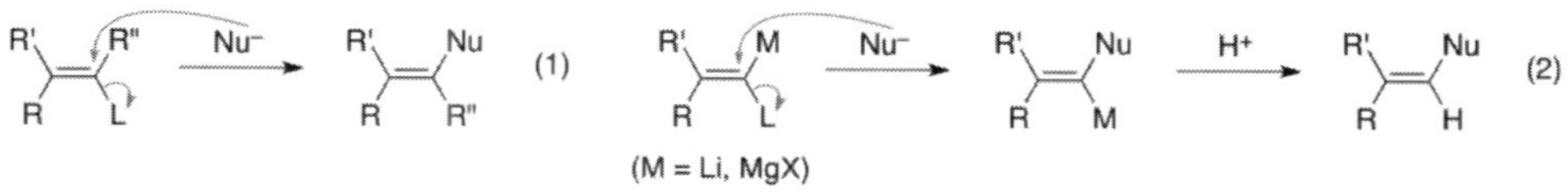

Figure 39. Substitution reaction at the vinylic carbon atom.

The authors took note of the vacant orbitals of a variety of vinylic compounds, including H_2C=CClLi (21) [79]. The LUMO+1 and LUMO+5 of lithium alkylidene carbenoid 21 corresponded to the C–Cl anti-bonding orbital and the π* orbital, respectively. They proposed that the nucleophiles would attack the LUMO+1 in the molecular plane and that the reaction proceeded with inversion.

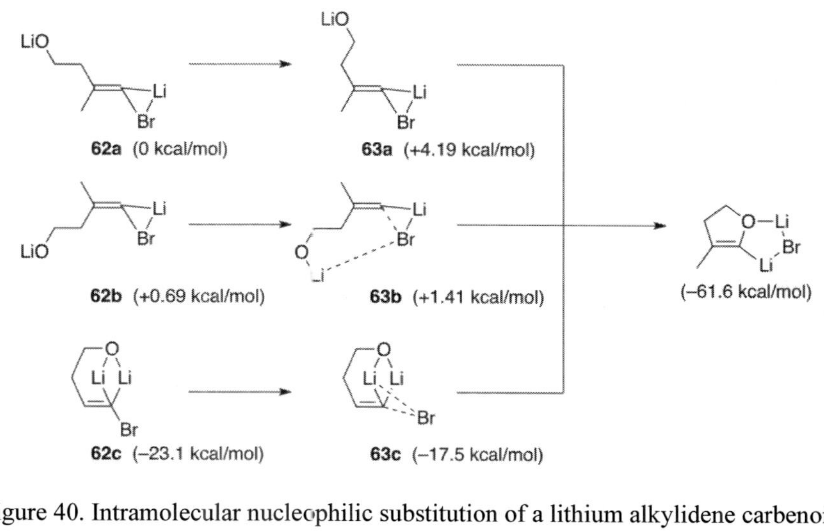

Figure 40. Intramolecular nucleophilic substitution of a lithium alkylidene carbenoid.

Narasaka and co-workers experimentally investigated the intramolecular substitution reaction of lithium alkylidene carbenoids with nucleophiles such as alkoxides and carbanions [80]. On the basis of detailed experimental results, an in-plane S_N2-type mechanism was proposed for the reaction. The authors computationally examined the possibility of the S_N2-type reaction of lithium alkylidene carbenoids (Figure 40) [53]. The activation energies for the intramolecular substitution reaction of three types of isomers 62a–c were calculated at the B3LYP/6-31+G(d) level. The cluster structure 62c was more stable than structures 62a and 62b by 23.1 and 23.8 kcal/mol, respectively, and the activation energies for 62a–c via the transition structures 63a–c were 4.2, 0.7, and 5.6 kcal/mol, respectively. From these results, the authors expected that the cyclization mainly proceeded from cluster 62c.

Subsequently, Ando reported the reaction mechanism of the intra- and intermolecular substitution of lithium alkylidene carbenoids in detail at the B3LYP/6-31+G* level (Figure 41) [81].

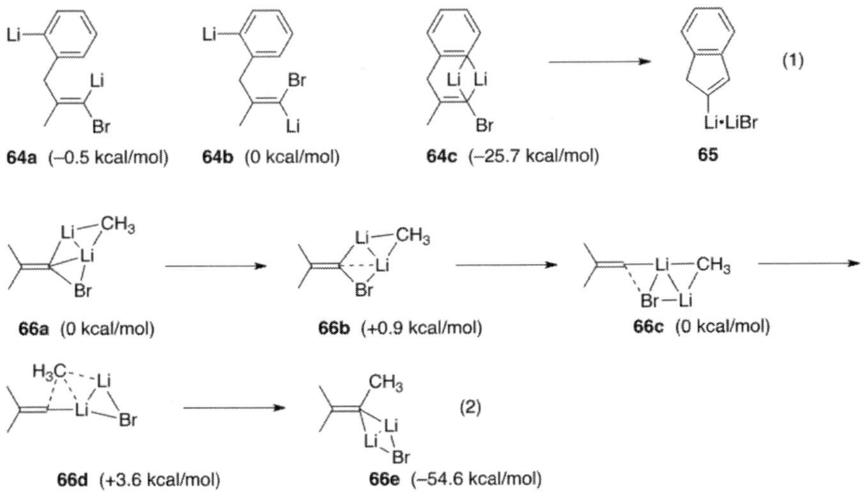

Figure 41. Intra- and intermolecular nucleophilic substitution of lithium alkylidene carbenoids.

For intramolecular substitution of the lithium alkylidene carbenoid 64 bearing an aryl lithium unit, structure 64c was more stable than 64a and 64b by 25.2 and 25.7 kcal/mol, respectively (Figure 41, Eq. 1). In addition, the isomerization of 64a to 64c was thought to readily occur. The activation energy for the substitution process from 64c to product complex 65 was estimated to be 5.0 kcal/mol. The intermolecular reaction with methyllithium initiated via C–Br bond cleavage in the reactant complex 66a with 0.9 kcal/mol activation energy to give the intermediate 66c (Figure 41, Eq. 2). The nucleophilic attack of the methyl anion on the vinylic carbon atom afforded product complex 66e. In this reaction, the methyl group was introduced to the product from the backside of the bromine atom. The activation energy for the geometrical isomerization of 66a was estimated to be 3.4 kcal/mol.

Satoh and co-workers extensively studied the reaction of magnesium alkylidene carbenoids with various nucleophiles [9,11]. Magnesium alkylidene carbenoids could react with aryllithiums, lithium acetylides, Grignard reagents, lithium amides, lithium thiolates, and lithium enolates to give multi-substituted alkenes. The authors took note of the hybridization of the carbenoid carbon atom in magnesium alkylidene carbenoid 40 (Figure 24) [44]. Natural bond orbital analysis of bis(dimethyl ether)-solvated (1-chlorovinyl)magnesium chloride 40 performed at the B3LYP/6-311++G(d,p) level showed that the molecule had an increased *s*-character of the carbenoid carbon atom in the C–Mg bond ($sp^{1.8}$) and an increased *p*-character of the carbenoid carbon atom in the C–Cl bond ($sp^{3.7}$) relative to those of the alkenyl carbon atom in chloroethene. The increased *p*-character in the C–Cl bond resulted in a weakening of the C–Cl bond and expanded C–C–Mg bond angles (129.4°). These factors provided a favorable circumstance for the S_N2 reaction at the alkenyl carbon atom of magnesium alkylidene carbenoids because the above-mentioned disadvantages in the S_N2 reaction at the alkenyl carbon atom were somewhat relieved.

3.5. Miscellaneous Reactions

Lithium alkylidene carbenoids can be generated from 1,1-dihaloalkenes via the halogen–lithium exchange reaction. Ando investigated the reaction of 1,1-dihaloalkenes with monomeric and dimeric methyllithium to generate lithium alkylidene carbenoids (Figure 42) [53,81]. The activation energy for the process from the reactant complex 67a to the product complex 67c through the transition state 67b was 15.6 kcal/mol at the B3LYP/6-31+G(d) level (Figure 42, Eq. 1). A similar process was investigated with a variety of substrates 68–72. The experimental *E*/*Z* stereochemistry results could be explained based on the computational results for the differences in the activation energies for the reaction with the *E*- and *Z*-isomers and the energy differences between the *E*- and *Z*-products (Figure 42, Eq. 2).

Davis and Liu described a theoretical study of the oxidative addition of tetrafluoroethylene to magnesium leading to the formation of (1,2,2-trifluorovinyl)magnesium fluoride 74 (Figure 43, Eq. 1) [82]. The activation energy for the reaction via the transition state 73, which lacked a planar geometry, was 19.0 kcal/mol at the MP2/6-31G**//SCF/MP2/6-31G** level, and the overall reaction was exothermic by 63.2 kcal/mol.

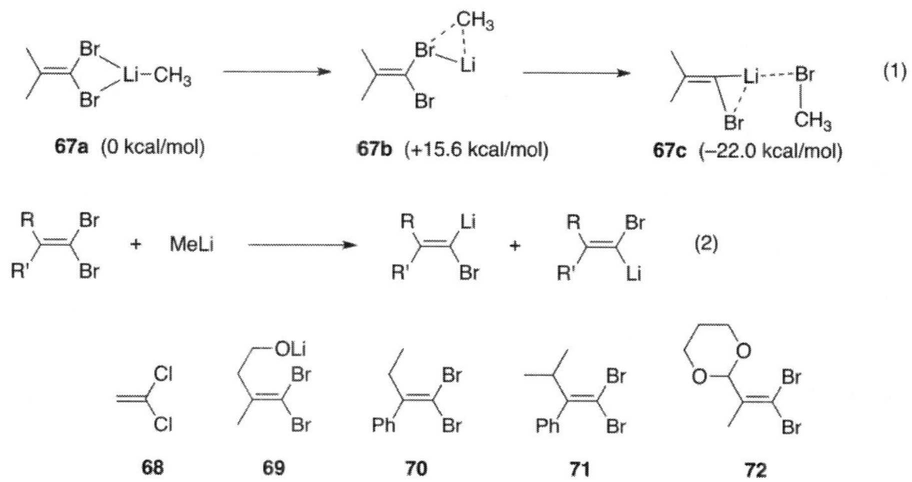

Figure 42. Halogen–lithium exchange reaction of 1,1-dihaloalkenes with methyllithium.

Metal carbenoids can act as nucleophiles (Figure 1, Eq. 4). Li and colleagues investigated the nucleophilic addition of lithium carbenoid 4 to formaldehyde (Figure 43, Eq. 2) [83]. The

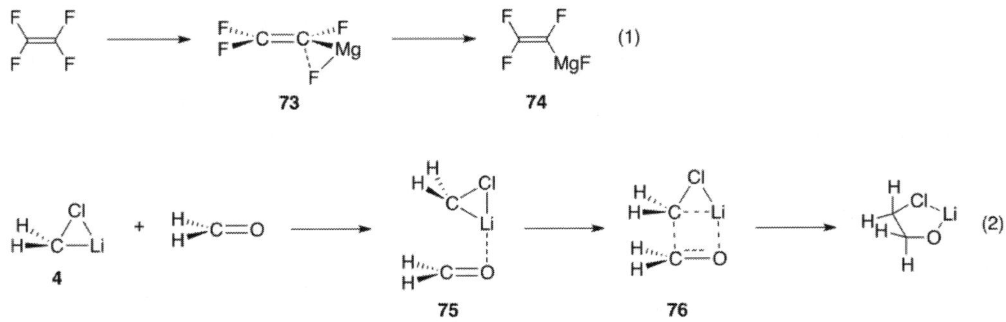

Figure 43. Generation and reaction of metal carbenoids.

activation energy for the addition reaction from reactant complex 75 to the product, lithium 2-chloroethanolate, via transition state 76 was 8.7 kcal/mol in THF solution at the HF/6-31G* level.

Conclusion

The chemistry of metal carbenoids has been an intensive area of research for some time. However, the precise nature of metal carbenoids is difficult to characterize experimentally due to the thermal instability of these compounds. Quantum chemical calculations have played an important role in the structural elucidation and mechanistic exploration of the reaction mechanisms of metal carbenoids. As described above, the RHF and MP2 methods were the mainstream methods in the quantum chemical studies of metal carbenoids. From the beginning of the twenty-first century, DFT studies have made a great contribution to the development of the chemistry of metal carbenoids. The DFT method is a useful tool because it has a reasonable computational cost and produces reliable results that are comparable to those derived from the MP2 method, as exemplified above. In particular, the vibrational frequency calculations of large molecules such as aggregates and solvated molecules using a large basis set at MP2 level are hard to treat from a practical point of view. The DFT calculations enabled us to calculate these systems with ease. Further theoretical studies of metal carbenoids using DFT calculations would continue to offer new insights into the chemistry of metal carbenoids.

References

[1] Köbrich, G. *Angew. Chem. Int. Ed.* 1967, *6*, 41–52.

[2] Köbrich, G. *Angew. Chem. Int. Ed.* 1972, *11*, 473–485.

[3] Siegel, H. *Top. Curr. Chem.* 1982, *106*, 55–78.

[4] Taylor, G. K. *Tetrahedron* 1982, *38*, 2751–2772.

[5] Braun, M. *Angew. Chem. Int. Ed.* 1998, *37*, 430–451.

[6] Boche, G.; Lohrenz, J. C. W. *Chem. Rev.* 2001, *101*, 697–756.

[7] Knorr, R. *Chem. Rev.* 2004, *104*, 3795–3849.

[8] Braun, M. In *The Chemistry of Organolithium Compounds*; Rappoport, Z.; Marek, I.; Eds.; John Wiley and Sons, Ltd: Chichester, UK, 2004; Vol. 2, pp 829–900.

[9] Satoh, T. *Chem. Soc. Rev.* 2007, *36*, 1561–1572.
[10] Capriati, V.; Florio, S. *Chem. Eur. J.* 2010, *16*, 4152–4162.
[11] Satoh, T. *Heterocycles* 2012, *85*, 1–33.
[12] Rodriquez, C. F.; Sirois, S.; Hopkinson, A. C. *J. Org. Chem.* 1992, *57*, 4869–4876.
[13] Apeloig, Y.; Schleyer, P. v. R.; Pople, J. A. *J. Am. Chem. Soc.* 1977, *99*, 1291–1296.
[14] Schaefer, H. F., III *Science* 1986, *231*, 1100–1107.
[15] Andrews, L.; Carver, T. G. *J. Phys. Chem.* 1968, *72*, 1743–1747.
[16] Hatzenbühler, D. A.; Andrews, L.; Carey, F. A. *J. Am. Chem. Soc.* 1975, *97*, 187–189.
[17] Seebach, D.; Siegel, H.; Gabriel, J.; Hässig, R. *Helv. Chim. Acta* 1980, *63*, 2046–2053.
[18] Seebach, D.; Hässig, R.; Gabriel, J. *Helv. Chim. Acta* 1983, *66*, 308–337.
[19] Boche, G.; Marsch, M.; Müeller, A.; Harms, K. *Angew. Chem. Int. Ed.* 1993, *32*, 1032–1033.
[20] Boche, G.; Harms, K.; Marsch, M.; Müeller, A. *J. Chem. Soc. Chem. Commun.* 1994, 1393–1394.
[21] Müeller, A.; Marsch, M.; Harms, K.; Lohrenz, J. C. W.; Boche, G. *Angew. Chem. Int. Ed.* 1996, *35*, 1518–1520.
[22] Dill, J. D.; Schleyer, P. v. R.; Pople, J. A. *J. Am. Chem. Soc.* 1976, *98*, 1663–1668.
[23] Collins, J. B.; Dill, J. D.; Jemmis, E. D.; Apeloig, Y.; Schleyer, P. v. R.; Seeger, R.; Pople, J. A. *J. Am. Chem. Soc.* 1976, *98*, 5419–5427.
[24] Clark, T.; Schleyer, P. v. R. *J. Chem. Soc. Chem. Commun.* 1979, 883–884.
[25] Clark, T.; Schleyer, P. v. R. *Tetrahedron Lett.* 1979, *51*, 4963–4966.
[26] Clark, T.; Schleyer, P. v. R. *J. Am. Chem. Soc.* 1979, *101*, 7747–7748.
[27] Abronin, I. A.; Zagradnik, R.; Zhidomirov, G. M. *J. Struct. Chem.* 1980, *21*, 1–5.
[28] Pross, A.; Radom, L. *J. Comput. Chem.* 1980, *1*, 295–300.
[29] Vincent, M. A.; Schaefer, H. F., III *J. Chem. Phys.* 1982, *77*, 6103–6108.
[30] Luke, B. T.; Pople, J. A.; Schleyer, P. v. R.; Clark, T. *Chem. Phys. Lett.* 1983, *102*, 148–154.
[31] Schleyer, P. v. R.; Clark, T.; Kos, A. J.; Spitznagel, G. W.; Rohde, C.; Arad, D.; Houk, K. N.; Rondan, N. G. *J. Am. Chem. Soc.* 1984, *106*, 6467–6475.
[32] Qiu, H.; Deng, C. *Chem. Phys. Lett.* 1996, *249*, 279–283.
[33] Hermann, H.; Lohrenz, J. C. W.; Kühn, A.; Boche, G. *Tetrahedron* 2000, *56*, 4109–4115.
[34] Kvíčala, J.; Štambaský, J.; Böhm, S.; Paleta, O. *J. Fluorine Chem.* 2002, *113*, 147–154.
[35] Rohde, C.; Clark, T.; Kaufmann, E.; Schleyer, P. v. R. *J. Chem. Soc. Chem. Commun.* 1982, 882–884.
[36] Bellagamba, V.; Ercoli, R.; Gamba, A.; Simonetta, M. *J. Chem. Soc. Perkin Trans. 2* 1985, 185–192.
[37] Canepa, C. Antoniotti, P.; Tonachini, G. *Tetrahedron* 1994, *50*, 8073–8084.
[38] Canepa, C.; Tonachini, G. *Tetrahedron* 1994, *50*, 12511–12520.
[39] Pratt, L. M.; Ramachandran, B.; Xidos, J. D.; Cramer, C. J.; Truhlar, D. G. *J. Org. Chem.* 2002, *67*, 7607–7612.
[40] Pratt, L. M.; Lê, L. T.; Truong, T. N. *J. Org. Chem.* 2005, *70*, 8298–8302.
[41] Pratt, L. M.; Merry, S.; Nguyen, S. C.; Quan, P.; Thanh, B. T. *Tetrahedron* 2006, *62*, 10821–10828.
[42] Pratt, L. M.; Phan, D. H. T.; Tran, P. T. T.; Nguyen, N. V. *Bull. Chem. Soc. Jpn.* 2007, *80*, 1587–1596.

[43] Schulze, V.; Löwe, R.; Fau, S.; Hoffmann, R. W. *J. Chem. Soc. Perkin Trans. 2* 1998, 463–466.

[44] Kimura, T.; Satoh, T. *J. Organomet. Chem.* 2012, *715*, 1–4.

[45] Satoh, T.; Ogino, Y.; Ando, K. *Tetrahedron* 2005, *61*, 10262–10276.

[46] Wang, B.; Deng, C.; Xu, L.; Tao, F. *Chem. Phys. Lett.* 1989, *161*, 388–392.

[47] Azizoglu, A.; Balci, M.; Mieusset, J.-L.; Brinker, U. H. *J. Org. Chem.* 2008, *73*, 8182–8188.

[48] Eccles, W.; Jasinski, M.; Kaszynski, P.; Zienkiewicz, K.; Stulgies, B.; Jankowiak, A. *J. Org. Chem.* 2008, *73*, 5732–5744.

[49] Wang, B.; Deng, C. *Chem. Phys. Lett.* 1988, *147*, 99–104.

[50] Wang, B.; Deng, C.; Xu, L.; Tao, F. *Youji Huaxue* 1988, *8*, 507–510.

[51] Zheng, S.-J.; Meng, L.-P.; Cai, X.-H.; Tang, T.-H. *J. Mol. Struct.* 1992, *255*, 327–333.

[52] Schoeller, W. W. *Chem. Phys. Lett.* 1995, *241*, 21–25.

[53] Yanagisawa, H.; Miura, K.; Kitamura, M.; Narasaka, K.; Ando, K. *Bull. Chem. Soc. Jpn.* 2003, *76*, 2009–2026.

[54] Barluenga, J.; González, J. M.; Llorente, I.; Campos, P. J.; Rodríguez, M. A.; Thiel, W. *J. Organomet. Chem.* 1997, *548*, 185–189.

[55] Campos, P. J.; Sampedro, D.; Rodríguez, M. A. *Organometallics* 1998, *17*, 5390–5396.

[56] Nelson, D. J.; Matthews, M. K. G. *J. Organomet. Chem.* 1994, *469*, 1–9.

[57] Nelson, D. J.; Hohmann-Sager, S. *Heteroat. Chem.* 1998, *9*, 623–630.

[58] Kvíčala, J.; Czernek, J.; Böhm, S.; Paleta, O. *J. Fluorine Chem.* 2002, *116*, 121–127.

[59] Pratt, L. M.; Nguyễn, N. V.; Lê, L. T. *J. Org. Chem.* 2005, *70*, 2294–2298.

[60] Mareda, J.; Rondan, N. G.; Houk, K. N.; Clark, T.; Schleyer, P. v. R. *J. Am. Chem. Soc.* 1983, *105*, 6997–6999.

[61] Hirai, A.; Nakamura, M.; Nakamura, E. *Chem. Lett.* 1998, 927–928.

[62] Nakamura, M.; Hirai, A.; Nakamura, E. *J. Am. Chem. Soc.* 2003, *125*, 2341–2350.

[63] Zhou, Y.-B.; Cao, F.-L. *J. Organomet. Chem.* 2007, *692*, 3723–3731.

[64] Wang, D.; Phillips, D. L.; Fang, W.-H. *Organometallics* 2002, *21*, 5901–5910.

[65] Li, Z.-H.; Ke, Z.; Zhao, C.; Geng, Z.-Y.; Wang, Y.-C.; Phillips, D. L. *Organometallics* 2006, *25*, 3735–3742.

[66] Ke, Z.; Zhao, C.; Phillips, D. L. *J. Org. Chem.* 2007, *72*, 848–860.

[67] Ke, Z.; Zhou, Y.; Gao, H.; Zhao, C.; Phillips, D. L. *Chem. Eur. J.* 2007, *13*, 6724–6731.

[68] Stiasny, H. C.; Hoffmann, R. W. *Chem. Eur. J.* 1995, *1*, 619–624.

[69] Pratt, L. M.; Trần, P. T. T.; Nguyễn, N. V.; Ramachandran, B. *Bull. Chem. Soc. Jpn.* 2009, *82*, 1107–1125.

[70] Ramachandran, B.; Kharidehal, P.; Pratt, L. M.; Voit, S.; Okeke, F. N.; Ewan, M. *J. Phys. Chem. A* 2010, *114*, 8423–8433.

[71] Zhang, X.-H.; Zhang, F.-L.; Geng, Z.-Y. *J. Chem. Sci.* 2010, *122*, 363–369.

[72] Azizoglu, A.; Yildiz, C. B. *Organometallics* 2010, *29*, 6739–6743.

[73] Azizoglu, A.; Yildiz, C. B. *J. Organomet. Chem.* 2012, *715*, 19–25.

[74] Chalifoux, W. A.; Tykwinski, R. R. *Chem. Rec.* 2006, *6*, 169–182.

[75] Jahnke, E.; Tykwinski, R. R. *Chem. Comm.* 2010, *46*, 3235–3249.

[76] Darses, B.; Milet, A.; Philouze, C.; Greene, A. E.; Poisson, J.-F. *Org. Lett.* 2008, *10*, 4445–4447.

[77] Pratt, L. M.; Nguyen, N. V.; Kwon, O. *Chem. Lett.* 2009, *38*, 574–575.

[78] Glukhovtsev, M. N.; Pross, A.; Radom, L. *J. Am. Chem. Soc.* 1994, *116*, 5961–5962.

[79] Lucchini, V.; Modena, G.; Pasquato, L. *J. Am. Chem. Soc.* 1995, *117*, 2297–2300.
[80] Chiba, S.; Ando, K.; Narasaka, K. *Synlett* 2009, 2549–2564.
[81] Ando, K. *J. Org. Chem.* 2006, *71*, 1837–1850.
[82] Davis, S. R.; Liu, L. *J. Mol. Struct.* 1994, *304*, 227–232.
[83] Li, J.; Sun, C.; Liu, S.; Feng, S.; Feng, D. *Sci. China: Chem.* 2000, *43*, 240–246.

Reviewed by Prof. Tsuyoshi Satoh (Tokyo University of Science).

In: Density Functional Theory
Editors: J. Morin and J.M. Pelletier, pp. 95-119
ISBN: 978-1-62417-954-9

Chapter 3

ON THE PREDICTION OF THERMOELECTRIC PROPERTIES OF LOW-DIMENSIONAL MATERIALS BY DENSITY FUNCTIONAL THEORY

***P. Boulet*[1] *and M.C. Record*[2,*]**
[1]MADIREL, Université Aix-Marseille,Marseille, France
[2]IM2NP, Université Aix-Marseille, Marseille, France

Abstract

Growing concerns regarding energy wastes and environmental changes have stimulated consciousness on the necessity to develop new, more friendly, energetic resources. Among the possibilities offered to find solutions, promoting the advances in materials and technologies for energy conversion appears as a promising route.

In thermoelectric modules the use of the so-called "Seebeck effect" allows to convert heat flow that crosses the module into electricity. The efficiency of a thermoelectric material is given by its figure of merit $ZT=(\alpha 2\sigma/\kappa)T$, where α is the Seebeck coefficient, σ and κ are the electrical and thermal conductivities, respectively, and T is the temperature. From this equation we see that improving ZT can be achieved either by increasing the powerfactor $\alpha 2\sigma$ or by reducing the thermal conductivity κ or both. However, for three-dimensional materials these physical properties are interrelated in a counter productive way with regards to our objectives. One way to circumvent this difficulty is to lower the dimensionality of the material down to the nanoscale. In this case, the parameters can be tuned, to some extend, independently from one another. However, the elaboration of nanomaterials, in particular nanowires and quantum dots, and the measurement of their properties are still a challenging task for chemists and physicists. Therefore, theoretical and quantum calculations, and numerical simulations are expected to play an important role for the understanding of the behaviour of these materials and for their developments. The density-functional theory (DFT) is a method of choice for calculating thermoelectric properties of low-dimensional materials. When combined with the semi-classical Boltzmann's transport equation, DFT can predict with reasonable accuracy thermoelectric properties. In addition, with recent developments of the density-functional perturbation theory (DFPT) it has been made easier to calculate properties

[*]E-mail address: m-c.record@univ-amu.fr. (Corresponding Author)

such as phonon spectra and electron-phonon interactions which are important properties in the field of thermoelectricity.

This chapter aims at delivering an updated picture of the recent advances in the theoretical investigation of thermoelectric properties of low-dimensional materials using DFT and related methods.

By converting waste heat into electricity through the thermoelectric power of materials without producing greenhouse gas emissions, thermoelectric generators could be an important part of the solution to today's energy challenge.

The thermoelectric effect refers to the phenomenon of the direct conversion of temperature gradients to electric voltage and vice versa. Thermoelectric generators can be used for converting heat originating from many sources, suchas solar radiation, automotive exhaust, and industrial processes, to electricity. On the other hand, thermoelectric coolers can find applications in refrigerators and other cooling systems. Considering the extremely high reliability in thermoelectric devices, since they are solid state devices without moving parts, they have wide applications in infrared sensors, computer chips and satellites. The inconvenience in these thermoelectric devices lies in their low efficiency, which limits them against wider applications. If their efficiency was significantly improved, thermoelectric devices could play an important role in the solution to today's energy challenge. Therefore, how to improve thermoelectric efficiency becomes the key issue in this research field.

There are three well-known effects involved in the thermoelectric phenomenon: the Seebeck, Peltier, and Thomson effects. In 1821, Thomas Johann Seebeck discovered that when a conductor is subjected to a temperature gradient, an electrochemical potential gradient appears in it [1]. This phenomenon, called the Seebeck effect after the name of its discoverer, can be expressed as

$$\frac{1}{q}\nabla\overline{\mu} = -\alpha\nabla T \tag{1}$$

where $q$is the charge of the carriers, $\overline{\mu}$ is the electrochemical potential, $\nabla T$is the temperature gradient, and $\alpha$is the so-called Seebeck coefficient (usually expressed in μV/K). The left-hand side part of equation (1) can equivalently be viewed as a voltage difference. On the opposite, when a conductor undergoes a voltage gradient a temperature difference appears between both its ends. This is the Peltier effect which corresponds to the reverse of the Seebeck effect. The ratio of the rate of heat $\dot{Q}$ (released or absorbed) to the electrical current I that traverses the conductor is called the Peltier coefficient (expressed in volts) that reads

$$\Pi = \frac{\dot{Q}}{I} \tag{2}$$

The Peltier coefficient can be either positive or negative depending on the sign of $\dot{Q}$. In 1854, Thomson (Lord Kelvin) unified these concepts by using an equilibrium thermodynamics approach. He showed that, the eponymous Thomson effect corresponds to the heat power absorbed by the external medium when an electrical current crosses a temperature gradient [1]:

$$\dot{Q} = -\beta \ I \ \nabla T \tag{3}$$

The coefficient of proportionality β is the Thomson coefficient. From non-equilibrium thermodynamics, it can be established that the Seebeck, Peltier and Thomson coefficients are interrelated kinetic coefficients. For instance, the first Kelvin relation reads,

$$\beta = \frac{d\Pi}{dT} - \alpha \tag{4}$$

and the second one reads,

$$\Pi = \alpha T \tag{5}$$

In addition, it appears that the α, β and Π transport coefficients are tensors, some useful mathematical objects when one is interested in anisotropic materials.

In 1957, in his work on thermoelements, Ioffe introduced a parameter to measure the material effectiveness [2]. From then on, the efficiency of thermoelectric devices is characterized by the thermoelectric material figure of merit *ZT*, which is a function of several coefficients:

$$ZT = \frac{\alpha^2 \sigma}{\kappa_e + \kappa_l} T \tag{6}$$

where σ is the electrical conductivity, α is the Seebeck coefficient, T is the mean operating temperature and κ is the thermal conductivity. The subscripts e and l in κ refer to electronic and lattice contributions, respectively. We note in passing that *ZT* bears no unit. The larger the figure of merit, the better the efficiency of the thermoelectric cooler or power generator. Therefore, there is significant interest in improving the figure of merit in thermoelectric materials for many industrial and energy applications.

In terms of applications, the history of thermoelectric materials is strongly associated with their efficiency. The earliest applications of the thermoelectric effect can be found in metal thermocouples, which have been used to measure temperature for many years. From the late 1950s, research on semiconducting thermocouples has been carried out, and semiconducting thermoelectric devices have been applied first to terrestrial cooling and power generation, and later to space power generation, due to their favourable, competitive energy conversion as opposed to other forms of small-scale electric power generators [3]. By the 1990s, many thermoelectric-based refrigerators have been developed and commercialised, and starting from about 2000, thermoelectric technology has been used to enhance some functions of automobiles such as thermoelectric cooled and heated seats [4]. However, because the device ZT is oftensmaller than 1, the low efficiency of thermoelectric devices has largely limited their application. It is with the discovery of new materials with increasing *ZT*values(e.g., $ZT > 1$) that the road to many new potential applications of thermoelectric technology have been open up.

According to Eq.(6), it seems obvious that, in principle, the direction to follow for increasing the figure of merit *ZT* is to increase the electrical conductivity and Seebeck coefficient and to decrease the thermal conductivity of the materials. However, the reality offers us a very different picture, and it is by no mean a straightforward task to improve *ZT* due to the fact that σ, α, and $\kappa_e$are all coupled with each other. As a typical example, we can refer to the famous Wiedemann-Franz law that connects the electrical conductivity and the electronic thermal conductivity together:

$$\kappa_e = \sigma L T \tag{7}$$

where L is the (temperature-dependent) Lorenz number. Therefore the first approach to increase *ZT* was to reduce the lattice thermal conductivity κ_l by favouring the point defect scattering or boundary scattering of heat carriers that are mainly phonons in the case of semiconductors.

By modifying an already promising compound, a successful strategy has been developed. It consisted in introducing point defects through the synthesis of isostructural solid solution alloys. The solid solutions feature atomic mass fluctuations in the crystal lattice (said differently, it features disorder) which causes strong phonon scattering, hence generally leading to a significant lowering of the thermal conductivity. The typical example of this is the Bi_2Te_3 system for which the $Bi_{2-x}Sb_xTe_3$ and $Bi_2Te_{3-x}Se_x$solid solutions display superior thermoelectric performance over the parent compound [5]. It is often observed, however, that solid solution alloying lead to deterioration in electronic performance (for instance, by decreasing the electrical carrier mobility), and an overall gain in ZT cannot be achieved. This is, for example, the case between the PbTe and $(PbTe)_{1-x}(PbSe)_x$systems.

About 15 year ago, Slack proposed his now famous idea of "phonon glass electron crystal" (PGEC) as a way to achieve outstanding record in low lattice thermal conductivities without compromising the electronic performances [6]. Following his idea, a PGEC material features cages or tunnels in its crystal structure inside which massive atoms reside. A particularly important property of these atoms is their small size with respect to the cage so that they are able to "rattle". "Rattling" frequencies are low and produce phonon damping which leads to a dramatic reduction of the lattice thermal conductivity. With this picture of the PGEC, a glass-like thermal conductivity may, in principle, coexist with electrical charge carriers bearing high mobility. The PGEC approach has stimulated numerous and significant new researches and has led to marked increases in the *ZT* values of several compounds, such as the clathrates [7-18].

To date, it has not been proven, though, that a PGEC material exists. In fact, for most of the purported PGEC materials, other factors coexist which can also account for the observed reduction in thermal conductivity.

Finally, through nanostructuring of bulk materials (especially by using "bottom up" approaches), significant reductions in the thermal conductivity of materials have been achieved. For example, several synthetic techniques have been applied to prepare nanostructured PbTe, with exceptional reductions in the thermal conductivity. At this stage, two major types of bulk nanostructured materials have emerged: (a) materials with self-formed inhomogeneities on the nanoscale driven by phase segregation phenomena such as spinodal decomposition and nucleation and growth; (b) materials which have been processed (e.g., ground) so as to be broken up into nanocrystalline pieces first, and subsequently sintered or pressed into bulk objects. Enhanced thermoelectric performance has been achieved in both types of materials, compared with the corresponding, non-nanostructured ones. For example, n-type nano polycrystalline Bi2Te3 material displays higher ZT (around1.25 at 420 K) than conventional, bulk Bi2Te3[5]. The improvement in *ZT* was assigned both to a slight increase in electrical conductivity and to a reduction by one fourth in thermal conductivity. However, the thermal stability of these systems can be an issue since at high temperatures grain growth and grain fusion will result in permanent loss of the nanocrystallinity. This inevitably leads to the reversion to the conventional alloy with concomitant loss of the high

ZT property. Therefore, it is likely that these systems are limited in operating temperatures below those that favours grain growth.

To assess the lower practical thermalconductivity limits of these systems, Kanatzidis [19] grew PbTe nanoparticles by inverse micelle "wet" chemical synthesis with narrow size dispersity and anaverage size of 9-12 nm. After complete characterization, these nanoparticles were processed into pellets by coldpressing, and their thermal conductivity was measured. The capping ligands used to stabilize these nanoparticles were either left on them or removed, resulting in a minor effect on the measured properties. Even though the pressed pellets did not reached the theoretical density, the measured total, room temperature thermal conductivity values ranged from0.40 to 0.50W/(m.K).Since these pellets were essentially electrically insulating, these values of κ originate from the lattice part of the thermal conductivity and correspond thus the lowest possible achievable value for any PbTe-based material. Even with such low values of thermal conductivity, the figure of merit does not go beyond 1.5. The figure-of-merit *ZT*, along with the ratio between the temperature of the hot and cold sides, determines the percentage of the Carnot efficiency a thermoelectric can attain. This can be assessed from the corresponding formula of the maximum efficiency a thermoelectric generator can reach [5]:

$$\Phi_{max} = \frac{T_H - T_C}{T_H} \frac{\sqrt{1+Z\langle T\rangle} - 1}{\sqrt{1+Z\langle T\rangle} + \frac{T_C}{T_H}} \quad (8),$$

in which we can recognize the Carnot efficiency. T_H and T_C are the temperatures at the hot and cold sides of the generator, respectively, and $\langle T\rangle$ is the mean average operating temperature. Ideally, *ZT* should be in the range of 2-3 to get an overall efficiency of about 25-30% if the hot side is kept at 500°C with the cold side maintained at ambient temperature. Recalling that current best thermoelectrics have *ZT* around 1-1.5, it seems extremely challenging, if not simply out of reach, to further reduce the thermal conductivity from that obtained after nanostructuration of the materials. As a consequence, it is reasonable to believe that, future large *ZT* enhancements will emerge from materials exhibiting substantial jumps of the power factor $\alpha^2\sigma$ from current levels. Moreover, when operating under fixed heat flow conditions, low-κ materials areunsuitable to maximize the electric power output since thermal shunts may be required to sustain the needed heat dissipation [20].

Table 1. Comparison of thermoelectric properties of metals, semiconductors and insulators at 300K (after reference [3])

Property	Metals	Semiconductors	Insulators
α (μV K^{-1})	5	200	10^3
σ (S cm^{-1})	10^6	10^3	10^{-12}
Z (T^{-1})	$3\ 10^{-6}$	$2\ 10^{-3}$	$5\ 10^{-17}$

The figure of merit of a material is influenced by its electronic structure. Indeed, from a microscopic point of view, σ, α and $\kappa_\varepsilon$can be determined from the electronic band structure. It is well known from electronics that, many materials can merely be classified into metals, semiconductors, and insulators. These three different classes of materials can be roughly characterized by zero, small and large band gaps, respectively, or alternatively, by theirfree-charge-carrierconcentration. The comparison of thermoelectric propertiesof metals, semiconductors and insulators at 300 K isshown in Table I and illustrated in Figure1.

Clearly, from Table 1, metals have very good electrical conductivity (~$10^6\Omega^{-1}cm^{-1}$). However, their very low Seebeck coefficient (~5 μVK^{-1}) and large thermal conductivity do not make them the most desirable materials for thermoelectric applications. For insulators, although they have large Seebeck coefficient (~1000 μVK^{-1}), their large band gap confers them an extremely low electrical conductivity (~$10^{-12}\Omega^{-1}cm^{-1}$) resulting in a small value of $\alpha^2\sigma$, andthus a small Z (~$5x10^{-17}$ K^{-1}), which is far smaller than that of metal (~$3x10^{-6}$ K^{-1}). The optimal thermoelectric materials, exhibiting a large value of $\alpha^2\sigma$,are to be found in the region near the crossover between semiconductor and metal (see Figure 1), with optimized carrier concentration of about $1x10^{19}$ cm^{-1}.

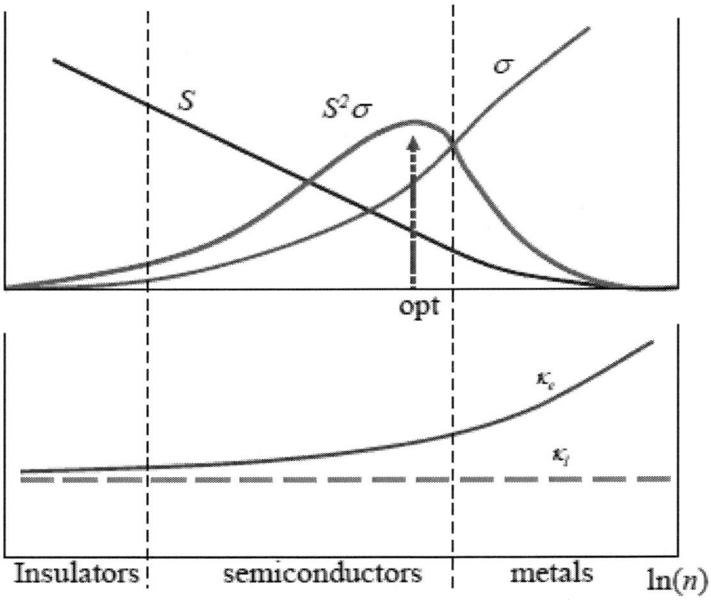

Figure 1. Seebeck coefficient α, electrical conductivity σ, power factor $\alpha^2\sigma$, and electronic (κ_e) and lattice (κ_l) thermal conductivities as a function of free-charge-carrier concentration n. The optimal carrier concentration is about $1x10^{19}$ cm^{-1}, which is indicated by an arrow (after References [2] and [3]).

By using the transport coefficients obtained from the resolution of Boltzmann's transport equation (BTE), and keeping all properties characterizing the material inside the transport distribution function (TDF), Mahan and Sofo [21] found that the best suited TDF, that is the one which yield the maximized figure of merit, is the one which takes the Dirac delta function. In other words, when the electronic density of state near the chemical potential has a sharp singularity the figure of merit can grow very large.By choosing some typical parameters forhigh-performance thermoelectric material, the authors [21]showed that a possible figure of merit as high as ZT=14can be obtained. Such high $ZT$is proposed to beachievable using rare-earth compounds. However, the authors also showed that when a background is added to the density of states, a drastic decrease of ZT is observed. Therefore, Mahan and Sofo's work pointed to some new indications for searching for good thermoelectric materials: (i) a very narrow distribution of energy carriers is required to have good Seebeck coefficient, (ii) a high carrier velocity in the direction of the applied electric field is necessary for high electrical conductivity, and (iii) a very small percentage (<1%) of background in density of states under a sharp peak is utterly desired to prevent the collapse of the power factor. In their theoretical study Mahan and Sofo did not consider the impact of the shape of the transport distribution function. The shape effect of the TDF on thermoelectric figure of merit was revisited by Fan et al. [22]. These authors confirmed both heuristically and numerically that among all the normalized TDF the Dirac delta function leads to the largest thermoelectric figure of merit. However, the delta function is unrealisable in practice, and more realistic shapes have been envisaged such as the Gaussian and the rectangular ones. Interestingly, the rectangular shape exhibits interesting properties, in particular it yields a *ZT* with a weak dependence on the background. For the case of TDF being bounded, a rectangular-shape distribution is again found to be the most favourable one. In addition, thermoelectric materials with high *ZT* should also violate the Wiedemann-Franz law to some appropriate degree. Another characteristic of the TDF is its asymmetric shape that should enhance the Seebeck coefficient, which could be achieved through various nanoroute.

Hicks and Dresselhaus published several papers in the 1990s in which they present the effect on the thermoelectric properties of nano confinement of the electronic structure. In these works, the authors solved a simplified Schrödinger equation obtained for devices exhibiting different symmetry that reflects the nano confinement.

In 1993, Hicks and Dresselhaus [23,24] showed that the use of confinement effects have the potential to significantly improve the thermoelectric figure of merit. In a one-dimensional wire made of Bi_2Te_3, a *ZT* of up to 14 was predicted for a wire with square cross section and width 5 Å. Similarly, using Bi_2Te_3 as a quantum-well structure, an increase of ZT by a factor of 13 compared to the bulk value was obtained, provided that the superlattice multilayers are made in a particular orientation.

Materials can be confined in one, two or three dimensions. A one-dimensional confinement can be obtained by creating a two-dimensional layer with thickness of the order of the nanometre, so then we get a 2D object. Alternatively, they can be confined in two dimensions, leaving a one-dimensional object with a nanoscale cross section (1D), or they can be confined in three dimensions, which effectively results in a zero-dimensional object (0D).

Devices with nanometre dimensions exhibit quantum mechanical effects [25]. This causes an alteration of the number of states (NOS) and density of states (DOS). The NOS (N(E)) and the DOS (D(E)), which can be defined as a function of energy E, affect the physical and chemical properties such as resistance, optical properties and thermal characteristics (Seebeck coefficient, thermal resistance, etc.). The NOS and DOS can be calculated either analytically, if simplified models are used, or numerically using quantum mechanical approaches. The 2D, 1D and 0D approximations can be utilized to analyse structures such as quantum wells, nanowires and nanoparticules, where quantum effects are observed for dimensions of less than a few nanometres. When treated classically, a free particle displays a continuous energy spectrum. However, this assumption is not valid for particles in confined systems, which must be described using quantum mechanics. Under such conditions, particles exist in discrete energy levels that form distinct bands. Energy levels outside these bands are forbidden.

When the number of bounded particles having an energy E is defined as $N(E)$, the density of states is defined as $D(E) = dN(E)/dE$. The DOS is then expressed as $D(E)dE$, which is the number of allowed energy levels per unit volume of the material, within the energy range $E$and $E+dE$. In three-dimensional materials (3D), the DOS is proportional to $E^{1/2}$and continuous in the E space. In two-dimensional materials (2D), where the structure is confined to quantum states in one dimension and effectively infinite in the other two, such as in the case of superlattices, the 2D DOS does no longer depend directly on the energy value in contrast to the case of the 3D DOS. Here, the DOS takes on a staircase, or Heaviside distribution function. For 1D materials, the DOS both depends on the inverse square-root of the energy and the Heaviside distribution function, whereas for 1D materials it corresponds to the Dirac delta distribution function. Calculated NOS and DOS for 0D, 1D, 2D and 3D materials as a function of energy are given in Table 2. The graphs for NOS and DOS of 0D, 1D, 2D, 3D structures as a function of E are presented in Figure 2.

Table 2. NOS and DOS for 0D, 1D, 2D, and 3D materials. $H(E\text{-}E_i)$ is the Heaviside function. $\delta(E\text{-}E_i)$ is the Dirac delta distribution function. K_0, K_1, K_2 and K_3 are coefficients, i is an integer. (We recall that, the derivative of the Heaviside distribution function is the Dirac distribution function)

Type	NOS (N(E))	DOS (D(E))
0D	$K_0\sum_i d_{0i}H(E-E_i)$	$K_0\sum_i d_{0i}\delta(E-E_i)$
1D	$K_1\sum_i d_{1i}(E-E_i)^{1/2}H(E-E_i)$	$\frac{1}{2}K_1\sum_i d_{1i}(E-E_i)^{-1/2}H(E-E_i)$
	$K_2\sum_i d_{2i}(E-E_i)H(E-E_i)$	$K_2\sum_i d_{2i}H(E-E_i)$
3D	$K_3E^{3/2}$	$\frac{3}{2}K_3E^{1/2}$

A critical issue for nanotechnology is that the components, structures, and systems are in a size regime for which we have little understanding of their fundamental behaviour. Theparticles are too small for direct measurements, though too large to be described by current rigorous first principle theoretical and computational approaches. They exhibit too many fluctuations in their chemical and physical properties to be treated monolithically in time and space, though they are too few to be described by a statistical ensemble. Fundamental understanding and highly accurate predictive methods are critical to successful manufacturing of nanostructured materials, devices, and systems.

Even though it is difficult today to fully model nanoscale systems, it is clear that theory, modelling and simulations is the key qualifying way for:

i. Reducing the time needed to design new materials;
ii. Developing nanoscale devices from the new materials;
iii. Increasing the reliability and predictability of the operation of the new devices;
iv. Designing and optimizing new nanoscale technologies.

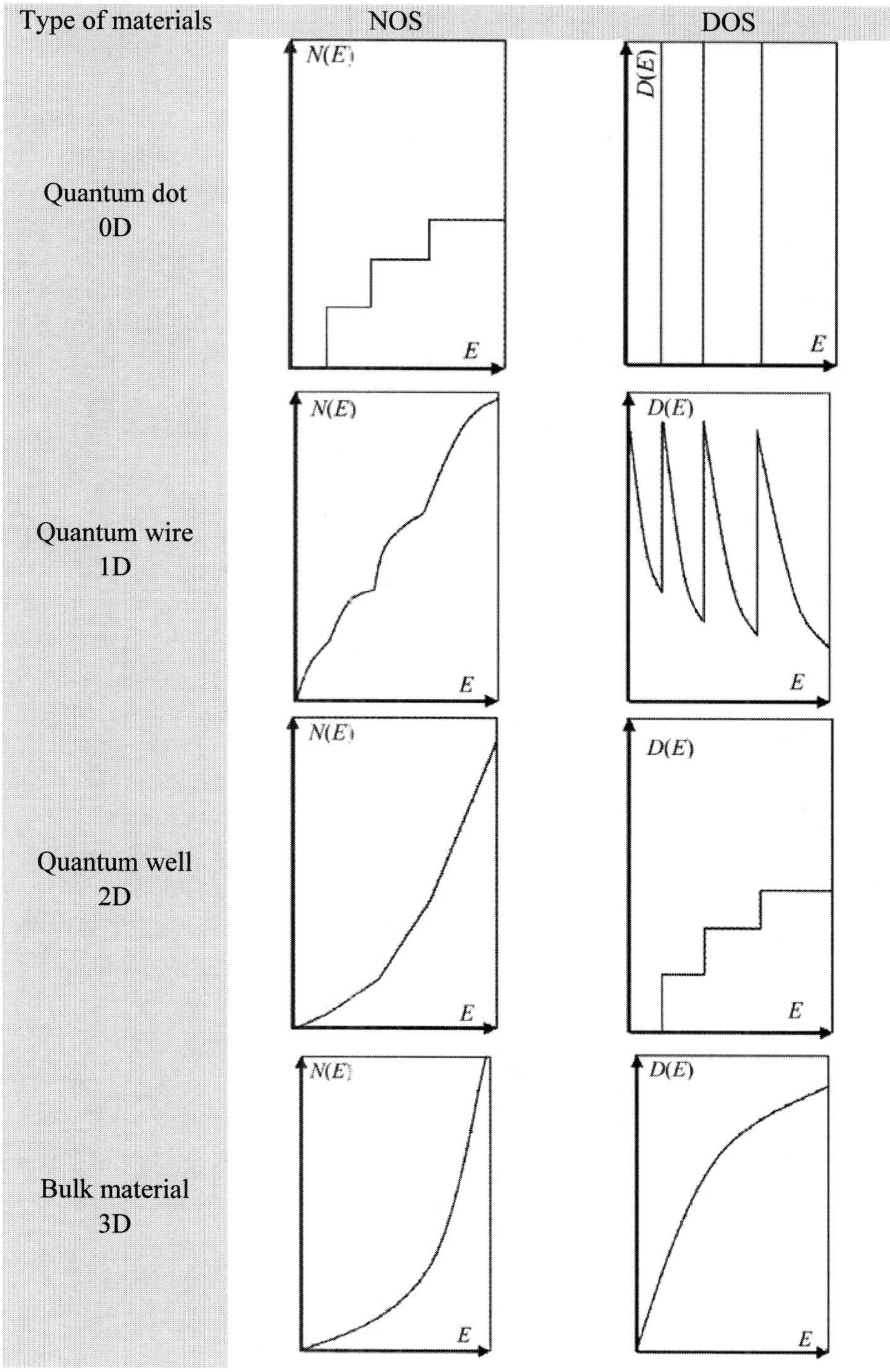

Figure 2. NOS (*N(E)*) and DOS (*D(E)*) of 0D, 1D, 2D, and 3D structures as a function of the energy *E*.

In recent years, ab initio or first-principles computations (i.e., methods based on solving the fundamental equations of quantum mechanics) have become ubiquitous in materials science, and by now it is difficult to find a material of scientific interest that has not been investigated from computational approach. The most popular approach to first-principles computations is by far the density-functional theory (DFT). Although the theoretical roots of modern DFT dates back to 1964 with the Hohenberg–Kohn theorems [26], the premises of the DFT can be found in the early models of Thomas (1927, [27]), Fermi (1927, [28]), Weizsacker (1935, [29]) and Dirac (1930, [30]) for the electron density-dependent kinetic and exchange functionals. The last two decades of DFT research have been particularly important to materials design due to the improved accuracy in the treatments of exchange and correlation quantum effects between electrons, the increased power of computational resources available to materials scientists, the improvement in the efficiency of numerical algorithms, and the introduction of numerous commercial or public-domain software packages for performing DFT static calculations and dynamic simulations (Quantum Espresso, VASP, WIEN2K, CRYSTAL, CPMD, SIESTA, CASTEP and CP2K, to cite a few) [31-33].

As said earlier, modern DFT has its foundation in the two Hohenberg-Kohn (H-K) theorems [26]:

i. The first one proves that all ground state properties of a system, including the total energy, are some functional of the ground-state electron density. This theorem can be demonstrated by a reducio ad adsurdum that shows that if one tries to apply two different external potentials $v(\vec{r})$ to get the same energy for the system of interest, the equations yield inconsistencies. Therefore, the external potential is determined, up to an additive constant, by the electron density of the system, which in turn determines the ground-state wave function;
ii. The second H-K theorem establishes the variational principle of the energy functional. It states that the true ground-state electronic density minimizes the energy functional. The H-K theorem implies that the ground-state for any system can be determined by varying the electronic density until the global minimum of the energy functional is found. Assuming that the energy functional $E_v[\rho]$ is differentiable, the constrained ground-state electronic density satisfies the stationary principle [34]:

$$\mu = \frac{\delta E_v[\rho]}{\delta \rho} = v(\vec{r}) + \frac{\delta F_{HK}[\rho]}{\delta \rho} \tag{9}$$

where μ is nothing else but the electronic chemical potential of the system, and $F_{HK}[\rho]$ is the Hohenberg-Kohn functional that can be expressed with respect to the kinetic energy and electron-electron interaction energy density-functionals:

$$F_{HK}[\rho] = \langle \Psi | \hat{T} + \hat{V}_{ee} | \Psi \rangle \tag{10}$$

Equation (9) is considered as the working DFT equation that has to be solved for the system of interest. We note in passing that, $F_{HK}[\rho]$ does not depend on the external potential $v(\vec{r})$ which confers the status of a universal functional to $F_{HK}[\rho]$.

The second theorem provides a justification for the earlier Thomas and Fermi electron density-based models. In fact, the energies found with those models are approximations to the true H-K energy.

With the density-functional theory we are at the point where the complex problem of solving the many-body Schrödinger equation of N electrons with 3N spatial coordinates is reduced to a simpler, though still difficult problem based on the electron density with only 3 spatial coordinates. However, while the H-K theorem guarantees in principle a correspondence between the ground-state electronic density and energy, the true functional denoted $F_{HK}[\rho]$ in equation (9) relating the electron density to the energy is unknown. It was Levy and Perdew who taught us how to find a solution to this problem through the constrained-search formulation [35-37]. It states that, if one searches among all the antisymmetric wave functions that yield the true electron density of the system, $F_{HK}[\rho]$ provides the minimum expectation value of $\langle \hat{T} + \hat{V}_{ee} \rangle$. The advantage of this formulation is that it also relaxes the constraint on the electron density to be *N*-representable rather than *v*-representable, and extends the domain of definition of $F_{HK}[\rho]$ from *v*-representable densities to *N*-representable ones [34]

For the vast majority of current applications of DFT, the variational problem of minimizing the energy functional is solved by considering the reformulation due to Kohn and Sham through the use of an auxiliary, noninteracting system [38]. The key point is to know how to find a proper wave function that yields the true electron density, that is, without resorting to use the approximate Thomas-Fermi-Dirac model. The way to do so is to use a fictitious system of noninteracting particles that would have the same electron density as the real system. By using this concept, Kohn and Sham reshaped the H-K equation (9) as:

$$\mu = v_{eff}(\vec{r}) + \frac{\delta T_s[\rho]}{\delta\rho(\vec{r})} \tag{11}$$

where $v_{eff}(\vec{r})$ is the effective Kohn-Sham potential that comprises the external, Coulomb and exchange-correlation potentials, and $T_s[\rho]$ is the kinetic energy of the noninteracting system. The exchange-correlation potential term comprises all the many-particle quantum interactions, including the part of the kinetic energy belonging to the interacting system, usually denoted as $T[\rho]-T_s[\rho]$,which is meant to be small. Still, the dependence of the kinetic energy of the noninteracting system on the electron density is an unknown functional of ρ, therefore a set of auxiliary wave functions ψ_i, called the Kohn-Sham wave functions, is used to construct $T_s[\rho]$:

$$T_s[\rho] = \sum_{i=1}^{N} \langle \psi_i | -\frac{1}{2}\nabla^2 | \psi_i \rangle \tag{12}$$

Finally, a set of Schrödinger-like Kohn-Sham equations has to be solved self-consistently to get the energies and wave functions of the particles:

$$\left[-\frac{1}{2}\nabla^2 + v_{eff}(\vec{r})\right]\psi_i = \varepsilon_i\psi_i \tag{13}$$

In practice, the exact energy functional for exchange and correlation $E_{xc}[\rho]$ and its functional derivative with respect to the electronic density (the exact potential $v_{xc}[\rho]$) are not known, and the development of approximate functionals that allow for the accurate calculation of physical quantities of interest, is a field of intensive research. One of the earliest density functionals, which is still in widespread use today in materials science, is the so-called local density approximation (LDA), where the functional depends only on the density at the coordinate where the functional is evaluated. More advanced approximations incorporate the gradient of the density, according to the generalized gradient approximation (GGA) [39,40], the laplacien of the electron density, and the kinetic energy density (meta-GGA functionals). The adiabatic connection allows to introduce a certain amount of Hartree-Fock exchange potential into the Kohn-Sham Hamiltonian, leading to the so-called hydrid functionals (see [34] page 186 and references therein). The functionals are parameters-dependent that can be fitted either on experimental data, which makes them some empirical functionals, or on the free electron gas for which the exact exchange and correlation functional is known. In the latter case, the corresponding functionals can be considered as ab initio functionals.

As we have seen, one of the main outcome when solving the Kohn-Sham equations for a material is its ground-state energy. A significant drawback with local and semi-local functionals, namely those that belong to the LDA, GGA and meta-GGA approximations, is that they contain a spurious electron self-interaction (SI) energy because the exchange potential is approximated and do not cancel exactly with the self-interaction term of the Coulomb part. This pitfall is absent in the Hartree-Fock theory. The spurious SI energy leads to considerable errors in reaction energies for which electrons are transferred between significantly different environments, such as a metal and a transition metal oxide. Intensive efforts have been done to alleviate the self-interaction issue. One popular approach which does not increase the computational burden compared to GGA or LDA, is to use a Hubbard U parameter that cancels part of the SI energy on the most problematic orbitals such as the localized *d*- or *f*-orbitals [41,42].

The main difficulty in using the DFT+U approach is that it requires the determination of an adequate U parameter. It can be done through a self-consistent linear response theory calculation [43] or by fitting the experimental formation energies of the materials of interest [44,45]. As mentioned before, an alternative way to DFT+U is to incorporate a fraction of the exact Hartree-Fock exchange through hybrid functionals as in the Heyd-Scuseria-Ernzerhof (HSE) functional, which is especially suited for solid-state applications [46-49]. These hybrid approaches are, however, computationally more expensive than the DFT+U one.

Another property of significant interest is the electronic band structure of the material. Solving the Kohn-Sham equation provides direct access to the charge density of a material which can give insight into charge localization and the nature of bonding or, in the case of spin-polarized systems, provide the magnetic polarization on different atoms. In addition, from the Kohn-Sham orbitals, band structures can easily be obtained. However, the accuracy of the band structures prediction using DFT is currently a major challenge. For instance, it is well-known that local and semi-local functionals systematically underestimate the energy gaps between occupied and unoccupied states in solids by 30-100% [50].

This underestimation has been attributed to the inherent lack of derivative discontinuity [51] and delocalization error [52]. The gap error can be partly cured using hybrid functionals.

Several approaches have been developed to address the limitations of semi-local functionals with respect to electronic structure properties. For example, the hybrid HSE functional, which incorporates a fraction of the exact Hartree-Fock exchange, has been shown to lead to much more accurate band gaps [48,53]. Excited-state methods such as GW [54,55]and time-dependent DFT [56] can also provide more reliable predictions of electronic structures. However, most of these methods are significantly more demanding in terms of computational power (at least an order of magnitude) than LDA or GGA, and current efforts are directed toward the development of cheaper (but often slightly less accurate) methods such as the Tran-Blaha potential [57], the GLLB-SC model potential [58, 59], or the Δ-sol method [53].

For what concerns us, namely the transport properties in semi-conducting materials, in particular the thermoelectric ones, several of these properties can be calculated from the energy bands and their derivatives. A widely used method is to feed the semi-classical Boltzmann transport equation (BTE) with the electronic band structure obtained from a DFT calculation. Madsen andSingh [60] developed such a code called BoltzTraP (Boltzmann Transport Properties) for solving the BTE and calculating these properties. This code relies on a Fourier expansion of the band energies where the space group symmetry is maintained by using star functions. The idea of the Fourier expansion is to use more star functions than band energies, but to constrain the fit so that the extrapolated energies are exactly equal to the calculated band-energies.The additional freedom in the star functions is used to minimize a roughness function, thereby suppressing oscillations in the extrapolated band energy curves between the high-symmetry *k*-points of the Brillouin zone [61-63].

Using this analytical representation of the bands, it then becomes a reasonably simple procedure to calculate band structure-dependent quantities. The method has been tested for several applications based on the Boltzmann theory, including the transport coefficients of intermetallic compounds [64], high TC superconductors [65] and thermoelectrics [66]. Furthermore the present code has already been applied to calculate the transport coefficients in a series of different clathrate structures [67] and a very good agreement was found with experimental values [68,69]. Good agreement was also found for the demanding Hall coefficient that depends on the second derivative of the bands [67,69]. Since the implementation of this code, a huge number of papers report on thermoelectric properties calculated by this code in bulk materials. By contrast, this code was scarcely used in the calculations of thermoelectric (TE) properties for low dimension materials [70,71].

Using the combined approach based on first-principles DFT and Boltzmann's semi-classical theory, Garcia-Fernandez et al. [70] calculated the transport and TE properties of the half-metallic, two-dimensional electronic structure confined in single $SrRuO_3$ layers of $SrTiO_3/SrRuO_3$ periodic superlattices. Close to the Fermi level, these authors found that the semiconducting majority-spin states display a very large in-plane component of the Seebeck tensor at room temperature, with α around1500 μV/K, but with a negligibly small electrical conductivity, whereas the metallic minority-spin states exhibit good in-plane conductivity, with σ=2.5 (mΩ.cm)$^{-1}$, but very small Seebeck coefficient. As a consequence, the total power factor and thermoelectric figure of merit for the undoped structure is too small for practical applications. Interestingly, the same conclusion holds for any reasonable levels of doping. These results support the idea that, the confinement of the electronic motion is not the only characteristic of the material that matters to describe the main features of the thermoelectric properties with respect to the chemical doping, butthe shape of the electronic density of states is also important. The evolution of the electronic structure, electrical conductivity, Seebeck coefficient, and power factor as a function of the chemical potential was explained by a simplified tight-binding model. The authors found that the electronic structure is composed by a pair of one-dimensional electron gases orthogonal to each other. Therefore, whereas the spatial confinement of the carriers is two-dimensional, the physical dimensionality of the electronic system (1D) can even be smaller.

Bi_2Te_3/Sb_2Te_3 superlattices TE properties were analyzed by means of DFT calculations and Boltzmann transport theory by Hinsche et al. [72]. The model of superlattices is made of quintuples, namely a staking of layered structures build from Bi_2Te_3 in which some atomic layers of Bi are replaced by Sb ones. This leads to a SL with composition $(Bi_2Te_3)_x(Sb_2Te_3)_{1-x}$. The results of the TE calculations show a noticeable anisotropy in the material. As a general trend, whatever the value of x and the nature of charge carriers, the in-plane electrical conductivity and power factor are larger than the out-of-plane ones. For the Seebeck coefficient, the out-of-plane is generally higher than the in-plane one, except for x=0.3 and x=0.7 in the case of electron doping. Experimentally, it was found a remarkably high out-of-plane figure of merit $ZT_{\perp}$ of 2.4 at room temperature [73]. However, from calculated data, the best estimate for the $ZT_{\perp}$ value amounts to 0.9 at room temperature. Hence, the DFT and BTE calculations fail to explain the high experimental value. TE transport is found to be best for hole-doped SL in the in-plane direction. For *p*-type SL the anisotropy of transport of the bulk material is retained whereas in the case of *n*-type SL, a larger anisotropy, caused by strong quantum well effect, is observed which prevents good cross-plane TE transport. In another publication, the authors report a detailed study of the behaviour of the anisotropic Lorenz function L [74]. The same theoretical approach was used as before and applied to Bi_2Te_3/Sb_2Te_3 SL and the bulk constituents at various charge carrier concentrations. Because of the quantum well effects mentioned earlier, large oscillatory deviations of L from the metallic limit L_0 can be observed, even at high doping. As a consequence, the determination of the electronic thermal conductivity can be in substantial error by a factor of two either from the low or the high side. This may also lead to wrong estimates of the lattice thermal contribution and figure of merit. Interestingly, a similar behaviour was observed in the case of *p*-type SiGe superlattices [75].

In 2010, Heinz et al. [71] used the combined approach based on the density-functional theory and the Boltzmann transport theory to examine the TE properties of Si and ZnO nanowires. Three H-passivated Si nanowires were used with diameters around 11Å and number of atoms comprised between 42 and 104. The authors found that the enhancement in the *ZT* value of Si nanowires is about eleven-fold compared to bulk silicon (*ZT*=0.007). It is worth noticing that the authors compared their results with those from Vo et al. who also used the same theoretical approach [76]. Vo et al. found *ZT* values for Si nanowires that are even larger than Heinz's ones by a factor of about 2.5. Therefore, we can see that transport and TE properties are very sensitive to both the calculation parameters (number of *k*-points, energy cut-off, surface termination) and to electron relaxation time used in the resolution of the BTE.

The difference in TE properties were compared between nanowires grown in different direction, namely [011], [111] and [001]. They were found to be quite different with respect to this feature. For instance, the electrical conductivity can vary from 1 to 3 times depending on the direction of growth. By contrast, the *ZT* value seems to be less sensitive and amount to around 0.07. For ZnO, the impact of the size of the nanowire on the TE properties was studied. The nanowires were grown in the direction perpendicular to the [0001] one and have diameters of 8 Å, 10 Å and 17 Å. The most noticeable effects of the 1D confinement are the decrease of the band gap when the diameter increases and the presence of a sharp peak in the electronic DOS for the small nanowire which disappears when the diameter of the nanowire gets larger. As with Si, it was found that ZnO nanowires exhibit an improvement in thermoelectric efficiency over bulk: depending on the value of the thermal conductivity used in the calculation of *ZT*, the minimum and maximum estimates for *ZT* increase by a factor of 3.14 and 28, respectively, compared to bulk ZnO. The latter value was obtained using a hypothetical latticecomponent of the thermal conductivity one order of magnitude less than the lattice component of thermal conductivity, which corresponds to the difference between the bulk and the $\left[10\bar{1}0\right]$ZnO nanowire values. Whereas the electrical conductivity of the nanowires considerably decreases compared to the bulk, the electrical thermal conductivity decreases too, by a factor of about 2, which is favourable to the *ZT* value. Overall, the increase is not significant enough toallow silicon or ZnO nanowire systems to be credible alternatives to novel technologies.

The thermoelectric properties of H-passivated silicon nanowires oriented in the direction [001] were also studied by Shi et al. [77]. The authors used a first-principle tight-binding electronic structure approach combined with the Boltzmann transport theory to investigate the size dependence of thermoelectric properties of silicon nanowires (SiNWs). The squared cross section of the nanowires was varied between 1 nm^2 and 18 nm^2. With increasing cross section area, the electrical conductivity and Seebeck coefficient evolve in opposite directions: while σ increases slightly, α decreases markedly (from 650 μV/K down to 500 μV/K). Therefore, the overall power factor decreases. In addition, the electronic thermal conductivity is observed to increase as the nanowire section grows larger. From experimental findings, the same behaviour is observed for the lattice thermal conductivity. As a consequence, *ZT* decreases when the section size of the nanowire increases. Moreover, the *ZT* value also decreases with the increase of the transverse size of the Si nanowire. These results demonstrate that in thermoelectric application, nanowires exhibiting a small diameter are preferred. In 2010, the same authors investigated the composition effects of silicon-germanium ($Si_{1-x}Ge_x$) nanowires (NWs) on the thermoelectric properties [78] using DFT electronic calculations and the BTE theory. As with the previous study on Si NW, the Si-Ge NW were oriented in a particular direction, namely the [110] one, and the NW exhibited a rectangular cross section. The evolution of σ and α with respect to the carrier concentration and the content in germanium x is very different. For *n*-type doping, whereas σ exhibits a minimum value around 50%Ge, α is constant over the 0-100% range of Ge composition. In the case of *p*-type doping, σ increases slightly with x whereas α decreases monotonically. Whatever the content in germanium, the seebeck coefficient is larger in the case of *p*-type doping (125 μV/K on average) than for *n*-doping (80 μV/K on average). The opposite is observed for the electrical conductivity. The power factor exhibited by the *n*-type $Si_{1-x}Ge_x$wires are much larger, by about 2 orders of magnitude, than that found in their *p*-type counterparts with the same Ge content and doping concentration, except when doping is very high (above 6x1020 cm^{-3}). For *n*-type doping, the curve of the maximum power factor with respect to the germanium content is a parabola with a minimum value located around x=0.5 and $\alpha^2\sigma$ amounting to 2 mW/(mK^2). For p-type doping the curve is relatively flat around 0.3 mW/(mK^2). According to the authors, the maximal obtainable figure of merit can be increased by a factor of 4.3 in*n*-type $Si_{0.5}Ge_{0.5}$ NWs, compared with the corresponding values in pure silicon nanowires. Given the fact that the measured ZT of *n*-type SiNW ranges between 0.6-1.0, the expected *ZT*value of *n*-type $Si_{1-x}Ge_x$NWs are estimated to be between 2.5 and 4.0.

Predictions for the thermoelectric properties of Si-nanowires were also made by Jauho et al. [79] from calculations based on the combination of density-functional theory, tight-binding, Tersoff empirical potential and nonequilibrium Green's functions approaches. They modelled Alkyl-decorated Si nanowires and nanotrees comprising a trunk and a branch. These nanostructures have been predicted to have interesting thermoelectric properties [80]. Since the electronic transport is coupled to the vibrations of the ions, local heating and energy relaxation of the material occur. In addition, the ionic vibrational motion itself is an important medium for heat transport. In this work, the authors report on the ballistic electronic and thermal transport properties of Si-nanowires, paying attention to nanostructuring, as mentioned before. Regarding the electrical transport, the electronic bands belonging to either the alkyl molecules or the lateral branch connected to the trunk are well separated from those belonging to the trunk. Therefore, the electrical transport capabilities of the materials, essentially assured by the trunk, are not affected by the nanostructuration of the Si nanowire. By contrast, the alkyl molecules and branches act as phonon scattering centers, so the thermal conductivity is observed to decrease when the number of scatterers increases. This trend is valid for both *n*-type and *p*-type doping. As a consequence, the *ZT* value in markedly increased (between 0.7 and 1.5, depending on the type of doping) with respect to bare SiNW and surface disordered SiNW in which surface vacancies are created.

Shelley and Mostofi [81] reported a detailed study on Si nanowires with axial Ge heterostructures using a combination of first principles density-functional theory, Tersoff interatomic potentials and Landauer-Buttiker (Green's functions) transport theory approaches. They considered nanowires with up to 8400 atoms and twelve Ge axial heterostructures spatially distributed according to various distributions namely, randomly, regularly or following the Fibonacci series. The length of the Ge wires followed the Gaussian or the uniform distribution. These calculations showed that the introduction of heterostructures reduces the power factor, but decreases the lattice thermal conductivity to such extend that, overall the figure of merit increases. Among all the systems considered, the nanowire oriented in the <111> direction with regularly distributed Ge heterostructures exhibits the highest figure of merit (*ZT*=3). It corresponds to an order of magnitude increase of *ZT* with respect to the equivalent pristine SiNW. Even when structural disorder is introduced in the NW, in the form of small variations in length of the heterostructures, *ZT* remains several times larger than that of SiNW. Nonetheless, when using a more realistic heterostructured nanowire, namely with a near-periodic distribution of Ge segments, the *ZT* value falls down to about 1.6. This result puts strong constraints on the experimental synthesis techniques. In conclusion, these calculations suggest that nanowires bearing axial heterostructure could be an interesting way to obtain good thermoelectric materials.

Chen et al. investigated the electronic, transport, and lattice vibrational behaviors of SiNWs by first-principles calculations to identify the physical origin of the high thermoelectric performance [82]. They used VASP code to relax the NW structure, solved the BTE to obtain transport coefficients, and calculated the phonon band dispersion patterns for two nanowires, one with 25 atoms the other with 57 Si atoms. The phonon bands were calculated within the harmonic approximation through PHONON package [83,84]. In the PHONON software, the phonon dispersion curves and phonon density spectra are calculated from a set of force constants calculated ab initio. It is noteworthy that, to date very few authors have tackled the burden of calculating phonon spectra of nanomaterials by ab initio methods. The authors report that the power factor of SiNWs are not significantly affected by the low dimensional structures, i.e., possessing the similar transport coefficients $\alpha$and σ to the bulk values. This contrasts with other prediction that shows that $\alpha^2\sigma$ is usually slightly smaller in nanowires. The authors assign this result to the fact that the bulk structure of Si, namely the tetrahedral environment of Si, is largely retained in the nanowires. In contrast, lattice heat transport in SiNWs is significantly decreased. This is attributed to both the low-lying optical phonon bands that mix with acoustic modes and the small slope of the acoustic modes near the center of the Brillouin zone that reduces the acoustic phonon velocities. The former effect leads to a decrease in the phonon mean relaxation time and the combination of both effects explains the reduced lattice thermal conductivity.

In the same scope, Demchenko et al. [85] analyzed the electronic and lattice contributions to the enhancement of *ZT* by using the DFT, the BTE theory and classical molecular dynamics (MD) simulations to investigate the thermoelectric properties of ZnO and Si nanowires. The authors showed that lattice thermal conductivity reduction is the prevailing factor in the improvement of thermoelectric properties in nanowires. The thermal conductivity driven by the lattice was calculated by means of MD simulations. From the trajectories, the heat current autocorrelation function is calculated, based on Green-Kuboformula. The LAMMPS code was used for this purpose. The interatomic interactions in ZnO was modelled using the Buckingham potential and included the Coulomb term. In the case of SiNW, the Stillinger-Weber potential was used. A sharp decrease, by a factor of 5 to 8, in thermal conductivity was observed in ZnO NW compared to the bulk structure case, and the smallest κ_l value was observed for the thinnest nanowire. This behaviour was attributed mainly to the scattering of phonons by the nanowires boundaries since the trend observed for κ_l follows the surface to volume ratio of the ZnO NW. It is again observed that the power factor decreases in nanowires compared to bulk ZnO. Itis worthwhile mentioning that whereas the lattice contribution to the enhancement of ZT grows larger with decreasing the size of nanowires, the benefit from the electronic contribution is very minor in the case of ZnO and even disadvantageous in the case of Si. Hence, the authors conclude thatlowering the dimensionality of the materials may not be sufficient to obtain good thermoelectrics.

The doping by gallium atoms of [0001]-oriented zinc oxide nanowires ($Zn_{1-x}Ga_xO$ NWs) have been investigated by Shi et al. [86]. From first-principle electronic structure calculations and BTE theory, the authors studied the thermoelectric property of the material. The cross section of the investigated NWs amounts to 0.7nm. Their results showed that the thermoelectric performance of the $Zn_{1-x}Ga_xO$ NWs is strongly dependent on the Ga contents. As expected, when x increases both σ and κ_e increase whereas α decreases. Therefore, the power factor reaches an optimal value of 908 μW/(mK^2) at x=0.04. The estimate of the $Zn_{1-x}Ga_xO$ NWs lattice thermal conductivity were done through non-equilibrium molecular dynamics simulations using Buckingham empirical potentials. Compared with the case of $Si_{1-x}Ge_x$ NWs for which κ_l is reduced by 60% at x=0.1 with respect to pure SiNWs, only 10% decrease is observed from ZnO NWs to Zn1−xGaxO NWs (with x=0.04). According to the authors, the small reduction is due to the low mass difference between Ga and Zn. As a consequence, compared to pure ZnO NWs, a maximum improvement by a factor 2.5 is achievable in room temperature *ZT* for $Zn_{1-x}Ga_xO$ NWs with x=0.04.

A core-shell structure of Ge/Si nanowires was studied by Chen et al. [87] to estimate the thermoelectric figure of merit. The authors used ab initio DFT electronic structure calculations coupled with the Boltzmann transport theory. Phonon dispersion curve calculations were also performed using the harmonic approximation. The models of nanowires used in the study contained 57 Si and Ge atoms and the nanowires were passivated with hydrogen atoms. The authors showed that the *ZT*value of the core-shell nanowire with *p*-type doping can reach 0.85 at 300 K, which is significantly larger than in pure SiNW for which, the experimental *ZT*value amounts to 0.36. As expected from the results described previously in this review, the authors concluded that the underlying mechanism for this enhanced *ZT*is ascribed to a drastic reduction of lattice thermal conductivity, which can be as small as 0.29 W/(mK), in the Ge/Si core-shell structure. Moreover, the authors suggested that appropriate tuning of the Ge content in the Ge/Si core-shell nanowire may further enhance the nanowire performance.

Vo et al. [88] performed ab initio DFT and BTE calculations to predict the thermoelectric figure of merit for silicon nanowires with diameters up to 3 nm thickness and grown in various directions, namely ([001], [011] and [111]. The authors found that σ, κ_l and α strongly depend on the growth direction. For instance, SiNW with diameter 1 nm and direction [011] bears a highly dispersive conduction band, therefore having electrons with small effective massand σ being high. The opposite is observed for SiNW with diameter 1nm grown in the [011] and [111] directions. By contrast, for SiNW of 3 nm, those grown in the [001] direction are more interesting for their electrical conductivity. Having determined these SiNW characteristics, the authors also predicted κ_l values for ZT being larger than 1. To conform with this value of *ZT*, SiNW with diameter 1 nm should have κ_l of the order of 2 W/(mK), whereas SiNW with diameter 1 nm should have κ_l of the order of 0.2 W/(mK) only, which is a very stringent condition from the synthesis side. Considering the fact that thermoelectric characteristics also depend on the carrier concentration that should be tuned properly, and on the NW surface roughness, optimization of the SiNW for practical purposes may be a very difficult task. The authors suggested that changing the growth direction and NW diameter, or modifying the surface structure by chemical etching or heat treatment, *ZT* value of SiNW could be tuned as will. They also suggested that further optimisation should be possible by alloying Si with other elements, or by functionalizing the surface of NWs.

Transport properties in SiGe nanowires were investigated by Amato et al. [89]. They performed first-principle calculations of the electron transport in the ballistic regime and they showed that for typical Ge concentrations the scattering of carriers is identical whether the NW is an alloyed SiGe NW or the NW corresponds to a Ge-segmented SiNW.

Murphy-Armando et al. [90] calculated from first-principle approach the deformation potentials of nanowires with various diameters and orientations ([100] and [110]), and having hydrogen and hydroxyl surface termination. The added value of this work is that the authors extended the deformation potential theory in order to obtain a model that incorporates the effects of low dimensionality in the electron-phonon scattering rate. From the scattering rates, the electron mobilities can be determined from which, the electronic conductivity enhancement of the NW can be estimated. The authors found that surface termination does not interfere much in the deformation potentials, unlike in the band structure, which is in agreement with previous papers [91,92]. The deformation potentials in the [110] direction are highly anisotropic, therefore those calculated for bulk Si or [100] wires, which are isotropic, cannot be transferred to NW oriented [110]. As a consequence, the scattering of electrons by the SiNW phonon breathing modes is suppressed which leads to a larger carrier mobility compared to the NW oriented [100]. Added to the fact that, in [110] SiNW, electrons bear a low effective mass and that enhanced scattering of the phonons along the direction of the nanowires decreases thermal conductivity, it is expected that a higher figure of merit should be observed for the [110] NW.

The influence of the diameter and passivation on electronic and thermoelectric properties of ZnO NWs bearing hexagonal cross-section have been investigated by means of ab initioelectronic structure calculations and one-dimensional Boltzmann transport equation solution by C. Liu and J. Li [93]. Without saturation of the dangling bonds a buckling effect is observed: the Zn atoms tend to go inward the NW whereas oxygen atoms go outward. This effect is suppressed when the NWs are passivated with hydrogen atoms. Other noticeable results are that, the surface states present in the bare NWs disappear in H-saturated NWs and the band gap opens up. Regarding the TE properties, σ and κ_e increase when the NW diameter grows large, but the effect is counteracted when surface hydrogen are added. Quantum confinement of the electrons and passivation enlarge the effective mass of the charge carriers, therefore increasing α. The largest figure of merit *ZT*achievable in thin NWs is larger than that in wide NWs. However, the size of NWs can not be reduced as will since the band gap widens and makes more difficult the doping of the NW. Here again, a compromise should be found. Moreover, bare NWs are better TE materials than passivated ones. In order to be effective enough to compete with conventionalTE materials, that is to reach *ZT* values higher than unity, ZnO NWs should exhibit lattice thermal conductivity at least 2 orders of magnitude lower than bulk value.

Li et al. [94] determined from first principle DFT calculations the vibrational and thermal properties for silicon nanowires bearing small diameter. Phonon spectrums of the SiNWs were calculated using the density functional perturbation theory. Four low-frequency modes were found with non-zero frequency, namely a dilatational mode, a torsional mode and two flexural modes. The heat-carrying acoustic branches exhibit bending, which is a consequence of a strong interaction between the acoustic phonon modes and the no-zero-frequency flexural ones. This observation has important implication since the phonon group velocity of the acoustic branches is reduced and hence, a lower lattice thermal conductivity of SiNWs should be observed.

TE properties of Si/Ge superlattice nanowires (SLNW) were investigated by DFT calculations combined with BTE theory [95]. Various models were used containing between 57 and 228 atoms, and the nanowires diameter was around 1.7 nm. According to the authors, Si/Ge SLNW is a promising candidate for thermoelectric applications. The TE properties are improved over those of pure Si NW. They predict *ZT* values for SLNWs larger than 1. The remarkable thermoelectric performance in Si/Ge SLNWs originates from the enhanced Seebeck coefficient, and depreciated lattice thermal conductivity due to additional interface scattering. According to the authors, further improvement of the TE performance of Si/Ge

SLNWs could be achieved by decreasing the nanowire diameter and superlattice period, or by tuning the constituent lengths and doping concentrations for a given superlattice period.

Ab initio DFT and tight-binding methods were applied by Bukala et al. for the study of the electronic properties, in particular the DOS of the material, for three kinds of PbTe-CdTe pseudo-binary systems [96]. The corresponding systems are PbTe NWs embedded in CdTe matrix, the CdTe quantum dots (QDs), and CdTe NWs in PbTe matrix. In this study, DFT calculations were used to geometrically relax the structures whereas the tight-binding technique was employed to calculate the DOS. As expected, the diameter and the concentration of the inclusion into the matrix influence the DOS. This effect is observed for instance in the case of CdTe anti-inclusions (A-NWs and A-QDs) in PbTe matrix. More interesting is that quantum confinement of PbTe nanowires leads to 1D sub-bands which changes the derivative of the electronic DOS at the Fermi level and hence, dramatically modifies the Seebeck coefficient. Both kinds of inclusions, namely PbTe clusters in CdTe matrix and CdTe clusters in PbTe, promote this effect. However, the changes are very abrupt and sensitive to the carrier concentration in the case of PbTe NWs. As a consequence, the authors conclude that the anti-structures are much more suitable from the perspective of controlled design.

Zhao et al. [97] performed a detailed investigation, combining experiments and DFT calculations, of the effect of CdS and ZnS as second phases on the thermoelectric properties of *p*-type PbS. They report a *ZT* values as high as 1.3 at 923 K for 2.5 at.% Na-doped *p*-type PbS with endotaxially nanostructured 3.0 at.% CdS. According to the authors such a high *ZT* value is obtainable because of the phonon scattering on multiple length scales, that reduces the lattice thermal conductivity, and the coherent interfaces between the PbS matrix and metal sulphide MS (M= Pb, Cd, Zn, Ca, and Sr) nanophase precipitates that favours the electronic transport by charge carriers. DFT calculations were mainly used to evidence the band alignment between PbS and the second phases in these materials. This alignment can be assessed using van de Walle and Neugebauer's findings [98]. These authors showed that a universal alignment exists for the electronic transition level of hydrogen in a wide range of materials, e.g., semiconductors and insulators. The alignment can be estimated using the following expression:

$$\Delta E^{f}(H^{q}) = E_{tot}(H^{q}) - E_{tot}(\text{bulk}) - 0.5E_{tot}(H_{2}) + q(E_{V} + \Delta E + E_{F}) \tag{14}$$

where E_v and E_F are valence band maximum and Fermi level, respectively, ΔE is an electrostatic potential correction term. $E_{tot}(\text{bulk})$ and $E_{tot}(H^q)$ are the total energy of the bulk system and total energy of the system with H defects with various charge states q. The hole transport is controlled by band offset minimization via valence bands alignment between the host PbS and the embedded second phase metal sulfide. The DFT calculations show that the smallest valence band offset was found between PbS and CdS and amounts to about 0.13 eV. This allows carrier transport between the endotaxially aligned components and minimises potential deterioration of the hole mobility and power factor. Therefore, embedding endotaxially nanostructured second phases into matrices can enhance TE performance by coupling phonon-blocking/electron-transmitting approach.

He et al. [99] investigated a set of nanostructured TE materials made of Sb, Bi, and Pb embedded in PbTe. Both experimental work using high-resolution transmission electron microscopy and theoretical calculations using density-functional perturbation theory (DFPT) was performed to study the relationship between nanostructures and lattice thermal conductivity. Due to the complexity of the PbTe-X (X = 2% Sb, Bi, or Pb) TE material, phonon scattering occursfrom many processes, namely strain, nanoscale precipitates, dislocations, boundaries and phonon-phonon interactions. In order to study the scattering by precipitates, DFPT calculations were performed and phonon spectra were analyzed through the Debye model and Slack [100] and morelli [101] models, according to which, the thermal conductivity depends mainly on the Debye temperature of acoustic phonons. The Debye temperature can hence be written as $\Theta_{a,i}= (h\omega_{a,i}/2\pi)/(k_B)$ for each acoustic phonon branch i. Under the condition that the phonon mode Grüneisen parameter is invariant with respect to a small strain, the expression of the lattice thermal conductivity can besimplified as $\kappa \propto \Theta^3_{a,i}$. The phonon properties of PbTe were calculated under various strain conditions. The results evidence that the presence of dislocations, precipitates, and strain in the TE material play an important role in increasing phonon scattering. As a consequence, the lattice thermal conductivity is strongly reduced. However, according to the authors, inclusion of properly selected precipitates is necessary to obtain this effect, namely precipitates that bring large mass contrast, and significant strain and dislocations.

In this chapter we have recalled the thermoelectric properties and the stringent constraints that a material has to meet to be a good thermoelectric material. Ideally, it has to exhibit large electrical conductivity and Seebeck coefficient and low electronic and lattice thermal conductivities. In general, increasing the electrical conductivity tends to deteriorate the Seebeck coefficient and, by virtue of the Wiedemann-Franz law, increases the thermal conductivity. Therefore, a compromise has to be found. A mean to decrease the thermal conductivity without depreciating the electrical one is to work on the lattice part only, namely to scatter phonons by, e.g., point defects, nano-inclusions or boundaries. Another route was opened by Slack with the "phonon glass electron crystal" concept of materials that contain rattling atoms. Nanostructuration of bulk materials can also significantly reduce the lattice thermal conductivity. Finally, it was shown from theoretical approaches that reducing the dimensionality of materials may drastically increase the Seebeck coefficient through quantum confinement of the charge carriers.

To date, because of technical hitches, only very few experimental investigations have been carried out to study low-dimensional materials, namely nanowires, superlattices and quantum dots, to estimate their thermoelectric performance. To circumvent these difficulties, many computational studies have been carried out and reported in the literature. Nevertheless they deal mainly with Si, Si-Ge and ZnO nanowires. Bi-Te/Sb-Te superlattices were also investigated. The density-functional theory is by far the most used methodology to study electronic properties of materials. We have briefly reviewed the history of the development of DFT from the early 1930s. However, using DFT to predict thermoelectric properties of nanomaterials such as nanowires or quantum dots is a challenging task that is still out of reach. Thus, the majority of the researchers relies on either the Boltzmann transport theory or the Green function methodology to calculate thermoelectric properties.

All the computational results described in the chapter show that the thermoelectric properties are very sensitive to the structure of the materials, e.g., whether it is passivated or not, isotropic or anisotropic in nature, or whether it may undergo surface reconstruction or not. The studies agree to say that, the main factor leading to the enhancement of the figure of merit is the decrease of the lattice thermal conductivities. However, it should be mentioned that, the computational estimation of the lattice thermal conductivity of materials is rarely done, but most often taken from the literature data. In addition, the phonon relaxation time is almost systematically fitted to experimental data or taken from corresponding bulk materials. This introduces large uncertainty in the relaxation time which is a crucial parameter to estimate the lattice thermal conductivity, and by consequence the figure of merit. The same remark holds for the relaxation time of charge carriers that appear in the expression of the electrical conductivity. Models exist for the relaxation time of charge carriers, but usually, experimental data are used or values are fitted from bulk data. Again, the uncertainty on the electrical conductivity may be large as well as for the figure of merit.

Finally, although the low-dimensionality of materials has demonstrated interesting features for improving TE properties of materials, it may not be the panacea. In the case of nanowires, for instance, although the reduction of the diameter lowers the thermal conductivity and increases the charge carrier confinement (and hence, the Seebeck coefficient), we have seen that doping very thin nanowires may be very challenging. In addition, the synthesis of such small nanowires is a difficult task. So far, few nanowires have a figureof merit greater than 1. Another example is that of quantum dots for which, we have seen that the inclusion of precipitates in matrices is not the only parameter that improves TE properties. Rather, the precipitates have to be chosen judiciously to obtain the target property, and this may not be always achievable.

Many parameters enter into the formula that yields the next promising thermoelectric material that remains to be discovered. Hopefully, it is from the mutual exchange and learning between experiments and computational investigations that one will find the long awaited good thermoelectric material.

References

[1] Pottier N., *Physique statistique hors d'équilibre*, EDP Sciences, 2007.

[2] Ioffe A. F.,*Semiconductor thermoelements and thermoelectric cooling*, Infosearch, London, 1957.

[3] WoodC., *Rep. Prog. Phys.*, 51 (1988) 459-539.

[4] Stabler F. R., *Mater. Res. Soc. Symp. Proc.*, 886 (2006) 0886-F01-04.1

[5] RoweD. M., *Thermoelectrics Handbook: Macro to Nano*; CRC/Taylor & Francis: Boca Raton, 2006; and references therein.

[6] SlackG. A., New Materials and Performance Limits for Thermoelectric Cooling. In *CRC Handbook of Thermoelectrics*; Rowe, D. M., Ed.; CRC Press: Boca Raton, 1995; pp 407-440.

[7] KimS. J., HuS. Q., UherC., HoganT., HuangB.Q., CorbettJ.D., KanatzidisM.G., *J. Solid State Chem.*, 153(2) (2000) 321–329.

[8] SchujmanS. B., NolasG.S., YoungR.A., LindC., WilkinsonA.P., SlackG.A., PatschkeR., KanatzidisM.G., UlutagayM., HwuS.J., *J. Appl. Phys.*, 87(3) (2000) 1529–1533.

[9] BeekmanM., NolasG.S., *Phys. B: Condens.Matter*, 383(1) (2006) 111–114.

[10] BeekmanM., NolasG.S., *J. Mater. Chem.*, 18(8) (2008) 842–851.

[11] BentienA., PachecoV., PaschenS., GrinY., SteglichF., *Phys. Rev.* B, 71(16) (2005) 165206.

[12] DongaJ., SankeyO.F., RamachandranG.K., McmillanP.F., *J. Appl. Phys.*, 87 (2000) 7726–7734.

[13] KimJ.-H., OkamotoN.L., KishidaK., TanakaK., InuiH., *Acta Mater.*, 54 (2006) 2057–2062.

[14] NolasG.S., Structure, Thermal Conductivity, and Thermoelectric Properties of Clathrate Compounds. In *Thermoelectrics Handbook: Macro- to Nano-StructuredMaterials*; Rowe, D. M., Ed.; CRC Press: Boca Raton, 2006; pp 34-1-13.

[15] NolasG.S., BeekmanM., GrykoJ., LambertonG., TrittT.M., McMillanP.F., *Appl.Phys. Lett.*, 82(6) (2003) 910–912.

[16] SalesB.C., ChakoumakosB.C., MandrusD., SharpJ.W., *J. Solid State Chem.*, 146(2) (1999) 528–532.

[17] SaramatA., SvenssonG., PalmqvistA.E.C., StieweC., MuellerE., D. Platzek, WilliamsS.G.K., RoweD.M., BryanJ.D., StuckyG.D., *J. Appl. Phys.*, 99 (2006) 023708-5.

[18] TrittT.M., BoettnerH., ChenL., *MRS Bull.*, 33(4) (2008) 366–368.

[19] KanatzidisM.G., *Chem. Mater.* 22 (2012) 648-659.

[20] Narducci D., *Appl. Phys. Lett.* 99 (2011) 102104.

[21] MahanG. D., and SofoJ. O., *Proc. Natl. Acad. Sci.* USA, 93 (1996) 7436-7439.

[22] FanZ., WangH-Q, ZhengJ-C, *J. Appl. Physics*, 109 (2011) 073713.

[23] HicksL.D., DresselhausM.S., *Physical Review* B 47 (1993) 16631–16634.

[24] HicksL.D., DresselhausM.S., *Physical Review* B 47 (1993) 12727–12731.

[25] GuthrieD. K., GaylordT. K., and GlytsisE. N., *Ieee Transactions on Education*,39 (1996) 465-470.

[26] HohenbergP., KohnW., *Phys Rev* B, 136 (1964) 864-871.

[27] ThomasL.H., *Proc. Cambridge Phil. Roy. Soc.*, 23 (1927) 542-548.

[28] FermiE., *Rend. Accad. Naz. Lincei*, 6 (1927) 602-607.

[29] von WeizsackerC.F., *Z. Phys.,* 96 (1935) 431-458.

[30] DiracP.A.M., *Proc. Cambridge Phil. Roy. Soc.,* 26 (1930) 376-385.

[31] ABINIT http://www.abinit.org/. Accessed 10 November 2012.

[32] Vienna Ab Initio Simulation Package (VASP) http://www.vasp.at/. Accessed 10 November 2012.

[33] Quantum Espresso. http://www.quantum-espresso.org/. Accessed 10 November 2012

[34] Parr R. G. and Yang W. in *Density-Functional Theory of Atoms and Molecules*, Oxford University Press, 1989.

[35] LevyM., Proc. *Nat. Acad. Sci.* USA, 76, 1979, 6062-6065.

[36] LevyM., Phys. *Rev.* A, 26 (1982) 1200-1208.

[37] LevyM., Perdew J. P. in *Density Functional Methods in Physics*, R. M. Dreizler and J. da Providencia, Ed., New York, Plenum, 1985.

[38] KohnW., ShamL.*J., Phys Rev* 140 (1965) A1133.

[39] PerdewJ.P., ParrR.G., LevyM., BalduzJ.L., *Phys Rev Lett*, 49 (1982) 1691-1694.

[40] PerdewJ.P., BurkeK., ErnzerhofM. *Phys Rev Lett* 77(18) (1996) 3865-3868.

[41] AnisimovV.I., AryasetiawanF., Lichtenstein A.I. *J Phys: Condens. Matter*9 (1997) 767-808.

[42] ZhouF., CococcioniM., KangK., Ceder G. *Electrochem Commun* 6 (2004) 1144-1148.

[43] CococcioniM., de GironcoliS., *Phys Rev* B, 71 (2005) 035105.

[44] WangL., MaxischT., CederG., *Phys Rev* B, 73(19) (2006) 195107.

[45] WangL., MaxischT., CederG., *Chem Mater*, 19(3) (2007) 543-552.
[46] HeydJ., ScuseriaG.E., ErnzerhofM., *J Chem Phys*, 124(21) (2006) 219906.
[47] HeydJ., ScuseriaG.E., ErnzerhofM., *J Chem Phys*, 118(18) (2003) 8207-8215.
[48] HeydJ., ScuseriaG.E., *J Chem Phys*, 121(3) (2004) 1187-1192.
[49] ChevrierV.L., OngS.P., ArmientoR., ChanM.K.Y., CederG., *Phys Rev*B, 82(7) (2010) 075122.
[50] WangC., PickettW., *Phys Rev Lett*, 51(7) (1983) 597-600.
[51] ShamL., SchlüterM., *Phys Rev Lett*, 51(20) (1983) 1888-1891.
[52] CohenA., Mori-SanchezP., YangW., *Phys Rev* B 77(11) (2008) 115123.
[53] ChanM., CederG., *Phys Rev Lett*, 105(19) (2010) 196403.
[54] HedinL., *Phys Rev*, 139(3A) (1965) A796-A823.
[55] AryasetiawanF., GunnarssonO., *Rep Progr Phys*, 61 (1998) 237-312.
[56] RungeE., GrossE. K. U., *Phys Rev Lett*, 52(12) (1984) 997-1000.
[57] TranF., BlahaP., *Phys Rev Lett*, 102(22) (2009) 226401.
[58] Kuisma M., Ojanen J., Enkovaara J., Rantala T. T., *Phys Rev* B, 82(11) (2010) 115106.
[59] GritsenkoO., van LeeuwenR., van LentheE., BaerendsE., *Phys Rev* A 51(3) (1995) 1944-1954.
[60] MadsenG. K. H., *J Am Chem Soc*, 128(37) (2006) 12140-12146.
[61] EuwemaR.N., StukelD.J., CollinsT.C., deWittJ.S., ShanklandD.G., *Phys. Rev.*, 178 (1969) 1419–1423.
[62] KoellingD.D., WoodJ.H., *J. Comput. Phys.*, 67 (1986) 253–262.
[63] PickettW.E., KrakauerH., AllenP.B., *Phys. Rev.* B, 38 (1988) 2721–2726.
[64] AllenP.B., SchulzW.W., *Phys. Rev.* B, 47 (1993) 14434–14439.
[65] AllenP.B., PickettW.E., KrakauerH., *Phys. Rev.* B, 37 (1988) 7482–7490.
[66] SinghD.J., Mazin I.I. *Phys. Rev.* B, 56 (1997) 1650–1653.
[67] MadsenG.K.H., SchwarzK., BlahaP., SinghD.J., *Phys. Rev.* B, 68 (2003) 125212.
[68] BentienA., PachecoV., PaschenS., GrinY., F. Steglich, *Phys. Rev.* B, 71 (2005) 165206.
[69] PachecoV., BentienA., Carrillo-CabreraW., PaschenS., SteglichF., Grin Y. *Phys.Rev.* B, 71 (2005) 165205.
[70] Garcıa-FernandezP., Verissimo-AlvesM., BilcD.I., GhosezP., and JunqueraJ., *Physical Review* B, 86 (2012) 085305.
[71] HeinzP.D., *Master of Science*, Texas State University-San Marcos (2010).
[72] HinscheN. F., YavorskyB. Y., GradhandM., CzernerM., WinklerM., KönigJ., BöttnerH., Mertig I. and ZahnP., *Phys. Rev.* B, 86 (2012) 085323.
[73] VenkatasubramanianR., Siivola E. and ColpittsT., O'QuinnB., *Nature*, 413 (2001) 597-602.
[74] Hinsche N. F., Mertig I. and Zahn P., *Journal of Electronic Materials*, in press DOI: 10.1007/s11664-012-2279-z.
[75] Hinsche N. F., Mertig I. and Zahn P., *J. Phys. Condens. Matter* 24 (2012) 275501.
[76] Vo T., Williamson A. J. and Galli G., *Physical Review* B 74 (2006) 045116.
[77] Shi L., Yao D., Zhang G., Li B., Appl., *Phys. Lett.*, 95 (2009) 063102.
[78] Shi L., Yao D., Zhang G., Li B., Appl., *Phys. Lett.*, 96 (2010) 173108.
[79] Jauho A. P., Engelund M., Markussen T. and Brandbyge M., Progress in nonequilibrium Green's functions IV, book series: *Journal of Physics ConferenceSeries*, 220, 012010, 2009.
[80] Blase X. and Fernandez-Serra M.-V., *Phys. Rev. Lett.*, 100 (2009) 046802.

[81] Shelley M. and Mostofi A. A., *Eur. Phys. Lett.*, 94 (2011) 67001.
[82] Chen X., Wang Y., Ma Y., Cui T., Zou G., *J. Phys. Chem.* C, 113 (2009) 14001-14005.
[83] Parlinski, K. PHONON Software; 2006.
[84] Parlinski, K.; Li, Z. Q.; Kawazoe, Y. *Phys. Rev. Lett.*1997, 78, 4063-4066.
[85] Demchenko D. O, Heinz P. D and Lee B., *Nanoscale Research Letters*, 6 (2011) 502.
[86] Shi L., Chen J., Zhang G., Li B., *Physics Letters* A 376 (2012) 978–981.
[87] Chen X., Wang Y., and Ma Y., *J. Phys. Chem.* C,114 (2010) 9096–9100.
[88] Vo T.T.M., Williamson A.J., Lordi V., Galli G., *Nanoletters*, 8-4 (2008) 1111-1114.
[89] Amato M., Ossicini S., R. Rurali, *Nano Lett.* 12 (2012) 2717–2721.
[90] Murphy-Armando F., Fagas G., and Greer J. C., *Nano Lett.*, 10 (2010) 869-873.
[91] Leu P.W., Shan B., Cho K., *Phys. Rev.* B, 73 (2006) 195320.
[92] Nolan M., O'Callaghan S., Fagas G., Greer J.C., Frauenheim T., *Nano Lett.* 7 (2007) 34-38.
[93] Liu C., Li J., *Physics Letters* A 375 (2011) 2878–2881.
[94] Li J. C., Wang C. L., Peng H., Wang M. X., Zhang R. Z., Wang H. C., Liu J., Zhao M. L. and Mei L. M., *Journal of applied Physics*, 108 (2010) 063702.
[95] Chen X., Wang Z., and Ma Y., *J. Phys. Chem.* C, 115 (2011) 20696–20702.
[96] Bukała M., Sankowski P., Buczko R., Kacman P., *Nanoscale Research Letters*, 6 (2011) 126.
[97] Zhao, J. L.-D.; He, S. Hao, C.-I. Wu, T. P. Hogan, C. Wolverton, V. P. Dravid, and M. G. Kanatzidis, *J. Am. Chem. Soc.*, 134 (2012) 16327−16336.
[98] van de Walle C.G., Neugebauer J., *Nature* 423 (2003) 626-628.
[99] He J., Sootsman J.R., Giradr S.N., Zheng J-C, Wen J., Zhu Y., Kanatzidis M. G. and Dravid V. P., *J. Am. Chem. Soc.* 132 (2010) 8669-8675.
[100] Slack G.A., *Solid State Phys.*, 34 (1979) 1-71.
[101] Morelli D.T., Heremans J.P., *Appl. Phys. Lett.*, 81 (2002) 5126.

In: Density Functional Theory
Editors: J. Morin and J.M. Pelletier, pp. 121-145
ISBN: 978-1-62417-954-9

Chapter 4

On the Use of DFT Computations to the Radical Scavenging Activity Studies of Natural Phenolic Compounds

Nikolaos Nenadis* and Maria Z. Tsimidou
Laboratory of Food Chemistry and Technology, School of Chemistry,
Aristotle University of Thessaloniki, Thessaloniki, Greece

Abstract

Theoretical methods have been recognized nowadays as a useful approach in antioxidant activity studies of phenolic compounds (AH). Toward this direction Density Functional Theory (DFT) calculations employing hybrid functionals such as the Becke's 3 and Lee Yang Parr (B3LYP) are most frequently applied solely or in combination with those obtained using less demanding semi-empirical methods. In this chapter, applications of the aforementioned computational methods in the field of food chemistry and life sciences regarding structure–antioxidant activity relationships of various classes of phenolic compounds (simple phenols, phenolic acids and related compounds, coumarins, naphthoquinones, xanthones, stilbenes, chalcones and flavonoids) are presented. Focus is mainly given on the computation of suitable molecular descriptors to characterize the radical scavenging activity of AH via hydrogen atom transfer, electron transfer or by sequential proton loss followed by electron transfer.

The computation of spin distribution on radicals formed by AH, when oxidized during the process, as a useful index of antioxidant activity is also acknowledged; whereas the contribution of other than phenolic hydrogens such as allylic ones to the radical scavenging efficiency is documented.

Except for calculations in the gas phase, simulation of real systems is approximated by the utilization of models that take into account implicitly and/or explicitly solvent effect. The potential of the above approaches, shortcomings and some key points for further studies are commented.

* E-mail address: niknen@chem.auth.gr

1. Introduction

Natural phenolic compounds (AH) are a diverse group of plant secondary metabolites bearing at least one aromatic ring with minimum one hydroxyl group [1].

These compounds (Figure 1) arise from various biosynthetic pathways and are widespread in the plant kingdom due to their multiple roles in the plants as e.g. the protection from herbivorous animals, insects, fungi and the oxidative damage caused by environmental ultraviolet radiation, the contribution to fruit coloration and signal transduction etc [2].

In plant derived foods (fruits, vegetables, cereals, wine, herbal teas etc) various AH occur from traces up to 8000 mg/kg [3, 4]. Consequently, AH are parts of the daily diet, with an estimated daily intake of few mg up to more than 1 g depending on the dietary habits [2, 3].

Over the last decades, the AH have attracted the interest of many researchers active in the field of Food Chemistry and Life Sciences. This was mainly due to their ability to retard oxidation of lipids and other valuable substrates (e.g. nucleic acids, proteins), when present at low concentration [5]. Oxidation of the aforementioned substrates, mediated mainly with the formation of oxygenated free radicals (commonly known as reactive oxygen species, ROS), has been associated with food deterioration and loss of nutritional value; whereas *in vivo*, with the formation of cardiovascular and other degenerative diseases [5, 6].

Thus, an upsurge was observed considering the amount of published work towards the examination of structure-antioxidant activity relationship (SAR) of phenols [e.g. 7]. The ultimate goal set was to recognize those structural criteria that enhance antioxidant performance, obtain valuable information to counteract oxidation or even a rationale to design novel antioxidants.

The AH can act via different mechanisms including inhibition of oxidative enzymes and metal chelation. Emphasis among researchers has been given on their ability to scavenge free radicals that is widely accepted as the principal mechanism of action [7]. Radical scavenging can be described by three different ways, namely via hydrogen atom transfer, HAT [R.1] towards free radicals (e.g. peroxyl, $ROO^\bullet$), single electron transfer followed by proton transfer, SET-PT [R.2] or by sequential proton loss followed by electron transfer, SPLET [R.3]:

$$AH + ROO^\bullet \rightarrow ROOH + A^\bullet \text{ [R.1]}$$

$$AH + ROO^\bullet \rightarrow [ROO^- \text{---} AH^{\bullet+}] \rightarrow ROOH + A^\bullet \text{ [R.2]}$$

$$AH + ROO^\bullet \rightarrow A^- + ROO^\bullet + H^+ \rightarrow ROO^- + A^\bullet + H^+ \rightarrow ROOH + A^\bullet \text{[R.3]}$$

The AH may react via all these three pathways concomitantly; nevertheless, SET-PT and SPLET are more favored in polar media which can facilitate charge separation [8, 9]. Regardless, of the pathway followed, the phenoxyradical ($A^\bullet$) or the radical cation ($AH^{\bullet+}$) are expected to be more stable than the original oxidizing species and less prone to promote oxidation.

The radical scavenging efficiency of phenolic compounds can be assessed using more than one *in vitro* methods which have been developed over the years [10 and ref therein].

Among the most popular ones frequently employed to elucidate SAR of AH are those based on monitoring the reduction of the level of synthetic radicals in solution. Assays using

	Class	Skeleton	Basic structure
I	Simple phenols	C_6	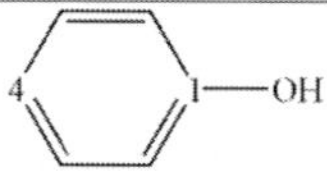
II	Benzoic acids	C_6-C_1	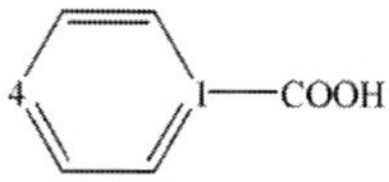
III	Phenylacetic acids	C_6-C_2	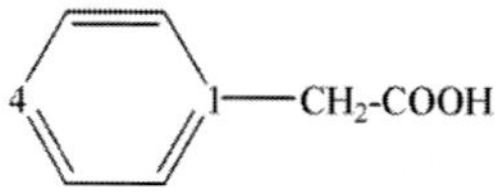
IV	Hydroxycinnamic acids	C_6-C_3	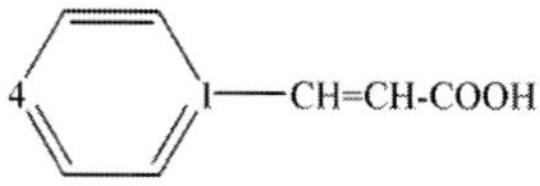
V	Coumarins	C_6-C_3	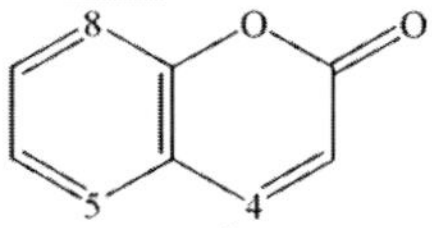
VI	Napthoquinones	C_6-C_4	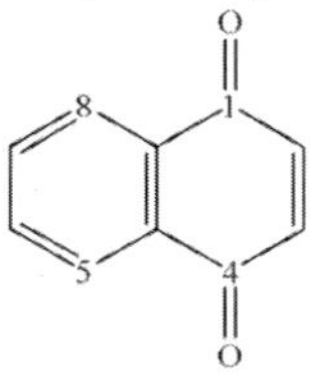
VII	Xanthones	C_6-C_1-C_6	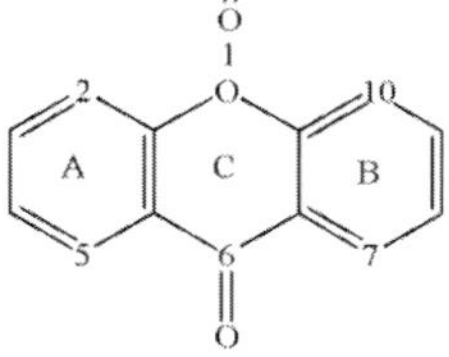
VIII	Stilbenes	C_6-C_2-C_6	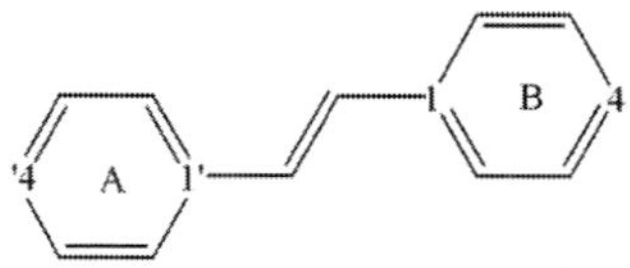
IX	Chalcones	C_6-C_3-C_6	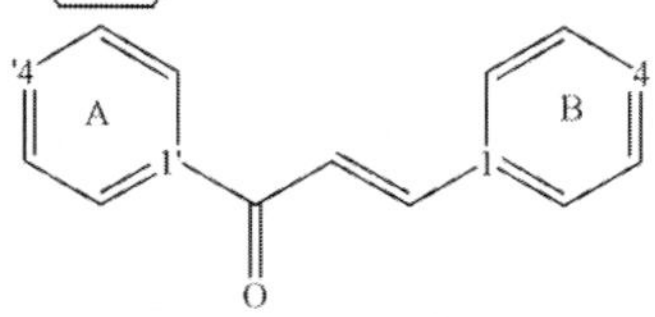
X	Flavonoids	C_6-C_3-C_6	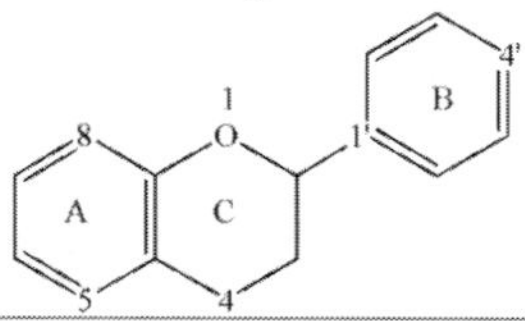

Figure 1. Some of the major classes of natural phenolic compounds (AH).

the 1,1-diphenyl-2-picrylhydrazyl, DPPH$^{\bullet}$ [11] and the 2,2-azinobis(3-ethylbenzothiazoline-6-sulfonic acid) diammonium salt radical cation, ABTS$^{\bullet+}$ [12] appear in hundreds of publications so far. The simple and rapid analytical protocols are based on decoloration of these radicals upon reaction with an AH and are applicable to both hydrophilic/lipophilic AH. Other assays employing scavenging of peroxyl (ROO$^{\bullet}$) radicals that play a key role in lipid oxidation are also common.

Among them the Crocin-Bleaching assay, CBA, [13] has been used in many SAR studies [e.g. 14, 15]. Peroxyl radicals are generated by the decomposition of a thermo-labile azo initiator whereas evaluation of activity is based on recording the decay of a coloured probe. The necessity to obtain meaningful data for the activity of the test AH in real systems led also to the utilization of numerous lipid models as substrates.

Bulk oils subjected to thermal stress at various temperatures are the simplest and most common ones [10]. Monitoring of oxidation is achieved by measuring primary and secondary oxidation products with various analytical techniques. Such tests can be time consuming depending on substrate unsaturation, storage conditions and antioxidant potency.

2. Quantum Chemical Methods

Thanks to the advances in computer science, quantum chemical calculations can be carried out at a reasonable length of time. Their utilization has been acknowledged as a green tool that can predict the activity of target compounds, and complement experimental findings providing insight in the chemistry involved.

The theoretical approach usually starts with quantum chemical calculations of the energetics of the parent compound and the respective oxidation products (phenoxy and/or cation radicals) [10]. In some cases, calculations are carried out also for the oxidizing species (free radicals involved) used experimentally [e.g. 16, 17]. A characteristic example is the calculations of the energetics of $DPPH^{\bullet}$ and its reduced forms after accepting a hydrogen atom (DPPH-H) or an electron ($DPPH^{-}$) in order to comment on AH activity and the preferred way of scavenging [e.g. 16].

2.1. Molecular Descriptors

According to HAT mechanism, the reactivity of an AH can be estimated by calculating the O-H bond dissociation enthalpy (BDE) values. SET-PT is approximated by the adiabatic ionization potential (IP) and O-H proton dissociation enthalpy (PDE) values, with the former one being more important since electron transfer is expected to be the rate-determining step.

Table 1. Common molecular descriptors related to the radical scavenging of phenolic compounds and the formulas for their calculation

Molecular descriptor	Formula
Bond dissociation enthalpy (BDE)	$H_r + H_h - H_p$
Adiabatic ionization potential (IP)	$E_{cr} - E_p$
Proton dissociation enthalpy (PDE)	$H_r + H_{pr} - H_{cr}$
Proton affinity (PA)	$H_a + H_{pr} - H_p$
Electron transfer energy (ETE)	$E_r - E_a$

H_r: enthalpy of phenoxy radical, H_h: enthalpy of the H-atom, H_p: enthalpy of the parent molecule, E_{cr}: energy of the cation radical, E_p: energy of the parent molecule, H_{pr}: enthalpy of the proton, H_{cr}: enthalpy of the cation radical, H_a: enthalpy of the anion, E_r: energy of the phenoxy radical, E_a: energy of the anion.

Regarding SPLET, the ability to form phenoxy anions can be predicted via calculation of proton affinity (PA), whereas electron transfer energy (ETE) can characterize the electron donating efficiency of the anion [18]. The formulas utilized for the calculation of these molecular descriptors are presented in Table 1.

Useful information regarding the stability of the phenoxy radical formed can be extracted via the computation of the spin distribution of the unpaired electron [19].

Nevertheless, to address more accurately the reactivity towards a specific radical, the BDE value of the reduced form of the radical after accepting a hydrogen atom (hydrogen atom affinity) or the IP value of its anion after accepting a single electron (electron affinity) by an AH should be considered [16, 17].

2.2. Methodology

So far in literature the values reported for phenols have been computed with a diversity of methods [e.g. 16, 18-20]. The approaches employed can be divided into three categories: (a) semiempirical methods (e.g. Austin Model 1, AM1), (b) advanced methods (e.g. *ab initio* and density functional theory, DFT) and (c) their combination (e.g. DFT/AM1). Selection among the three approaches is based on a variety of factors, such as theoretical background of researchers, software availability and computer resources. A critical factor is the computational time. The length of the latter depends on various factors as i.e. the approach selected, the size of the test compound, as well as the size of the basis set that is the group of functions used to describe the molecular orbitals. All these eventually affect the number of integrals that have to be solved and consequently have an impact on the duration of the calculation. For example, the effect of the size of the basis set on the CPU time required for the calculation is reported to scale N^4 times, meaning that utilization of a 2-fold larger set of functions will make the duration of the calculation to increase by 16-fold [21].

To consider a method as suitable for calculations the relative difference between computed values for bonds and angles of some of the test compounds and experimental ones with X-ray analysis (where available) should be less than 2-3% [19, 22]. Other approach to consider the method efficiency is to compute a gas-phase BDE (BDE_g) value for the simple phenol comparable to the experimental one [e.g. 23]. The latter value has been reported in the range 85.1-90.2 kcal/mol [24, 25 and ref therein] with a consensus of a mean value of 88.7 kcal/mol[26]. In cases when a statistically high correlation between the values of a computed molecular descriptor and experimental data on activity for a group of AH is obtained, the method is also considered as proper [e.g. 22].

Density functional theory (DFT) calculations in combination or not with a semi-empirical structure optimization step, dominate antioxidant activity studies [26]. Contrary to semi-empirical methods, which take into consideration only the valence electrons, DFT ones model all electrons in a molecule. The energy of the test compound can be determined from the electron density instead of a wavefunction. In this way a substantial decrease of computation time in comparison to i.e. HartreeFock calculations is succeeded and with comparable accuracy to that of more demanding methods (e.g., Møller-Plesset, MP2) [21]. This is related to the improvements made to model the Coulombic interactions between the electrons of a system, achieved with the use of various approximations such as those included in hybrid functionals. Among those functionals available, the Becke 3 and Lee Yang Parr, B3LYP is the most frequently employed one in DFT calculations of antioxidant activity due to e.g the accuracy of results on geometry optimisation, as well as the prediction of bond energy such as the -O–H of phenols [26]. Less frequently employed is the Becke 3Perdew 86 denoted as B3P86 [17, 27, 28]. Lately few studies on AH radical scavenging activity have been published using the recently developed functional M05-2X, claiming to perform better than B3LYP [29-31].

The methodology adopted in most of the studies using DFT/B3LYP is to calculate all relevant indices of radical scavenging activity in the gas-phase which is considered to provide primary information on the activity of an AH in short time [26].

As reactions take place usually in solution, solute-solvent interactions can also be considered. In the available approaches, the solvent may be introduced implicitly and/or explicitly. Implicit solvent models, instead of using individual solvent molecules use a continuum with a dielectric constant value equal to that of the solvent; whereas, the studied compound is located in a cavity, which in its simplest form is consisted of spheres around each atom [21].

The dielectric field can cause polarization to the compound which in turns polarizes the field (polarized continuum model). The outcome is described by a charge distribution on the surface of the cavity. The solvation energy (ΔG_{sol}), is usually the sum of the electrostatic component (ΔG_{el}), the dispersion-repulsion interaction (ΔG_{dr}) and the cavity formation (ΔG_{cav}). A wide variety of models available that differ in the approximations are detailed by Tomasi and co-workers [32].

These models usually provide a consistent description of solvation at the expense of a substantial increase in computational time when compared to gas-phase calculations. Explicit solvent effects are considered important in cases where formation of intermolecular hydrogen bonds with phenolics (e.g. water) is favoured, having, thus, an impact on hydrogen atom donation [33]. The approximations include the addition in the input file of individual molecules of solvent (e.g. water) so as to form hydrogen bonds with the active sites of the AH in the gas-phase [e.g. 34] or concomitantly with the use of a continuum model to include implicit solvent effects as well [34, 35]. The later approximation increases dramatically the cost of calculation and consequently it is not preferred in the literature.

3. DFT Examination of Structure-Radical Scavenging Activity Relationship of Natural Phenolic Compounds

In the following paragraphs the applicability of DFT calculations to radical scavenging activity is described for various classes of natural phenolic compounds (Figure 1, I-X). Pure DFT calculations or in combination with a semi-empirical or Hartree-Fock geometry optimization step in the gas-phase or even after considering solvation, and basis sets ranging from 6-31G up to 6-311G++(3df, 3pd) have been used. In the paragraphs 3.1-3.8, the different effects (aromatic ring substitution, planarity etc) that can be examined using theoretical calculations are presented for selected case studies.

3.1. Simple Phenols

Simple phenols (Figure 1, I) usually constitute part of the structure of other classes of AH. In this sense these compounds are often used as a reference in SAR studies for other classes of natural phenols [22, 36, 37]. The role of substitution (hydroxyl, methoxy) to the efficiency of hydrogen atom or electron donation has been extensively examined. Thus, pyrogallol (1,2,3-trihydroxy benzene), and catechol (1,2-dihydroxy benzene), are predicted to be the most efficient hydrogen atom donors among the simple phenols, in line with experimental findings [11, 38]. Their efficiency is related to the donation of more than one hydrogen atoms from the existing hydroxyl groups, as well as the further stabilization of the derived phenoxy radical via formation of intramolecular hydrogen bonds (IHB). For example, the active site of pyrogallol, is the –OH group at C-2. Each of the two IHBs formed with the –OH groups at C-1 and C-3 lower the BDE_g value by 7-8 kcal/mol respectively. Consequently, the difference in BDE_g values (ΔBDE_g) between pyrogallol and phenol is almost -15 kcal/mol [22],though such a difference depends on the methodology employed [39]. The importance of the presence of hydroxyl groups with electron donating properties and the formation of IHB is better illustrated when comparisons are made among catechol (1,2-dihydroxy benzene), hydroquinone (1,4-dihydroxy benzene), resorcinol (1,3-dihydroxybenzene) and phloroglucinol (1,3,5-trihydroxy benzene). In the first two compounds the presence of a second –OH group with electron donating properties at C-2 or C-4 results in a significant HAT activity (ΔBDE_g = -9.4 and -5.2 kcal/mol) when compared to phenol. On the contrary, the presence of –OH groups with electron-withdrawing effect (C-3 and/or C-5) does not improve activity since BDE_g values for resorcinol and phloroglucinol were comparable to that of phenol [40]. This observation was in accordance to experimental ones using the $DPPH^•$ for these four AH [36].

Theoretical outcomes regarding the contribution of a methoxy group at an *o*-position to the phenolic –OH group (guaiacol) are influenced by various factors. Thus, according to gas-phase calculations using different basis sets -6-31+G(,p) up to 6-311+G(2d,2p)- it was suggested that its presence may not have an impact in activity granted that the ΔBDE_g values between guaiacol and phenol were in the range -1.0 to + 1.6 kcal/mol [23]. In polar media (methanol, water) using a PCM model [22] it was shown that introduction of a methoxy group at C-2 enhanced HAT efficiency as the BDE value in the liquid phase (BDE_l) was 4.7 kcal/mol lower than that of phenol. The findings were in line with $DPPH^•$ experimental data for these two compounds [17]. With regards to the electron donating efficiency, there is no clear trend as that previously reported for HAT. Pyrogallol and catechol are predicted to be of equal potency with a difference in IP values (ΔIP), both in gas and in the liquid phase, of -6.4 to -9.8 kcal/mol than that of phenol [22]. The corresponding compounds are expected to be better electron donors than resorcinol (ΔIP= -2.2 to -3.1 kcal/mol). Comparisons made with guaiacol showed that the respective monophenol could donate an electron more easily by 4.2 and ~6 kcal/mol respectively, according to IP_g values [22]. Yet, as pointed out by the same authors this is not expected in polar solvents such as methanol and water. The three compounds, using a PCM model, were predicted to be equipotent electron donors; the PDE value, characterizing the deprotonation efficiency of the cation radical was lower in liquid phase (PDE_l) than that of phenol (by ~2 kcal/mol) only for pyrogallol. So, the latter phenol was found to be more efficient electron donor when SET-PT pathway is favoured in a polar solution.

3.2. Phenolic Acids and Related Compounds

Phenolic acids (Figure 1, II-IV) are characterized by the presence of a carboxyl group either attached to an aromatic ring (benzoic acids) or at the end of a short side chain (phenylacetic acids or hydroxycinnamic acids). The interest of researchers in these classes of compounds is mainly due to their antioxidant potential but also because they are associated with the biosynthesis of flavonoids and their *in vivo* degradation during digestion of plant food [41-43]. Phenolic acids have been investigated in order to elucidate the SAR both experimentally and theoretically. Nevertheless, theoretical studies are rather limited to certain compounds.

BenzoicAcids

In accordance to findings presented for simple phenols, the HAT activity of benzoic acids (Figure 1, II) is governed by the hydroxylation pattern in the aromatic ring. Thus, gallic (3,4,5-trihydroxy) acid, followed by protocatechuic (3,4-dihydroxy) are predicted to be the most effective ones as reflected by their lower BDE_g and BDE_l values (by ~5-9 kcal/mol). Vanillic (4-hydroxy, 3-methoxy) acid bearing a guaiacol moiety should be more efficient than *p*-hydroxybenzoic acid both in gas and liquid phase (ΔBDE = -0.7 to -6 kcal/mol) [36, 44, 45].

An additional structural characteristic studied in this class is the role of a syringol moiety (4-hydroxy, 3,5-dimethoxy) present in the acid (syringic). Experimental findings with various *in vitro* radical scavenging activity assays [15, 46] suggest that the latter is more efficient than vanillic acid, and in certain cases more potent even than protocatechuic acid. Theoretical calculations clearly support the superior activity of syringic acid over that of vanillic due to easier donation of a hydrogen atom both in gas and liquid phase (ΔBDE = -2.5 to -4.3 kcal/mol) [44, 45, 47]. A higher activity for syringic acid over that of protocatechuic acid could not be supported in the gas-phase (ΔBDE_g = 1.4 to 5 kcal/mol) [44, 45]. When solvation was taken into account, in methanol and water (regardless of the ionization of the –COOH), syringic acid was predicted to be more potent (ΔBDE_l = -0.9 to -2.5 kcal/mol) [47]. Within the frame of SET, the syringol moiety enhances electron donation in benzoic acids as pointed out by comparison to protocatechuic acid in the gas (ΔIP_g~ -12 kcal/mol) and liquid (ΔIP_l -3 to -4.5 kcal/mol) phase [47].

Further studies within the class of benzoic acids have been carried for various dihydroxy derivatives to examine the effect of –OH location on the radical scavenging activity.

Theoretical data are limited to those presented by Nenadis et al. [36] regarding the 2,3-, 2,4-, 2,5- and 3,4- dihydroxy counterparts. Computations in ethanol to mimic the environment of radical scavenging assays also used by these authors ($DPPH^{\bullet}$, $ABTS^{\bullet+}$) clearly demonstrated that the one (2,4-) bearing the –OH groups in *meta* position was the least effective hydrogen and electron donor.

The respective compound presented in ethanol higher BDE_l (by 10-11 kcal/mol) and IP_l (by ~5-10 kcal/mol) values than the other dihydroxy benzoic acids tested. The most efficient hydrogen atom donor was predicted to be the 3,4-dihydroxy counterpart (ΔBDE_l = -1.5 and - 2.4 kcal/mol) than the 2,5- and 2,3- dihydroxy benzoic acids.

The lower efficiency predicted for the latter acid, despite the presence of a catechol moiety, is due to the formation of an IHB with the carboxylic group. In case of electron donation, however, the 2,5- configuration is proposed to be the most beneficial than that of the 2,3- and 3,4- dihydroxy compounds (ΔIP_l = -4.5 and -5.1 kcal/mol). Comparison with experimental findings ($DPPH^{\bullet}$, $ABTS^{\bullet+}$) indicated an agreement with the trend based on IP_l values rather than that on the BDE_l ones [36].

PhenylaceticAcids

The only available theoretical data reported on the radical scavenging activity of these compounds (Figure 1, III) are those presented by Ordoudi et al. [37] regarding the HAT efficiency of homoprotocatechuic (3,4-dihydroxy) and homovanillic (4-hydroxy, 3-methoxy) acids. The BDE_g values verified the importance of the catechol moiety (ΔBDE_g= -7.9 kcal/mol) in accordance with $DPPH^{\bullet}$ data presented by the authors.

HydroxycinnamicAcids

This group of phenolic acids (Figure 1, IV) has been extensively studied experimentally and theoretically [7]. For HAT efficiency, it seems that similar SAR are expected to those reported in benzoic acids according to the BDE_g calculated values for common members [19]. Similarly to syringic acid, there was an interest to examine theoretically the role of a second methoxy group attached to the aromatic ring by comparing the activity of sinapic acid (4-hydroxy, 3,5-dimethoxy) and the 3,4- dihydroxy derivative, known as caffeic acid. Data in gas-phase and polar media [19, 36, 45, 48] were consistent with those previously reported for the syringic-protocatechuic acids pair. Based on these calculations it seems that hydrogen atom donation of catechol moiety is affected by the polarity of the solvent, whereas this is not the case for compounds bearing a syringol moiety. Experimental data on radical scavenging activity are rather controversial [19, 36, 49] suggesting that the conditions as well as the oxidizing radical may affect the results. Thus, theoretical approaches [e.g. 50] taking into account in the calculations the radical species scavenged experimentally, may be proved more informative.

An important structural aspect within the class of hydroxycinnamic acids studied was the extension of conjugation in the propenoic side chain. Emphasis has been given on the activity of caffeic-dihydrocaffeic acid counterparts. The latter is expected to be an active *in vivo* form [51]. Absence of a double bond in the side chain affects molecular planarity since the saturated side chain is almost perpendicular to the plane with a dihedral angle of 84 ^{o}C according to gas-phase computations [52]. In addition to this, the derived phenoxy radical is less stabilized and the spin density of the unpaired electron is mainly located on the oxygen atom of the radical (0.34); whereas in caffeic acid the spin density on the same atom is lower (0.29) due to the extended delocalization [19, 52]. Such a structural change does not seem to affect BDE_g values as differences observed were less than 1 kcal/mol [37, 52]. This finding was related to the presence of the catechol moiety that could mask the role of other structural characteristics. Nenadis et al. [53] using a series of radical scavenging assays (e.g. $DPPH^{\bullet}$, $ABTS^{\bullet+}$, bulk oil oxidation) showed that dihydrocaffeic acid was more efficient than caffeic one. Therefore, the BDE_g values for the most active –OH group did not support these experimental data [52]. Calculations with the aid of a PCM model showed that in ethanol dihydrocaffeic acid was more efficient hydrogen atom (ΔBDE_l= -1.3 kcal/mol) and electron (ΔIP_l= -5 kcal/mol) donor than caffeic acid [36]. These predictions may justify an increased activity in solution but not in a non-polar medium such as an oil substrate where HAT dominates [54]. A re-examination of the HAT activity of the two compounds suggested that the observed differences could be due to the ease of quinone formation by dihydrocaffeic acid and the contribution of allylic hydrogen atoms from its propanoic side chain (Nenadis and Tsimidou un-published data). This is better illustrated in Figure 2.

Recently, Leon-Carmona et al. [31] following a different approach (DFT/ M05-2X) and using a recently developed continuum model [55] commented that caffeic acid should react faster with free radicals than dihydrocaffeic at pH 7.4. According to the information presented, the increased activity was proposed to stem from the contribution of phenoxide anions. The latter were estimated to account for ~7% of the concentration of caffeic at a physiological pH, while in the case of dihydrocaffeic acid not more than 1%.

The Effect of the Carboxylic Group

To comment on the effect of the carboxyl group when not directly attached to the aromatic ring, comparisons of the radical scavenging activity have been carried out between benzoic acids and members from the other two groups of phenolic acids. Mohajeri et al. [45] examined a series of hydroxycinnamic acids and their corresponding hydroxybenzoic ones, whereas Ordoudi et al. [37] presented information on catechol and guaiacol derivatives from all the three sub-classes of phenolic acids.

In the latter study, catechol and guaiacol were also considered as references. Gas-phase calculations by Mohajeri et al. [45] showed that the hydroxycinnamic acids were predicted to be more efficient hydrogen atom (ΔBDE_g = -2.5 to -3.9 kcal/mol) and electron (ΔIP_g= -7 to -15.4 kcal/mol) donors than the corresponding benzoic ones. Therefore, the electron withdrawing effect of the carboxylic group as denoted by the positive (0.42) Hammet-type value (σp) [56] is attenuated by the propenoic side chain. These observations were in accordance to information obtained from various *in vitro* radical scavenging activity assays [7]. The higher HAT efficiency of hydroxycinnamic acids may, partially at least, attributed to the better stabilization of the derived phenoxy radical due to extended delocalization of the unpaired electron.

This was supported by the gas-phase spin density values computed at the oxygen of the respective phenoxy radicals. Those values were in the range 0.30-0.36 for hydroxycinnamic acids, whereas for the hydroxybenzoic ones ranged from 0.33 to 0.43 [45]. The negative impact of the direct introduction of a carboxylic group (benzoic acids) was verified by the theoretical data of Ordoudi et al. [37]. Comparison with simple phenols (guaiacol, catechol) indicated a rising in the BDE_g values by ~2-2.7 kcal/mol. Increasing the distance from the aromatic ring with a $-CH_2-$ group (phenylacetic acids) led to BDE_g values similar to that of the simple phenols.

Those computed for the corresponding hydroxycinnamic acids were even lower (by ~ 1 kcal/mol) probably due to the extended conjugation. Though these data are reasonable in terms of electronic phenomena, the higher radical scavenging activity observed experimentally for the 3,4- dihydroxyphenylacetic acid over that of caffeic with the $DPPH^{\bullet}$ assay could not be justified [37].

The enhanced activity of the phenylacetic acid could partially at least attributed to a similar mechanism previously mentioned for dihydrocaffeic-caffeic acids pair. Thus, the non-planar homoprotocatechuic acid can form a phenoxy radical at the expense of 76.4 kcal/mol and a quinone at 73.9 kcal/mol in the gas-phase, whereas the BDE_g value for allylic hydrogen atom abstraction is predicted to be 77.3 kcal/mol (Nenadis and Tsimidou unpublished data, see Figure 2 for comparison).

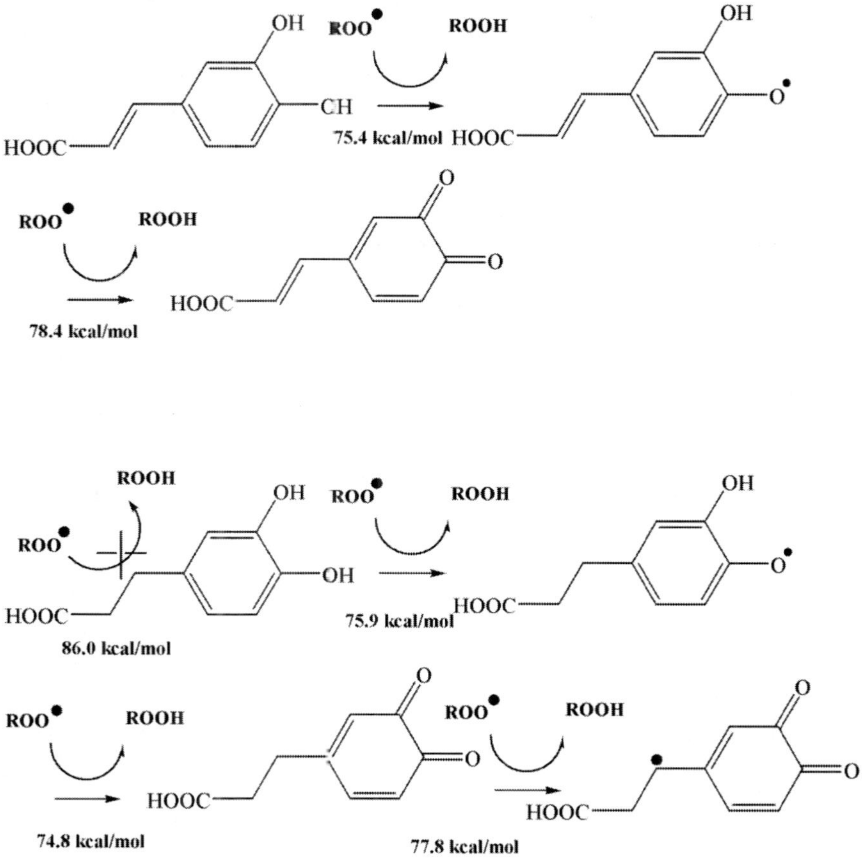

Figure 2. Proposed mechanism of radical (ROO•) scavenging activity of caffeic-dihydrocaffeic acids. BDE values were computed at the B3LYP/6-311++G(2d,2p)//B3LYP/6-31 G level of theory in the gas phase (Nenadis and Tsimidou un-published data).

The carboxylic moiety is prone to ionization, thus, at a physiological pH the acids are expected to bear a carboxylate anion. This group presents electron donating properties (σp = 0.0) in comparison to the –COOH group (σp = 0.42) [56] and is expected to promote both HAT and SET. This is characteristically demonstrated by the computations carried out in water for a series of hydroxybenzoicacids [47].

Comparisons made for parent compounds and the corresponding ions in water showed that the latter presented lower BDE_l values by 0.8-2.7 kcal/mol and IP_l ones by ~6.8-8.0 kcal/mol. The importance of carboxylic group deprotonation in radical scavenging activity was also commented in the past by Kong et al. [57].

The Effect of the Carboxylic Group Esterification

Except for being ionized, the carboxylic group may be also esterified. Characteristic examples are the chrologenic acid (ester of caffeic acid with quinic acid) widely found in natural products [51] or the methylated ester of gallic acid which can be formed *in vivo* [58]. Theoretical examination of the radical scavenging activity of these forms in comparison to the respective acids [36, 58] indicated that esterification may improve hydrogen atom (ΔBDE ~ -1 kcal/mol) and electron donation (ΔIP ~ -3 kcal/mol). Nevertheless, in case the ionization of the carboxylic group is facilitated the opposite trend can be expected as e.g. shown for gallic acid-gallic acid methylester [59]. At this point it should be stressed that theoretical findings based on the aforementioned computed molecular descriptors should be interpreted with caution when referred to esterification. This is due to the fact that the size of the esterified moiety as i.e. presented for methyl, ethyl, propyl, octyl and dodecyl gallate [e.g. 58], though not greatly affecting the hydrogen atom (ΔBDE_g< -0.5 kcal/mol) or electron donating efficiency (ΔIP_g< -3 kcal/mol) have a significant impact on polarity of compounds- dodecyl gallate was found ~30 times more lipid soluble than methyl- and can additionally impose steric hindrance. These characteristics are very important for the activity in a real system [10]. Thus, the effect on radical scavenging activity of polarity and steric phenomena of AH cannot be addressed when the proposed molecular descriptors are solely used.

PhenolicAcidRelated Compounds

Except for phenolic acids, other compounds, biosynthetically related to them and of potential interest as antioxidants can be found. Characteristic examples are hydroxybenzaldehydes, also known for their antimicrobial and flavouring properties [60, 61]. A series of natural ones was recently examined with the aid of DFT [47] and results were compared to those obtained for the respective acids. Thus, introduction of an aldehyde group instead of a carboxylic one was not found to affect the BDE_g values. A small increase (< 3.0 kcal/mol) was observed for the IP_g ones. This was probably due to the comparable electronic phenomena induced by the two groups according to Hammet-type values (σp), which are 0.42 and 0.45, respectively [56]. In solution, however, the mechanism of action seems to become more complicated. The carboxylic group is prone to ionization, whereas the aldehyde one may facilitate ionization of hydroxyl groups to those AH bearing catechol, syringol and guaiacol moiety. This was supported by literature pKa values, as well as by computed proton affinity (PA_l) ones in methanol [47]. Thus, in case of ionization in water (pH > 3.0) the carboxylate anion ($\sigma p = 0.00$) [56] is predicted to cause a decrease in the BDE_l (by 0.8-2.7 kcal/mol), as well as the IP_l (by 6.8-8.0 kcal/mol) values. Nevertheless, aldehydes, could form phenoxide anions in methanol or at physiological pH with lower IP_l value of ~ 25 kcal/mol than the parent AH. This is due to the strong electron donating character of the $-O^-$ ($\sigma p = -0.81$) over that of the –OH ($\sigma p = -0.37$) [56]. Hence, computational data and comparison with available experimental ones on radical scavenging activity (CBA, $DPPH^\bullet$, $ABTS^{\bullet+}$, bulk oil oxidation) for the test AH [15] indicated that in polar media the aldehydes may rather react faster via SPLET or SET. Whereas, the corresponding acids with HAT, even in an aqueous environment at physiological pH value.

These data verified previous suggestions on the preferable mechanism of action of some 4-hydroxybenzoic acids made by Tyrakowska et al. [62] with the aid of gas-phase calculations and the $ABTS^{•+}$ assay. In case HAT dominates as i.e. in bulk oil the superior activity of protocatechuic acid to that of the corresponding aldehyde [15] was ascribed to the ease of quinone formation of the former (ΔBDE_g = -1.7 kcal/mol) [47]. The higher HAT efficiency of acids over that of aldehydes in both polar and non-polar media was also reported by the theoretical study of Galano et al. [63] comparing the radical scavenging activity of vanilic acid and vanillin.

The same authors also studied the activity of vanillic alcohol (flavoring compound) bearing an electron donating CH_2OH group (σp = 0.00) [56].

According to their findings the alcohol was more efficient hydrogen atom donor in non-polar media. Nevertheless in water (pH 7.4), the activity of the acid was improved due to ionization of the carboxylic group. The importance of electron donating substituents [56] even not directly attached to the aromatic ring has been pointed out in the past by Nenadis et al. [64]. They examined the activity of a series of ferulic acid (4-hydroxy, 3-methoxy hydroxycinnamic acid) related compounds and found that coniferylacohol bearing a CH_2OH group (σp = 0.00) and isoeugenol bearing a $-CH_3$ group (σp= -0.17) were more efficient hydrogen atom (ΔBDE_g= -2.4 to -3.2 kcal/mol) and electron (ΔIP_g= -5.6 to -14.5 kcal/mol) donors than the respective acid (σp = 0.45), aldehyde (σp = 0.42) or ethylester (σp = 0.45). These findings were in accord to those obtained experimentally using the $DPPH^{•}$ assay and a bulk oil oxidation experiment [64].

3.3. Coumarins

Coumarins are natural phenolic compounds that bear a benzopyranone skeleton (Figure 1, V) and derive biosynthetically from hydroxycinnamic acids [65]. Various natural ones have been tested toward the scavenging of free radicals with *in vitro* assays [66-68]. Nevertheless, the only available theoretical study on HAT and SET efficiency is that of Zhang and Wang [69]. These authors studied the activity of six coumarins, namely the 7-monohydroxy one (umbelliferon), scopoletin bearing a guaiacol moiety (7-hydroxy, 6-methoxy-), three dihydroxy ones (5,7-, 6,7-, 7,8-) with a methyl group at C-4 as well and daphnetin (7,8-dihydroxycoumarin). Comparison of BDE_g and IP_g values with literature data on peroxyl radical scavenging at physiological pH [70] suggested that the respective compounds should react via HAT. The BDE_g value computed for umbelliferon was found higher than that of phenol by 1 kcal/mol.

Consequently, the pyranone ring seems to have an electron withdrawing effect since it does not facilitate hydrogen atom donation. In line with information for simple phenols (3.1) those coumarins bearing a catechol moiety were the most efficient radical scavengers. Comparison of daphnetin with 4-methyl-daphnetin indicated that the methyl group at A ring did not affect activity, in agreement with experimental data [70]. These two compounds were however predicted to be less efficient than the 6,7-dihydroxy coumarin. This was related to the fact that the most stable conformers for these two compounds were those bearing two IHBs, one being formed between the C-8 hydroxy and the –O– groups.

As a consequence the BDE_g value increased by ~ 4 kcal/mol. However, this finding was not in accordance with experimental data which pointed that 7,8-dihydroxy coumarins were ~ 1.4 more efficient [70]. This discrepancy was explained by the fact that the active –OH group in the case of 6,7-dihydroxy coumarin is free to form a strong intermolecular hydrogen bond with water, whereas in daphnetin and 4-methyl daphnetin this should be avoided due to the two IHBs [69].

3.4. Naphthoquinones

Alkannins and shikonins are naturally occurring derivatives of the dihydroxynaphthoquinone (5,8-dihydroxy) naphthazarin (Figure 1, VI) bearing an isohexenylside chain attached to C-3. These compounds used as food colorants in some countries, are of great interest in the field of life sciences due to the broad spectrum of biological properties [71, 72]. Considering the –OH configuration it seems that naphtoquinones could hardly donate a hydrogen atom due to the formation of strong intramolecular hydrogen bond.

This was clearly demonstrated for a series of natural ones studied by Ordoudi et al. [73] according to BDE_g values. Thus, for all six compounds examined (deoxyshikonin, alkannin, shikonin, acetyloshikonin, isovalerylshikonin, *β,β*-dimethylacrylshikonin) the reported BDE_g values were 6.7-8.4 kcal/mol higher than that of phenol.

On the contrary, the respective IP_g ones were lower by 4.5-15 kcal/mol suggesting a possible electron donating ability. However, deoxyshikonin, isovalerylshikonin, *β,β*-dimethylacrylshikonin were found inactive using the CBA, and $DPPH^{\bullet}$ assays [73]. The lack of efficiency observed for the latter two was rather related to solubility issues and streric hindrance. The experimental trend of activity for the rest of the test compounds could not be explained in terms of either –OH BDE_g or IP_g values.

Importance of AllylicHydrogens

Alkanin/shikonnin and related compounds bear an isohexenyl side chain. Consequently, allylic hydrogen atoms, proposed to add to the activity of other phenolic compounds in past studies [74] could affect activity.

Calculation of C-H BDE_g values for deoxyshikonin, shikonin and acetyloshikonin indicated that abstraction was feasible at C-1' of the side chain. The lowest BDE_g value (by 8-10 kcal/mol) was computed for shikonin pointing out the positive contribution of the –OH group attached at C-1'. The latter was ascribed to the better stabilization of the derived carbon radical as evidenced by the gas-phase computed spin density value on C-1'atom for shikonin (0.28) and deoxyshikonin (0.53) [73].

3.5. Xanthones

Limited are also the studies for natural xanthones (Figure 1, VI). Gas-phase computations [15] for a series of natural ones isolated from the stem bark of *Garciniavieillardii*P. [75] namely, 6-*O*-methyl-2-deprenylrheediaxanthone B, vieillardixanthone, forbexanthone, buchanaxanthone and isocudraniaxanthone A, indicated that from a thermodynamic point of view the tested AH could scavenge the $DPPH^{\bullet}$ via the HAT mechanism. The IP_g values for all of these xanthones were higher than that of the $DPPH^{-}$ so it was concluded that SET mechanism was not favoured. From the compounds examined it was evidenced that the presence of an –OH at C-2 and C-3 (vieillardixanthone, isocudraniaxanthone A) enhanced the activity despite the formation of IHB between the –OH group at C-2 and the oxygen located in the C ring. The latter bond was rather week as the BDE_g values were comparable to that of catechol. Substitution of one –OH group with a methoxy one resulted in higher BDE_g values (by ~4-13 kcal/mol), as the electron donating effect of the methoxy group was counterbalanced by the IHB. Such an increase was however lower (~ 4 kcal/mol) when the methoxy group was situated at C-3 position in 6-*O*-methyl-2-deprenylrheediaxanthone B instead of C-2 (~13 kcal/mol) in buchanaxanthone. This was due to the fact that in the former case the HAT is favoured by the electron donating properties of the –O– located in C ring. Comparison with a model compound bearing only A and C rings and the same substitution pattern of the most active xanthones (vieillardixanthone, isocudraniaxanthone A) showed that the B ring had a negligible effect on the radical scavenging efficiency. Martinez et al. [76] studying theoretically the radical scavenging activity of 20 natural xanthones isolated from the pericarp of *Garciniamangostana* verified that for the neutral ones electron donation was not possible thermodynamically. However, for those which deprotonation is feasible at physiological pH, predictions supported that the respective phenoxide anions could scavenge free radicals via electron donation.

3.6. Stilbenes

Stilbenes are natural AH bearing two aromatic rings linked with an ethylidene bridge (Figure 1, VIII). The most known member is *trans*-resveratrol (4,3',5'-trihydroxystilbene) which is found mainly in grapes and is acknowledged for its health benefits [77]. This compound though bearing a resorcinol moiety in one ring and a single –OH in the other one seems to be a potent hydrogen atom donor when compared to phenol (ΔBDE_g= -5.6 kcal/mol) [26]. The hydrogen atom donation is proposed to take place from the single –OH group at C-4' instead of the others (ΔBDE_g= -5.8 to -6.3 kcal/mol) [78]. This is due to the better stabilization of the radical formed through resonance [79, 80]. Despite being less efficient hydrogen atom donor than known antioxidants, the IP_g value for the neutral compound was reported to be 30.7 kcal/mol lower than that of phenoland significantly lower than many other AH, suggesting action through electron donation [26]. However, Iuga et al. [30] using DFT/M05-2X proposed that *trans*-resveratrol could scavenge hydroperoxyl radicals only via hydrogen atom donation. Instead, SET-PT pathway or an adduct formation was possible at physiological pH when the oxidizing species was the highly active hydroxyl radical.

Examining the activity of other monomeric hydroxystilbenes, Mikulski and Molski [81] reported that *trans*-astrigin bearing an additional –OH group at C-3' was more efficient hydrogen atom (ΔBDE_g= -9.2 kcal/mole) and electron (ΔIP_g= -7.5 kcal/mole) donor than resveratrol. A higher potency than that of *trans*-resveratrol was also predicted for *trans*-4,4'-dihydroxystilbene. This finding indicated that the presence of a single –OH group at C-4 instead of a resorcinol moiety results in the availability of two almost equipotent –OH groups (ΔBDE_g = -1.02 kcal/mol), as well as a higher electron donating efficiency (ΔIP_g= -4.8 kcal/mol) than that of *trans*-resveratrol. Such an observation may also be due to a better stabilization of the derived phenoxy radicals as predicted by the gas-phase computed spin density values [81]. The higher efficiency of both of these compounds over that of *trans*-resveratrol, is supported by existing experimental $DPPH^{\bullet}$ findings [82, 83]. Formation of a glucoside or glucuronide derivative at C-3 has a rather small effect to the BDE_g value of the –OH group at C-4' when compared to *trans*-resveratrol (ΔBDE_g= 0.2 and -0.6 kcal/mol). Even so, formation of such derivatives renders the –OH group at C-3 unavailable for radical scavenging, decreasing thus, the available active sites. This may, partially at least, explain the higher efficiency (~2.7-fold) of *trans*-resveratrol over that of *trans*-piceid acid (3-O glucoside of resveratrol) with the $DPPH^{\bullet}$ assay [82]. In the case of electron donation, comparison between resveratrol and the aforementioned derivatives, indicated that a glucoside moiety but mainly a glucuronide one was beneficial both in the gas-phase (ΔIP_g ~ -7 and -14 kcal/mol) and water (ΔIP_l~ -1.4 and -3.0 kcal/mol) [81].

3.7. Chalcones

Chalcones bearing two aromatic rings connected via a three-carbon unsaturated chain with a carbonyl group (Figure 1, IX) are natural antioxidants widely distributed in nature that can also be transformed via enzymatic cyclization to flavonoids [1]. Kozlowski et al. [17] examining the antioxidant activity of a series of 11 chalcones using both $DPPH^{\bullet}$ assay and computational methods pointed out the importance of catechol moiety in the B ring for a high activity, as well as the contribution of an -OH group at C-6'. Nevertheless, the effect of the latter is rather masked when the compound bears also a catechol moiety. Absence of the double bond (dihydrochalcones) resulted in a decrease of radical scavenging activity, as BDE_g and BDE_l values increased by 2.3 and 4 kcal/mol. Computations for the $DPPH^{\bullet}$ and DPPH-H proposed that for an efficient scavenging via HAT a BDE value of ~82 kcal/mol is sufficient. Additional calculations with the aid of a PCM model for other radicals such as $^{\bullet}OH$, $CH_3{}^{\bullet}CHOH$, and $CH_3OO^{\bullet}$, indicated that chalcones can also react with them via electron donation [17]. The authors suggested that except for HAT, the SET pathway should not be overlooked in radical scavenging studies, as it may be also important in biological systems.

3.8. Flavonoids

Flavonoids are characterized by two aromatic rings that are linked to a three-carbon aliphatic chain condensed to form a pyran ring (Figure 1, X). It is the most populated class of phenolic compounds in the plant kingdom (some thousands of members) and separated into different sub-classes based on the structural characteristics of the C ring (Figure 3).

SAR studies for flavonoids are numerous involving both experimental and theoretical approaches. Despite the fact that in nature flavonoids are usually found as glucosides, research on their antioxidant activity focuses mainly on the corresponding aglycons [7].

B Ring Structural Features

The B ring is considered the active site of flavonoids according to experimental findings and its hydroxylation pattern seem to have a great impact to the radical scavenging activity [7]. As already mentioned for the other classes of phenolics the presence of a catechol or a pyrogallol moiety is beneficiary for HAT enhancement as reflected on BDE_g values. Hence, Lespade and Bercion [84] reported for some selected flavonols that myricetin bearing a pyrogallol moiety in B ring (3',4',5',5,7pentahydroxy) is expected to donate a hydrogen atom more easily (ΔBDE_g= -4.6 kcal/mol) than quercetin (3',4'5,7 tetrahydroxy). The respective compound was predicted to be significantly more active than kaempferol (4',5,7trihydroxy) and morin (2',4'5,7 tetrahydroxy), (ΔBDE_g= -12.6 kcal/mol). In the same study, the importance of the catechol moiety was also shown by comparing the flavones luteolin (3',4',5,7tetrahydroxy) and apigenin (4',5,7 tetrahydroxy), with a ΔBDE_g value of ~ -8 kcal/mol. These finding are in line with various experimental data on radical scavenging activity of flavonols and flavones [7]. Additional to these, the beneficial effect in activity of a pyrogallol moiety in B ring was also commented by Zhang and Wang, [74] for epigallocatechin-epicatechin pair (ΔBDE_g= -3.5 kcal/mol).

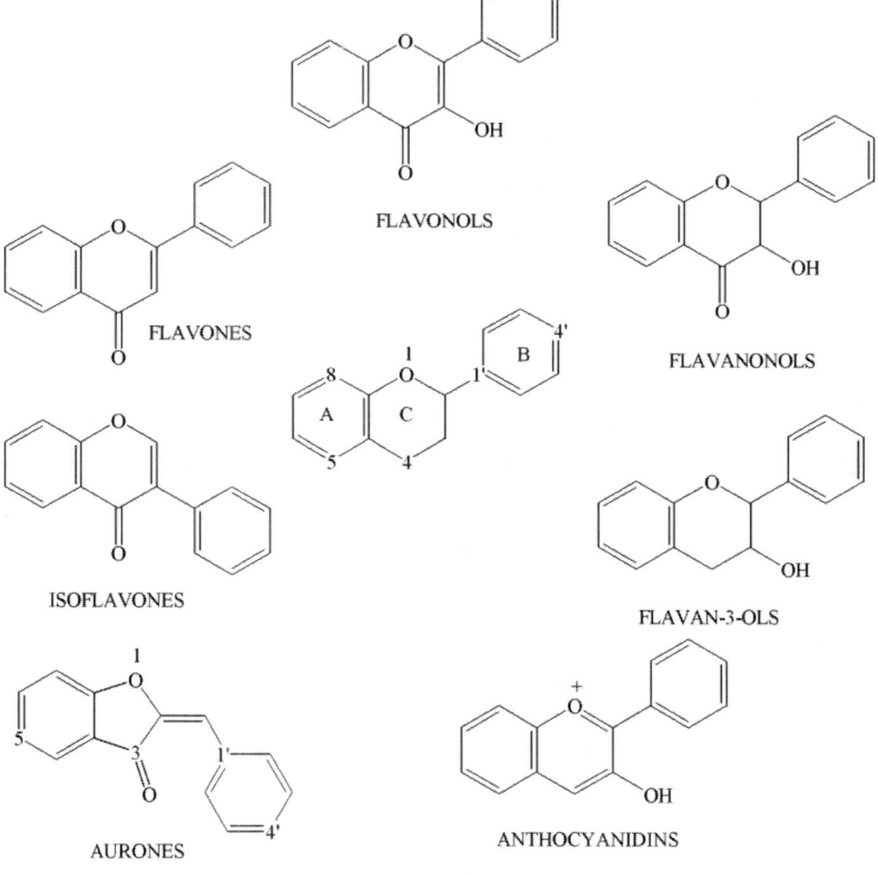

Figure 3. Flavonoid sub-classes.

With regards to electron donating efficiency as this was estimated via computation of IP_l values in water [84], small differences in the activity of the neutral compounds were reported. Thus, myricetin and quercetin were predicted to be of equal potency. Kaempferol presented a higher IP_l value by only 1.3 kcal/mol, whereas morin, bearing a resorcinol moiety, was the least active with an IP_l value higher by 3.1 kcal/mol.

The ionization, as pointed out by Lemanska et al. [85] experimentally and theoretically may positively affect the antioxidant efficiency of flavonoids by increasing the HAT, but mainly the electron donating ability. Therefore quercetin and kaempferol anions were predicted to be more potent hydrogen atom (ΔBDE_g= -4.9 and -12.7 kcal/mol) and electron donors (ΔIP_g= ~ -100 kcal/mol) than their neutral forms. Similar observations were also made for morin [84]. However, the magnitude of potency and the relative trend in activity in a polar environment will be eventually modulated by the ability of the test compounds to deprotonate. Quercetin seems to be more prone to anion formation (pK_{a1}=7.03) followed by morin (pK_{a1}=7.5) and kaempferol (pK_{a1}=8.2). As a consequence, quercetin should be possibly more active in case SPLET is facilitated [84, 86].

C Ring Structural Features

The structural characteristics of ring C, as stated previously, are those that differentiate flavonoids into sub-classes. To elucidate SAR studies focus is usually on the comparative examination of quercetin, luteolin, taxifolin and catechin/epicatechin. Based on information presented by Leopoldini et al. [26] quercetin is predicted to be more potent hydrogen atom donor, as the BDE_g value of the –OH group located at C-4' was ~ 2 kcal/mol lower than that of the other three flavonoids. Analogous were the findings of Vaganek et al. [86] and Antonczak [87] though a 0.7-1.5 kcal/mol lower BDE_g value was computed for quercetin. The above results suggest that the hydrogen atom donating efficiency of the catechol moiety in B ring is not greatly affected by the structural characteristics of C ring. Nevertheless, C ring configuration is crucial for molecular planarity, as well as the stability of the phenoxy radical formed [87]. Furthermore, as pointed out by computed BDE_g values for other –OH groups in C and A rings, the configuration in the former ring may have an impact on the activity of the compounds. As shown in the past [27] for quercetin-taxifolin pair and recently [86] for quercetin, luteolin, taxifolin and epicatechin, the –OH group at C-3 can probably act as a hydrogen atom donor in the case of quercetin. This stems from the fact that the respective BDE_g value was reported to be ~24 and ~17 kcal/mol lower than that of taxifolin and epicatechin respectively. Consequently, findings reporting a higher activity of quercetin in comparison to that of luteolin lacking the respective –OH group can be justified [7]. In addition, it may also explain the opening of C ring in quercetin as an increased gas-phase spin density (0.44) located at C-2 atom after reaction of the C-3 –OH group can attract further a radical attack [27]. The importance of the respective –OH group has been pointed out also by Aliaga and Lissi [88] who studying the activity of quercetin and rutin with the $ABTS^{\bullet +}$ assay and theoretical calculations commented that glucosilation in the case of the latter compound leads to a decrease in activity due to less "reactive centres". Similar theoretical findings were also presented [84] for other glucosides of quercetin (quercitrin) and kaempferol (astragalin).

The presence of a -C=O group at C-4 has a negative impact to HAT efficiency of compounds. This was evident by the ease of HAT from the –OH group at C-5 of epicatechinin comparison to that of taxifolin and luteolin (ΔBDE_g = -17 and -21 kcal/mol respectively). This was clearly due to the lack of IHB which hampers HAT [86]. Thus, epicatechin should be more active than taxifolin and also luteolin despite the extended conjugation of the latter. This is further justified buy the possible contribution of allylichydrogens from C-2 and/or C-4 of the C ring as proposed by Zhang and Wang, [74].

Examination of the electron donating ability showed that the lowest IP_g value was obtained for quercetin when compared to epicatechin (ΔIP_g= -1.9 kcal/mol), luteolin (ΔIP_g= -9.5 kcal/mol) and taxifolin (ΔIP_g= -10.3 kcal/mol). These findings highlight the importance of the concomitant presence of an –OH group at C-3 and the 2,3 double bond for an enhanced activity [86]. Even so, the absence of double bond in epicatechin is almost compensated by the lack of the carbonyl group.

Limited are the data on the role of the structural features of C ring on ionization of flavonoids and the activity of their ions. Martins et al. [89] using both experimental and theoretical approach reported the following acidity trend for the four flavonoids: luteolin>quercetin>taxifolin>catechin. From these compounds only in taxifolin the most acidic –OH group was predicted to be that at C-7 instead of C-4'. As evident the presence of 2,3 double bond in combination with the C=O at C-4 is necessary for high deprotonation efficiency. Differences in acidity as mentioned in previous paragraphs may have a consequence in radical scavenging efficiency when reactions take place in a polar environment which favours SPLET or SET-PT. Thus, in methanol luteolin could possible react faster than the rest of the selected flavonoids. However in an aqueous environment and at a pH value facilitating ionization of all these flavonoids, luteolin would be rather more potent only from taxifolin[86]. This is postulated by the higher gas-phase ETE (ETE_g) value computed for the corresponding anion at C-4' than that of quercetin (4.8 kcal/mol) and catechin (~18 kcal/mol) C-4' anions respectively.

Other Structural Features of C Ring

Anthocyanidins are the aglycones of anthocyanins, which are common natural pigments. The difference in their structure from the aforementioned flavonoids is the double bonds in C ring that facilitate conjugation throughout the three rings, as well as the positive charge at the –O– of C ring (Figure 3). The order of activity among common members based on BDE_g values is in accordance with the hydroxylation pattern in B ring [90]. However, the structure of C ring seems to significantly affect the activity when compared to that of other flavonoid sub-classes. Thus, cyanidin (3',4',5,7 tetrahydroxy) seems to be a poor HAT donor since the BDE_g value for the catechol group in B ring is close to that of the –OH group at C-4 of keampferol (4',5,7 trihydroxyflavonol) and apigenin (4',5,7-trihydroxyflavone), known to be of low potency [7]. In fact, the active site of cyanidin, in case HAT is feasible, should be the –OH group at C-3 [26, 86]. The computed BDE_g value was reported to be ~2 kcal/mol lower than that for the respective group of quercetin [86]. Similarly to high BDE value, cyanidin presents very high IP value (higher by 70-80 kcal/mol) when compared to qurcetin, taxifolin and epicatechin. This is rather expected since cyanidin already bearing charge is difficult to donate a second electron to form a less stable divalent cation radical. Contrary to this, cyanidin is the most prone to deprotonation. This may have a significant impact on activity in case SPLET or SET is facilitated considering e.g. that the ETE_g value after deprotonation is almost 9.5 kcal/mol lower than that of the neutral quercetin [86]. This may justify a high $ABTS^{\bullet+}$ activity reported by Rice-Evans et al. [91].

Introduction of a furan (aurones) C-ring instead of a pyran one (Figure 3), as shown for auresidin (3',4'5,7 tetrahydroxyaurone) and luteolin [92] results in a ~1.5 kcal/mol decrease in BDE_g value of the catechol moiety in ring B. Even so, a significant impact was reported on the HAT from the –OH group at C-5 (ΔBDE_g= -15 kcal/mol). The latter difference is clearly related to the weaker IHB between the –OH at C-5 and the –C=O of C ring formed in auresidin. The enthalpy for such a bond was estimated to be ~8 kcal/mol for the aurone, whereas in luteolin was ~14 kcal/mol. The C ring configuration in aurones seems also to benefit the electron donation when comparisons were made for the aforementioned two compounds (ΔIP_g= -6 and ΔIP_l= -6.5 kcal/mol).

Isoflavonoids is a known sub-class of flavonoids where B ring is attached to C-3 of C ring instead of C-2 (Figure 3). Zhang et al. [93] studied with the aid of DFT the radical scavenging activity of some natural compounds. The authors found that B ring was the active centre and proposed HAT as the major pathway followed. Nevertheless, the authors did not make any comparison with members from the other flavonoid sub-classes. Thus, the only available theoretical data are those regarding some natural aurones and isoaurones [92]. However in this case due to the presence of a pentacyclic ring in isoaurones the carbonyl group is displaced adjacent to the oxygen of the ring. Computation of BDE_g values indicated that such a structural differentiation is not expected to greatly influence the HAT ability from B ring. A significant effect, though, was found on the HAT efficiency from the –OH group located in the A ring. Calculations showed that the BDE_g and BDE_l were ~8 kcal/mol lower for the –OH group at C-7. Such a decrease was attributed to the extended conjugation in the molecule of isoaurones which permitted the spin density delocalization throughout the whole structure of the AH. On the other hand in aurones, the delocalization is restrained only to rings A and C [92]. The neutral isoaurones are also expected to be efficient electron donors (ΔIP_g~ -0.8 to -6 kcal/mol, and ΔIP_l= -3 to- 6.2 kcal/mol) than the corresponding aurones, as well as luteolin (ΔIP_g ~ -11.3 kcal/mol and ΔIP_l= -7 to - 8.7 kcal/mol).

A Ring Structural Features

The substitution pattern in A ring has been reported to have a minor effect on antioxidant activity, except for cases where no available –OH groups are located in B ring [7]. Indeed this is evident from the emphasis given so far on the role of other structural features of the flavonoids. Despite this, the -OH groups can have some effect as mentioned in the previous paragraphs in case this is favoured by the C ring configuration. Theoretical calculations have pointed out however some exceptions. Characteristic examples are predicted to be the natural auronesmaritimetin (3',4',6,7tetrahydroxyaurone) and licoagroaurone [3',4',6 trihydroxy, 7-(3-methyl-2-butenyl) aurone], as well as the isoauroneisoaurostatin (4',6 dihydroxyisoaurone) [22, 92]. In the first one, according to both BDE_g and BDE_l values in various solvents, it was shown that the active site was located in A ring. Therefore, the respective BDE value was predicted to be 1.1-1.7 kcal/mol lower than that for the catechol moiety of B ring. On the other hand, in licoagroaurone bearing one –OH group at A ring, allylic hydrogen atom abstraction was predicted from the 3-methyl-2-butenyl group located at C-7 in A ring. The computed BDE_g value for this hydrogen was comparable to that of catechol. In isoaurostatin, the only –OH group located in ring A was predicted to present 1.5-2.4 kcal/mol lower BDE_g and BDE_l value than that found for the –OH group at C-4'.

Conclusion

DFT calculations can provide useful information on the contribution of structural characteristics of phenolic compounds to their radical scavenging activity. This is possible in a reasonable length of time especially if gas-phase calculations are used. Besides the recognition that solvation models can provide more reliable information, improvements are expected in order implicit and explicit solvent effects to be assessed concomitantly and more accurately as well as in a reasonable time window.

The available studies point out BDE as the most valuable molecular descriptor. Nevertheless other such as PA and ETE can be useful as well when ionization phenomena take place. The presence of a pyrogallol or a catechol moiety is of pivotal importance for the HAT radical scavenging activity. Electron donating substituents and extended conjugation are also important. The role of allylic hydrogen atoms, often neglected in SAR studies cannot be overlooked.

At present, a clear limitation of theoretical predictions concern assessment of the radical scavenging activity in complex environments (e.g. dispersion systems, liposomes). This is due to the fact that AH properties such as lipophilicity, the ability to cross membranes, ionic interactions and steric hindrance can affect the overall activity. Thus, quantitative structure-activity relationship (QSAR) models, build up on the basis of experimental results and various computed descriptors, including those giving information on e.g. lipophilicity or binding ability to sites with *in vivo* significance, should gain more importance in the near future.

The availability of various alternative protocols to compute the molecular descriptors using DFT and the constant improvement in approximations impose the necessity for standardization of protocols to facilitate literature comparison. Such a target already set in experimental assays, seems imperative. Computer facilities render another serious limitation. Still, there is a need for a consensus on the use of a series of antioxidants as references to assist data evaluation and comparison among researchers. Last but not least, an extension of research on the applicability of computed molecular descriptors to the activity of other categories of antioxidants (e.g. bearing –NH, -SH groups) is also required.

References

[1] Harborne, J. B. Plant phenolics; Day, P. M. and Harborne, J. B.; Eds, Academic Press: London, 1989.

[2] Crozier, A.; Jaganath, I. B.; Clifford, M. N. In *Plant secondary metabolites: Occurrence, structure and role in the human diet*; Crozier, A.; Clifford, M. N.; Ashihara, H.; Eds, Blackwell Publishing Ltd: Oxford, 2006; pp. 1-24.

[3] King, A.; Young, G. *J. Am. Diet Assoc*. 1999, 99, 213-218.

[4] Määttä-Riihinen, K. R.; Kamal-Eldin, A.; Törrönen, A. R. *J. Agric. Food Chem*. 2004, 52, 6178-6187.

[5] Halliwell, B.; Aeschbacht, R.; Loliger J.; Aruoma, O. I. *Food Chem. Toxicol.* 1995, 33, 601-617.

[6] Min, D. B.; Choe, E.O. *Crit. Rev. Food Sci. Nutr.*2006, 46, 1-22.

[7] Tsimidou, M. Z.; Nenadis, N.; Zhang, H. Y. In *Antioxidant plant phenols: Sources, structure-activity relationship, current trends in analysis and characterization*; Boskou, D.; Gerothanasis, I.; Kefalas, P.; Eds, Reasearch Signpost: Kerala, India, 2006; pp. 29-68.
[8] Wright, J. S.; Johnson, E. R.; DiLabio, G. A. *J. Am. Chem. Soc.* 2001, 123, 1173-1183.
[9] Litwinienko, G.; Ingold, K. U. J. *Org. Chem.* 2003, 68, 3433-3438.
[10] Nenadis, N.; Tsimidou, M. Z. In *Oxidation in foods and beverages and antioxidant applications: Understanding mechanisms of oxidation and antioxidant activity*, part 2 antioxidants in foods and beverages; Decker, E. A.; Elias, R.; McClements, D. J.; Eds, Woodhead Ltd: Cambridge, UK; 2010; Vol. 1., pp. 332-367.
[11] Brand-Williams, W.; Cuvelier, M. E.; Berset, C. Lebensm. *Wiss. Technol.* 1995, 28, 25-30.
[12] Re, R.; Pellegrini, N.; Proteggente, A.; Pannala, A.; Yang, M.; Rice- Evans, C. *Free Rad. Biol. Med.* 1999, 26, 1231-1237.
[13] Bors, W.; Michel, C.; Saran, M. *Biochim. Biophys. Acta*1984, 796, 312-319.
[14] Ordoudi, S. A.; Tsimidou, M. Z. J. Agric. *Food Chem.* 2006, 54, 9347-9356.
[15] Bountagkidou, O. G.; Ordoudi, S. A.; Tsimidou, M. Z. *Food Res. Inter.* 2011, 43, 2014-2019.
[16] Ji, H. F.; Tang, G. Y.; Zhang, H. Y. *QSAR CombinatorialSci.* 2005, 24, 826-830.
[17] Kozlowski, D.; Trouillas, P.; Calliste, C.; Marsal, P.; Lazzaroni, R.; Duroux, J. L. *J. Phys. Chem. A* 2007, 111, 1138-1145.
[18] Justino, G. C.; Vieira, A. J. S. C. *J. Mol. Model.* 2010, 16, 863-876.
[19] Bakalbassis, E. G.; Chatzopoulou, A.; Melissas, V. S.; Tsimidou, M.; Tsolaki, M.; Vafiadis, *A. Lipids* 2001, 36, 181-190.
[20] Zhang, H. Y. J. *Am. Oil Chem. Soc.* 1998, 75, 1705-1709.
[21] Young, D. C. *Computational chemistry: A practical guide for applying techniques to real world problems*, John Wiley and Sons, Inc.: N Y, 2001.
[22] Nenadis, N.; Sigalas, M. P. *J. Phys. Chem.* A 2008, 112, 12196-12202.
[23] Bakalbassis, E. G.; Lithoxoidou, A. T.; Vafiadis, A. P. *J. Phys. Chem.* A 2003, 107, 8594-8606.
[24] Wayner, D. D. M.; Lusztyk, E.; Pagé, D.; Ingold, K. U.; Mulder, P.; Laarhoven, L. J. J.; Aldrich, H. S. *J. Amer. Chem. Soc.* 1995, 117, 8737–8744.
[25] Borges dos Santos, R. M.; Martinho Simoes, J. A. *J. Phys. Chem. Ref.* Data 1998, 27, 707-740.
[26] Leopoldini, M.; Russo, N.; Toscano, M. *Food Chem.* 2011, 125, 288-306.
[27] Trouillas, P.; Marsal, P.; Siri, D.; Lazzaroni, R.; Duroux, J. L. *Food Chem.* 2006, 97, 679-688.
[28] Anouar, E.; Calliste, C. A.; Košinová, P.; Di Meo, F.; Duroux, J. L.; Champavier, Y.; Marakchi, K.; Trouillas, P. *J. Phys. Chem.* A 2009, 113, 13881-13891.
[29] Marković, Z. S.; Dimitrić Marković, J. M.; Milenković, D.; Filipović, N.*J. Mol. Model.* 2011, 17, 2575-2584.
[30] Iuga, C.; Alvarez-Idaboy, J. R.; Russo, N. *J. Org. Chem.* 2012, 77, 3868-3877.
[31] León-Carmona, J. R.; Alvarez-Idaboy, J. R.; Galano, A. *Phys. Chem. Chem. Phys.* 2012, 14, 12534-12543.
[32] Tomasi, J.; Mennucci, B.; Cammi, R. *Chem. Rev.* 2005, 105, 2999-3093.
[33] Litwinienko, G.; Ingold, K. U. *Accounts Chem. Res.* 2007, 40, 222-230.

[34] Guerra, M.; Amorati, R.; Pedulli, G. F. *J. Org. Chem.* 2004, 69, 5460-5467.
[35] Kozlowski, D.; Marsal, P.; Steel, M.; Mokrini, R.; Duroux, J. L.; Lazzaroni, R.; Trouillas, P. *Rad. Res.* 2007, 168, 243-252.
[36] Nenadis, N.; Wang, L. F.; Tsimidou, M.; Zhang, H. Y. J. *Agric. Food Chem.* 2004, 52, 4669-4674.
[37] Ordoudi, S. A.; Tsimidou, M. Z.; Vafiadis, A. P.; Bakalbassis, E. G. *J. Agric. Food Chem.* 2006, 54, 5763-5768.
[38] Hotta, H.; Nagano, S.; Ueda, M.; Tsujino, Y.; Koyama, J.; Osakai, T. *Biochim. Biophys. Acta*2002, 1572, 123-132.
[39] Ji, H. F.; Zhang, H. Y. *New J. Chem.* 2005, 29, 535-537.
[40] Thavasi, V.; Leong, L. P.; Bettens, R. P. A. *J. Phys. Chem. A* 2006, 110, 4918-4923.
[41] Aura, A. M.; O'Leary, K. L.; Williamson, G.; Ojala, M.; Bailey, M.; Puupponen-Pimia, R.; Nuutila, A. M.; Oksman-Caldentey, K. M.; Poutanen, K. *J. Agric. Food Chem.* 2002, 50, 1725-1730.
[42] Rechner, A. R.; Kuhnle, G.; Bremner, P.; Hubbard, G. P.; Moore, K. P.; Rice-Evans, C. A. *Free Rad. Biol. Med.* 2002, 33, 220-235.
[43] Manach, C.; Donovan, J. L. *Free Rad. Res.* 2004, 38, 771-785.
[44] Vafiadis, A. P.; Bakalbassis, E. G. *J. Amer. Oil Chem. Soc.* 2003, 80, 1217-1223.
[45] Mohajeri, A.; Asemani, S. S. *J. Mol. Strut. THEOCHEM* 2009, 930, 15-20.
[46] Natella, F.; Nardini, M.; Di Felice, M.; Scaccini, C. *J. Agric. Food Chem.*1999, 47, 1453-1459.
[47] Nenadis, N.; Tsimidou, M. Z. *Food Res. Inter.* 2012, 48, 538-543.
[48] Lithoxoidou, A. T.; Bakalbassis, E. G. *J. Amer. Oil Chem. Soc.* 2004, 81, 799-802.
[49] Pekkarinen, S. S.; Stockmann, H.; Schwarz, K.; Heinonen, I. M.; Hopia, I. A. *J. Agric. Food Chem.* 1999, 47, 3036-3043.
[50] Nsangou, M.; Dhaouadi, Z.; Jaidane, N.; Ben Lakhdar, Z. *J. Mol. Str. THEOCHEM* 2008, 862, 53-59.
[51] Clifford, M. N. *J. Sci. Food Agric.* 2000, 80, 1033-1043.
[52] Bakalbassis, E. G.; Nenadis, N.; Tsimidou, M. *J. Amer. Oil Chem. Soc.* 2003, 80, 459-466.
[53] Nenadis, N.; Boyle, S.; Bakalbassis, E. G.; Tsimidou, M. *J. Amer. Oil Chem. Soc.* 2003, 80, 451-458.
[54] Frankel, N. E. *Lipid oxidation*. The Oily Press Ltd: Dundee, Scotland, 1998.
[55] Marenich, A. V.; Cramer, C. J.; Truhlar, D. G. *J. Phys. Chem.* B 2009, 113, 6378-6396.
[56] Hansch, C.; Leo, A.; Taft, R. W. *Chem. Rev.* 1991, 91, 165-195.
[57] Kong, L.; Sun, Z. L.; Wang, L. F.; Zhang, H. Y.; Yao, S. D. *Helv. Chim. Acta* 2004, 87, 511-515.
[58] Lu, Z.; Nie, G.; Belton, P. S.; Tang, H.; Zhao, B. *Neurochem. Inter.* 2006, 48, 263-274.
[59] Ji, H. F.; Zhang, H. Y.; Shen, L. *Bioorg. Med. Chem. Lett.* 2006, 16, 4095-4098.
[60] Mangas, J.; Rodriguez, R.; Moreno, J.; Suarez, B.; Blanco, D. J. Agric. *Food Chem.* 1996, 44, 3303-3307.
[61] Friedman, M.; Henika, P. R.; Mandrell, R. E. *J. Food Prot.* 2003, 66, 1811-1821.
[62] Tyrakowska, B.; Sofers, A. E. M. F.; Szymusiak, H.; Boeren, S.; Boersma, M. G., Lemanska, K., Vervoot, J.; Rietjens, I. M. C. M. *Free Rad. Biol. Med.* 1999, 27, 1427-1436.

[63] Galano, A.; León-Carmona, J. R.; Alvarez-Idaboy, J. R. *J. Phys. Chem.* B 2012, 116, 7129-7137.
[64] Nenadis, N.; Zhang, H. Y.; Tsimidou, M. Z. J. *Agric. Food Chem.* 2003, 51, 1874-1879.
[65] Bourgaud, F.; Hehn, A.; Larbat, R.; Doerper, S.; Gontier, E.; Kellner, S.; Matern, U. *Phytochem. Rev.* 2006, 5, 293-308.
[66] Paya, M.; Goodwin, P. A.; De lasHeras, B.; Hoult, J. R. S. *Biochem. Pharmacol.* 1994, 48, 445-451.
[67] Rehakova, Z.; Koleckar, V.; Cervenka, F.; Jahodar, L.; Saso, L.; Opletal, L.; Jun, D.; Kuca, K. Toxicol. *Mechanisms Methods* 2008, 18, 413-418.
[68] Thuong, P. T.; Hung, T. M.; Ngoc, T. M.; Ha, D. T.; Min, B. S.; Kwack, S. J.; Kang, T. S.; Choi, J. S.; Bae, K. *Phytother. Res.* 2010, 24, 101-106.
[69] Zhang, H. Y.; Wang, L. F. *J. Mol. Struct. THEOCHEM* 2004, 673, 199-202.
[70] Foti, M.; Piattelli, M.; Baratta, M. T.; Ruberto, G. *J. Agric. Food Chem.* 1996, 44, 497-501.
[71] Andrikopoulos, N. K.; Kaliora, A. C.; Assimopoulou, A. N.; Papapeorgiou, V. P. *Phytother. Res.* 2003, 17, 501-507.
[72] Assimopoulou, A. N.; Papageorgiou, V. P. *Phytother. Res.* 2005, 19, 141-147.
[73] Ordoudi, S. A.; Tsermentseli, S. K.; Nenadis, N.; Assimopoulou, A. N.; Tsimidou, M. Z.; Papageorgiou, V. P. *Food Chem.* 2011, 124, 171-176.
[74] Zhang, H. Y.; Wang, L. F. *J. Amer. Oil Chem. Soc.* 2002, 79, 943-944.
[75] Hay, A. E.; Aumond, M. C.; Mallet, S.; Dumontet, V.; Litaudon, M.; Rondeau, D.; Richomme, P. J. *Nat. Prod.* 2004, 67, 707-709.
[76] Martínez, A.; Hernández-Marin, E.; Galano, A. *Food Funct.* 2012, 3, 442-450.
[77] Soobrattee, M. A.; Neergheen, V. S.; Luximon-Ramma, A.; Aruoma, O. I.; Bahorun, T. *Mut. Res. Fund. Mol.* M. 2005, 579, 200-213.
[78] Leopoldini, M.; Marino, T.; Russo, N.; Toscano, M. *J. Phys. Chem.* A 2004, 108, 4916–4922.
[79] Cao, H.; Pan, X.; Li, C.; Zhou, C.; Deng, F.; Li, T. *Bioorg. Med. Chem. Lett.* 2003, 13, 1869-1871.
[80] Queiroz, A. N.; Bruno, A. Q.; Moraes, M. W.; Borges, R. S. *Eur. J. Med. Chem.* 2008, 44, 1644-1649.
[81] Mikulski, D.; Molski, M. *Eur. J. Med. Chem.* 2010, 45, 2366-2380.
[82] Fauconneau, B.; Waffo-Teguo, P.; Huguet, F.; Barrier, L.; Decendit, A.; Merillon, *J. M. Life Sci.* 1997, 61, 2103-2110.
[83] Li-Shuang, L. V.; Gu, X.; Ho, C. T.; Tang, J. *J. Food Lipids* 2006, 13, 131-144.
[84] Lespade, L.; Bercion, S. *Free Rad. Res.* 2012, 46, 346-358.
[85] Lemańska, K.; Szymusiak, H.; Tyrakowska, B.; Zieliński, R.; Soffers, A. E.; Rietjens, I. M. 2001, *Free Rad. Biol. Med.* 2001, 31, 869-881.
[86] Vagánek, A.; Rimarcˇík, J.; Lukeš, V.; Klein, E. *Comp. Theor. Chem.* 2012, 991, 192-200.
[87] Antonczak, S. *J. Mol. Struct. THEOCHEM* 2008, 856, 38-45.
[88] Aliaga, C.; Lissi, E. A. *Canadian J. Chem.* 2004, 82, 1668-1673.
[89] Martins, H. F. P.; Leal, J. P.; Fernandez, M. T.; Lopes, V. H. C.; Cordeiro, M. N. D. S. *J. Amer. Soc. Mass Spec.* 2004, 15, 848-861.
[90] Guzmán, R.; Santiago, C.; Sánchez, M. *J. Mol. Struct. THEOCHEM* 2009, 935, 110-114.

[91] Rice-Evans, A. C.; Miller, N. J.; Paganga, G. *Free Rad. Biol. Med.* 1996, 20, 933-956.
[92] Nenadis, N.; Sigalas, M. P. *Food Res. Inter.* 2011, 43, 2014-2019.
[93] Zhang, J.; Du, F.; Peng, B.; Lu, R.; Gao, H.; Zhou, Z. *J. Mol. Struct. THEOCHEM* 2010, 955, 1-6.

In: Density Functional Theory
Editors: J. Morin and J.M. Pelletier, pp. 147-169
ISBN: 978-1-62417-954-9

Chapter 5

POLARIZABILITY OF C_{60}/C_{70} FULLERENE [2+1]- AND [1+1]-ADDUCTS: A DFT-PROGNOSIS

Denis Sh. Sabirov *
Institute of Petrochemistry and Catalysis of Russian Academy of Sciences, Ufa, Russia

Abstract

The paper is devoted to the application of DFT methods to calculation of static polarizabilities of fullerene derivatives. The comparison of the calculated and experimental values has been performed for polarizabilities of C_{60}, C_{70} and some fluoro[60]fullerenes. As it has shown, DFT methods can be effectively used for the prediction of fullerene derivatives polarizabilities, which are necessary for the description of the functioning of fullerene-based nanosystems but have not been measured by the moment.

A model of polarizability for two types of fullerene derivatives $C_{60}X_n$ ([2+1]- and [1+1]-adducts) as a function of the number of added groups has been developed. DFT-calculations show that mean polarizabilities of both classes of fullerene derivatives do not grow up linearly with the increase of the number of addends n and are characterized by negative deviations from the additive scheme, *i.e.* depression of polarizability takes place. General formula for calculation of mean polarizability of the functionalized C_{60} fullerene has been derived based on polarizability depression. Its applicability to the related compounds (*e.g.*, C_{70} fullerene derivatives) has been shown.

Predicted by us with DFT methods, this interesting phenomenon may be important in the design of fullerene-containing nanostructures with regulated polar characteristics.

Keywords: Fullerene C_{60}, fullerene C_{70}, fullerene cycloadducts, fullerene halogenides, DFT methods, additive scheme, polarizability, depression of polarizability

1. Introduction

Applied quantum chemistry have reached a new level with the development of density functional theory methods, which allowed achieving a seemingly unattainable goal – to

* E-mail address: diozno@mail.ru

improve the accuracy of quantum chemical calculations with simultaneous reducing the required time costs. The current situation can be truly called the age of DFT because many works in the field of organic and physical chemistry are followed by DFT-calculations. Their high predictive ability raised quantum chemistry from the level of abstract theory (when calculated data are compared with experimental more often in qualitative aspect) to a completely independent research method, reproducing experimentally-measured characteristics of substances and processes with high accuracy. It is noteworthy that many molecular properties can be obtained from the calculated wave functions [1]. These facts allowed replacing the experiment by DFT-calculations in the case when the use of experimental techniques is difficult or impossible.

The study on polarizability of fullerenes and their derivatives is one of the problems with the limited experimental data. Polarizability is a molecule's ability to acquire induced dipole moment in electric fields (including electric fields of other molecules). In the case of low fields, it is defined as a ratio of the induced dipole moment of a molecule μ_{ind} to the electric field E that produces this dipole moment:

$$\mu = \alpha E, \tag{1}$$

where α is polarizability tensor (a 3×3 matrix, symmetric about the main diagonal). Its trace is invariant under coordinate system [2]. Mean polarizability is calculated as the arithmetic mean of the diagonal elements of α tensor:

$$\alpha = \frac{1}{3}\left(\alpha_{xx} + \alpha_{yy} + \alpha_{zz}\right). \tag{2}$$

Mean polarizability determines many physical properties such as dielectric constant, refractive index, van der Waals constant, ion mobility in gas, *etc*. (See Table 1.1 in Ref. [3]). Therefore, information concerning to polarizability is useful for analysis of the structure, physical and chemical properties of molecular and atomic clusters [4].

Polarizability has the dimension of volume and can be interpreted as the degree of the filling the space by molecule's electronic cloud. Therefore, molecular systems with a large number of electrons should demonstrate high α values. Fullerenes are high-polarizability molecules [5–18] (Table 1): the measured mean polarizability values of C_{60} and C_{70} are ~80 and ~105 Å^3 [5]. High α values of fullerenes have been used for explaining the distinctiveness of physical and chemical processes in fullerene-containing systems, such as the anomalously effective quenching of electronically-excited states of organic compounds by C_{60} and C_{70} [16], propensity for aggregation [19], formation of donor–acceptor complexes [20], reactivity of higher fullerenes in 1,3-dipolar cycloaddition [18]. Moreover, polarizability is considered as a reactivity index of fullerenes in reactions of cycloaddition [21].

Only the mean polarizability of C_{60} and C_{70} has been measured (Table 1). Its estimation for the other fullerenes (discovered later and less available) has been performed utilizing diverse quantum chemical methods. The most exciting results in this field have been obtained *via* research of the quantum-size effect on the carbon fullerenes polarizability that nonlinearly depends on the fullerene radius [22]. Development of the computational methodology and equipment makes possible to perform theoretical studies on polarizability of bulky molecular systems, *e.g.*, static and dynamic polarizabilities of the giant C_{540} fullerene [14]; so far, this

Table 1. Static polarizabilities of fullerenes, a comparison between theoretical methods and experimental data (Å^3)

Molecule	Calculated data			Experimental data
	B3LYP/Λ1	PBE/3ζ	Other theoretical estimations	
C_{60} (I_h)	80.3[a]	82.7[b]	78.8 (CPHF/(7s4p)[3s2p]),[c] 75.1 (HF/ 6-31 ++G),[d] 83.0 (topological model),[e] 77.5 (point dipole interaction model),[f] 82.1 (PBE/NRLMOL),[g] 81.6 (PBE0/SVPD),[h] 78.4 (VWN/DZVP/GEN-A2)[i]	76.4±8.0,[j] 88.9±6.0[k]
C_{70} (D_{5h})	100.7[a]	102.7[l]	93.2 (CPHF/(7s4p)[3s2p]),[c] 89.8 (HF/ 6-31 ++G),[d] 97.8 (VWN/DZVP/GEN-A2),[i] 103.0 (PBE/NRLMOL)[m]	101.9±13.9,[j] 108.5±8.2[k]
C_{76} (D_2)	–	112.3[n]	–	–
C_{78} (D_3)	–	115.3[n]	–	–
C_{84} (D_{2d})	–	124.1[n]	113.3 (CPHF/(7s4p)[3s2p])[c]	–
C_{540} (I_h)	–	–	1192.6 (PBE/NRLMOL),[g] 1243.8 (PBE0/SVPD),[h] 1254.0 (VWN/DZVP/ GEN-A2)[i]	–

[a] Taken from [6]. [b] Taken from [7]. [c] Taken from [8]. [d] Taken from [9]. [e] Taken from [10]. [f] Taken from [11]. [g] Taken from [12]. [h] Taken from [13]. [i] Taken from [14]. [j] Taken from [5]. [k] Taken from [15]. [l] Taken from [16]. [m] Taken from [17]. [n] Taken from [18].

paper represents the largest dynamic polarizability calculation, ever presented in the literature.

The accumulated data on fullerenes polarizability stimulate theoretical studies on polarizability of endofullerenes with encapsulated atoms of noble gases [23, 24], metals [25] or another fullerene [26] and giant fullerene-containing onions [27]. Nanostructures, listed above, are topological compounds, promising for molecular nanotechnologies [28]. For these applications, it is important to know the ease, with which the endohedral atom can be manipulated using an applied electric field. Theoretical studies [23–27] show that the mean polarizability of an endohedral complex cannot be calculated as the sum of the polarizabilities of carbon cage and a guest-molecule. In the case of endofullerenes with encapsulated noble gases, this violation of the additivity (called the exaltation of polarizability) can be either positive or negative, depending on the size of a carbon cage [24]. As estimated [29], C_{60} fullerene acts effectively as a small Faraday cage, with only 25% of the electric field penetrating the interior of the molecule. Thus, influencing the endo-atom is difficult in the case of the negative exaltation of polarizability, but as a qubit the endohedral atom should be well shielded from environmental electrical noise [29].

There is a lack of researches on the polarizability of exohedral fullerene derivatives. Just few earlier theoretical works deal with the mean polarizabilities of the short list of the functionalized fullerenes: fullerenols $C_{60}(OH)_x$ [30], fullerene halogenides $C_{60}F_{18}$ [31, 32], $C_{60}F_{17}CF_3$ [33], $C_{60}Cl_{30}$ [31], $C_{58}F_{18}$, $C_{58}F_{17}CF_3$ [34], $C_{50}Cl_{10}$ [35], N-methyl-3,4-fulleropyrrolidine [36], PBCM [37], and C_{56} fullerene derivatives [38]. Most of the articles contain the results of quantum chemical calculations of fullerene derivatives polarizability without an explanation of the obtained data. However, the search of the relations between the structure of compounds and their physical properties is interesting in a fundamental aspect.

Fullerenes willingly attach various molecules and highly-reactive particles (*e.g.*, radicals, carbenes, nitrenes), resulting in the exohedrally-functionalized carbon cages. The number for exohedral fullerene derivatives includes diverse fullerene adducts with carbo- and heterocycles, products of radical reactions $R_kH_lC_{60}$ (R is alkyl, perfluoroalkyl, *etc.*), fullerene-based polymers and dendrimers [39]. Many of these compounds are perspective derivatives for nanomaterials, pharmaceuticals and molecular devices [40, 41].

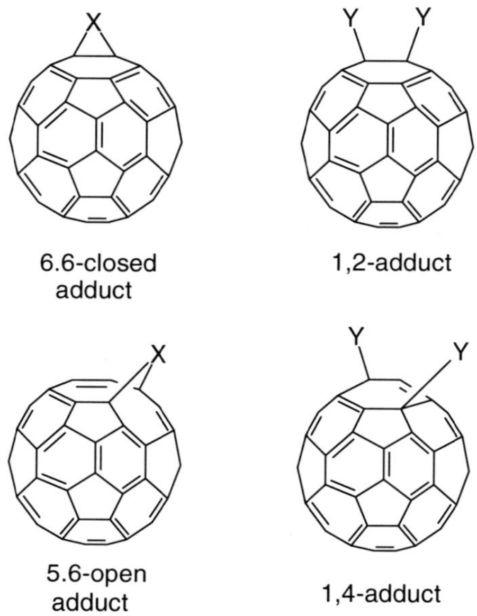

Figure 1. General structures of [2+1]- and [1+1]-adducts of C_{60} fullerene (left and right, respectively).

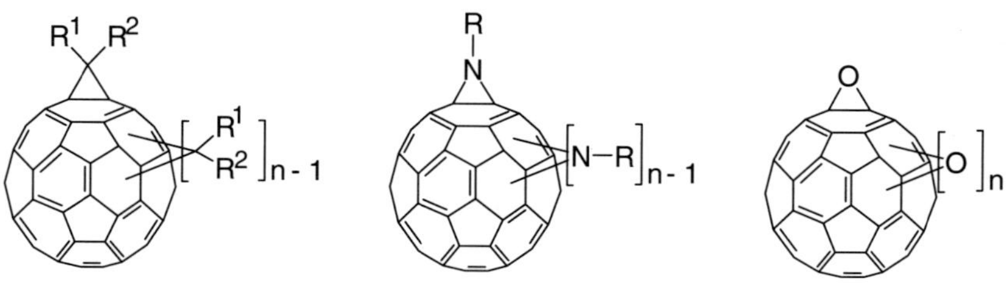

1 **2** **3**

Figure 2. General formulae of [2+1]-cycloadducts under study: cyclopropa- (1), aziridino- (2) and epoxyfullerenes (3).

Intermolecular interactions underlie the functioning of the mentioned applications of fullerene-containing systems and, while the polarizability determines the intermolecular interactions, its investigations are the basis to understand the mechanisms of processes, in which fullerenes and their derivatives take part.

There are two wide classes of fullerene adducts, which are [2+1]- and [1+1]-adducts (Figure 1).

[2+1]-Adducts are precursors of various fullerenes derivatives, including C_{60} conjugates with natural compounds. Moreover, X fragment can bridge the natural compound moieties with a fullerene cage in such conjugates [42, 43]. Cyclopropa- (1) and aziridinofullerenes (2) are well-known representatives of [2+1]-adducts. Their variety is explained by diverse qualities, *n* number, as well as the relative positions of substituents R, R^1, and R^2. Nowadays, methods for the synthesis of $C_{60}(CR^1R^2)_n$ and $C_{60}(NR)_n$ with various *n* are being developed [44, 45].

Constructed with carbon and oxygen atoms, fullerene epoxides (3) are another type of [2+1]-cycloadducts. Oxidative functionalization of fullerenes leads to the change of their initial polar characteristics [46]. As shown earlier, epoxides $C_{60}O_x$, $C_{70}O_y$ ($x = 1–6$, $y = 1–4$) can be easily produced by liquid-phase ozonation of a respective fullerene [47, 48].

Halofullerenes (halogenofullerenes, fullerene halogenides) $C_{60}Hal_n$ with Hal = F, Cl, Br are [1+1]-adducts of primary interest. These perspective compounds wait for applications in various branches of molecular electronics and nanotechnology. For example, polyfluorofullerenes are blocks for donor–acceptor diads and molecular devices, based on the long-life charge separation [49]. Since the great experimental material on the methods of synthesis, chemical and physical properties of halofullerenes has been accumulated in reviews and books (See, *e.g.*, [50, 51]), here we just note their interesting properties: different reactivity in the presence and in the absence of electron donor molecules [52], chemiluminescence upon ozonation, which depends on the number of halogen atoms added [53], tendency to form intermolecular complexes [20], electrochemical behavior, which is determined by the type and the number of atoms, decorated a fullerene cage [54].

Functionalization of C_{60} (and other fullerenes) changes the initial physicochemical characteristics. Even if the number of the attached addends is minimal, the electronic structure is transformed. As consequence, the effectiveness of the use of initial fullerene and its derivative in the same technology may differ. For example, B3LYP/6-31+G* calculations predict $C_{60}X_2$ to be superior to C_{60} in the electron-injection property but to be inferior in the electron-transport property [55]. To foresee what changes may occur in the synthesized fullerene derivative, various theoretical approaches are worked out (for example, the computational design of perspective 6.6-closed and 5.6-open $C_{60}CR^1R^2$ with improved electronic properties [56] and fullerene-based compounds with desirable static dielectric constants [57]).

In the present paper, we summarize our recent theoretical investigations on the fullerene derivatives polarizability, performed by modern DFT methods. The elucidation of the fullerene derivatives polarizability as a function of number of addends was the final goal of this research. The following compounds have been in the focus of the study:

1) [2+1]-adducts: epoxyfullerenes (**3**) and the simplest cyclopropa- (**1**) and aziridinofullerenes (**2**) with $R = R^1 = R^2 = H$;
2) [1+1]-adducts: fluoro-, chloro- and bromofullerenes.

2. DFT Methods and Theoretical Models for Calculations of Fullerenes Polarizability

2.1. DFT Methods Applied to Fullerenes Polarizability

Currently, various DFT methods are applied to quantum chemical studies on fullerenes and their derivatives. The choice of a method is determined by a research task. Choosing a method for polarizability calculations is often explained by good reproduction of experimental values of $\alpha(C_{60})$ and $\alpha(C_{70})$ (See Table 1; one of the exhaustive comparison between diverse semiempirical and DFT methods has been performed in [58]).

Among the other methods, we prefer Perdew–Burke–Ernzerhof (PBE) density functional theory method [59] with the 3ζ split-valence basis set [60] (Table 2), implemented in Priroda program [61]. This program suite was developed for parallel computing on multiprocessor computer systems with shared or distributed memory, driven by the Unix-like operation systems. In our case, these are clusters based on dual processor nodes connected by a high-speed network.

Table 2. Orbital basis sets for calculation of electronic configuration in PBE/3ζ method

Element	Contracted Gaussian-type functions	Uncontracted Gaussian-type functions
H	(5s1p)/[3s1p]	5s2p
C, O, N, F	(11s6p2d)/[6s3p2d]	10s3p3d1f
Cl	(15s11p2d)/[10s6p2d]	14s3p3d1f1g
Br	(18s14p9d)/[13s10p5d]	18s3p3d1f1g
I	(21s17p12d)/[15s12p8d]	23s3p3d1f1g

Table 3. Comparison between the PBE/3ζ-calculated [7, 64] and experimental IR spectra [67] for C_{60} and C_{70} (cm^{-1})

C_{60}		C_{70}	
PBE/3ζ	Experimental data	PBE/3ζ	Experimental data
		457	457
		529	533
		565	563
526	527	574	572
576	576	642	643
1179	1183	674	670
1432	1429	793	745
		1131	1130
		1414	1406
		1430	1430
		1460	1463

PBE/3ζ method reproduces structures and physicochemical characteristics of fullerenes and their derivatives with high accuracy [6, 16, 18, 21, 24, 32, 62–66], including the measured mean polarizabilities of C_{60} and C_{70} (See Table 1). To demonstrate the accuracy, we have imposed on each other the calculated and the experimental IR spectra of C_{60} and C_{70} fullerenes (Table 3).

Due to the paucity of experimental data on fullerene derivatives polarizability, in addition to PBE/3ζ, we usually use an auxiliary method, which has to be based on another functional and/or basis set. For example, B3LYP/Λ1 [68, 69] density functional method allows obtaining trustworthy α values for C_{60} and C_{70} (Table 1) and key structural parameters of the testing set of fullerene halogenides (See Supplementary material to [6]). However, B3LYP functional both in our [6] and in the previous calculations [31] has led to understatement of the components of polarizability tensors and, as consequence, to lower α in comparison with the respective PBE/3ζ-values. It is caused by the fact that B3LYP functional underestimates polarizability of the pristine C_{60}. Nevertheless, polarizabilities, calculated by PBE and B3LYP functionals, are comparable qualitatively, and their joint use is able to elucidate general trends of fullerene derivatives polarizability with a high degree of reliability. In the present paper, the results, obtained by PBE/3ζ method, are presented. After DFT-optimizations and vibration modes solving (to prove that all the stationary points, respective to the molecules under study, are minima of the potential energy surfaces) using standard techniques, the components of polarizability tensors α have been calculated in terms of finite field approach as the second order derivatives of the total energy *E* with respect to the homogenous (*i.e.*, the field gradient and higher derivatives are zero) external electric field *F*:

$$\alpha_{ij} = -\frac{\partial^2 E}{\partial F_i \partial F_j} \tag{3}$$

Tensors α have been calculated in the arbitrary coordinate system and then diagonalized. Their eigenvalues have been used for the calculation of the mean polarizabilities by Equation (2) and anisotropy of polarizability:

$$a^2 = \frac{1}{2}\left((\alpha_{yy} - \alpha_{xx})^2 + (\alpha_{zz} - \alpha_{yy})^2 + (\alpha_{zz} - \alpha_{xx})^2\right). \tag{4}$$

2.2. Additive Scheme for Calculation of Fullerene Adducts Polarizability

Currently, many quantum chemical methods are suitable for calculation of polarizability. Nevertheless, evaluation of polarizability in terms of additive schemes has not lost its relevance. Thus, a comparison of the quantum-chemically obtained values with the additive ones is significant for studies of correlation of the polarizability with molecular structure and the exploration of the mutual influence of a molecule's fragments at each other. Since one of the first additive schemes of polarizability was worked out by Le Fèvre [70], two trends of their improvement have been developed [2]:

1) Particularization of the additive scheme, taking into account the dependence of the parameters of atoms and bonds polarizabilities of on the types of the nearest molecular fragments (this approach should result in the use of polarizabilities of large structural fragments [2]) as the increments, which allows taking into account all the interactions within them).

2) Development of a strict additive scheme with a description of any deviations from it as a manifestation of interatomic interactions.

The additive scheme of the first type has been efficiently applied to fullerene derivatives [6, 66]. It involves the partitioning of a molecule on (n + 1) subunits of two types: a fullerene cage and n addends attached. For example, mean polarizabilities of C_{60} [2+1]-cycloadducts are estimated by this scheme within a chosen DFT method as:

$$\alpha_{\text{add}}(C_{60}X_n) = \alpha(C_{60}) + n\alpha_X, \tag{5}$$

where

$$\alpha_X = \alpha(C_{60}X) - \alpha(C_{60}) \tag{6}$$

are increments (X are bivalent chemical moieties, *e.g.*, >CH_2, >NH, >O). In the case of halofullerenes ([1+1]-adducts), $C_{60}Hal_2$ is a simplest derivative, so the increments and additive polarizabilities are determined as:

$$\alpha_{\text{Hal}} = \frac{1}{2}\left(\alpha(C_{60}Hal_2) - \alpha(C_{60})\right), \tag{7}$$

$$\alpha_{\text{add}}(C_{60}Hal_n) = \alpha(C_{60}) + n\alpha_{\text{Hal}}. \tag{8}$$

Increments α_X and α_{Hal} describe a change of polarizability upon the addition of one X (or Hal) fragment (atom), accompanied by disappearance of π-component of one 6.6 bond (α_X and $\alpha_{Hal} > 0$).

3. Polarizability of Fullerene Mono- and Bisadducts. Influence of Isomery on Polarizability and Its Anisotropy

At first, polarizabilities of [2+1]-adducts $C_{60}X$ with X = O (4), NH (5), CR^1R^2 (6–13), and $C_{70}O$ (14, 15) have been studied in detail (Figure 3) [6, 64, 66] and collected in Table 4, which also includes mean polarizabilities of some substituted cyclopropa[60]fullerenes:

1) Compounds 7–11, whose formation we have studied theoretically previously (7–11 have been synthesized *via* catalytic reactions [71, 72]);
2) Compound 12, a carboxylic derivative of $C_{60}CH_2$; 12 and its esters, having a complex of physiological activities [73, 74] as well as the ability to generate singlet oxygen, are considered to be perspective for nanomedicine [75];
3) Phenyl-C_{61}-butyric acid methyl ester (PCBM) (13), used in solar cells as a material of *n*-type [51].

For all the studied compounds 4–13, α values are higher than those of pristine C_{60}. Higher α values in comparison with the pristine fullerene also characterize C_{70} epoxides 14 and 15.

It is well-known that [2+1]-addition to C_{60} and C_{70} leads to two types of adducts, *viz.* 6.6-closed (addition to 6.6 bond) and 5.6-open (addition to 5.6 bond with its simultaneous cleavage) (Figure 1) [39]. Often, 5.6-open derivatives are formed in a mixture with their 6.6-

closed isomers in the same reactions and then convert to them spontaneously or under thermal treating.

If an addend, being attached to fullerene, is asymmetric, two types of 5.6-open adducts are possible, differing by the relative positions of R^1 and R^2 substituents. As an example, the substituted homofullerenes $C_{60}CR^1R^2$ (19–22) theoretically have two isomers (Figure 3).

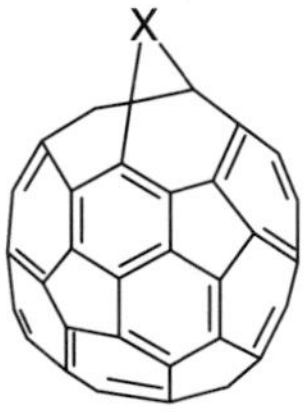

4 X = O

5 X = NH

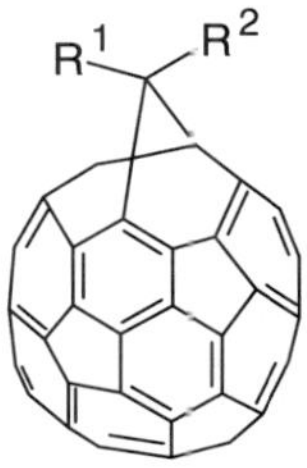

6 - 13

	R^1	R^2
6	H	H
7	CH_3	$COOCH_3$
8	H	$(CH_2)_3CH_3$
9	CH_3	$(CH_2)_3CH_3$
10	CH_2CH_3	$(CH_2)_3CH_3$
11	$(CH_2)_2CH_3$	$(CH_2)_3CH_3$
12	COOH	COOH
13	C_6H_5	$(CH_2)_3COOCH_3$

14

15

Figure 3. Structures of the studied 6.6-closed [2+1]-cycloadducts.

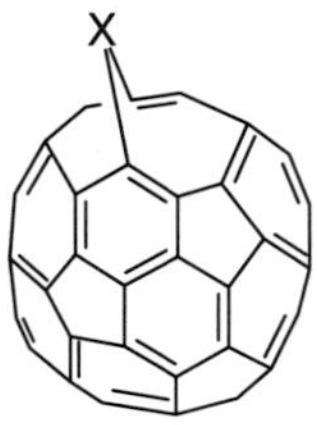

16 X = O

17 X = NH

18 - 22

	R^1	R^2
18	H	H
19a	H	$(CH_2)_3CH_3$
19b	$(CH_2)_3CH_3$	H
20a	CH_3	$(CH_2)_3CH_3$
20b	$(CH_2)_3CH_3$	CH_3
21a	CH_2CH_3	$(CH_2)_3CH_3$
21b	$(CH_2)_3CH_3$	CH_2CH_3
22a	$(CH_2)_2CH_3$	$(CH_2)_3CH_3$
22b	$(CH_2)_3CH_3$	$(CH_2)_2CH_3$

23

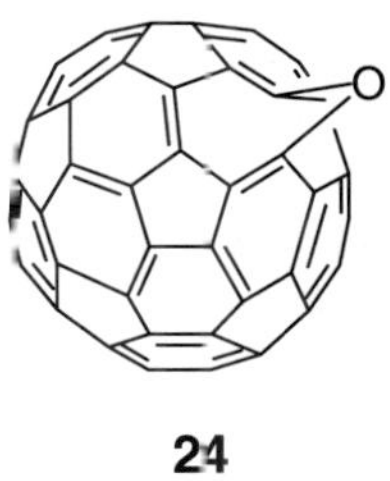

24

25

26

Figure 4. Structures of the studied 5.6-open [2+1]-cycloadducts.

According to the previous quantum chemical calculations, isomers 19b–22b, in which the bulkier R is placed under the nearest pentagon, are more stable thermodynamically. The corresponding a and b isomers for each of 19–22 compounds have almost the same mean polarizabilities. In the case of both C_{60} and C_{70} adducts (Figure 4, Table 4), the mean polarizability of a 5.6-open isomer exceeds the polarizability of its 6.6-closed counterpart on ~0.5 Å^3. It is explained with the contribution of π-electronic system to polarizability of the studied molecules: because all 6.6 double bonds remain unbroken in 5.6-open derivatives (*i.e.* the initial π-electronic system does not change significantly), they are characterized with higher α values than respective 6.6-closed cycloadducts.

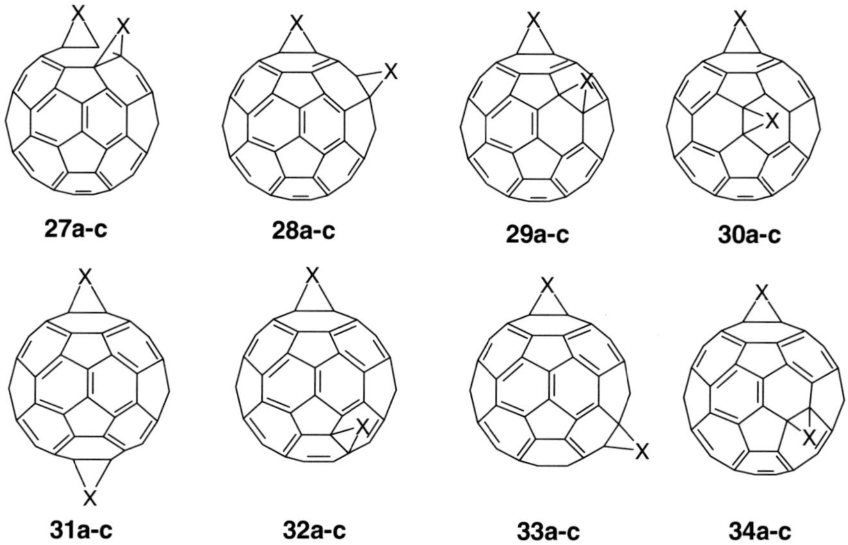

Figure 5. Structures of eight isomeric bisadducts: **a** – X = O, **b** – X = NH, **c** – X = CH_2.

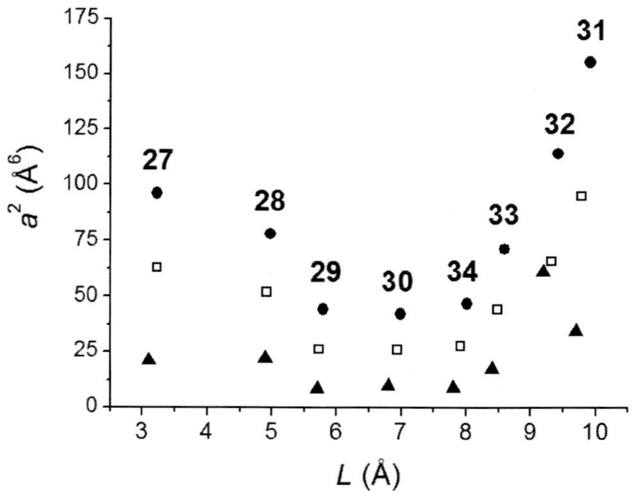

Figure 6. Dependences of anisotropy of polarizability on the distance between central atoms of addends in $C_{60}X_2$ (27–34): ● – X = CH_2, □ – X = NH, ▲ – X = O.

Mean polarizabilities of isomeric bisepoxy-, bisaziridino- and biscyclopropa[60]fullerenes 27–34 are approximately equal (~84, ~86 and ~87 $Å^3$, respectively). More significant differences are observed for anisotropy of their polarizability (a^2). Dependences of a^2 on the internuclear distance between the central atoms of the attached moieties *L* are shown in Figure 6.

Table 4. Mean polarizabilities of C_{60} and C_{70} [2+1]-cycloadducts, calculated by PBE/3ζ ($Å^3$)

Compound	α	Compound	α
4	83.2^a	16	83.9^a
5	84.3^b	17	84.8^a
6	85.0^b	18	85.5^a
7	94.8^c	19a (19b)	$94.4\ (94.8)^c$
8	93.9^c	20a (20b)	$96.5\ (96.7)^c$
9	96.1^c	21a (21b)	$98.6\ (98.7)^c$
10	98.2^c	22a (22b)	$100.8\ (100.9)^c$
11	100.3^c	23	103.8^a
12	91.7^c	24	103.8^a
13	$108.4^c\ (85.6^d)$	25	104.4^a
14	103.2^a	26	103.5^a
15	102.6^a	35 (36)	$109.3\ (109.9)^c$

[a] Taken from [64]. [b] Taken from [6]. [c] Calculated specially for this paper by PBE/3ζ according to standard methodology [6]. [d] B3LYP/6-31G(d) calculations from [37].

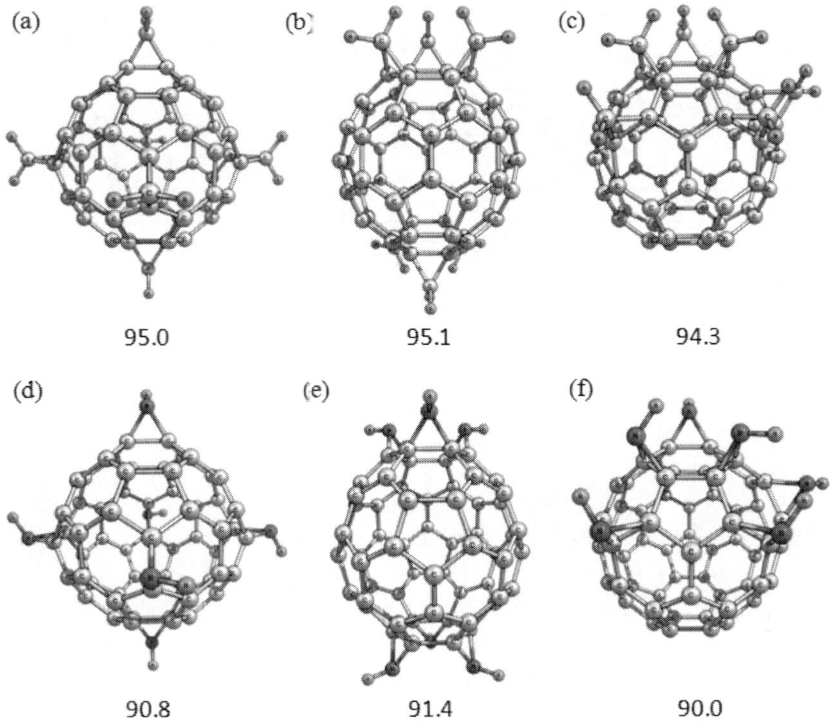

Figure 7. Hexakisadducts $C_{60}(CH_2)_6$ (a–c) and $C_{60}(NH)_6$ (d–f) with uniform (a, d), focal (b, e) and compact (c, f) distributions of X groups on a fullerene cage and their mean polarizabilities ($Å^3$).

Regioisomers are characterized by the different a^2 values. In the case of X = CH_2 and NH, the highest values of anisotropy are typical for *trans*-1-$C_{60}X_2$ (31), and the smallest ones correspond to equatorial bisadducts *e*-$C_{60}X_2$ (30) (bisepoxy[60]fullerenes fall out of this trend).

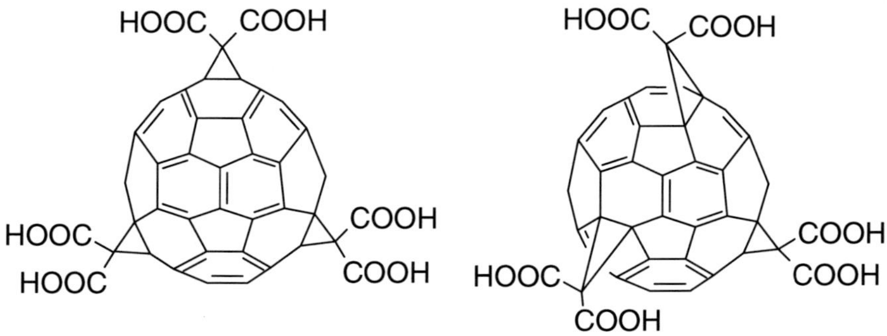

35 **36**

Figure 8. D_3- (**35**) and C_3-symmetry isomers (**36**) of '*carboxyfullerene*'.

Thus, isomeric fullerene derivatives demonstrate approximately the same mean polarizabilities and differ by anisotropies. It is also true for C_{60} [2+1]-adducts with greater number of addends and for [1+1]-adducts. According to DFT-calculations, the mean polarizabilities of $C_{60}X_6$ (X = CH_2, NH) isomers with compact, focal or uniform distributions of X moieties on a fullerene framework do not differ significantly (Figure 7).

Table 5. Mean polarizabilities of halofullerenes, obtained by PBE/3ζ method pure quantum-chemically and in terms of additive scheme (Å^3) [6]

Molecule	α	α_{add}	Molecule	α	α_{add}
1,2-$C_{60}F_2$	84.0	84.0	1,2-$C_{60}Cl_2$	89.6	89.6
C_s-$C_{60}F_{16}$	87.8	93.1	C_s-$C_{60}Cl_6$	100.7	103.4
C_{3v}-$C_{60}F_{18}$	87.4	94.4	T_h-$C_{60}Cl_{24}$	140.2	165.5
D_{5d}-$C_{60}F_{20}$	88.4	95.7	C_1-$C_{60}Cl_{28}$	149.4	179.3
T_h-$C_{60}F_{24}$	84.7	98.3	D_{3d}-$C_{60}Cl_{30}$	161.0	186.2
C_1-$C_{60}F_{36}$	88.8	106.1	C_2-$C_{60}Cl_{30}$	150.4	186.2
C_3-$C_{60}F_{36}$	88.8	106.1	1,2-$C_{60}Br_2$	92.9	92.9
T-$C_{60}F_{36}$	89.0	106.1	C_{2v}-$C_{60}Br_6$	109.9	113.4
D_3-$C_{60}F_{48}$	90.5	113.9	C_s-$C_{60}Br_8$	119.6	123.7
S_6-$C_{60}F_{48}$	90.4	113.9	T_h-$C_{60}Br_{24}$	172.6	205.6

In the case of the substituted cyclopropa[60]fullerenes, mean polarizability remains regardless of positional relationship of the addends attached. We have demonstrated it on the example of two '*carboxyfullerenes*' – *t,t,t*- (**35**) и *e,e,e*-tris(dicarboxymethano)fullerenes (**36**), produced *via* reaction of fullerene with malonic ester derivatives [76] (these species are of a great interest due to their physiological activity, *e.g.*, inhibition activity towards some enzymes [77]).

In spite of the different patterns of additions in 35 and 36 compounds (Figure 8), their mean polarizabilities are equal (Table 4). The analogous situation is typical for isomeric halofullerenes (Table 5). Most of them have been obtained as mixtures of several isomers.

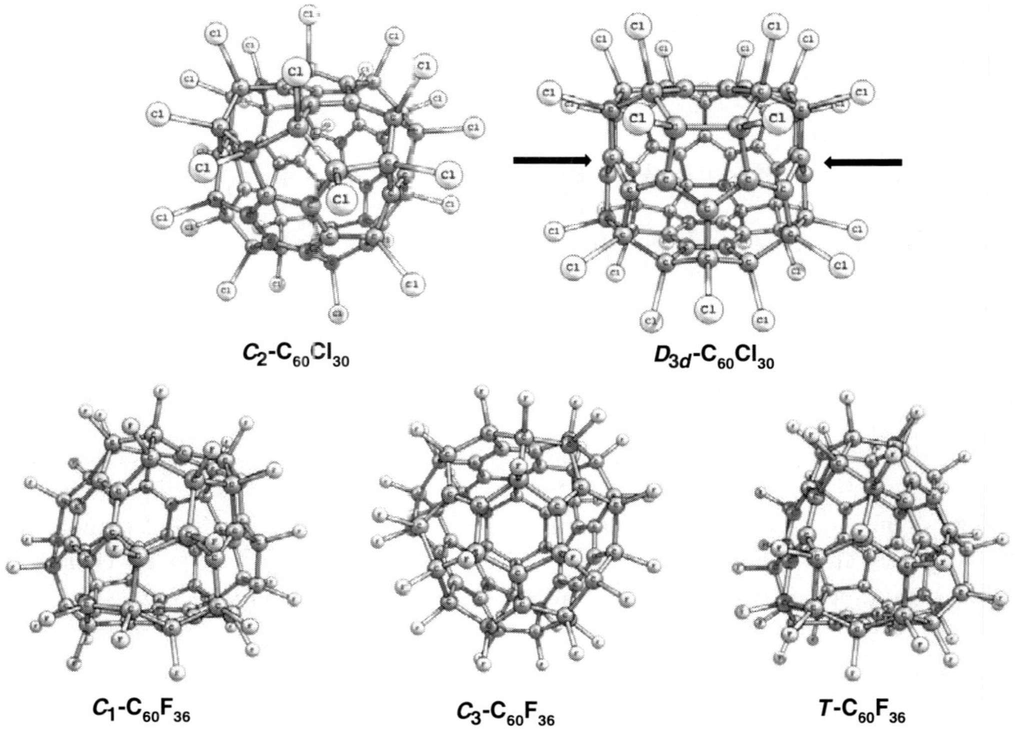

Figure 9. Isomeric $C_{60}Cl_{30}$ and $C_{60}F_{36}$ halofullerenes. Trannulene equatorial belt in D_{3d}-$C_{60}Cl_{30}$ is marked by arrows.

Moreover, isomerization processes are able to take place in halogen-containing fullerene derivatives. For example, a slow room-temperature interconversion of C_1 and C_3 isomers of $C_{60}F_{36}$ occurs in the presence of ambient atmosphere [78]. Carbon skeletons of fluorinated matters changes insignificantly upon the isomerization [79]:

$$C_1\text{-}C_{60}F_{36} \leftrightarrow C_3\text{-}C_{60}F_{36} \quad (9)$$

Isomers of $C_{60}Hal_n$, which differ by the relative Hal positions, are characterized by approximately the same values of mean polarizability [6], *e.g.*, for all $C_{60}F_{36}$ isomers α equals to ~89 Å^3 (Table 5). However, the third isomer T-$C_{60}F_{36}$ has the carbon skeleton, differing significantly from the mentioned above meantime it has the same polarizability.

Simultaneously, the difference in mean polarizability achieves the largest value for two $C_{60}Cl_{30}$ isomers (~10 Å^3) with dissimilar geometries of carbon frameworks. Also the structure of isomer with higher polarizability D_{3d}-$C_{60}Cl_{30}$ is characterized by the trannulene equatorial belt, which made up by facilely-polarizable conjugated double bonds (Figure 9).

In general, isomery should effect negligibly on polarizability of fullerene adducts if there are no bright contrasts in their carbon cage structures.

4. Depression of Polarizability of C_{60}/C_{70} Fullerene [2+1]- and [1+1]-Adducts

4.1. General Formula for Calculation of C_{60} Adducts Polarizability

Nowadays, the situation in the fullerene synthetic chemistry might be described as follows: generally, a fullerene derivative with the required number of addends and their relative position is available *via* modern synthetic strategies. Therefore, relation between fullerene derivative polarizability and number of addends is of the applied interest.

Such dependence has been studied in detail for the simplest representatives of cyclopropa- $C_{60}(CH_2)_n$ and aziridinofullerenes $C_{60}(NH)_n$ (**1** and **2** with $R = R^1 = R^2 = H$) [66]. Initially, we have proposed that polarizability should linearly increase with the increase of n value according to the additive scheme (Equation (5)). The additive polarizabilities have been calculated to adducts with n up to 30 (which is a number of double bonds in C_{60} molecule). Then, we have calculated $C_{60}(CH_2)_n$ and $C_{60}(NH)_n$ polarizabilities pure quantum-chemically. For this purpose, the only randomly chosen isomer has been selected for each n because, as it has been shown before, mean polarizability does not depend on the positional relationship of functional groups and are determined mainly by their number (except $C_{60}X_{29}$ and $C_{60}X_{30}$, presented by the single isomers).

As it has turned out, the differences between $\alpha(C_{60}X_n)$ and $\alpha_{add}(C_{60}X_n)$ become greater with n increase. According to Equation (5), polarizabilities of $C_{60}(CH_2)_n$ and $C_{60}(NH)_n$ should enlarge linearly if $n \rightarrow 30$. However, we observe the depression of polarizability $\Delta\alpha$, *i.e.* the negative deviation of $\alpha(C_{60}X_n)$ from $\alpha_{add}(C_{60}X_n)$:

$$\Delta\alpha = \alpha_{\text{add}}\left(C_{60}X_n\right) - \alpha\left(C_{60}X_n\right) \tag{9}$$

For both classes of cycloadducts, $\Delta\alpha$ achieves maximal value at $n = 30$ (Figure 10). Based on mathematical induction, analysis of the computed data allowed obtaining a fitting function [66], which unites mean polarizability and number of addends in a fullerene derivative molecule:

$$\alpha\left(C_{60}X_n\right) = \alpha\left(C_{60}\right) + n\alpha\left(X\right) - \frac{\Delta\alpha\left(C_{60}X_{n_{\max}}\right)n^2}{n_{\max}^2}, \tag{10}$$

where $\Delta\alpha\left(C_{60}X_{n_{\max}}\right)$ is a depression of polarizability of the totally-functionalized fullerene derivative. Fitting functions (10), render the quantum-chemically obtained values of [2+1]-cycloadducts polarizability with high accuracy (Figure 10). Derived strictly for $C_{60}(CH_2)_n$ and $C_{60}(NH)_n$ cycloadducts, this formula reproduces well the DFT-calculated mean polarizabilities of other derivatives of C_{60}: polyepoxy-, fluoro-, chloro- and bromo[60]fullerenes (Figure 11).

Parameters of fitting functions (10) for the studied fullerene derivative are collected in Table 6. Depression of polarizability of fullerene derivatives is a unique property of three-dimensional (3D) molecular systems. It has been proved by calculating mean polarizabilities of two series of chloro-derivatives of naphthalene $C_{10}H_{8-n}Cl_n$ and anthracene $C_{14}H_{10-n}Cl_n$

(these are examples of 2D halocarbons) [6]. Their mean polarizabilities are successfully described in terms of additive scheme, analogous to the applied to fullerene derivatives (respective values of α and α_{add} for halocarbons do not differ significantly). It means that in 2D case, there is no depression of polarizability, so it is a prerogative of fullerene derivatives (3D molecular systems).

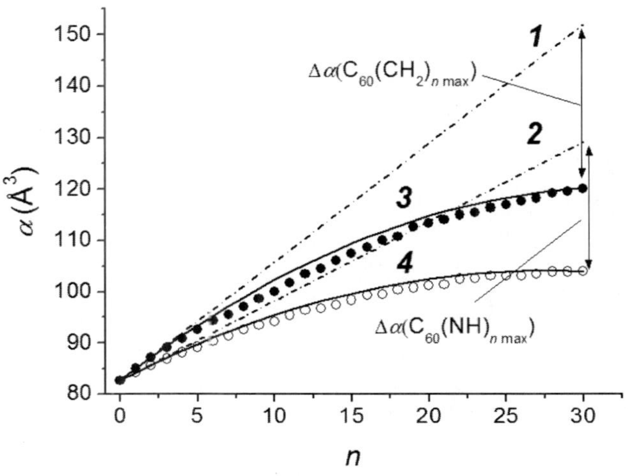

Figure 10. Dependences of α on n values, obtained by PBE/3ζ method in terms of additive scheme without (1 – $C_{60}(CH_2)_n$, 2 – $C_{60}(NH)_n$) (Equation (5)) and with the correction on the depression of polarizability (3 – $C_{60}(CH_2)_n$, 4 – $C_{60}(NH)_n$) according to Equation (10). Black and white circles correspond to pure quantum-chemically calculated α values of $C_{60}(CH_2)_n$ and $C_{60}(NH)_n$, respectively.

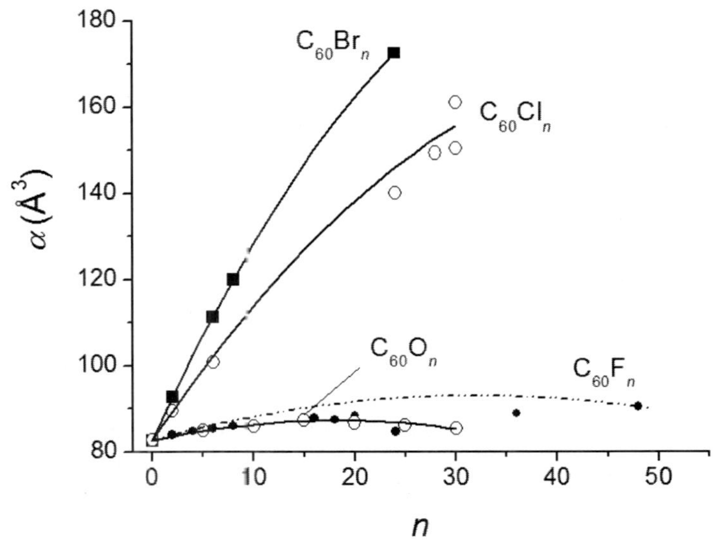

Figure 11. Dependences of fullerene derivatives mean polarizabilities *vs*. number of attached atoms in a molecule, obtained by PBE/3ζ method. Symbols correspond to the pure quantum-chemically calculated mean polarizabilities; lines show dependences α *vs*. n, obtained by the use of fitting function (10).

Table 6. Parameters of Equation (10) for calculation of polarizability, applied to various fullerene adducts

Fullerene adducts	α_X (Å^3)	$\Delta\alpha\left(C_{60}X_{n_{max}}\right)$(Å^3)	n_{max}
$C_{60}(CH_2)_n$	2.31	31.8	30
$C_{60}(NH)_n$	1.55	25.2	30
$C_{60}O_n$	0.50	12.3	30
$C_{60}F_n$	0.65	23.5	48
$C_{60}Cl_n$	3.45	30.5	30
$C_{60}Br_n$	5.12	33.0	24
$C_{60}I_n$	8.17	47.8	24
$C_{60}H_n$	0.525	17.6	36
$C_{70}Cl_n$	3.78	41.6	28
$C_{70}[OOC(CH_3)_3]_n$	11.51	8.7	10

Is there a physical meaning to the Equation (10)? This question will be answered by further theoretical and experimental investigations. Mathematically obtained, the expression for depression of polarizability

$$\Delta\alpha\left(C_{60}X_n\right)=\frac{\Delta\alpha\left(C_{60}X_{n\max}\right)n^2}{n_{\max}^2} \qquad (11)$$

can be rewritten as

$$\Delta\alpha\left(C_{60}X_n\right)=\frac{\Delta\alpha\left(C_{60}X_{n\max}\right)}{n_{\max}}\frac{n}{n_{\max}}n. \qquad (12)$$

Here,

$$\Delta\alpha\left(C_{60}X_{n\max}\right)^{spec}=\frac{\Delta\alpha\left(C_{60}X_{n\max}\right)}{n_{\max}}, \qquad (13)$$

$$\phi=\frac{n}{n_{\max}}, \qquad (14)$$

where $\Delta\alpha\left(C_{60}X_{n\max}\right)^{spec}$ is a specific depression of polarizability, corresponding to deviation from the additive scheme per each addend, and □ is a degree of functionalization. Thus, the resulting depression is proportional to the two adverted values.

The disadvantage of our explanation of depression of polarizability and the derived general formula should be noted. One of the parameters, determining Δα value according to Equation (11), is a maximal number of addends (n_{max}), which are able to be attached to fullerene skeleton. The n_{max} value is difficult to determine exactly (both experimentally and theoretically), as it is significant for the effective use of the general formula (10). Though the stability of some totally-functionalized fullerene derivatives has been clearly shown (e.g., polyepoxide $C_{60}O_{30}$ [80]), $n_{max} = 30$ is rather hypothetical value. Notwithstanding, the use of the theoretically possible maximal value ($n_{max} = 30$) for [2+1]-cycloadducts and n_{max} values for [1+1]-adducts, experimentally-known at the moment (Table 6), demonstrates good agreement between fitting function and DFT-calculations.

4.2. Depression of Polarizability of C_{60} Derivatives in Experimental Studies

There is the only experimental work where mean polarizabilities of two halofullerenes have been measured [81]. The experimental technique is based on Kapitza–Dirac–Talbot–Lau interferometry [82, 83]. The obtained experimental data on $C_{60}F_{36}$ and $C_{60}F_{48}$ correspond to the mixtures of isomers [81]. Nevertheless, as follows from the calculations (Table 7), the structure does not influence on the static mean polarizability of a halofullerene (though it remains significant for evaluation of dipole moments).

Despite the fact that DFT methods, used by us, overestimate the respective measured values, both experimental and theoretical studies indicate the equality of $C_{60}F_{36}$ and $C_{60}F_{48}$ polarizabilities. A mismatch between the measured and calculated values may occur because the computation represents static polarizabilities while the experiment [81] yields the optical polarizability at 532 nm laser wavelength.

The equal mean polarizabilitites of fullerene fluorides with different numbers of addends in a molecule [81] is the first confirmation of our theoretical assumptions about depression polarizability [6, 64, 66].

Table 7. Mean polarizabilities of polyfluoro[60]fullerenes[a] (Å^3)

Molecule	OLYP/Λ1	BLYP/Λ1	PBE/3ζ	B3LYP/Λ1
C_1-$C_{60}F_{36}$	78.2	79.5	88.8	93.2
C_3-$C_{60}F_{36}$	78.2	79.5	88.8	93.3
T-$C_{60}F_{36}$	78.5	79.8	89	93.7
Average	78.3	79.6	88.9	93.4
Experimental[b]	60.3±7.7			
D_3-$C_{60}F_{48}$	79.2	80.6	90.5	95.3
S_6-$C_{60}F_{48}$	79.1	80.5	90.4	95.1
Average	79.2	80.6	90.5	95.2
Experimental[b]	60.1±7.5			

[a] PBE/3ζ and B3LYP/Λ1 are taken from [6]; OLYP/Λ1 and BLYP/Λ1 calculations have been performed specially for this paper according to the standard computational methodology [6]. [b] Experimental data are taken from [81].

4.3. Phenomenon of Polarizability Depression for the Other Fullerene Derivatives

To demonstrate applicability of the theoretical model, described above, to congenerous classes of fullerenes derivatives, we have studied mean polarizabilities of the following species: experimentally-obtained chloro- and *tert*-buthylperoxy-derivatives of C_{70} [84, 85], and C_{60} derivatives (hydrides [86] and hypothetical iodo[60]fullerenes $C_{60}I_n$). Types of fullerene or functional groups added are varied in these classes of fullerene derivatives. Formula (10) accurately reproduces values of $C_{70}X_n$ and $C_{60}X_n$ mean polarizabilities obtained by a DFT method (Figure 12) [6, 87]. So, it can be recommended for an accurate calculation of polarizability for various types of fullerene derivatives without performing of resource-intensive quantum chemical calculations.

Previously, theoretical studies for $C_{60}H_2$ [88] and $C_{60}H_4$ [89] have shown that electronic properties of fullerene core are changed significantly if even 2 and 4 hydrogen atoms are added. It makes the hydrogenation of C_{60} an effective method for modifying the hole-transport property, and some isomers of $C_{60}H_2$ have potential utility as hole-transport materials. Based on DFT-calculations, we can ascertain that, in spite of the mentioned changes of electronic properties of C_{60} hydrides compared to original C_{60} [88, 89], there are no bright contrasts in mean polarizability upon the number of hydrogen atoms increasing [87].

In our works, we pay particular attention to high-polarizability molecules, for example iodo[60]fullerenes, which has not been obtained at the moment [51]. As known, there are no examples of fullerene derivatives with C–I bonds, possibly due to the constraints, arising between the voluminous iodine atoms. It makes the reaction of iodine with fullerene core thermodynamically unfavorable [90].

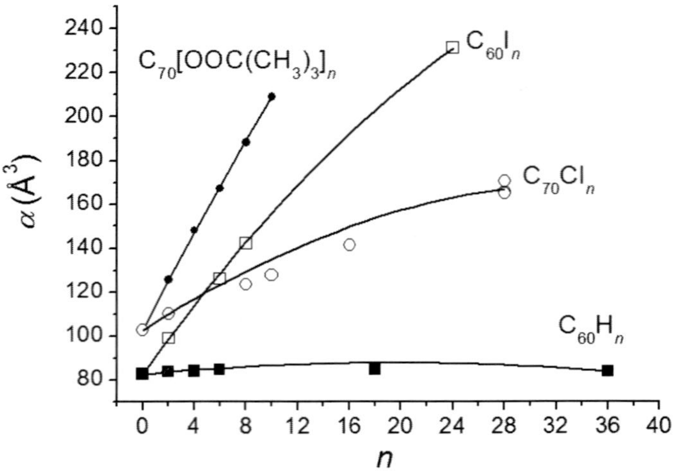

Figure 12. Dependences of mean polarizabilities of C_{60} and C_{70} derivatives *vs.* number of the attached groups in a molecule (PBE/3ζ calculations). Symbols correspond to the pure quantum-chemically calculated values; lines show dependences α *vs.* *n*, obtained by the use of fitting function (10) with the parameters, taken from Table 6.

Table 8. Calculated mean polarizability and depression of polarizability (in parentheses) of $C_{70}X_n$ (Å^3) [87]

X	$C_{70}X_8$	$C_{70}X_{10}$
H	101.9 (1.1)	102.1 (1.0)
CH_3	116.9 (8.6)	120 (11.3)
C_6H_5	196 (12.4)	217.8 (17.09)
Cl	123.6 (9.3)	128 (12.5)
Br	137.5 (10.8)	144.8 (14.9)
$OOC(CH_3)_3$	188.1 (6.6)	209.1 (8.7)

However, the iodination of C_{60} fullerene should lead to easily-polarizable compounds, which have a strongly expressed response to external electric fields, because $C_{60}I_n$ have the highest mean polarizabilities among the other C_{60} derivatives. The works, intended to the synthesis of the iodinated C_{60}, are being performed [91]. Also polarizabilities of C_{70} derivatives $C_{70}X_8$ and $C_{70}X_{10}$ with the same pattern of addition have been calculated (Table 8) [87]. All the classes of the compounds under study demonstrate $\Delta\alpha \neq 0$. Values of $\Delta\alpha$ for the fullerene derivatives, decorated by highly-polarizable groups or atoms, are large enough to be experimentally observed (*e.g.*, for X = Br and C_6H_5).

Conclusion

The performed systematic DFT study on the fullerene derivatives polarizability allows formulating guidelines for molecular design of fullerene-based molecules with regulated polarizability:

1. Polarizabilities of exohedral fullerene derivatives are higher than those of a respective pristine fullerene.
2. Polarizabilities of 5.6-open isomers of fullerene derivatives are higher than those of respective 6.6-closed isomers.
3. Regioisomers of $C_{60/70}X_n$ are characterized with the approximately equal mean polarizabilities and differ with the anisotropy of polarizability.
4. Depression of polarizability of polyadducts $C_{60/70}X_n$ with $n = 3$ appears ($\Delta\alpha > 0$) and increases with the increase of n values.

Depression of polarizability seems to be the most interesting item of the listed above. Conjecturing its potential importance both for fundamental and applied studies, we have derived the general formula for calculation of depression polarizability, which have demonstrated high accuracy for description of the mean polarizabilities of fullerene derivatives with diverse structure.

Writing this paper, we have found the first experimental work, devoted to mean polarizabilities of polyfluorofullerenes. The measured data confirms our DFT-based theoretical constructions and makes us confident that both theoretical and experimental studies on the depression of polarizability will continue and lead to the useful information for materials science of fullerenes and their derivatives.

Acknowledgments

The work was supported by the Presidium of Russian Academy of Sciences (Program No. 24 'Foundations of Basic Research of Nanotechnologies and Nanomaterials').

References

[1] Helgaker, T.; Coriani, S.; Jørgensen, P.; Kristensen, K.; Olsen, J.; Ruud, K. *Chem. Rev.* 2012, *112*, 543−631.

[2] Vereshchagin, A. N. *Polarizability of Molecules*, Nauka, 1980 (in Russian).

[3] Bonin, K. D.; Kresin, V. V. *Electric-Dipole Polarizabilities of Atoms, Molecules and Clusters*, World Scientific, 1997.

[4] Broyer, M.; Antoine, R.; Benichou, E.; Compagnon, I.; Dugourd, Ph. *C. R. Physique* 2002, *3*, 301−317.

[5] Compagnon, I.; Antoine, R.; Broyer, M.; Dugourd, Ph.; Lermé, J.; Rayane, D. *Phys. Rev.* A 2001, *64*, 025201.

[6] Sabirov, D. Sh.; Garipova, R. R.; Bulgakov, R. G. *Chem. Phys. Lett.* 2012, *523*, 92–97.

[7] Sabirov, D. Sh.; Khursan, S. L.; Bulgakov, R. G. *J. Mol. Graph. Model.*, 2008, *27*, 124–130.

[8] Fuchs, D.; Rietschel, H.; Michel, R. H.; Fischer, A.; Weis, P.; Kappes, M. M. *J. Phys. Chem.* 1996, *100*, 725–729.

[9] Jonsson, D.; Norman, P.; Ruud, K.; Ågren, H.; Helgaker, T. *J. Chem. Phys.* 1998, *109*, 572–578.

[10] Luzanov, A. V.; Bochevarov, A. D.; Shishkin, O. V. *J. Struct. Chem.* 2001, 42, 296–300.

[11] Jensen, L.; Åstrand, P.-O.; Mikkelsen, K. V. *J. Phys. Chem.* B 2004, *108*, 8226–8233.

[12] Zope, R. R.; Baruah, T.; Pederson, M. R.; Dunlap, B. I. *Phys. Rev.* B 2008, *77*, 115452.

[13] Rappoport, D.; Furche, F. *J. Chem. Phys.* 2010, *133*, 134105.

[14] Calaminici, P.; Carmona-Espindola, J.; Geudtner, G.; M. Köster, A. M. *Int. J. Quantum Chem.* 2012, *112*, 3252–3255.

[15] Berninger, M.; Stefanov, A.; Deachapunya, S.; Arndt, M. *Phys. Rev.* A 2007, 76, 013607.

[16] Bulgakov, R. G.; Galimov, D. I.; Sabirov, D. Sh. *JETP Lett.* 2007, *85*, 632–635.

[17] Zope, R. R. *J. Phys. B: At. Mol. Opt. Phys.* 2007, *40*, 3491–3496.

[18] Sabirov, D. Sh.; Bulgakov, R. G.; Khursan, S. L.; Dzhemilev, U. M. *Dokl. Phys. Chem.* 2009, *425*, 54–56.

[19] Bezmel'nitsyn, V. N.; Eletskii, A. V.; Okun', M. V. *Phys. Usp.* 1998, *41*, 1091–1114.

[20] Konarev, D. V.; Lyubovskaya, R. N. *Russ. Chem. Rev.* 1999, *68*, 19–38.

[21] Sabirov, D. Sh.; Bulgakov, R. G.; Khursan, S. L. *ARKIVOC* 2011, *part 8*, 200–224.

[22] Gueoruiev, G. K.; Pacheco, J. M.; Tománek D. *Phys. Rev. Lett.* 2004, *92*, 215501.

[23] Yan, H.; Yu, Sh.; Wang, X.; He, Y.; Huang, W.; Yang, M. *Chem. Phys. Lett.* 2008, *456*, 223–226.

[24] Sabirov, D. Sh.; Bulgakov, R. G. *JETP Lett.* 2010, *92*, 662–665.

[25] Torrens, F. *Microelectron. Eng.* 2000, *51–52*, 613–626.

[26] Zope, R. R. *J. Phys. B: At. Mol. Opt. Phys.* 2008, *41*, 085101.

[27] Langlet, R.; Mayer, A.; Geuquet, N.; Amara, H.; Vandescuren, M.; Henrard, L.; Maksimenko, S.; Lambin, Ph. *Diamond Relat. Mater.* 2007, *16*, 2145–2149.

[28] Liu, S.; Sun, S. *J. Organomet. Chem.* 2000, *559*, 74–86.

[29] Delaney, P.; Greer, J. C. *Appl. Phys. Lett.* 2004, *84*, 431–433.

[30] Rivelino, R.; Malaspina, Th.; Fileti, E.E. *Phys. Rev. A* 2009, *79*, 013201.

[31] Tang, Sh.-W.; Feng, J.-D.; Qiu, Y.-Q.; Sun, H.; Wang, F.-D.; Chang, Y.-F.; Wang, R.-Sh. *J. Comput. Chem.* 2010, *31*, 2650–2657.

[32] Sabirov, D. Sh.; Bulgakov, R. G. *Fullerenes Nanotubes Carbon Nanostruct.* 2010, *18*, 455–457.

[33] Tang, Ch.; Zhu, W.; Zou, H.; Zhang, A.; Gong, J.; Tao, Ch. *Comput. Theoret. Chem.* 2012, *991*, 154–160.

[34] Tang, Ch.; Zhu, W.; Deng, K. *Chin. J. Chem.* 2010, *28*, 1355–1358.

[35] Yang, Y.; Wang, F.-H.; Zhou, Y.-S. *Phys. Rev. A* 2005, *71*, 013202.

[36] Zhang, C. R.; Liang, W. Zh.; Chen, H. Sh.; Chen, Y. H.; Wei, Zh. Q.; Wu, Y. Zh. *J. Mol. Struct. THEOCHEM* 2008, *862*, 98–104.

[37] Zhang, C. R.; Chen, H. Sh.; Chen, Y. H.; Wei, Zh. Q.; Pu, Zh. Sh. *Acta Phys. Chim. Sin. (Wuli Huaxue Xuebao)* 2008, *24*, 1353–1358.

[38] Tang, Sh.-W.; Feng, J.-D.; Qiu, Y.-Q.; Sun, H.; Wang, F.-D.; Su, Zh.-M.; Chang, Y.-F.; Wang, R.-Sh. *J. Comput. Chem.* 2011, *32*, 658–667.

[39] Thilgen, C.; Diederich, F. *Chem. Rev.* 2006, *106*, 5049–5135.

[40] Guldi, D. M.; Prato, M. *Chem. Commun.* 2004, 2517–2525.

[41] Wang, Sh.; Gao, R.; Zhou, F.; Selke, M. *J. Mater. Chem.* 2004, *14*, 487–493.

[42] Yurovskaya, M. A.; Trushkov, I. V. *Russ. Chem. Bull. Int. Ed.* 2002, *51*, 367–443.

[43] Tuktarov, A. R.; Dzhemilev, U. M. *Russ. Chem. Rev.* 2010, *79*, 585–610.

[44] Tuktarov, A. R.; Korolev, V. V.; Khalilov, L. M.; Ibragimov, A. G.; Dzhemilev, U. M. *Russ. J. Org. Chem.* 2009, *45*, 1594–1597.

[45] Akhmetov, A. R.; Tuktarov, A. R.; Dzhemilev, U. M.; Yarullin, I. R.; Gabidullina L. A. *Russ. Chem. Bull. Int. Ed.* 2011, *60*, 1885–1887.

[46] Heymann, D.; Weisman, R. B. *C. R. Chimie* 2006, 9, 1107–1116.

[47] Bulgakov, R. G.; Nevyadovskii, E. Yu.; Belyaeva, A. S.; Golikova, M. T.; Ushakova, Z. I.; Ponomareva, Yu. G.; Dzhemilev, U. M.; Razumovskii, S. D.; Valyamova, F. G. *Russ. Chem. Bull. Int. Ed.* 2004, *53*, 148–159.

[48] Bulgakov, R. G.; Ponomareva, Yu. G.; Muslimov, Z. S.; Tuktarov, R. F.; Razumovsky, S. D. *Russ. Chem. Bull. Int. Ed.* 2008, *57*, 2072–2080.

[49] Goryunkov, A. A.; Ovchinnikova, N. S.; Trushkov, I. V.; Yurovskaya, M. A. *Russ. Chem. Rev.* 2007, *76*, 289–312.

[50] Boltalina, O. V.; Galeva, N. A. *Russ. Chem. Rev.* 2000, *69*, 609–621.

[51] Troshin, P. A.; Troshina, O. A.; Lyubovskaya, R. N.; Razumov, V. F. *Functional fullerene derivatives, methods of synthesis and perspectives for use in organic electronics and biomedicine*, Ivanovo, 2008, (in Russian).

[52] Yoshida, Yu.; Otsuka, A.; Drozdova, O. O.; Saito, G. *J. Am. Chem. Soc.* 2000, *122,* 7244–7251.

[53] Bulgakov, R. G.; Galimov, D. I.; Mukhacheva, O. A.; Goryunkov, A. A. *Russ. Chem. Bull. Int. Ed.*, 2010, *59*, 1843–1845.

[54] Troshin, P. A.; Troshina, O. A.; Peregudova, S. M.; Yudanova, E. I.; Buyanovskaya, A. G.; Konarev, D. V.; Peregudov, A. S.; Lapshina, A. N.; Lyubovskaya, R. N. *Mendeleev Commun.* 2006, *16*, 206–208.

[55] Tokunaga, K. *Chem. Phys. Lett.* 2009, *476*, 253–257.

[56] Wang, Y.; Seifert, G.; Hermann, H. *Phys. Stat. Sol.* A. 2006, *203*, 3868–3872.

[57] Tokunaga, K. Computational Design of New Organic Materials: Properties and Utility of Methylene-Bridged Fullerenes C_{60}. In: *Handbook on Fullerene: Synthesis,*

Properties and Applications (Editors Verner, R. F.; Benvegnu, C.), Nova Science Publishers, Inc. 2011, 517–537.

[58] Alparone, A.; Librando, V.; Minniti, Z. *Chem. Phys. Lett.* 2008, *460*, 151–154.

[59] Perdew, J. P.; Burke, K.; Ernzerhof, M. *Phys. Rev. Lett.* 1996, *77*, 3865–3868.

[60] Laikov, D. N. *The development of saving approach to calculation of molecules by a density functional method, its application to the complicated chemical problems,* PhD thesis, Moscow State University, 2000 (in Russian).

[61] Laikov, D. N.; Ustynyuk, Yu. A. *Russ. Chem. Bull. Int. Ed.* 2005, *54*, 820–826.

[62] Sabirov, D. Sh.; Bulgakov, R. G. *Comput. Theoret. Chem.* 2011, *963*, 185–190.

[63] Tulyabaev A. R.; Khalilov, L. M. *Comput. Theoret. Chem.* 2011, *976*, 12–18.

[64] Sabirov, D. Sh.; Bulgakov, R. G. *Chem. Phys. Lett.* 2011, *506*, 52–56.

[65] Pankratyev, E. Yu.; Tulyabaev, A. R.; Khalilov, L. M. *J. Comp. Chem.* 2011, *32*, 1993–1997.

[66] Sabirov, D. Sh.; Tukhbatullina A. A.; Bulgakov, R. G. *Comput. Theoret. Chem.* 2012, *993*, 113–117.

[67] Sokolov, V. I.; Stankevich, I. V. *Russ. Chem. Rev.* 1993, *62*, 419–435.

[68] Becke, A. D. *J. Chem. Phys.* 1993, *98*, 5648–5653.

[69] Laikov, D. N. *Chem. Phys. Lett.* 2005, *416*, 116–120.

[70] Le Fèvre, R. J. W. *Adv. Phys. Org. Chem.* 1965, *3*, 1–90.

[71] Tuktarov, A. R.; Akhmetov, A. R.; Sabirov, D. Sh.; Khalilov, L. M.; Ibragimov, A. G.; Dzhemilev, U. M. *Russ. Chem. Bull. Int. Ed.* 2009, *58*, 1724–1730

[72] Tuktarov, A. R.; Korolev, V. V.; Sabirov, D. Sh.; Dzhemilev U. M. *Russ. J. Org. Chem.* 2011, *47*, 41–47.

[73] Satoh, M.; Matsuo, K.; Kiriya, H.; Mashino, T.; Hirobe, M.; Takayanagi, I. *Gen. Pharmacol.* 1997, *29*, 345–351.

[74] Okuda, K.; Hirota, T.; Hirobe, M.; Nagano, T.; Mochizuki, M.; Mashino, T. *Fullerene Sci. Technol.* 2000, *8*, 127–142.

[75] Hamano, T.; Okuda, K.; Mashino, T.; Hirobe, M.; Araka, K.; Ryu, A.; Mashiko, S.; Nagano, T. *J. Chem. Soc. Chem. Commun.* 1997, 21–22.

[76] Lamparth, I.; Hirsch, A. *J. Chem. Soc. Chem. Commun.* 1994, 1727–1728.

[77] Wolff, D. J.; Barbieri, C. M.; Richardson, C. F.; Schuster, D. I.; Wilson, S. R. *Arch. Biochem. Biophys.* 2002, *399*, 130–141.

[78] Avent, A. G.; Taylor, R. *Chem. Commun.* 2002, 2726–2727.

[79] Avdoshenko, S. M.; Ioffe, I. N.; Sidorov, L. N. *J. Phys. Chem. A* 2009, *113*, 10833–10838.

[80] Ren, X.-Y.; Liu, Z.-Y. *J. Mol. Graph. Model.* 2007, *26*, 336–341.

[81] Hornberger, K.; Gerlich, S.; Ulbricht, H.; Hackermüller, L.; Nimmrichter, S.; Goldt, I. V.; Boltalina, O.; Arndt, M. *New J. Phys.* 2009, *11*, 043032.

[82] Gerlich, S.; Eibenberger, S.; Tomandl, M.; Nimmrichter, S.; Hornberger, K.; Fagan, P. J.; Tüxen, J.; Mayor, M.; Arndt, M. *Nature Commun.* 2011, *2*, 263.

[83] Gerlich, S.; Hackermüller, L.; Hornberger, K.; Stibor, A.; Ulbricht, H.; Goldfarb, F.; Savas, T.; Müri, M.; Mayor, M.; Arndt, M. *Nature Phys.* 2007, *3*, 711–715.

[84] Xiao, Z.; Wang, F.; Huang, Sh.; Gan, L.; Zhou, J.; Yuan, G.; Lu, M.; Pan, J. *J. Org. Chem.* 2005, *70*, 2060–2066.

[85] Troshin, P. A.; Lyubovskaya, R. N. *Russ. Chem. Rev.* 2008, *77*, 305–349.

[86] Goldshleger, N. F.; Moravsky, A. P. *Russ. Chem. Rev.* 1997, *66*, 323–342

[87] Sabirov, D. Sh.; Garipova, R. R.; Bulgakov, R. G. *Fullerenes Nanotubes Carbon Nanostruct.* 2012, *20*, 386–390.

[88] Tokunaga, K.; Ohmori, Sh.; Kawabata, H.; Matsushige, K. *Jpn. J. Appl. Phys.* 2008, *47*, 1089–1093.

[89] Tokunaga, K.; Kawabata, H.; Matsushige, K. *Jpn. J. Appl. Phys.* 2008, 47, 3638–3642.

[90] Zhou, O.; Cox, D. E. *J. Phys. Chem. Solids* 1992, 53, 1373–1390.

[91] Tuktarov, A. R. (unpublished data).

In: Density Functional Theory
Editors: J. Morin and J.M. Pelletier, pp. 171-186
ISBN: 978-1-62417-954-9

Chapter 6

THE APPLICATION OF DENSITY FUNCTIONAL THEORY TO CALCULATION OF PROPERTIES OF ENVIRONMENTALLY IMPORTANT SPECIES DI- AND TRIMETHYLNAPHTHALENES

Bojana D. Ostojić* and Dragana S. Đorđević
Institute of Chemistry, Technology and Metallurgy,
University of Belgrade, Belgrade, Serbia

Abstract

Alkylated naphthalenes are persistent environmental pollutants but there is still little information on their characteristics. In this chapter we present the application of Density Functional Theory (DFT) methods for obtaining structural parameters of dimethylnaphthalenes (DMNs) and trimethylnaphthalenes (TMNs) through fully geometry optimization, their vibrational frequencies, IR intensities, Raman activities, and the assignment of vibrational modes in the ground electronic state. Almost all of the investigated molecules are characterized by a planar equilibrium geometry. The knowledge of aromaticity of these molecules based on nucleus-independent chemical shifts (NICS) can lead to a more sophisticated understanding of the reactivity of these molecules. The condensed electrophilic Fukui function calculated at the B3LYP/cc-pVTZ level of theory show the localization of the most positive parts susceptible to electrophilic attack. These regions are particularly important from the point of view of dioxigenation reaction on the aromatic rings of DMNs and TMNs by the naphthalene dioxigenase (NDO) enzymes in the process of biodegradation. The obtained equilibrium geometries and the transition state (TS) geometries enable further investigation of the π^*-σ^* hyperconjugation effects and their influence on the methyl group torsional barriers in the ground electronic state (S_0). Obtained results present basis for the investigation of the conformational flexibility of the aromatic rings of these molecules using *ab initio* techniques which can give important information about geometry relaxation of these molecules in possible intermolecular interactions.

* E-mail address: bostojic@chem.bg.ac.rs (Corresponding Author)

Introduction

In the last two decades Density Functional Theory (DFT) method has become a computationally attractive alternative to wave function techniques. Comparative studies have shown that DFT methods can give equal or even better results than some *ab initio* techniques. The geometries, energies, vibrational frequencies, NMR shifts, and variety of other important molecular properties can be obtained with a high level of accuracy with the help of DFT methods. A lot of books, special issues, and articles in international journals on fundamental and computational as well as conceptual aspects of DFT and its application are available [1-25]. The literature on DFT is rich in excellent reviews and overviews. In this chapter we put emphasis on the application of DFT methods for calculation of properties of some alkylated naphthalenes as species of great environmental concern.

Polycyclic aromatic hydrocarbons (PAHs) are a class of organic compounds composed of two or more fused aromatic rings. They have toxic, mutagenic, and carcinogenic properties and are of great importance for the environment. These compounds are released into the environment through the incomplete burning of fuels, garbage or other organic substances resulting from human activities. The main anthropogenic sources include processes such as wood preservation, petroleum refining, transport, power and heat generation, lignite pyrolysis sites, municipal waste incineration, hazardous waste landfills, traffic *etc*. The U.S. Environmental Protection Agency [26] defined 16 priority pollutants among PAHs: low molecular weight PAHs (2–3-ring PAHs) including naphthalene, acenaphthylene, acenaphthene, fluorene, phenanthrene, and anthracene and high molecular weight PAHs (4–6-ring PAHs) including fluoranthene, pyrene, benz[a]anthracene, chrysene, benzo[b]fluoranthene, benzo[k]fluoranthene, benzo[a]pyrene, dibenz[a,h]anthracene, indeno[1,2,3-c,d]pyrene, and benzo[g,h,i]perylene. Due to their ubiquitous occurrence, recalcitrance, bioaccumulation potential and carcinogenic activity, the PAHs have gathered significant environmental concern. Naphthalene (NPH) is the PAH with the structure of two fused benzene rings. Naphthalene and alkylated naphthalenes are semi-volatile, present in the atmosphere mostly in the gas phase. Trimethylnaphthalenes (TMNs) as well as other alkylated naphthalenes such as methyl- and dimethylnaphthalenes (DMNs) are constituents of diesel fuel. Various DMNs and TMNs are among alkylated naphthalenes which constitute main volatile PAHs of bitumen emissions. However, until recently, there has not been much information available on alkyl homologues of naphthalene [27-39].

This chapter reports the results of the DFT calculation of the full geometry optimization of DMN and TMN isomers, their vibrational frequencies, IR intensities, Raman activities, and the assignment of the vibrational modes. The characteristic features of Raman spectra of TMNs are presented and the simulated Raman spectra of selected TMN isomers are reported. The rotational constants and dipole moment values of all TMN are given. In the second part of the chapter the calculated vibrational wavenumbers of two DMN isomers in the ground electronic state are compared with the experimental values. The magnetic properties of all DMN isomers based on nucleus-independent chemical shifts (NICS) are presented. The results were obtained using the Gaussian 09 [40] suite of programs.

Trimethylnaphthalenes

Changes of Geometrical Parameters Introduced by Methylation in 1,2,6-TMN and 1,2,7-TMN

It is widely recognized that methylation can induce, enhance, reduce or even abolish the toxic, carcinogenic, or mutagenic properties of PAHs. Methyl substitution in PAHs can alter their carcinogenic activity depending on the number of substituents and on the position of substitution [41]. Small changes in structure can have a considerable effect on carcinogenic activity. Therefore it is essential to investigate changes of structural and electronic properties of PAHs that methyl substituents cause.

The fully optimized molecular geometries of the 1,2,6- and 1,2,7-TMN isomers at the B3LYP/cc-pVTZ level, along with the labeling of atoms, are displayed in Figure 1. The frequency calculations were carried out at the same level as the geometry optimization using analytic evaluation of the second derivatives of energy with respect to the nuclear displacements. Bauschlicher and Langhoff investigated substituted PAHs and their ions and observed computational problems associated with symmetry breaking [42].

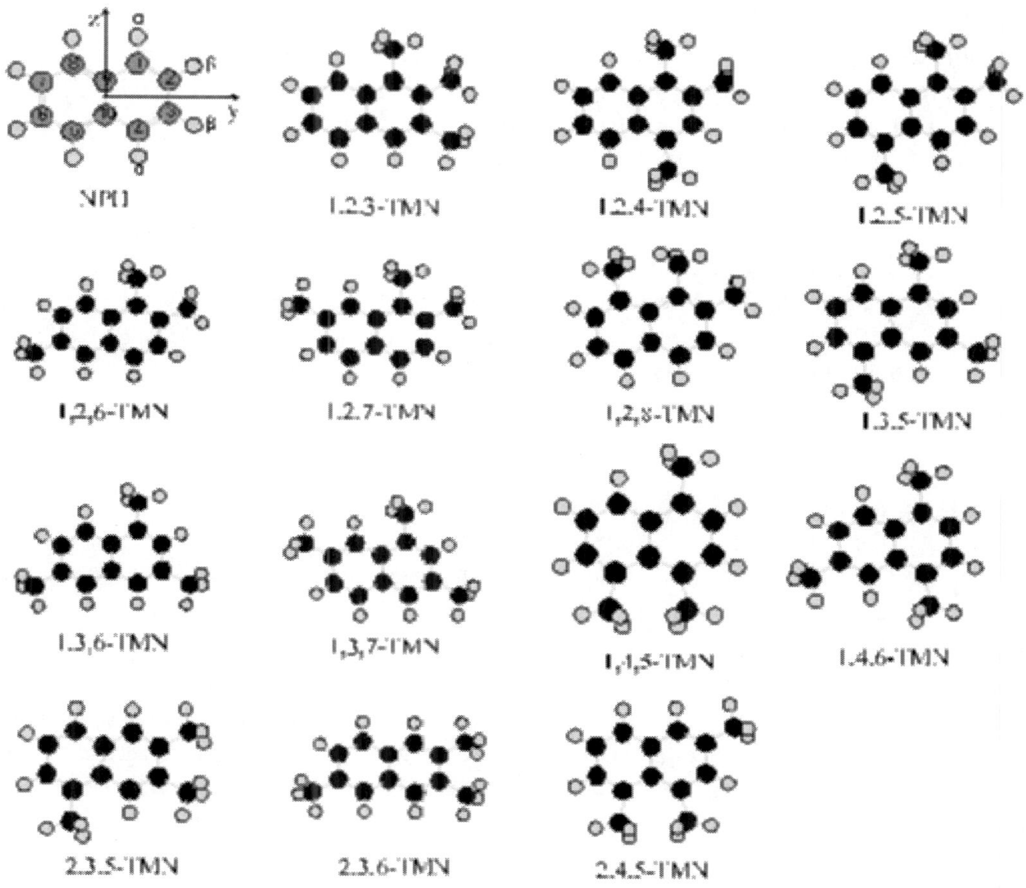

Figure 1. The equilibrium geometries of TMN isomers and atom numbering.

Within expected C_s symmetry of the 2-methylnaphthalene molecule, the results of their calculations at the B3LYP [43,44] level showed one imaginary frequency. They noticed that many of the methyl PAHs have lower symmetry than expected. In some cases the B3LYP approach appears to break symmetry, resulting in a less-accurate vibrational spectrum and running the BP86 [45,46] approach can detect an unexpected symmetry breaking in the B3LYP approach [47]. Therefore, we have performed the calculations within C_1 symmetry. In addition to the frequency calculations at the B3LYP level of theory which showed no imaginary values, implying that the structures of these two isomers are minima on the potential energy surfaces, we also checked the results using the BP86 functional. The geometry optimizations at the BP86 level have been carried out using the cc-pVTZ basis set [48].

The distortion of the benzene rings is obvious from the increase of the bond lengths C_1–C_2 (1.385 Å) and C_1–C_9 (1.430 Å) from C_3–C_4 (1.366 Å). Three methyl groups introduce naphthalene rings distortions. The bond length between C_1 and C_2 (the substitution places of two methyl groups) is 0.015 Å larger than the corresponding bond length (1.370 Å) in the naphthalene molecule fully optimized at the same level of theory. Another major structural change is found for the bond length between C_9 and the carbon atom bound to the methyl group in the α position, C_1. The computed bond length C_1–C_9 in both isomers is 0.014 Å larger than the corresponding bond length (1.416 Å) in the naphthalene molecule. The most remarkable changes of the bond angles from those in the naphthalene molecule can be noticed for the angles C_8–C_9–C_1 and C_4–C_{10}–C_5. The C_4–C_{10}–C_5 bond angle of 1,2,6-TMN and 1,2,7-TMN is smaller (121.66° and 121.92°, respectively), than the corresponding angle in naphthalene, 122.33°. On the other hand, compared to the C_8–C_9–C_1 bond angle in the naphthalene molecule (122.33°), the corresponding angles in the 1,2,6-TMN and 1,2,7-TMN isomers are 122.66° and 122.41°, respectively. Methylation of the rings in naphthalene results in an increase in the area of rings of 0.0340Å^2 and 0.0356Å^2 for 1,2,6-TMN and 1,2,7-TMN, respectively, compared to the corresponding value for the rings of the naphthalene molecule. The area of the first ring where two methyl groups are attached is 0.0215Å^2 and 0.0208Å^2 larger for 1,2,6-TMN and 1,2,7-TMN, respectively, than the area of the second ring with one methyl group.

Molecular Constants and Dipole Moments of TMNs

In our study on TMNs reported previously [37] it was shown that most TMNs are weakly or moderately polar molecules. Due to the small value of the dipole moment for some of TMNs (*e.g.* 1,3,6-TMN) the rotational spectrum is expected to be weak. However, 1,2,8-TMN, 2,3,5-TMN, and 1,2,3-TMN have sufficiently large dipole moments for rotational spectroscopy and the results that will be presented in this section complement the data reported previously [37]. McNaughton et al. reported molecular constants of benzanthrone obtained from the Fourier Transform Microwave (FT-MW) spectroscopy (13-18 GHZ) of a supersonic rotationally cold molecular expansion [49]. They showed that DFT calculations at the B3LYP/6-311+G(d,p) level of theory closely predict the experimental rotational and centrifugal distortion constants. The calculations at the B3LYP/6-311+G(d,p) level of theory shown that they can predict closely the ground state rotational constants and centrifugal distortion constants of phenanthridine, acridine, 1,10-phenanthroline, quinazoline, quinoxaline, and phthalazine [50,51]. The predicted dipole moments at the B3LYP/cc-pVTZ level of theory can give values close to experimentally derived values [50,52]. We compared the rotational constants of naphthalene calculated at the B3LYP/6-311+G(d,p) level of theory (Table 1) with those accurately determined by Baba *et al.* employing the ultrahigh-resolution laser spectroscopy [53]. To our knowledge the rotational constants of TMNs are not published so far. Given these predictions, one might expect the B3LYP/6-311+G(d,p) rotational constants and the B3LYP/cc-pVTZ dipole moment values of TMNs presented in Table 1 to provide reasonable estimates.

Table 1. The rotational constants (cm^{-1}) of TMNs and NPH calculated at the B3LYP/6-311+G(d,p) level, dipole moment values (Debye) calculated at the B3LYP/cc-pVTZ level and the comparison with experimental values

Molecule	*A*	*B*	*C*	*μ*
1,3,6-TMN	0.0471355	0.0194096	0.0138957	0.11
1,4,5-TMN	0.0362647	0.0286138	0.0161377	0.31
1,3,7-TMN	0.0514134	0.0190566	0.0140117	0.31
1,2,6-TMN	0.0613322	0.0179414	0.0139889	0.37
1,4,6-TMN	0.0389104	0.0240948	0.0150046	0.39
2,3,6-TMN	0.0662923	0.0162392	0.0131391	0.44
1,3,5-TMN	0.0373055	0.0254034	0.0152406	0.44
1,2,5-TMN	0.0481965	0.0219550	0.0152115	0.45
1,2,4-TMN	0.0387614	0.0258521	0.0156436	0.51
1,2,7-TMN	0.0534306	0.0194752	0.0143871	0.56
2,4,5-TMN	0.0421092	0.0235947	0.0152500	0.66
1,2,8-TMN	0.0435755	0.0244408	0.0158672	0.86
2,3,5-TMN	0.0493236	0.0204601	0.0145784	0.89
1,2,3-TMN	0.0474333	0.0220361	0.0151731	0.95
NPH	0.104274 0.104052[a]	0.0411281 0.0411269[a]	0.0294947 0.0294838[a]	0.00

[a] exp. data from ultrahigh-resolution laser spectroscopy [53].

Raman Spectra of TMN Isomers

It is well known fact that the accurate computation of the IR and Raman intensities is not an easy task and requires the large basis sets with diffuse functions. For small molecules DFT calculations using the aug-cc-pVTZ and Sadlej pVTZ basis sets can give results in very good agreement with the observed values from IR and Raman experiments in the gas-phase [54]. In this chapter we present the results of the calculations of Raman spectra of TMNs using the B3LYP/6-311+G(2df,p) level of theory. Planarity was not assumed during full geometry optimization and all the calculations have been done within C_1 symmetry.

The vibrational frequencies obtained using the DFT calculations are proven to be in good agreement with experimental ones if scaled to compensate for the basis set truncation effects, the incompleteness of electron correlation, and the neglect of anharmonicity [55-57]. The wavenumbers have been corrected using the scaling factor of 0.9686. The scaling factor used for frequencies were determined by Merrick et al. [57] for the hybrid functional B3LYP in conjunction with the 6-311+G(2df,p) basis set.

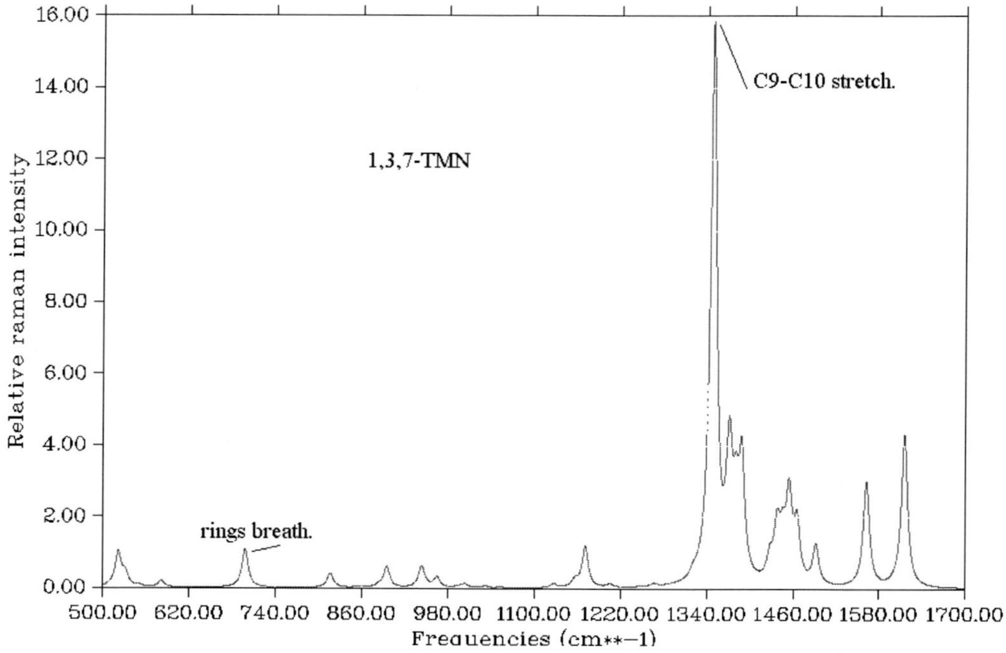

Figure 2. The calculated Raman spectrum of the 1,3,7-TMN isomer in the 500-1700 cm^{-1} region obtained at the B3YP/6-311+G(2df,p) level of theory. Wavenumbers have been corrected by 0.9686 scaling factor reported by Merrick *et al.* [57]. Lorentz lineshapes with FWHM of 10 cm^{-1} are used.

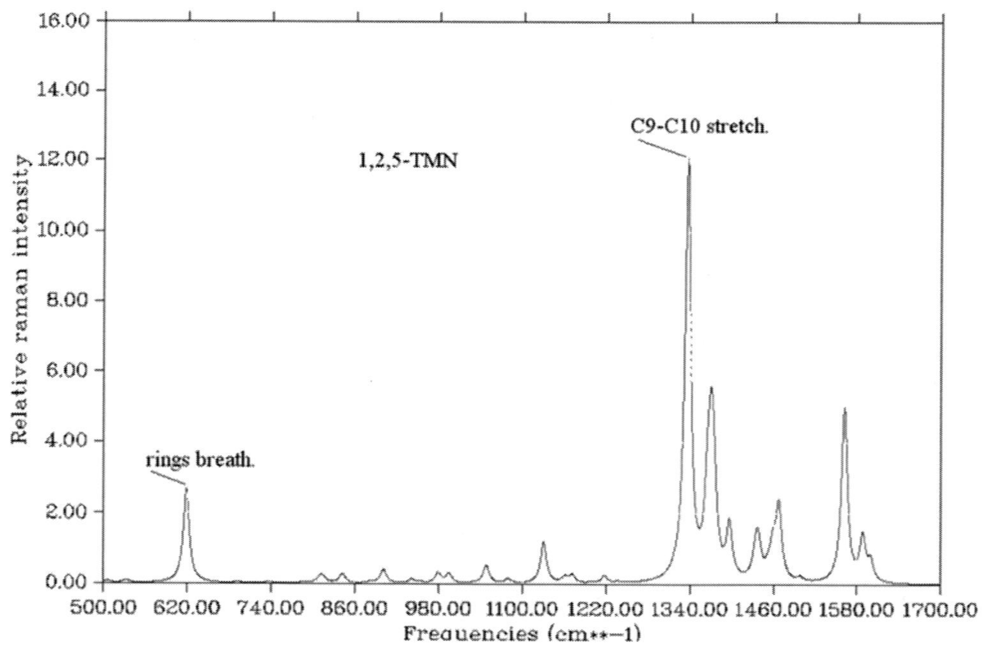

Figure 3. The calculated Raman spectrum of the 1,2,5-TMN isomer in the 500-1700 cm^{-1} region obtained at the B3YP/6-311+G(2df,p) level of theory. Wavenumbers have been corrected by 0.9686 scaling factor reported by Merrick *et al.* [57]. Lorentz lineshapes with FWHM of 10 cm^{-1} are used.

The assignment of vibrational modes was carried out on the basis of normal modes as displacements in redundant internal coordinates and through the visualization software Avogadro [58]. The Raman spectra were computed in the double harmonic approximation. The simulated Raman spectra (Figures 2 and 3) are obtained with the help of the Multiwfn 2.3 program [59]. The variation of the line width as a function of mode is not taken into account and simulated spectra are represented by convoluting the computed frequency values with pure Lorentzian band shape employing a full width at half maximum (FWHM) of 10 cm^{-1}.

Table 2. Harmonic vibrational wavenumbers ω (cm^{-1}) and scaled wavenumbers[a] ω_{sc} (cm^{-1}) for the characteristic vibrational modes in the spectra of TMNs calculated at the B3LYP/6-311+G(2df,p) level of theory and the comparison with the available exp. data for 1,5-DMN and NPH

Molecule	methyl group sym. C-H stretching mode		C_9-C_{10} stretching mode		rings "breathing" mode	
	ω	ω_{sc}	ω	ω_{sc}	ω	ω_{sc}
2,3,6-TMN	3019	2924	1403	1359	752	728
2,3,5-TMN	3019	2924	1396	1352	731	708
1,2,3-TMN	3027[f,g] 3020[g]	2932[f,g] 2925[g]	1392	1348	692	670
1,3,7-TMN	3019	2924	1392	1348	721	698
1,2,7-TMN	3023	2928	1391	1347	726	704
1,2,4-TMN	3023	2928	1391	1347	715	692
1,3,6-TMN	3020	2925	1391	1347	786	761
1,2,6-TMN	3023	2928	1390	1347	728	705
1,4,6-TMN	3019	2924	1385	1342	722	700
1,3,5-TMN	3020	2925	1382	1338	571	553
1,2,5-TMN	3023	2929	1379	1336	639	619
1,2,8-TMN*	3042[g] 3019[g]	2946[g] 2924[g]	1375	1332	620	601
2,4,5-TMN	3050[g] 3019[g]	2954[g] 2924[g]	1364	1321	602	583
1,4,5-TMN	3051	2955	1363	1320	617	597
1,5-DMN	3020	2926 2920[b]	1379	1336 1360[b]	641	621 620[b]
NPH			1394 1397[e]	1350 1353[e] 1353[c]	774 775[e]	750 750[e] 758[d]

[a]The scaling factor 0.9686 reported by Merrick *et al.* is used for correction of the wave numbers [57].
[b]The experimental values obtained from the FTIR and FT Raman [35].
[c]FTIR and Raman [60]
[d][61]
[e]The values obtained at the B3LYP/cc-pVTZ level of theory. The scaling factor 0.9682 taken reported by Merrick *et al.* [57] is used for correction of the wave numbers.
[f]The C-H stretching vibration in methyl group is coupled to the C-C stretching vibration of the C-CH_3 bonds.
[g]Both bands corresponding to symmetric C-H stretching vibrations in methyl group appear with high Raman activity.
*The 1,2,8-TMN is the isomer whose rings geometry deviates from planarity.

In the Raman spectra of TMNs three characteristic regions with pronounced peaks important for the identification of these species can be recognized (Figures 2 and 3). This pattern bears a resemblance to the pattern in the simulated spectra of DMNs as Alparone and Librando pointed out [33]. The first one is in the high-frequency end of the spectrum, in the region 2900-3050 cm^{-1} that is assigned to the C-H stretching in methyl group of TMN. The peak is attributed to symmetric stretching vibration of the CH_3 groups (Table 2).

In some TMNs (1,2,3-TMN, 12,8-TMN, and 2,4,5-TMN) two bands corresponding to symmetric C-H stretching vibrations in methyl group appear with high Raman activity. The second pronounced peak in the Raman spectra of TMNs located between 1300-1400cm^{-1} corresponds predominantly to a stretching vibration of the central C_9-C_{10} bond which connects two rings. The third peak ascribed mainly to rings breathing vibration is located between 550 and 770 cm^{-1}. The bands in this region are informative and may serve for distinguishing the TMNs.

Although further experimentally observed spectra with more features over the previously reported spectra as well as isotope substitution analysis of the data are needed, we believe that the results presented will be of help in identification of these TMN isomers in an unknown mixture by spectroscopic methods as well as in investigation of intermolecular interactions of these molecules with other biomolecules.

Dimethylnaphthalenes

IR and Raman Spectra of DMNs

In this chapter we compare the theoretically calculated frequencies of 1,5-DMN and 2,3-DMN with those obtained experimentally. The gas phase IR spectra of three dimethylnaphthalenes (1,5-DMN, 1,6-DMN, and 2,6-DMN) at 0.5 cm^{-1} resolution have been

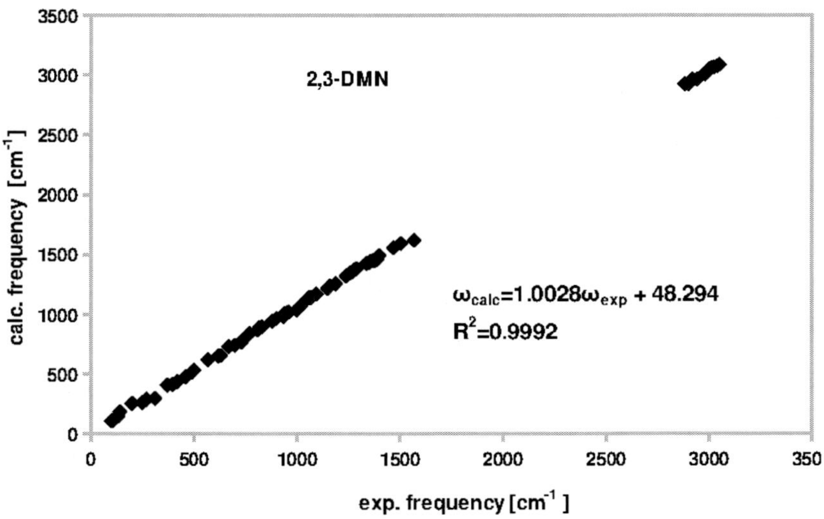

Figure 4. The comparative graph of calculated frequencies at the B3LYP/cc-pVTZ level of theory with experimental values reported by Prabhu *et al.* [36] for the 2,3-DMN isomer.

reported by Prasanta *et al.*[34]. Nagabalasubramanian and Periandy [35] reported the results of the FTIR and FT Raman measurements of 1,5-DMN in the range 3600-10 cm^{-1} in the solid phase. Prabhu *et al.* recorded FTIR and FT Raman spectra of 2,3-DMN in the region 4000-100 cm^{-1} [36]. We performed the full geometry optimization of 1,5-DMN and 2,3-DMN employing the B3LYP/6-311+G(2df,p) and B3LYP/cc-pVTZ levels of theory, respectively, and frequency calculations in the double harmonic approximation. The 1,5-DMN isomer is characterized with two distant methyl groups while 2,3-DMN possesses two closely interacting methyl groups. The previous work on the IR and Raman spectra of these two isomers [33, 35, 36] shows that vibrational frequencies scaled by a single factor can give the results in good accordance with the experimental data. Despite limitations employed, it can be concluded that DFT calculations at the B3LYP/6-311+G(2df,p) and B3LYP/cc-pVTZ levels of theory are able to closely predict the experimental vibrational frequencies of 1,5- and 2,3-DMN (Figures 4 and 5) and to guide the initial spectral search of other DMNs.

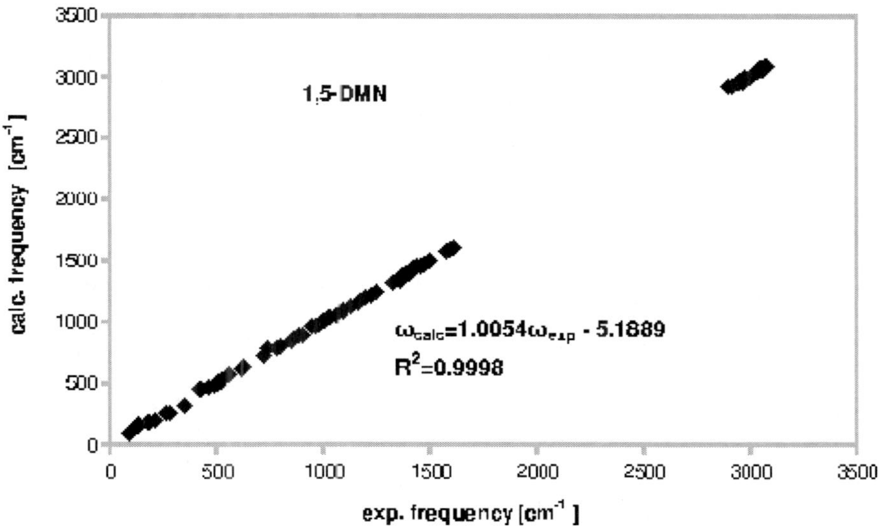

Figure 5. The comparative graph of calculated frequencies at the B3LYP/6-311+G(2df,p) level of theory with experimental values reported by Nagabalasubramanian and Periandy [35] for the 1,5-DMN isomer.

NICS values of DMNs

The knowledge of aromaticity [62] of DMNs based on NICS values [63] as the most widely used method to discuss aromaticity can lead to a more sophisticated understanding of the reactivity of these molecules. All ten isomers of DMN are characterized by a planar equilibrium geometry of the aromatic system. The NICS values of DMNs are calculated as the total isotropic shielding at the point 1Å above the ring centroid (NICS(1)) employing the GIAO-B3LYP [64,65] level of theory. Another magnetic index suggested by Fowler and Steiner [66,67], based on the total MO contribution to the zz component of the NICS tensor ($NICS_{zz}$) was also presented. $NICS_{zz}$ was computed 1Å above the ring centroids ($NICS(1)_{zz}$). A good performance of the $NICS(1)_{zz}$ index for planar aromatic rings compared to the most refined NICS aromaticity index NICS(0)π_{zz} which employs localized molecular orbitals (LMO) or canonical molecular orbitals (CMO) dissection for selection of only the π contribution to the zz component of the tensor[67-69] was demonstrated by Fallah-Bagher-Shaidaei *et al.* [70]. We employed one of Jensen's polarization-consistent pcS-n basis sets, pcS-2, designed specifically for the calculation of the nuclear magnetic shielding constants [71]. The pcS-2 basis set could be described as a triple-zeta quality basis set. The number of contracted basis functions for the NICS calculation of a DMN isomer was 564.

The methyl substitution causes the enlargement of the area of rings compared to the naphthalene molecule. It can be noticed that the methyl substitution in the α position has larger effect to the ring area than the methyl substitution in the β position [39]. The area of the first ring of the 1,4-DMN isomer is the largest and the area of the first ring of the 2,6-DMN and 2,7-DMN is smallest among DMN isomers. If both methyl groups are in the α position and in the close proximity like in the 1,8-DMN isomer it causes the steric strain of the methyl groups on rings and it is reflected in the enlargement of the ring are. The 1,4-DMN and 1,8-DMN characterized by the presence of the bulky methyl groups in the α position possess the largest area of rings among DMNs. The results of calculations of the NICS–based magnetic indices (Table 3) show that the presence of two methyl groups does not change the degree of aromaticity of the naphthalene rings to a large extent. For all ten isomers of DMN the NICS values are negative indicating diatropicity associated with aromatic character.

Table 3. The NICS data (ppm) for both rings of DMN isomers and NPH obtained using the GIAO method and B3LYP level of theory with the pcS-2 basis set. The fully optimized geometries at the B3LYP/cc-pVTZ level of theory are employed

Molecule	1. ring		2. ring	
	NICS(1)	NICS(1)$_{zz}$	NICS(1)	NICS(1)$_{zz}$
NPH	-10.46	-29.55	-10.46	-29.55
1,2-DMN	-10.17	-27.16	-10.32	-28.72
1,3-DMN	-9.81	-26.21	-10.36	-28.93
1,4-DMN	-9.73	-26.03	-10.46	-29.32
1,5-DMN	-10.12	-27.66	-10.12	-27.66
1,6-DMN	-9.99	-27.33	-10.19	-27.85
1,7-DMN	-10.00	-27.42	-10.14	-27.77
1,8-DMN	-10.01	-27.55	-10.01	-27.55
2,3-DMN	-10.28	-27.67	-10.26	-28.76
2,6-DMN	-10.04	-27.51	-10.04	-27.51
2,7-DMN	-10.02	-27.45	-10.02	-27.45

Other Properties of TMNS

Methyl Group Rotation, Hyperconjugation, Conformational Flexibility, Aromaticity, and Fukui Functions of TMNs

The obtained equilibrium geometries and the TS structures of TMNs using the DFT methods enabled further investigation of other important properties of TMNs that required *ab initio* techniques. It is well known that internal rotation in molecules may have a significant effect on the non-covalent interactions. Since each of the three methyl groups in a TMN can have a different effect it is interesting to investigate the methyl group torsion in these molecules. The methyl group torsion is closely related to the π*-σ* hyperconjugation (HC) effects. Nakai and Kawai discovered a type of orbital interaction named π*-σ* hyperconjugation. It appears between an unoccupied benzene π* orbital and an unoccupied σ* orbital of the C-H bond in the methyl group [72]. The model was successfully applied to explain the stability of the highest occupied molecular orbital (HOMO) and the lowest unoccupied molecular orbital (LUMO) of substituted toluenes and methylnaphthalenes [72-74]. We investigated the π*-σ* hyperconjugation effects in 1,4,6-trimethylnaphthalene and their influence on the torsional barriers in the ground electronic state [38]. In the HOMO and LUMO of 1,4,6-TMN the π*-σ* hyperconjugation is also observed. The barrier heights for the methyl group torsion in the positions 1 and 4 of 1,4,6-TMN are higher than the corresponding barrier height of 1-methylnaphthalene. The barrier height for the methyl group torsion in the position 6 of 1,4,6-TMN is lower compared to the barrier height of 2-methylnaphthalene. The variations of the rotational barriers in the S_0 state of 1,4,6-TMN are shown to be directly connected with the stability of HOMO and the orbitals stability is determined by the π*-σ* hyperconjugation [38].

Conformational flexibility is important source of relaxation of geometry of molecules in intra- and intermolecular interactions. Its role in the interactions between nucleic acid bases and water [75-77] is particularly interesting. It was demonstrated that naphthalene [78] and azanaphthalenes [79] possess high conformational flexibility. Our results showed that 1,4,6-TMN possesses high conformational flexibility of aromatic system [38]. The conformational deformability of 1,4,6-TMN calculated at the MP2/cc-pVDZ level of theory is larger compared to the naphthalene molecule. The deformational energy of the aromatic system in 1,4,6-TMN is about 0.3 kcal/mol lower compared to the deformational energy of the NPH rings. Its aromaticity based on NICS value calculated at the GIAO-B3LYP/6-311+G** level of theory using the B3LYP/cc-pVTZ optimized geometry is lower compared to NPH but this TMN isomer keeps a high degree of π-electron delocalization within the rings and remains aromatic [38].

A number of bacterial species are known to degrade PAHs. The group of enzymes responsible for the biodegradation of PAHs by bacteria is aromatic naphthalene dioxigenase and they catalyze oxidation of many aromatic compounds [80,81]. Naphthalene 1,2-dioxigenase (NDO) is one of these enzyme systems. According to the analysis presented by Wammer and Peters [82], the biodegradation reaction involves an electrophilic attack on alkylated naphthalenes by the NDO enzyme. They pointed out to four regions on the naphthalene rings susceptible to electrophilic attack based on ionization potential maps. We present here the results of the calculations of the condensed Fukui function [83] for two TMN isomers. They are calculated employing the natural population analysis (NPA)[84] and Bader's topological population analysis scheme [85] at the B3LYP/cc-pVTZ level of theory as implemented in the Gaussian 09 package [40]. The Fukui function $f^-(r)$ was obtained as a finite difference approximation from the population in an N-electron and N-1 system using the equilibrium geometry of the N-electron system. A good performance of the B3LYP functional and the cc-pVTZ basis set for the calculation of the condensed Fukui function $f^-(r)$ has been reported by De Proft *et al.* [86]. The Figures 6 and 7 were obtained using the Multiwfn 2.5 program [59]. The most positive parts of the $f^-(r)$ are localized on the C_1-C_2, C_3-C_4, C_5-C_6, and C_7-C_8 bonds, indicating that these bonds are susceptible to electrophilic attack [39]. In accordance with the findings of Wammer and Peters for naphthalene, these regions are particularly important from the point of view of dioxigenation reaction on the aromatic rings of TMNs by the NDO enzymes in the process of biodegradation [87].

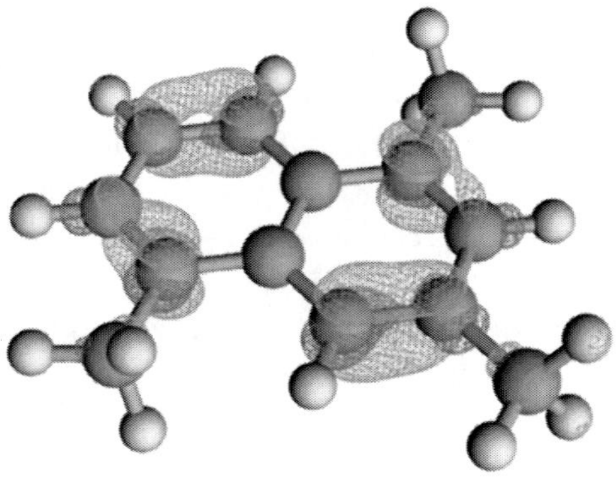

Figure 6. The electrophilic Fukui function for the 1,3,5-TMN isomer calculated at the B3LYP/cc-pVTZ level of theory (the isovalue is 0.004 a.u.).

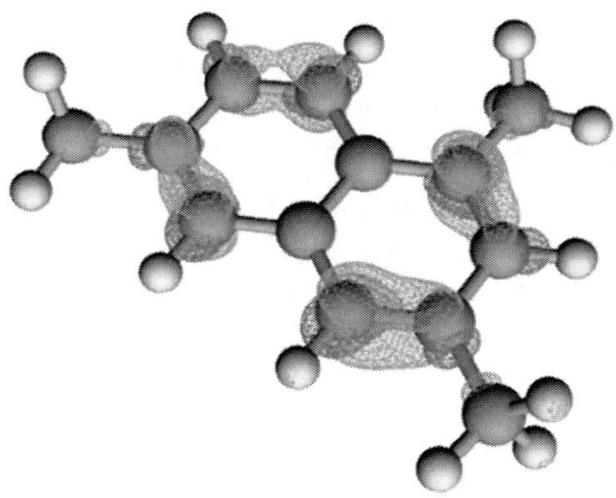

Figure 7. The electrophilic Fukui function for the 1,3,6-TMN isomer calculated at the B3LYP/cc-pVTZ level of theory (the isovalue is 0.004 a.u.).

Another important property of TMNs for biodegradation is their static dipole polarizability. Librando and Alparone [30] proposed polarizability values of the DMN isomers to be predictors of their biodegradation rates. They concluded that the average static dipole polarizabilities of DMN isomers are dependent on the positions of the methyl substituents. Librando and Alparone reported that the polarizabilities of DMNs increase in the order α,α-DMN < α,β-DMN < β,β-DMN and are in excellent linear relationship with the observed first-order biomass-normalized rate coefficients observed by Wammer and Peters [87].

Based on the results of ionization potentials (IP) and electron affinities (EA) of TMNs presented in our previous study [37], the IP and EA values remain almost constant along the series of TMN isomers suggesting that oxidative and reductive mechanism arguments may not be used to explain their different activities related to naphthalene 1,2-dioxigenase. We assumed that the binding of a particular TMN isomer to the active site of the NDO enzyme will be controlled by intermolecular interactions similar to the DMN isomers. The results of our calculations show that the average static dipole polarizabilities of TMNs in gas phase as well as in water solution increase in the order α,α,α-TMN < α,α,β-TMN < α,β,β-TMN < β,β,β-TMN [37]. It can be expected that the biodegradation rates of TMNs by bacterial enzyme NDO will follow the same trend as the polarizability changes. Consequently, the calculated polarizability values can be used as predictors of the biodegradation rates of TMNs with the NDO enzyme.

Conclusion

Alkylated naphthalenes constitute a group of widespread pollutants of great environmental interest. There have not been many papers published on alkylated naphthalenes. For TMNs modern spectroscopic studies are not known and current data can be used for distinguishing and identification of TMNs. The comparison with available experimental data on DMNs shows that DFT methods are able to predict closely the experimental vibrational frequencies and to guide the initial spectral search of other DMNs. The results of calculations of the NICS–based magnetic indeces for all ten isomers of DMN show that the NICS values are negative indicating diatropicity associated with aromatic character. The DFT calculations in conjunction with *ab initio* techniques enabled the investigation of methyl group torsion, the π*-σ* hyperconjugation effects, the conformational deformability of the aromatic rings and other important properties of TMN molecules. We hope that the results presented will encourage further experimental studies of these molecules and will be of help in their identification and understanding their binding to the cavity of enzymes and intermolecular interactions with biomolecules and other molecules in the environment.

Acknowledgment

We acknowledge the financial support of the Ministry of Education and Science of the Republic of Serbia (Contract No. 172001).

References

[1] Parr, R. G. *Annu Rev Phys Chem* 1983, 34, 631-656.
[2] Parr, R. G.; Yang, W. *Density Functional Theory of Atoms and Molecules*; Oxford University Press and Clarendon Press: New York and Oxford, 1989.
[3] Jones, R. O.; Gunnarsson, O. *Rev Mod Phys* 1989, 61, 689-746.
[4] Kryachko, E. S.; Ludena, E. V. *Density Functional Theory of Many Electron Systems*; Kluwer: Dordrecht, 1990.

[5] *Density Functional Theory: An Approach to the Quantum Many Body Problem*; Dreizler, R. M.; Gross, E. K. U.; Eds.; Springer-Verlag: Berlin, 1990.
[6] Perdew, J. P. In *Density Functional Theory of Many Fermion Systems*; Trickey, S. B.; Ed.; Advances in Quantum Chemistry 21; Academic Press: San Diego, 1990; pp.1-405.
[7] Ziegler, T. *Chem Rev* 1991, 91, 651-667.
[8] *Density Functional Methods in Chemistry*; Labanowski, J. K., Andzelm, J. W., Eds.; Springer-Verlag: New York, 1991.
[9] March, N. H. *Electron Density Theory of Atoms and Molecules*; Academic Press: London, 1992.
[10] *Modern Density Functional Theory, A Tool for Chemistry, Theoretical and Computational Chemistry*; Politzer, P., Seminario, J. M., Eds.; Elsevier: Amsterdam, 1995; Vol. 2.
[11] Chong, D. P. *Recent Advances in Density Functional Methods*; World Scientific: Singapore, 1995.
[12] *Density Functional Theory*; Nalewajski, R. F., Ed.; Topics in Current Chemistry 179-182; Springer: Berlin, 1996.
[13] *Theoretical and Computational Chemistry*; Seminario, J., Ed.; Recent Developments and Applications of Modern Density Functional Theory 4; Elsevier: Amsterdam, 1996.
[14] *Metal-Ligand Interactions: Structure and Reactivity*; Russo, N.; Salahub, D. R., Eds.; Kluwer: Dordrecht, 1996.
[15] Geerlings, P.; De Proft, F.; Martin, J. M. L. *In Recent Developments and Applications of Modern Density Functional Theory*; Seminario, J. M., Ed.; Theoretical and Computational Chemistry 4; Elsevier: Amsterdam, 1996; p 773.
[16] *DFT Methods in Chemistry and Material Science*; Springborg, M., Ed.; Wiley: New York, 1997.
[17] *DFT, a Bridge between Chemistry and Physics*; Geerlings, P.; De Proft, F.; Langenaeker, W.; Eds.; VUB Press: Brussels, 1998.
[18] Nagy, A. *Phys Rep* 1998, *298*, 1-79.
[19] Special Issue on Progress in Density Functional Theory, Frenking, G. *J Comput Chem* 1999, 20 (1)
[20] *A Chemist's Guide to Density Functional Theory*; Koch, W.; Holthausen, M. C.; Eds.; Second Edition, Wiley-VCH Verlag, Weinheim, 2001.
[21] *Recent Advances in Density Functional Methods,* Part III; Barone, V.; Bencini, A.; Fantucci, P.; Eds.; World Scientific: Singapore, 2002.
[22] *A Primer in Density Functional Theory*; Fiolhais, C; Nogueira, F.; Marques, M.; Eds.; Springer-Verlag, Berlin Heidelberg, 2003.
[23] Geerlings, P.; De Proft, F.; Langenaeker, W. *Chem Rev* 2003, 103, 1793-1873.
[24] van Mourik, T. *J Chem Theory Comput* 2008, 4, 1610-1619.
[25] L. Goerigk; Grimme, S. *Phys Chem Chem Phys*, 2011, 13, 6670-6688.
[26] U.S. *Environmental Protection Agency*, (2000). Superfund sites. Available: http://www.epa.gov/superfund/sites/index.htm.
[27] Mosby, W.L., *J Am Chem Soc* 1952, 74, 2564-2569.
[28] Mayer, T. J.; Duswalt, J. M. *J Chem Eng Data* 1973, 18, 337-344.
[29] Mora-Diez, N.; Boyd, R. J.; Heard, G. L., *J Phys Chem A* 2000, 104, 1020-1029.
[30] Librando, V.; Alparone, A. *Environ Sci Technol* 2007, 41, 1646-1652.
[31] Librando, V.; Alparone, A. *Polycycl Arom Comp.* 2007, 27, 65-94.

[32] Alparone, A.; Librando, V. *Comput Theor Chem* 2011, 965, 244-245.
[33] Alparone, A.; Librando, V. *Struct Chem* 2012, 23, 1467-1474.
[34] Das, P.; Arunan, E.; Das, P. K. *Vib Spectrosc* 2008, 47, 1-9.
[35] Nagabalasubramanian, P. B.; Periandy, S. *Spectrochim Acta A* 2010, 77, 1099-1107.
[36] Prabhu, T.; Periandy, S.; Mohan, S. *Spectrochim Acta A* 2011, 78 566-574.
[37] Ostojić, B. D.; Đorđević, D. S. *Chemosphere* 2012, 88, Issue 1, 91-97.
[38] Ostojić, B. D.; Đorđević, D. S. *Chem Phys Lett* 2012, 536, 19-25.
[39] Ostojić, B. *et al.*, unpublished data
[40] Frisch, M. J.; Trucks, G. W.; Schlegel, H. B.; Scuseria, G. E.; Robb, M. A.; Cheeseman, J. R.; Scalmani, G.; Barone, V.; Mennucci, B.; Petersson, G. A.; Nakatsuji, H.; Caricato, M.; Li, X.; Hratchian, H. P.; Izmaylov, A. F.; Bloino, J.; Zheng, G.; Sonnenberg, J. L.; Hada, M.; Ehara, M.; Toyota, K.; Fukuda, R.; Hasegawa, J.; Ishida, M.; Nakajima, T.; Honda, Y.; Kitao, O.; Nakai, H.; Vreven, T.; Montgomery, Jr., J. A.; Peralta, J. E.; Ogliaro, F.; Bearpark, M.; Heyd, J. J.; Brothers, E.; Kudin, K. N.; Staroverov, V. N.; Keith, T.; Kobayashi, R.; Normand, J.; Raghavachari, K.; Rendell, A.; Burant, J. C.; Iyengar, S. S.; Tomasi, J.; Cossi, M.; Rega, N.; Millam, J. M.; Klene, M.; Knox, J. E.; Cross, J. B.; Bakken, V.; Adamo, C.; Jaramillo, J.; Gomperts, R.; Stratmann, R. E.; Yazyev, O.; Austin, A. J.; Cammi, R.; Pomelli, C.; Ochterski, J. W.; Martin, R. L.; Morokuma, K.; Zakrzewski, V. G.; Voth, G. A.; Salvador, P.; Dannenberg, J. J.; Dapprich, S.; Daniels, A. D.; Farkas, O.; Foresman, J. B.; Ortiz, J. V.; Cioslowski, J.; Fox, D. J.; Gaussian 09, Revision B.01, Gaussian, Inc., Wallingford CT, 2010.
[41] Jones, D.W.; Matthews, R.S. In *Progress in Medicinal Chemistry*; Ellis, G. P.; West, G. B.; Eds.; Publisher: North-Holland, Amsterdam, NL, 1974; Vol. 10. pp. 159-203.
[42] Bauschlicher, C.W.; Langhoff, S. R. *Spectrochim Acta A* 1997, 53, 1225 -1240.
[43] Lee, C.; Yang, W.; Parr, R. G. *Phys Rev B* 1988, 37, 785-789.
[44] Becke, A. D. *J Chem Phys* 1993, 98, 5648-5652.
[45] Becke, A. D. *Phys Rev A* 1988, 38, 3098-3100.
[46] Perdew, J. P. *Phys Rev B* 1986, 33, 8822-8824 and eratum ibid. 34, 7406-7406.
[47] Bauschlicher, C. W.; Hudgins, D. M.; Allamandola, L. J. *Theor Chem Acc* 1999, 103, 154-162.
[48] Dunning, Jr., T. H. *J Chem Phys* 1989, 90, 1007-1023.
[49] McNaughton, D.; Godfrey, P. D.; Grabow, J. -U. *J Mol Spectrosc* 2012, 274, 1–4.
[50] McNaughton, D.; Godfrey, P. D.; Brown, R. D.; Thorwith, S.; Grabow, J.-U. *Astrophys J* 2008, 678, 309-315.
[51] McNaughton, D.; Godfrey, P. D.; Grabow, J. -U.; Jahn, M. K.; Dewald, D. A. *J Chem Phys* 2011, 134, 154305.
[52] Thorwirth, S.; McCarthy, M. C.; Dudek, J. B.; Thaddeus, P. *J Mol Spectrosc* 2004, 225, 93-95.
[53] Baba, M.; Kowaka, Y.; Nagashima, U.; Ishimoto, T.; Goto, H.; Nakayama, N. *J Chem Phys* 2011, 135, 054305
[54] Zvereva, E. E.; Shagidullin, A. R.; Katsyuba, S. A. *J Phys Chem A* 2011, 115, 63–69.
[55] Halls, M. D.; Velkovski, J.; Schlegel, H. B. *Theor Chem Acc* 2001, 105, 413- 421.
[56] Sinha, P.; Boesch, S. E.; Gu, C.; Wheeler, R. A.; Wilson, A. K. *J Phys Chem A* 2004, 108, 9213-9217.
[57] Merrick, J. P.; Moran, D.; Radom, L. *J Phys Chem A* 2007, 111, 11683-11700.

[58] Hanwell, M. D.; Curtis, D. E.; Lonie, D. C.; Vandermeersch, T.; Zurek, E.; Hutchison, G. R. *J Cheminf* 2012, 4:17 doi:10.1186/1758-2946-4-17, Avogadro, 2011. Avogadro: an open-source molecular builder and visualization tool. Version 1.0.3;
[59] Lu, T.; Chen, F. *J Comp Chem* 2012, 33, 580-592.
[60] Srivastava, A.; Singh, V.B. *Ind J Pure App Phys* 2007, 45, 714-720.
[61] Foing, B. H.; Ehrenfreud, P. *Nature* 1994, 369, 296-298.
[62] Krygowski, T. M.; Cyranski, M. K.; Czarnocki, Z.; Häfelinger, G.; Katritzky, A. R. *Tetrahedron* 2000, 56, 1783-1796.
[63] Chen, Z.; Wannere, C. S.; Corminboeuf, C.; Puchta, R.; Schleyer, P. v. R. *Chem Rev* 2005, 10, 3842-3888.
[64] Wolinski, K.; Hilton, J. F.; Pulay, P. *J Am Chem Soc* 1990, 12, 8251-8260.
[65] Cheeseman, J. R.; Trucks, G. W.; Keith, T. A.; Frisch, M. J. *J Chem Phys* 1996, 104, 5497-5509.
[66] Fowler, P. W.; Steiner, E. *Mol Phys* 2000, 98, 945-953.
[67] Steiner, E.; Fowler, P. W.; Jenneskens, L. W. *Angew Chem Int Ed* 2001, 40, 362-366.
[68] Cernusak, I.; Fowler, P. W.; Steiner, E. *Mol Phys* 2000, 98, 945-953.
[69] Corminboeuf, C.; Heine, T.; Seifert, G.; Schleyer, P. v. R. *Phys Chem Chem Phys* 2004, 6, 273-276.
[70] Fallah-Bagher-Shaidaei, H.; Wannere, C. S.; Corminboeuf, C.; Puchta, R.; Schleyer, P. v. R. *Org Lett* 2006, 8, 863-866.
[71] Jensen, F. *J Chem Theory Comput* 2008, 4, 719-727.
[72] Nakai, H.; Kawai, M. *J Chem Phys* 2000, 113, 2168-2174.
[73] Nakai, H.; Kawai, M. *Chem Phys Lett* 1999, 307, 272-276.
[74] Nakai, H.; Kawamura, Y. *Chem Phys Lett* 2000, 318, 298-304.
[75] Shishkin, O. V.; Gorb, L.; Leszczynski, J. *Struct Chem* 2009, 20, 743-749.
[76] Palamarchuk, G. V.; Shishkin, O. V.; Gorb, L.; Leszczynski, J. *J Biomol Struct Dyn* 2009, 26, 653-661.
[77] Sukhanov, O. S.; Shishkin, O. V.; Gorb, L.; Podolyan, Y.; Leszczynski, J. *J Phys Chem B* 2003, 107, 2846-2852.
[78] Zhigalko, M. V.; Shishkin, O. V.; Gorb, L.; Leszczynski, J. *J Mol Struct* 2004, 693, 153-159.
[79] Zhigalko, M. V.; Shishkin, O. V. *J Struct Chem* 2006, 47, 823-830.
[80] Kauppi, B.; Lee, K.; Carredano, E.; Parales, R.E.; Gibson, D.T.; Eklund, H.; Ramaswamy, S. *Structure* 1998, 6, 571–586.
[81] Gibson, D.T.; Parales, R.E. *Curr Opin Biotechnol* 2000, 11, 236–243.
[82] Wammer, K. H.; Petters, C. A. *Environ Toxicol Chem* 2006, 25, 912-920.
[83] Parr, R. G.; Yang, W. T. *J Am Chem Soc* 1984, 106, 4049-4050.
[84] Reed, A. E.; Curtiss, L. A.; Weinhold, F. *Chem Rev* 1988, 88, 899-926.
[85] Bader, R. F. W. *Atoms in Molecules: A Quantum Theory*, Oxford University Press, Oxford, UK, 1990.
[86] De Proft, F.; Martin, J. M.L.; Geerlings, P. *Chem Phys Lett* 1996, 256, 400-408.
[87] Wammer, K. H.; Petters, C. A. *Environ Sci Technol* 2005, 39, 2571-2578.

In: Density Functional Theory
Editors: J. Morin and J.M. Pelletier, pp. 187-225
ISBN: 978-1-62417-954-9

Chapter 7

TRANSPORT OF ORGANIC MATERIALS FROM MOLECULES TO ORGANIC SEMICONDUCTORS

Kenji Hirose
Green Innovation Research Laboratories, NEC Corporation

Abstract

We review our recent works on transport calculations of organic materials, which include single molecules, molecular wires, and organic semiconductors. First we describe the transport calculation methods based on the nonequilibrium Green's function (NEGF) and the density-functional theory (DFT) in the localized basis sets and in the plane-wave expansion. Then we apply the method to the transport properties of single molecules and molecular wires. The problems such as conductance and its eigenchannel analysis, bias drop in the molecules, and local heating at the contact to electrodes are explored and presented.

Next we describe the charge transport of organic semiconductors, which are regarded as molecular systems assembling weakly with van der Waals interactions. Recent advances of single-crystal organic transistors with high mobility require us to understand the fundamental transport mechanisms in these mechanically flexible organic materials. Using the time-dependent wave-packet diffusion (TD-WPD) method combined with classical molecular dynamics (MD) with parameters taken from the DFT-D calculations which include the van der Waals energy, we show the temperature dependence of charge transport with strong electron-phonon interaction in polaronic states and treat the crossover from the localized hopping-transport to delocalized band-transport.

Keywords: Molecular electronics, NEGF method, organic semiconductors, flexible device, van der Waals interaction, TD-WPD method, electron-phonon coupling, polaron

1. Introduction

In this review article, we present transport calculations based on the density-functional theory (DFT) for two types of electron devices constructed by using organic materials.

1.1. Molecular Electronics — Ultimately Tiny Functional Devices

The idea of molecular electronics is to utilize single molecules to operate as ultimately tiny functional electron devices. The recent rapid progress of the nanometer-size technologies drives much interest in this field. Presently various applications such as sensors, diodes, switches, circuits, the next generation of transistors are proposed. The technology how to fabricate the atomic-scale contacts for the control of electron transport properties of single molecules becomes important to realize the quantum devices for molecular electronics.

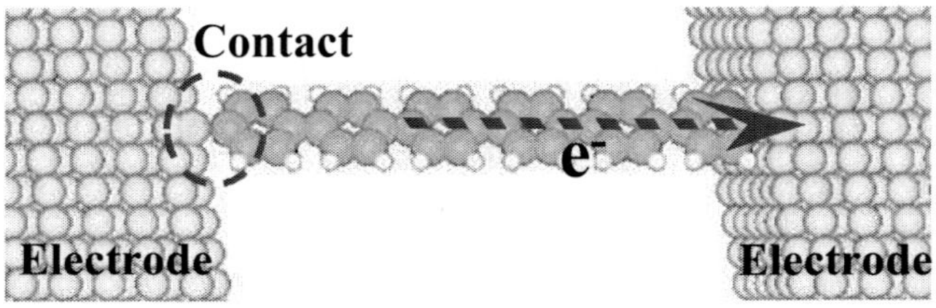

Figure 1. Schematic view of the molecular device. Single molecule is sandwiched between electrodes where electrons go through, creating ultimately tiny functional devices.

1.2. Organic Semiconductor Electronics — Flexible Devices

Organic semiconductors are crystals composed of organic materials such as organic oligomers, polymers, and small organic molecules, which are regarded as the assemblies of single molecules bonded with weak van der Waals interaction. Recently much attention has been paid to the field of organic semiconductor electronics such as field-effect transistors (FETs), light-emitting diodes (LEDs), electronic paper, and solar cells due to their structural flexibility and large area coverage using inexpensive printing processes. The understanding of their charge-transport mechanisms has been strongly required.

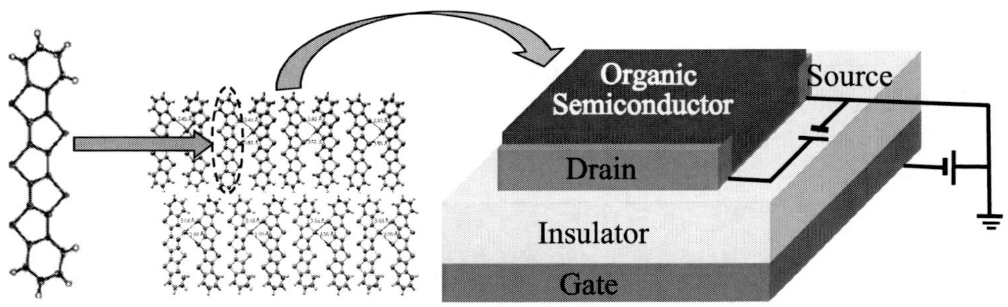

Figure 2. Schematic view of the organic semiconductor device. Molecules bonded weakly by van der Waals interaction are utilized for channel materials, creating flexible transistors.

2. Transport Calculations for Single Molecule

In this section, we describe the calculation method for the transport using the nonequilibrium Green's function (NEGF) method. First we briefly describe the fundamental derivations of the equations [1]. Then we explain how the formulas are utilized for the nanometer-scale systems sandwiched between electrodes. We also describe the basis sets to construct an effective Hamiltonian by numerical atomic orbital basis sets and by the plane-wave expansion using the recursion-transfer-matrix (RTM) method.

2.1. Nonequilibrium Green's Function (NEGF)

When we apply a bias voltage to a system, it becomes a nonequilibrium situation, where the system is no more described by the periodic boundary condition as in the usual band calculations. Instead we need to develop more complicated formulas with the open boundary condition. To describe the electronic states and transport properties, the nonequilibrium Green's function (NEGF) provides us with a powerful computational method. Since the nonequilibrium system is not restored to the original system for asymptotically large times, it is not sufficient to describe the nonequilibrium system only by the conventional time-ordered Green's function, which assumes that the system only changes its phase factor for the large times. However, this assumption is no more applicable for the nonequilibrium system, since the system does not return to its ground state as $t \to +\infty$ and irreversible effects occur in the process of $t = -\infty$ and $t = +\infty$.

In the NEGF formalism, this problem is circumvented by taking the time ordering such that the system evolves from $t = -\infty$ then continues the time evolution up to t_0 and returns back to $t = -\infty$. The contour-ordering operator is expressed by T_C, which follows the contour C coming back after a long time to the original point at $t = -\infty$. Then we define the contour-ordering Green's function by

$$G(\mathbf{r}, t : \mathbf{r}', t') = -i\langle T_C \left\{ \hat{\psi}(\mathbf{r}, t)\hat{\psi}^\dagger(\mathbf{r}', t') \right\}\rangle, \tag{1}$$

where $\hat{\psi}(\mathbf{r}, t)$ is the field operator. According to the formal theory of the nonequilibrium Green's functions, the perturbation expansion in the interaction picture using the contour-ordered S-matrix has the same structure as the equilibrium expansion. The contour-ordering Green's function obeys the same Dyson equation as the usual Green's function and the perturbation expansion is possible along the contour C [2].

Usually we introduce the 2×2 Green's function matrices, which requires at least four Green's functions corresponding to two time branches to describe the nonequilibrium Green's function [3]. In addition to the time-ordered Green's function $G^t(\mathbf{r}, t : \mathbf{r}', t')$ and the anti-time-ordered Green's function $G^{\tilde{t}}(\mathbf{r}, t : \mathbf{r}', t')$ we define the lesser and the greater Green's functions as

$$\begin{aligned} G^<(\mathbf{r}, t : \mathbf{r}', t') &= i\langle \hat{\psi}^\dagger(\mathbf{r}', t')\hat{\psi}(\mathbf{r}, t)\rangle \\ G^>(\mathbf{r}, t : \mathbf{r}', t') &= -i\langle \hat{\psi}(\mathbf{r}, t)\hat{\psi}^\dagger(\mathbf{r}', t')\rangle. \end{aligned} \tag{2}$$

These constitute four Green's functions.

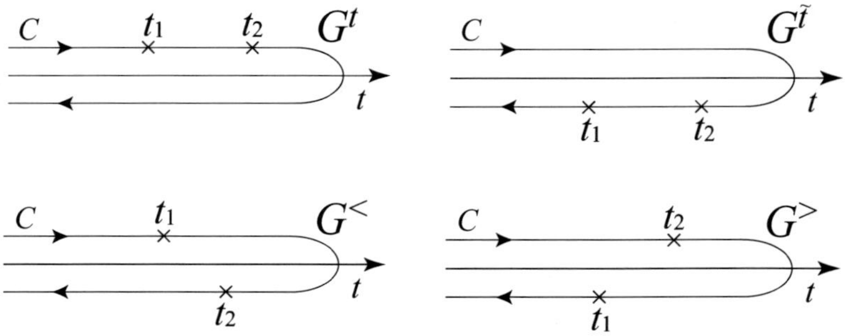

Figure 3. Time contour C for the four Green's functions G^t, $G^{\tilde{t}}$, $G^<$, and $G^>$. The system evolves from $t = -\infty$ and continues the evolution up to t_0. Then it returns back to $t = -\infty$.

It is convenient to define the retarded and the advanced Green's functions

$$\begin{aligned} G^r(\mathbf{r}, t : \mathbf{r}', t') &= -i\theta(t - t')\langle\left[\hat{\psi}(\mathbf{r}, t), \hat{\psi}^\dagger(\mathbf{r}', t')\right]\rangle \\ G^a(\mathbf{r}, t : \mathbf{r}', t') &= i\theta(t' - t)\langle\left[\hat{\psi}(\mathbf{r}, t), \hat{\psi}^\dagger(\mathbf{r}', t')\right]\rangle, \end{aligned} \tag{3}$$

which are expressed using the lesser and the greater Green's functions as

$$\begin{aligned} G^r(\mathbf{r}, t : \mathbf{r}', t') &= \theta(t - t')\left[G^>(\mathbf{r}, t : \mathbf{r}', t') - G^<(\mathbf{r}, t : \mathbf{r}', t')\right] \\ G^a(\mathbf{r}, t : \mathbf{r}', t') &= -\theta(t' - t)\left[G^>(\mathbf{r}, t : \mathbf{r}', t') - G^<(\mathbf{r}, t : \mathbf{r}', t')\right]. \end{aligned} \tag{4}$$

Note that, since $\theta(t - t') + \theta(t' - t) = 1$, two relations connect six Green's functions

$$G^r - G^a = G^> - G^<, \quad G^t + G^{\tilde{t}} = G^> + G^<, \tag{5}$$

thus we have effectively four independent Green's functions.

Then the electron density is expressed by

$$\hat{\rho}(\mathbf{r}, t) = -iG^<(\mathbf{r}, t : \mathbf{r}, t) \tag{6}$$

and the current density is written as

$$\begin{aligned} \hat{\mathbf{j}}(\mathbf{r}, t) &= \left\langle \frac{e\hbar}{2mi}\left[\hat{\psi}(\mathbf{r})^* \frac{\partial \hat{\psi}(\mathbf{r})}{\partial \mathbf{r}} - \hat{\psi}(\mathbf{r}) \frac{\partial \hat{\psi}(\mathbf{r})^*}{\partial \mathbf{r}}\right] \right\rangle \\ &= \left\langle \frac{e\hbar}{2mi} \lim_{\mathbf{r} \to \mathbf{r}'} \left(\nabla_{\mathbf{r}} - \nabla_{\mathbf{r}'}\right) \hat{\psi}^*(\mathbf{r}')\hat{\psi}(\mathbf{r}) \right\rangle \\ &= \frac{e\hbar}{2m} \lim_{\mathbf{r}' \to \mathbf{r}} \left(\nabla_{\mathbf{r}'} - \nabla_{\mathbf{r}}\right) G^<(\mathbf{r}, t : \mathbf{r}', t). \end{aligned} \tag{7}$$

The spectral function is expressed by the lesser and greater Green's function as

$$A(k, \epsilon) = i\left[G^r(k, \epsilon) - G^a(k, \epsilon)\right] = i\left[G^>(k, \epsilon) - G^<(k, \epsilon)\right]. \tag{8}$$

In equilibrium, we can show $G^<(\epsilon) = if(\epsilon)A(\epsilon)$ and $G^>(\epsilon) = -i\left(1 - f(\epsilon)\right)A(\epsilon)$, which represent the fluctuation-dissipation theorem for electron $G^<$ and hole $G^>$ propagations. This shows that the width of the spectrum is related to the decay of the system.

2.2. Electric Current Expression of Junction Systems in NEGF Formalism

We consider the system composed of the center region (C) connected to the left (L) and right (R) electrodes with different chemical potentials $\mu_{L,R}$ [4] described by the Hamiltonian,

$$\hat{H} = \sum_k \epsilon_{kL}\hat{c}^\dagger_{kL}\hat{c}_{kL} + \sum_k \epsilon_{kR}\hat{c}^\dagger_{kR}\hat{c}_{kR} + \sum_n \epsilon_n \hat{d}^\dagger_n \hat{d}_n$$
$$+ \sum_{k,n}\left[V_{kL,n}\hat{c}^\dagger_{kL}\hat{d}_n + V^*_{kL,n}\hat{d}^\dagger_n\hat{c}_{kL}\right] + \sum_{k,n}\left[V_{kR,n}\hat{c}^\dagger_{kR}\hat{d}_n + V^*_{kR,n}\hat{d}^\dagger_n\hat{c}_{kR}\right]. \quad (9)$$

Here $\hat{c}^\dagger_k\hat{c}_k$ represents the electron in the electrodes where the electronic states are continuous ϵ_k labeled by k, while $\hat{d}^\dagger_n\hat{d}_n$ represents the electron in the center region where the electronic states are ϵ_n labeled by n.

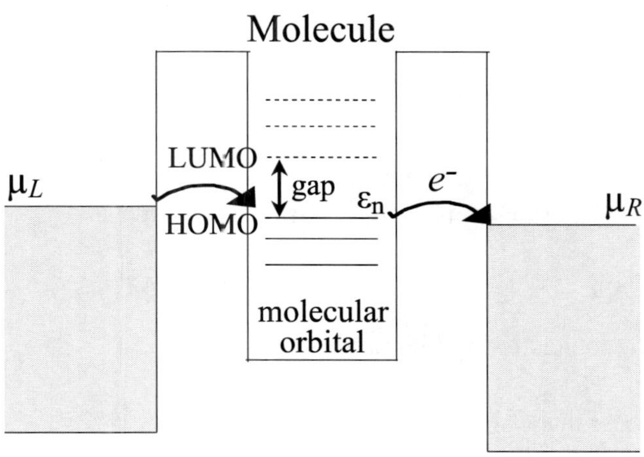

Figure 4. Schematic view of the band diagram of the junction system between electrodes. Electrons go through the resonant states of molecular orbitals.

The electrical current is expressed from the number of electrons $\hat{N}_L = \sum_k \hat{c}^\dagger_{kL}\hat{c}_{kL}$ by

$$I_L = -e\langle\frac{d\hat{N}_L}{dt}\rangle = -\frac{ie}{\hbar}\langle\left[\hat{H},\hat{N}_L\right]\rangle = \frac{ie}{\hbar}\sum_{k,n}\left[V_{kL,n}\langle\hat{c}^\dagger_{kL}\hat{d}_n\rangle - V^\dagger_{kL,n}\langle\hat{d}^\dagger_n\hat{c}_{kL}\rangle\right]. \quad (10)$$

For the steady state, we define the nonequilibrium Green's functions as

$$G^<_{n,kL}(t,t') = i\langle\hat{c}^\dagger_{kL}(t')\hat{d}_n(t)\rangle \quad G^<_{kL,n}(t,t') = i\langle\hat{d}^\dagger_n(t')\hat{c}_{kL}(t)\rangle, \quad (11)$$

which satisfies the properties $G^<_{kL,n}(t=t') = -G^<_{n,kL}(t=t')$. Then the current becomes

$$I_L = \frac{2e}{\hbar}\int\frac{d\epsilon}{2\pi}\mathrm{Re}\left(\sum_{k,n}V_{kL,n}G^<_{n,kL}(\epsilon)\right). \quad (12)$$

From the equation of motion, the Green's function for the coupling is expressed by the multiples of the Green's function for the electrodes and those for the center region [1]

$$G^<_{n,kL}(\epsilon) = \sum_m V^*_{kL,m}\left[G^r_{nm}(\epsilon)g^<_{kL}(\epsilon) + G^<_{nm}(\epsilon)g^a_{kL}(\epsilon)\right]. \tag{13}$$

Here the nonequilibrium Green's functions of the center region are defined by

$$G^r_{n,m}(t,t') = -i\theta(t-t')\langle\left[\hat{d}_n(t), \hat{d}^\dagger_m(t')\right]\rangle, \quad G^<_{n,m}(t,t') = i\langle\hat{d}^\dagger_n(t')\hat{d}_n(t)\rangle, \tag{14}$$

and those of electrodes are defined by

$$g^<_{kL}(t,t') = i\langle\hat{c}^\dagger_{kL}(t')\hat{c}_{kL}(t)\rangle, \quad g^a_{kL}(t,t') = i\theta(t'-t)\langle\left[\hat{c}_{kL}(t), \hat{c}^\dagger_{kL}(t')\right]\rangle. \tag{15}$$

With use of these variables, the current I_L becomes

$$I_L = \frac{2e}{\hbar}\int\frac{d\epsilon}{2\pi}\mathrm{Re}\left(\sum_{k,n,m} V_{kL,n}V^*_{m,kL}[G^r_{n,m}(\epsilon)g^<_{kL}(\epsilon) + G^<_{n,m}(\epsilon)g^a_{kL}(\epsilon)]\right). \tag{16}$$

Here we assume that the electrons are noninteracting inside the electrodes keeping a steady state, then the Green's functions remain in a Fourier space such as

$$g^<_{kL}(\epsilon) = 2\pi i f_L(\epsilon)\delta(\epsilon-\epsilon_{kL}) \quad g^a_{kL}(\epsilon) = i\pi\delta(\epsilon-\epsilon_{kL}). \tag{17}$$

Defining the coupling to electrodes with multiple modes as

$$\Gamma^L_{n,m}(\epsilon) = \sum_k V_{kL,n}(\epsilon)V^*_{m,kL}(\epsilon)\delta(\epsilon-\epsilon_{kL}), \tag{18}$$

we have

$$I_L = \frac{ie}{\hbar}\int \mathrm{Tr}\left\{\Gamma^L(\epsilon)\left[G^<(\epsilon) + f_L(\epsilon)\left(G^r(\epsilon) - G^a(\epsilon)\right)\right]\right\}d\epsilon. \tag{19}$$

Since in a steady state $I_R = -I_L$, the total current is obtained from $I = (I_L - I_R)/2$ and we have the general expression as

$$I = \frac{ie}{h}\int \mathrm{Tr}\left\{\left[\Gamma^L(\epsilon) - \Gamma^R(\epsilon)\right]G^<(\epsilon) + \left[f_L(\epsilon)\Gamma^L(\epsilon) - f_R(\epsilon)\Gamma^R(\epsilon)\right]\left[G^r(\epsilon) - G^a(\epsilon)\right]\right\}d\epsilon. \tag{20}$$

We note that the assumption used is that the electrons are noninteracting inside the electrodes and the electrodes keep steady states, even if electrons lose their energies via various inelastic processes in the electrodes. We can see that, in equilibrium $f_L(\epsilon) = f_R(\epsilon) = f(\epsilon)$, by using the fluctuation-dissipation relation $G^<(\epsilon) = if(\epsilon)A(\epsilon) = -f(\epsilon)\left(G^r(\epsilon) - G^a(\epsilon)\right)$, we confirm that the current is vanishing and $I = 0$.

[1]We use the multiplication rule for the time loop as $A^< = B^rC^< + B^<C^a$, known as the Langreth theorem.

Let us consider the electron transport in the mean-field approximation as density-functional theory (DFT) where the current flows through the Kohn-Sham orbital levels [2]. To obtain the current, we need $G^{<}(\epsilon)$ and $G^{r}(\epsilon) - G^{a}(\epsilon)$, which are given by

$$G^{<}(\epsilon) = G^{r}(\epsilon)\Sigma^{<}(\epsilon)G^{a}(\epsilon) = G^{r}(\epsilon)\left(if_L(\epsilon)\Gamma^L(\epsilon) + if_R(\epsilon)\Gamma^R(\epsilon)\right)G^{a}(\epsilon)$$
$$G^{r}(\epsilon) - G^{a}(\epsilon) = -iG^{r}(\epsilon)\left(\Gamma^L(\epsilon) + \Gamma^R(\epsilon)\right)G^{a}(\epsilon). \quad (21)$$

Then we have

$$I = \frac{2e}{h}\int (f_L(\epsilon) - f_R(\epsilon))\,\mathrm{Tr}\left[G^{r}(\epsilon)\Gamma^L(\epsilon)G^{a}(\epsilon)\Gamma^R(\epsilon)\right] d\epsilon, \quad (22)$$

which is called as Fisher-Lee relation [5] and corresponds to the Landauer formula.

2.3. Modeling for Molecular System between Electrodes

In the previous section, we show a general formula of the electric current flowing between semi-infinite electrodes.Here we derive the methodology for the numerical computations to obtain the electric current of the molecular systems between electrodes, expressed by the effective Hamiltonian based on the density-functional theory (DFT).

As in the previous section, we divide the system into three regions, the left (L) and right (R) semi-infinite electrodes and the center (C) region between electrodes. When we set the boundaries deep inside the electrodes, the Hamiltonian is partitioned in the boundaries as

$$\hat{H} = \begin{pmatrix} \hat{H}_L & \hat{V}_{LC} & 0 \\ \hat{V}_{LC}^{\dagger} & \hat{H}_C & \hat{V}_{CR}^{\dagger} \\ 0 & \hat{V}_{CR} & \hat{H}_R \end{pmatrix}. \quad (23)$$

Then the Kohn-Sham equation is written as a matrix equation

$$\begin{pmatrix} \hat{H}_L & \hat{V}_{LC} & 0 \\ \hat{V}_{LC}^{\dagger} & \hat{H}_C & \hat{V}_{CR}^{\dagger} \\ 0 & \hat{V}_{CR} & \hat{H}_R \end{pmatrix} \begin{pmatrix} |\psi_L\rangle \\ |\psi_C\rangle \\ |\psi_R\rangle \end{pmatrix} = \epsilon\hat{I} \begin{pmatrix} |\psi_L\rangle \\ |\psi_C\rangle \\ |\psi_R\rangle \end{pmatrix}. \quad (24)$$

We note that it is possible to express the Hamiltonian in the present form when the potentials $\hat{V}_{LC}$ and $\hat{V}_{CR}$ are short range in the interface regions.

Expanding the wavefunctions with the localized basis sets $|\psi_{L,C,R}\rangle$, we have the coupled equations as

$$\left(\epsilon\hat{I} - \hat{H}_L\right)|\psi_L\rangle = \hat{V}_{LC}|\psi_C\rangle, \quad \left(\epsilon\hat{I} - \hat{H}_R\right)|\psi_R\rangle = \hat{V}_{CR}|\psi_C\rangle \quad (25)$$

for the electrode regions (L, R). When we define the Green's functions $\hat{g}_{L,R}(\epsilon)$ as

$$\left(\epsilon\hat{I} - \hat{H}_L\right)\hat{g}_L(\epsilon) = \hat{I}, \quad \left(\epsilon\hat{I} - \hat{H}_R\right)\hat{g}_R(\epsilon) = \hat{I}, \quad (26)$$

[2] We need TDDFT approach for the rigorous treatment of electron transport within the DFT formalism.

the equation for the center region (C) becomes

$$\begin{aligned}\left(\epsilon\hat{I} - \hat{H}_C\right)|\psi_C\rangle &= \hat{V}_{LC}^{\dagger}|\psi_L\rangle + \hat{V}_{CR}^{\dagger}|\psi_R\rangle, \\ &= \hat{V}_{LC}^{\dagger}\hat{g}_L(\epsilon)\hat{V}_{LC}|\psi_C\rangle + \hat{V}_{CR}^{\dagger}\hat{g}_R(\epsilon)\hat{V}_{CR}|\psi_C\rangle \\ &= \hat{\Sigma}_L(\epsilon)|\psi_C\rangle + \hat{\Sigma}_R(\epsilon)|\psi_C\rangle \end{aligned} \tag{27}$$

where we define self-energies $\hat{\Sigma}_{L,R}(\epsilon)$ for electrodes L and R by

$$\begin{aligned}\hat{\Sigma}_L(\epsilon) &= \hat{V}_{LC}^{\dagger}\hat{g}_L(\epsilon)\hat{V}_{LC} = \hat{V}_{LC}^{\dagger}\left(\epsilon\hat{I} - \hat{H}_L\right)^{-1}\hat{V}_{LC} \\ \hat{\Sigma}_R(\epsilon) &= \hat{V}_{CR}^{\dagger}\hat{g}_R(\epsilon)\hat{V}_{CR} = \hat{V}_{CR}^{\dagger}\left(\epsilon\hat{I} - \hat{H}_R\right)^{-1}\hat{V}_{CR}\end{aligned} \tag{28}$$

and then we obtain

$$\left[\epsilon\hat{I} - \hat{H}_C - \hat{\Sigma}_L(\epsilon) - \hat{\Sigma}_R(\epsilon)\right]|\psi_C\rangle = 0. \tag{29}$$

This equation means that the effects from the semi-infinite electrode regions are separated as the self-energies $\hat{\Sigma}_{L,R}(\epsilon)$ [6, 7, 8]. It should be noted that, to construct the self-energies $\hat{\Sigma}_{L,R}(\epsilon)$, we need the matrix elements of the electrode regions and those of the boundary parts only, and we do not need the matrix elements of the center region. Usually, the atomic configurations for the electrodes are taken as the bulk ones and thus we treat a small number of atoms to describe their electronic states. We calculate the self-energies $\hat{\Sigma}_{L,R}(\epsilon)$ only once for the beginning of the calculation procedure, since we assume that they are not affected situated deep inside the electrodes.

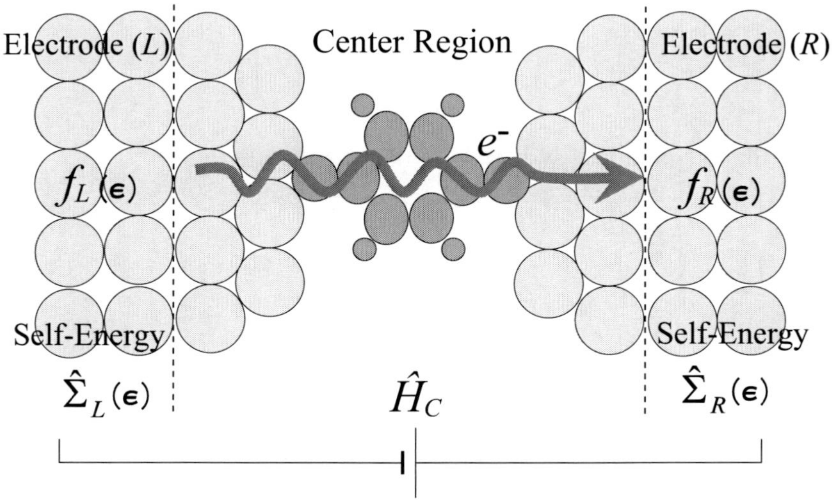

Figure 5. Schematic illustration of the junction systems in the nonequilibrium Green's function approach. The total system is divided into three parts at the boundaries. To construct the effective Hamiltonian in the nanosystem, the effects from semi-infinite electrodes are included in the Hamiltonian of the center region $\hat{H}_C$ as the self-energies $\hat{\Sigma}_{L,R}$.

After we obtain the self-energies for electrodes $\hat{\Sigma}_{L,R}(\epsilon)$, we solve the Green's function in the center region C, taking the inverse of the following matrix with an expansion of the localized basis sets of $|\psi\rangle$

$$\hat{G}^{r,a}(\epsilon) = \left[\epsilon\hat{I} - \hat{H}_C - \hat{\Sigma}_L(\epsilon) - \hat{\Sigma}_R(\epsilon) \pm i0^+\right]^{-1}. \tag{30}$$

Note that the dimension of this matrix $\langle\psi_i|\hat{G}^{r,a}|\psi_j\rangle$ is finite. However, as the center region for the nanostructures become larger and larger, this matrix inversion procedure would become much more time-consuming, since the scaling of the matrix inversion increases as $O(N^3)$ with the number of basis sets to represent the Hamiltonian in the center region.

When we have the retarded Green's functions $\hat{G}^r(\epsilon)$ for the center region, we can construct the electron density $\rho(\mathbf{r})$ from it. For the nonequilibrium situation, we need to obtain the lesser Green's function $\hat{G}^<(\epsilon)$ for the electron density where we have different chemical potentials μ_L and μ_R in the electrodes. Using the coupling constant $\hat{\Gamma}^{L(R)}(\epsilon)$, an imaginary part of the self-energy $\Sigma_{L(R)}(\epsilon)$ defined by

$$\hat{\Gamma}^{L(R)}(\epsilon) = i\left(\hat{\Sigma}_{L(R)}(\epsilon) - \hat{\Sigma}^*_{L(R)}(\epsilon)\right), \tag{31}$$

the lesser Green's function $\hat{G}^<(\epsilon)$ is obtained from

$$\hat{G}^<(\epsilon) = \hat{G}^r(\epsilon)\left(if_L(\epsilon)\hat{\Gamma}^L(\epsilon) + if_R(\epsilon)\hat{\Gamma}^R(\epsilon)\right)\hat{G}^a(\epsilon), \tag{32}$$

where we assign different $\mu_{L,R}$ for the Fermi distribution functions. The total electron density $\rho(\mathbf{r}) = -iG^<(\mathbf{r},\mathbf{r})$ in the center region (C) is obtained from

$$\rho(\mathbf{r}) = \frac{1}{2\pi i}\int G^<(\mathbf{r},\mathbf{r},\epsilon)d\epsilon. \tag{33}$$

We note that $\rho(\mathbf{r})$ can be divided as

$$\rho(\mathbf{r}) = -\frac{1}{\pi}\mathrm{Im}\int_Z f(z)G^r(\mathbf{r},\mathbf{r},z)dz + \frac{1}{2\pi i}\int_{\mu_R}^{\mu_L} G^<(\mathbf{r},\mathbf{r},\epsilon)d\epsilon \tag{34}$$

where Z is the upper complex contour energy plane. For obtaining the first term, we use the following relation to satisfy in the equilibrium region $f(\epsilon) = f_L(\epsilon) = f_R(\epsilon)$,

$$\begin{aligned}\hat{G}^<(\epsilon) &= if(\epsilon)\hat{G}^r(\epsilon)\left[\hat{\Gamma}^L(\epsilon) + \hat{\Gamma}^R(\epsilon)\right]\hat{G}^a(\epsilon)\\ &= f(\epsilon)\left[\hat{G}^r(\epsilon) - \hat{G}^a(\epsilon)\right] = -2if(\epsilon)\mathrm{Im}\hat{G}^r(\epsilon).\end{aligned} \tag{35}$$

Then the first term in Eq. (34) is the electron density in the equilibrium situation where we can use the imaginary contour integral technique for the retarded Green's function $G^r(\epsilon)$ as usual. The second term in Eq. (34) is the electron density in the nonequilibrium situation where we need a fine mesh for the direct energy integration.

Based on the density-functional theory, we can construct the effective potential for $\hat{H}_C[\rho(\mathbf{r})]$. These procedures are repeated as usual until we obtain the self-consistent solutions for $\rho(\mathbf{r})$ and the effective potential in $\hat{H}_C[\rho(\mathbf{r})]$. After we obtain the self-consistent solutions, we can calculate the current flowing between electrodes as

$$I = \frac{2e}{h}\int_{\mu_R}^{\mu_L}(f_L(\epsilon) - f_R(\epsilon))\,\mathcal{T}(\epsilon)d\epsilon \tag{36}$$

where the transmission is

$$\mathcal{T}(\epsilon) = \mathrm{Tr}\left[\hat{G}^r(\epsilon)\hat{\Gamma}^L(\epsilon)\hat{G}^a(\epsilon)\hat{\Gamma}^R(\epsilon)\right] \tag{37}$$

and the conductance G in the zero bias limit is obtained from

$$G = -\left[\frac{\partial I}{\partial V}\right]_{V\to 0} = \frac{2e^2}{h}\mathrm{Tr}\left[\hat{G}^r(\epsilon)\hat{\Gamma}^L(\epsilon)\hat{G}^a(\epsilon)\hat{\Gamma}^R(\epsilon)\right] = \frac{2e^2}{h}\mathrm{Tr}\left[\hat{\mathcal{T}}^\dagger\hat{\mathcal{T}}\right]. \tag{38}$$

The matrix $\hat{\mathcal{T}}^\dagger\hat{\mathcal{T}}$ formed from a product of the transmission matrices $\hat{\mathcal{T}}$ is diagonalized using a unitary matrix $\hat{U}$ to transform the original channels into eigenchannels as

$$G = \frac{2e^2}{h}\mathrm{Tr}(\hat{U}^\dagger\hat{\mathcal{T}}^\dagger\hat{\mathcal{T}}\hat{U}) = \frac{2e^2}{h}\sum_i \mathcal{T}_i. \tag{39}$$

The total conductance is expressed as a sum of the eigenchannel transmissions $\mathcal{T}_i$ at the Fermi energy ϵ_F. In this basis, the scattering state in each channel is independent of other states without interchannel scattering. We note that this picture known as the eigenvalue decomposition is effective for the ballistic transport case.

2.4. DFT Calculations Based on NEGF Method

2.4.1. Localized Basis Sets

To construct the effective Hamiltonian for transport, we use the basis sets for the wavefunctions. Here we describe the numerical atomic orbital basis sets. As localized basis sets, analytical atomic orbitals such as Gaussian and Wannier functions are often used. To reduce the number of elements for the Hamiltonian matrix, the use of numerical pseudo-atomic orbitals for the pseudopotential is effective as localized basis sets $\phi(\mathbf{r})$, where its radial part is numerically constructed to confine in a finite range at the cutoff radius r_c^l as

$$R_l^{ps}(r) = \begin{cases} \text{finite} & r \le r_c^l \\ 0 & r > r_c^l. \end{cases} \tag{40}$$

Here l is the angular momentum with $l = 0, 1, 2, \cdots$. Then the Green's function is expanded using the localized radial part of the pseudo-wavefunction $R_l^{ps}(r)$ with the spherical harmonic function $Y_{lm}(\hat{\Omega})$ $(m = -l, -l+1, \cdots, l-1, l)$ as

$$G^{r,a}(\mathbf{r}, \mathbf{r}', \epsilon) = \sum_{ij} \phi_i(\mathbf{r}) G_{ij}^{r,a}(\epsilon)\phi_j^*(\mathbf{r}') = \sum_{ij} G_{ij}^{r,a}(\epsilon)\rho_{ij}(\mathbf{r}, \mathbf{r}') \tag{41}$$

with

$$\phi_i(\mathbf{r}) = R_l^{ps}(r - r_\alpha)Y_{lm}(\hat{\Omega}_\alpha), \quad \rho_{ij}(\mathbf{r}, \mathbf{r}') = |\phi_i\rangle\langle\phi_j| = \phi_i(\mathbf{r})\phi_j^*(\mathbf{r}') \tag{42}$$

for localized basis set and and the density matrix. Here α denotes atom and we use the composite index $\{\alpha lm\} \to i$. Then the matrix elements of the Green's functions in the center region are obtained from

$$\sum_j \left[\epsilon S_{ij} - (H_C)_{ij} - \Sigma_L(\epsilon) - \Sigma_R(\epsilon) \pm i0^+\right] G_{jk}^{r,a}(\epsilon) = \delta_{ik} \tag{43}$$

where the overlap $S_{ij} = \langle \phi_i(\mathbf{r}) | \phi_j(\mathbf{r}) \rangle$ is calculated from

$$S_{ij} = 4\pi i^{l_i - l_j} \sum_{l=|l_i-l_j|}^{l_i+l_j} i^l c_l(l_i m_i, l_j m_j) Y_{l,-m_i+m_j}(\hat{\Omega}_{\alpha_i - \alpha_j}) \times \int q^2 j_l(q|\mathbf{r}_{\alpha_i} - \mathbf{r}_{\alpha_j}|) R_i^{ps}(q) R_j^{ps}(q) dq. \tag{44}$$

Here $c_l(l_i m_i, l_j m_j)$ is the Clebsh-Gordon coefficients and $j_l(x)$ is the spherical Bessel functions. The Hamiltonian in the center region H_C is divided into the kinetic part T and the potential parts V. The matrix elements for T and V are

$$T_{ij} = 4\pi i^{l_i - l_j} \sum_{l=|l_i-l_j|}^{l_i+l_j} i^l c_l(l_i m_i, l_j m_j) Y_{l,-m_i+m_j}(\hat{\Omega}_{\alpha_i - \alpha_j}) \times \int (q^4/2) j_l(q|\mathbf{r}_{\alpha_i} - \mathbf{r}_{\alpha_j}|) R_i^{ps}(q) R_j^{ps}(q) dq \tag{45}$$

$$V_{ij} = \int \rho_{ij}(\mathbf{r}, \mathbf{r}) V_{\text{eff}}(\rho(\mathbf{r})) d\mathbf{r} + \int\int \rho_{ij}(\mathbf{r}, \mathbf{r}') V_{\text{non-local}}(\mathbf{r}, \mathbf{r}') d\mathbf{r} d\mathbf{r}' \tag{46}$$

with local effective potential in the DFT formalism as

$$V_{\text{eff}}(\rho(\mathbf{r})) = \sum_{\alpha} V_{\text{loc}}^{ps}(\mathbf{r} - \mathbf{r}_\alpha) + \int \frac{\rho(\mathbf{r}')}{|\mathbf{r} - \mathbf{r}'|} d\mathbf{r}' + V_{\text{xc}}[\rho(\mathbf{r})] \tag{47}$$

and separable non-local pseudopotential with its corresponding pseudo-wavefunction ϕ_{lm}

$$V_{\text{non-loc}}(\mathbf{r}, \mathbf{r}') = \sum_l \sum_{m=-l}^{l} \frac{|\phi_{lm}(\mathbf{r}) \delta V_l^{ps}(\mathbf{r})\rangle \langle \delta V_l^{ps}(\mathbf{r}') \phi_{lm}(\mathbf{r}')|}{\langle \phi_{lm} | \delta V_l^{ps} | \phi_{lm} \rangle}. \tag{48}$$

The local potential part is calculated in the Fourier space using the relation of

$$\rho_{ij}(\mathbf{q}) = \sqrt{\frac{2}{\pi}} \sum_{l=|l_i-l_j|}^{l_i+l_j} (-i)^l c_l(l_i m_i, l_j m_j) Y_{l,-m_i+m_j}(\hat{\Omega}_{\mathbf{q}}) \times \int r^2 j_l(qr) R_i^{ps}(r) R_j^{ps}(r) dr, \tag{49}$$

which is effective for the accurate treatments of the long-range Coulomb potentials. We note that, in the zero bias case, the matrix elements of the Hamiltonian are constructed in the localized basis sets to perform O(N) calculations.

2.4.2. Plane-Wave Basis Sets

The electronic states calculations based on the DFT formalism are possible to perform using the Laue representation in plane-wave expansions as

$$\psi_i^{L/R}(\mathbf{r}) = e^{i\mathbf{k}_{||}\mathbf{r}_{||}} \sum_j e^{i\mathbf{g}_{||}^j \mathbf{r}_{||}} \varphi_i^{L/R}(\mathbf{g}_{||}^j, z) \tag{50}$$

where the wavefunctions are classified into two categories of scattering states, one coming from left (L) and one from right (R) electrodes. Then we solve the Kohn-Sham equation

$$\left[-\frac{\hbar^2}{2m}\nabla^2 + V_{\text{eff}}(\rho(\mathbf{r}))\right]\psi_i^{L/R}(\mathbf{r}) + \int V_{\text{non-loc}}(\mathbf{r},\mathbf{r}')\psi_i^{L/R}(\mathbf{r}')d\mathbf{r}' = \epsilon\psi_i^{L/R}(\mathbf{r}). \tag{51}$$

The coefficients $\varphi_i^{L/R}(\mathbf{g}_{||}^j, z)$ construct a matrix $U^{L/R}(z)$ with the elements in $\mathbf{g}_{||}^j$ as column and those in $\mathbf{k}_z^i(=\sqrt{2m\epsilon/\hbar^2 - |\mathbf{k}_{||}+\mathbf{g}_{||}^j|^2})$ as row vectors.

We solve $U^{L/R}(z)$ numerically for each energy ϵ. In the recursion-transfer-matrix (RTM) method [9], the discretized matrix equation for $U^{L/R}(z_p)$ is solved recursively as

$$U(z_{p-1}) = [b_p - a_p U(z_p)]^{-1} c_p \tag{52}$$

with matrix coefficients of

$$a_p = I - \frac{1}{6}\Delta^2 V_{\text{eff}}(z_{p+1}),\ \ b_p = 2I + \frac{5}{3}\Delta^2 V_{\text{eff}}(z_p),\ \ c_p = I - \frac{1}{6}\Delta^2 V_{\text{eff}}(z_{p-1}) \tag{53}$$

for the local potential $V_{\text{eff}}(z_p)$ first. This method is shown to be accurate and computationally stable for obtaining scattering states with open boundary conditions. Then we solve the matrix equation to include the separable non-local pseudopotential $V_{\text{non-local}}(\mathbf{r},\mathbf{r}')$ [10]. Since the obtained wavefunctions for each energy ϵ constitute the orthonormal functions, we can construct the retarded Green's function directly using these wavefunctions. Then we use the NEGF technique to construct the charge density and the electric current.

2.5. Experiments toward Molecular Devices

In order to realize a molecular device, we need to form nanometer-scale gap and to arrange single molecules between the gap. Presently two methods, mechanically controllable break junction technique and use of electromigration, are effective for the creation of nanogap.

When we assume that the connecting atoms have strong chemical bonding to both electrodes, since a molecule has electronic states with an energy gap of several eV between the highest occupied molecular orbital (HOMO) and the lowest unoccupied molecular orbital (LUMO), electric current flows when one of electronic states of the molecule align with the chemical potentials through resonant tunnelings. Thus we observe peak structures in the differential conductance $G = dI/dV$ in the transport experiments.

2.6. Calculations of Single Molecule Transport

Here, we take the benzene-dithiol molecule, Bucy ball C_{60}, and the poly-thiophene molecular wires to see the molecule transport properties. We assume that the molecules have ideal contacts to electrodes.

2.6.1. Benzene-1,4-Dithiol Molecule between Au(111) Electrodes

Fig. 6 (left) shows the atomic configuration for the molecule between semi-infinite Au(111) electrodes. The anchoring sulfur atom is located at a hollow site. The transmission spectrum shown in Fig. 6 (right) indicates the resonant tunneling behavior through occupied molecular orbitals hybridized with electrodes at the Fermi level, closer to the HOMO between the gap of the molecule, which means that this molecule works for an p-type material of hole transport for Au(111) electrodes.

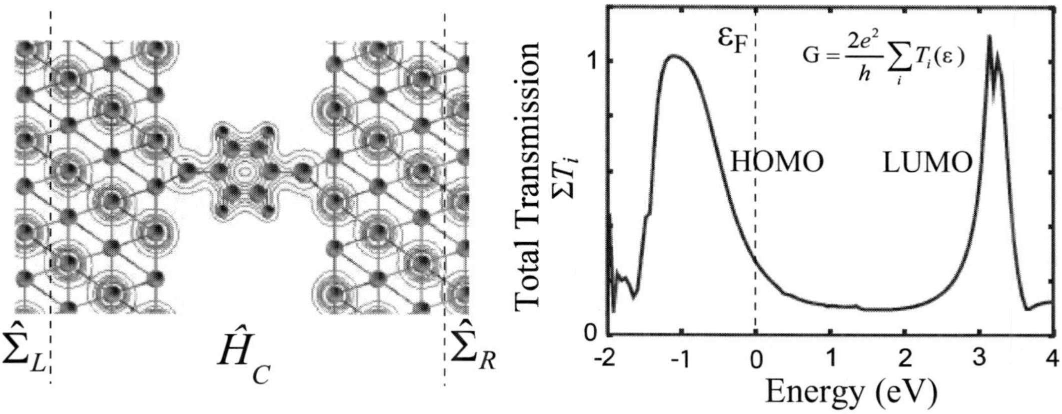

Figure 6. (Left) Atomic configuration for benzene-1,4-dithiol molecule sandwiched between semi-infinite Au electrodes. Contour lines show charge density. (Right) Total transmission spectrum of benzene-1,4- dithiolate molecule between Au(111) electrodes.

2.6.2. Bucky Ball between Al(111) Electrodes

Next let us see transport properties of Bucky ball C_{60}. Figure 7 (left) shows the atomic configuration for C_{60} between semi-infinite Al(111) 4×4 electrodes with protrusions. The electrode spacing is taken to be 10 Å and the centers of five-membered ring of C_{60} are contacted to the electrodes.

We see that there are number of peaks in the total transmissions shown in Fig. 7 (right), which indicate sharp resonant tunneling behaviors. The splitting in the degenerate states are clearly seen because of the breaking of symmetry due to the connection to the electrodes. The isolated C_{60} has the five-fold degenerate h_u HOMOs and the three-fold degenerate t_{1u} LUMOs, but the degeneracy is lifted into two states. From the channel decomposition analysis, we see that three eigenchannels contribute to the transport at ϵ_F. Also there are sharp peaks originating in individual molecular orbitals. The resonant widths at a higher energy are wider than those at a lower energy, since the wavefunctions of LUMOs spread out and the hybridization between the C_{60} molecule and electrode is large [11].

It is important to note that the electron transfer occurs from Al electrodes to C_{60}, and thus the relative Fermi energy position is not in the energy gap but is located inside the conduction band (LUMOs). Therefore, C_{60} acts as a conducting molecule and electrons move through the states originating from the unoccupied molecular orbitals of C_{60}. This

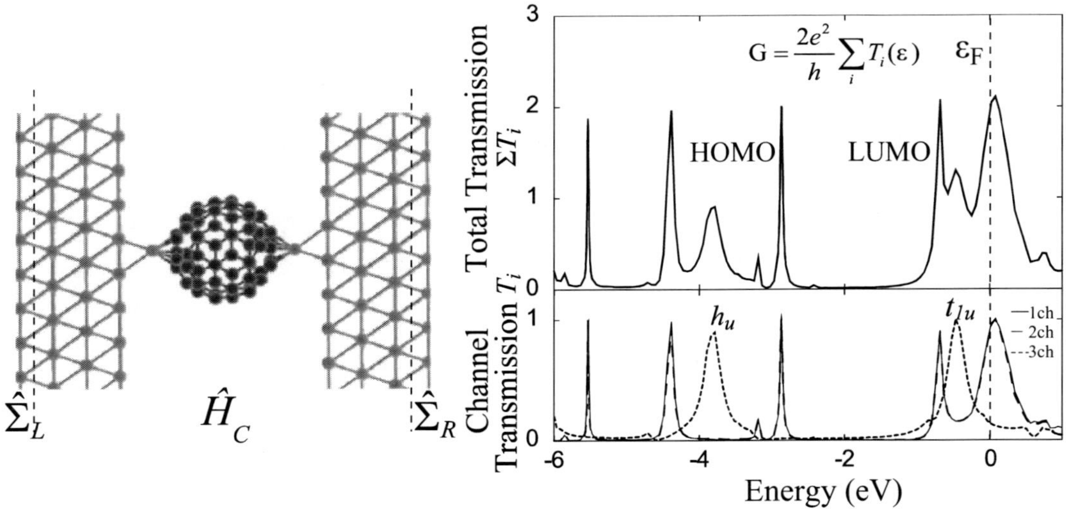

Figure 7. (Left) Schematic picture of atomic configuration for a Buck ball C_{60} sandwiched between semi-infinite Al electrodes. (Right) Total transmission (top) and eigenchannel transmission (bottom) of C_{60} between Al electrodes.

indicates that the Bucky ball C_{60} acts as an accepter for Al electrodes and works for an n-type material of electron (not hole) transport. We note that C_{60} is an important material as an acceptor in the organic devices. The molecular orbitals contributing to the charge transport depend on the interaction and the charge transfer between molecules and electrodes.

2.7. Energy Dissipation — Contact Problem

The contact problem as how to form strong chemical bonds between a tiny single molecule and electrodes is one of the most difficult and essential problems for the realization of molecular electronics. It is very important to understand how the contact to electrodes affect the electron transport and heating problem, which becomes more and more important for constructing tiny electron devices. Here we study the contact effects of molecules to electrodes to see how the bad contact affects the energy dissipation of electrons due to inelastic electron-phonon scatterings.

2.7.1. Phonon Emission at Contact

We consider the system where a molecule is sandwiched between jellium electrodes with the left contact is not well formed. In this case, the vibration of phonon with small frequency $\hbar\omega$ is induced between the gap region [12].

Fig. 8 represents a schematic picture of the present situation. Electron coming from the left electrode loses its energy at the contact to induce the spontaneous phonon emission, which becomes a source of local heating of the system.

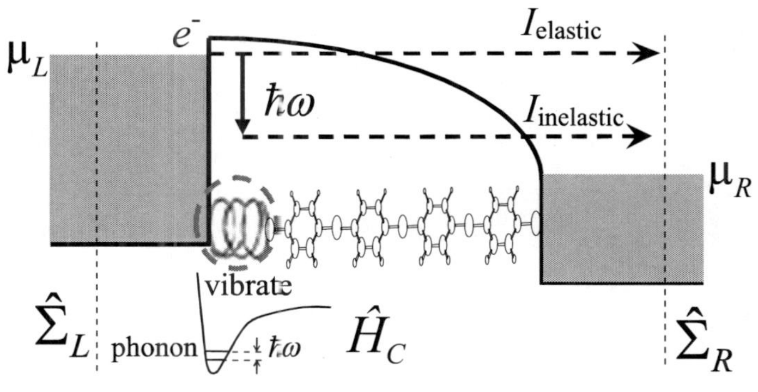

Figure 8. Schematic pictures of a molecule between electrodes with the left contact is not well formed, where the vibration with small frequency $\hbar\omega$ is induced within the gap region.

2.7.2. Formalism for Inelastic Transport

Let us consider to formulate the elastic and inelastic transport of electrons in the junction system with phonon emission. Here we treat the scatterings from phonons in the weak coupling case. Then we can limit the small displacement of atomic positions where the Hamiltonian for electron-phonon coupling is described by [13]

$$\hat{H} = \hat{H}_C + \sum_{\nu,j} \left. \frac{\partial \hat{H}_{\text{ion}}}{\partial Q_{\nu,j}} \right|_{Q=0} Q_{\nu,j}. \tag{54}$$

Here $\mathbf{Q} = \mathbf{R} - \mathbf{R_0}$ is displacement variables around equilibrium positions of atoms R_0 where the variables are $\nu = x, y, z$ and j runs all nuclei. We calculate the normal coordinates $\hat{Q}$ from the classical Newton's equation with frequencies ω_q ($q = \nu, j$) as

$$M_q \frac{\partial^2 R_q(t)}{\partial t^2} = -\frac{\partial E(\{R\})}{\partial R_q}, \quad Q_q(t) = R_q(t) - R_{q,0} = Q_q e^{i\omega_q t} \tag{55}$$

and are expressed quantum-mechanically by $\hat{Q}_q = \sqrt{\hbar/2M_q\omega_q}\left(\hat{a}_q^\dagger + \hat{a}_q\right)$. Here $\hat{a}^\dagger$ and $\hat{a}$ are the creation and annihilation operators of phonons, which obey the bosonic commutator relations with the occupation number of

$$N_q = \langle \hat{a}_q^\dagger \hat{a}_q \rangle = \frac{1}{e^{\hbar\omega_q/k_BT} - 1} \left(\approx \frac{k_BT}{\hbar\omega_q} \quad \text{for large } T \right). \tag{56}$$

The Hamiltonian for electron-phonon interaction is written in the weak coupling limit as

$$\hat{H}_{\text{e-pf}} \approx \sum_{n,q} \sqrt{\frac{\hbar}{2M_q\omega_q}} \left\langle \frac{\partial V_{\text{ion}}}{\partial Q_q} \right\rangle \hat{d}_n^\dagger \hat{d}_n \left(\hat{a}_q^\dagger + \hat{a}_q\right). \tag{57}$$

When an electron interacts with phonons, it dissipates energy from inelastic electron-phonon scatterings. Including the effect into Green's functions through self-energies

$$G^<(\epsilon) = G^r(\epsilon)\left(if_L(\epsilon)\Gamma^L(\epsilon) + if_R(\epsilon)\Gamma^R(\epsilon) + \Sigma^<_{\text{e-ph}}(\epsilon)\right) G^a(\epsilon)$$
$$G^r(\epsilon) - G^a(\epsilon) = -iG^r(\epsilon)\left(\Gamma^L(\epsilon) + \Gamma^R(\epsilon) + \Gamma_{\text{e-ph}}(\epsilon)\right) G^a(\epsilon), \tag{58}$$

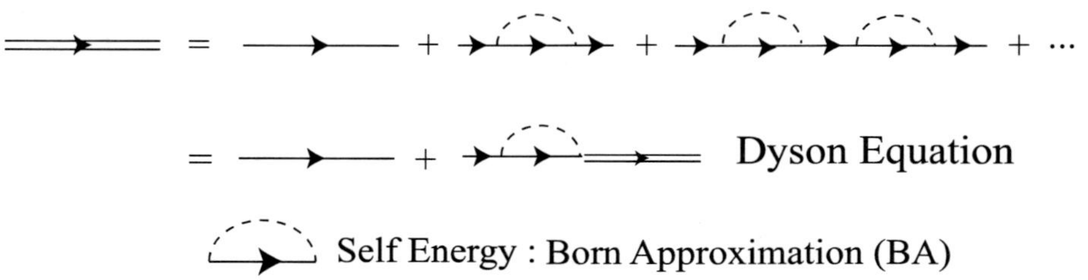

Figure 9. Diagrammatic summation of Green's function for one-phonon diagram.

we have two contributions for the current as

$$
\begin{aligned}
I_{\text{elastic}} &= \frac{2e}{h}\int (f_L(\epsilon)-f_R(\epsilon))\,\mathrm{Tr}\left[G^r(\epsilon)\Gamma^L(\epsilon)G^a(\epsilon)\Gamma^R(\epsilon)\right]d\epsilon \\
I_{\text{inelastic}} &= \frac{e}{h}\int \mathrm{Tr}\left\{i\left[\Gamma^L(\epsilon)-\Gamma^R(\epsilon)\right]G^r(\epsilon)\Sigma^<_{\text{e-ph}}(\epsilon)G^a(\epsilon)\right. \\
&\left. +\left[f_L(\epsilon)\Gamma^L(\epsilon)-f_R(\epsilon)\Gamma^R(\epsilon)\right]G^r(\epsilon)\Gamma_{\text{e-ph}}(\epsilon)G^a(\epsilon)\right\}d\epsilon, \qquad (59)
\end{aligned}
$$

where $G^r(\epsilon)$ with electron-phonon scatterings is obtained from the Dyson equation by

$$G^r(\epsilon) = G_0^r(\epsilon) + G_0^r(\epsilon)\Sigma^r_{\text{e-ph}}(\epsilon)G^r(\epsilon). \qquad (60)$$

The first equation has the same form for a non-interacting case describing the elastic current, while the second equation describes the inelastic current which causes energy relaxation.

The self-energies are evaluated from the bubble diagrams shown in Fig. 9 as $\Sigma_{\text{e-ph}}(t,t') = i\sum_q |g_q|^2 G_0(t,t')D(t,t')$, where $D(t,t')$ is free-phonon Green's function and we abbreviate coupling constant as $g_q = \sqrt{\hbar/2M_q\omega_q}\,\langle\partial V_{\text{ion}}/\partial Q_q\rangle$. The Fourier transforms are obtained using the Keldysh contour techniques by

$$
\begin{aligned}
\Sigma^<_{\text{e-ph}}(\epsilon) &= \frac{i}{2\pi}\sum_q\int |g_q|^2 G_0^<(\epsilon-\epsilon')D^<(q,\epsilon')d\epsilon' \\
\Sigma^r_{\text{e-ph}}(\epsilon) &= \frac{i}{2\pi}\sum_q\int |g_q|^2\left[G_0^<(\epsilon-\epsilon')D^r(q,\epsilon')\right. \\
&\left. +G_0^r(\epsilon-\epsilon')D^<(q,\epsilon') + G_0^r(\epsilon-\epsilon')D^r(q,\epsilon')\right]d\epsilon', \qquad (61)
\end{aligned}
$$

where phonon Green's functions are

$$
\begin{aligned}
D^<(q,\hbar\omega) &= -2\pi i\left[(N_q+1)\delta(\omega+\omega_q) + N_q\delta(\omega-\omega_q)\right] \\
D^r(q,\hbar\omega) &= \frac{1}{\hbar\omega-\hbar\omega_q+i0^+} - \frac{1}{\hbar\omega+\hbar\omega_q+i0^+}. \qquad (62)
\end{aligned}
$$

The calculations proceed as follows. Here we use the plane-wave expansion for the electronic states calculations with finite bias voltage. We construct the bare Green's functions $G_0^{r,<}(\epsilon)$ without electron-phonon scatterings and obtain the self-energies for phonon $\Sigma^{r,<}_{\text{e-ph}}(\epsilon)$. These are used to obtain the dressed Green's function $G^{r,<}(\epsilon)$ for the charge density for self-consistent DFT calculations. Then the currents are obtained from Eq. (59).

2.7.3. Elastic vs. Inelastic Transport

Here we present the results for molecular wires when the left contact S atom is set at 4.0 a.u. from the edge of jellium electrode [3] . Using the Newton's equation analysis in Eq. (55) for the contact atom, we find the frequency of $\hbar\omega_0 = 8.6$ meV for the vibration at contact.

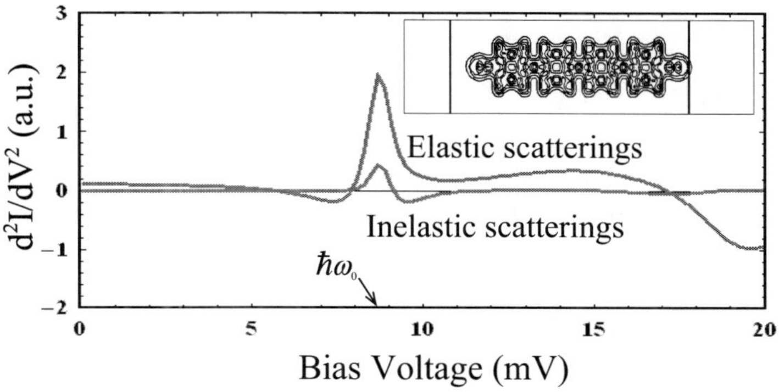

Figure 10. Second derivatives of elastic and inelastic electric currents for an applied bias voltage. For both cases, there are peaks corresponding to the phonon energy at the contact. (inset) Charge density distribution $\rho(\mathbf{r})$ of the molecular wire between jellium electrodes.

Fig. 10 shows the second derivatives of electric currents as a function of bias voltage. For both elastic and inelastic scatterings, there are peaks corresponding to the phonon energy at contact $\hbar\omega_0$. We can see that I_{elastic} is larger than $I_{\text{inelastic}}$, which is the source of local heating. It should be noted that these second derivative structures of the electric current can be observed by the inelastic electron tunneling spectroscopy (IETS) experiments.

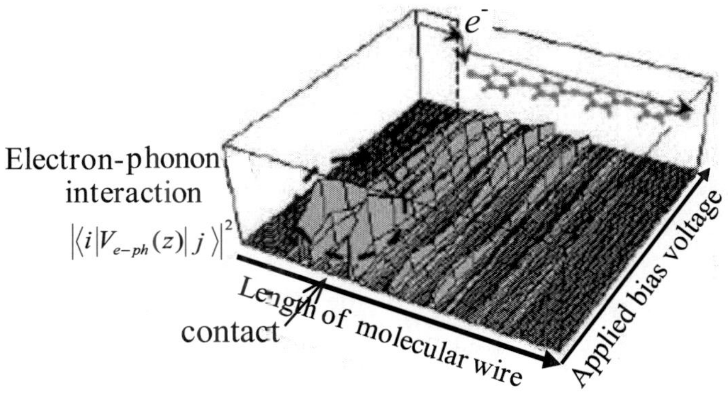

Figure 11. Electronphonon interaction as a function of the phonon energy across the molecular wire when the left contact to electrode is not well formed.

[3]The equilibrium distance from the edge of jellium electrode to the contact S atom (right side) is 1.9 a.u.

To observe where specifically the local heating due to phonon emission occurs, let us see the electron-phonon interaction strength. Fig. 11 shows the electron-phonon interaction as a function of phonon energy across the molecular wire. Here we evaluate the electron-phonon interaction by

$$|\langle i|V_{e-ph}(z)|j\rangle|^2 \approx \left|\sum_{m,n} \int d\mathbf{r}_{||}\varphi_i^L(\mathbf{g}_{||}^m, z)\nabla V_{loc}(\mathbf{r})\varphi_j^R(\mathbf{g}_{||}^n, z)\right|^2, \tag{63}$$

where there is an energy difference of eV between i and j states. Since the effective potential changes very rapidly at the left contact, we see that the electron-phonon interaction becomes very large there, which induces a phonon emission and bias drop [4]. This shows that the electron interacts with the phonon at contact to pass through the molecular wire.

In view of the power of local heating by energy dissipation, we can evaluate it from $\langle W\rangle \sim \hbar\omega_q \sum_q \int (2\pi/\hbar)(1+\langle n_q\rangle)g(\omega_q)\left|\langle i|V_{e-ph}|j\rangle\right|^2 \nu^L(\epsilon)\nu^R(\epsilon-\hbar\omega_q)d\epsilon$ where $g(\omega_q), \nu(\epsilon)$ are the density of states for phonon and electrons, because the transition rate by phonon scatterings becomes $\gamma_{i,j} = g(\epsilon_j - \epsilon_i)\left|\langle i|V_{e-ph}|j\rangle\right|^2 (n_q+1)$ for phonon emission [5]. Since $n_q \propto T/\omega_q$ as in Eq. (56), we can say that there are a number of interactions with phonons and correspondingly an electron loses its energy mostly at contact [6].

3. Transport Calculations for Organic Semiconductors

The method described in the previous section is difficult to apply to the transport properties of organic semiconductors, since the size of the system is too large to treat and also there is "strong" interaction of electron transport with vibrational modes to create the polaronic states. The binding energy of polaron (reorganization energy) has similar energetic orders of transfer integrals and thermal excitation energy $k_BT \simeq 25$ meV at room temperature. To clarify the charge transport properties from an atomistic viewpoint, we use a semi-classical approach for phonons assuming that the nuclear motions of the lattice is treated classically via Born-Oppenheimer approximation, whereas the carrier dynamics is evaluated purely based on the quantum mechanical approach. This assumption has been frequently taken in the studies of conductive polymers. The electron motion in materials is described directly by coupling the quantum- mechanical time-evolution calculation of an electron wave packet with the classical molecular lattice dynamics simulations.

In this section, we describe the calculation method using the electron wave-packet, called as time-dependent wave-packet diffusion (TD- WPD) method [14]. First we describe the basic derivations of the equations based on the Kubo formula. Then we explain how the formulas are utilized for the charge transport with strong electron-phonon couplings.

[4]There is a regime where such local heating occurs at contact. When the distance is small, the electron-phonon interaction is too small to induce phonon emissions. On the other hand, when the distance is too large, there is no interatomic force between molecule and electrodes, thus no phonon emission.

[5]The transition rate for phonon absorption is $\gamma_{j,i} = g(\epsilon_j - \epsilon_i)\left|\langle i|V_{e-ph}|j\rangle\right|^2 n_q$, which satisfies the detailed balance relation of $\gamma_{i,j}/\gamma_{j,i} = \exp(-\hbar\omega_q/k_BT)$. The power would be estimated from $W = -\int \epsilon \times I(\epsilon)d\epsilon$.

[6]Vibrational frequencies of internal modes of molecules are much higher than that of the mode at contact.

3.1. Time-Dependent Wave-Packet Diffusion (TD-WPD) Method

We consider the conductivity σ of organic materials with the volume Ω from the time-dependent form of linear-response Kubo formula as

$$\sigma = \lim_{t \to +\infty} \int_{-\infty}^{+\infty} d\epsilon \left(-\frac{df(\epsilon - \epsilon_F)}{d\epsilon} \right) \times e^2 \mathrm{Tr} \left[\frac{\delta(\epsilon - \hat{H}_e)}{\Omega} \cdot \frac{(\hat{x}(t) - \hat{x}(0))^2}{t} \right], \tag{64}$$

where $\mathrm{Tr}[\delta(\epsilon - \hat{H}_e)/\Omega] \equiv \nu(\epsilon)$ is the density of states (DOS). The dynamical change of electronic states by the lattice distortion due to the polaron formation is included in the time- dependent Hamiltonian $\hat{H}_e(t)$.

The mobility μ of a carrier with the elementary charge q is defined as the conductivity of Eq. (64) divided by the charge density

$$\mu = \frac{\sigma}{qn} \tag{65}$$

where the charge density is defined by $qn = q \int f(\epsilon - \epsilon_F)\nu(\epsilon)d\epsilon$. We note that when we evaluate the diffusion coefficient D as the long-time limit of the time-dependent diffusion coefficient,

$$D = \lim_{t \to +\infty} \frac{1}{t} \frac{\int d\epsilon f(\epsilon - \epsilon_F) \mathrm{Tr}[\delta(\epsilon - \hat{H}_e)\,(\hat{x}(t) - \hat{x}(0))^2]}{\int d\epsilon f(\epsilon - \epsilon_F) \mathrm{Tr}[\delta(\epsilon - \hat{H}_e)]}, \tag{66}$$

we can extract the well-known Einstein relation of

$$\mu = \frac{qD}{k_B T}. \tag{67}$$

Here we approximate the Fermi-distribution function as $f(\epsilon - \epsilon_F) \simeq \exp\{-(\epsilon - \epsilon_F)/k_B T\}$ since we consider the low carrier states of the semiconductor. The mean free path is also obtained from the diffusion coefficient

$$l_{\mathrm{mfp}} = \frac{D}{v}, \tag{68}$$

where v is the carrier velocity defined by $v = \lim_{t \to 0} \sqrt{D(t)/t}$.

Here we consider how this formula derives various transport regimes. (1) In the extended (diffusive) regime where electrons scatter many times with impurities and phonons to converge to a constant value $D_{\max}(\epsilon, t \to +\infty) \to D(\epsilon)$. This is because the velocity correlation vanishes after the relaxation time τ with $\langle v(s)v(s')\rangle = \langle v^2\rangle \exp(-|s - s'|/\tau)$. Thus $D(\epsilon) = \langle v^2\rangle\tau$ and the resistance is proportional to the system length L and is inversely proportional to S. This shows the classical Ohm's law. (2) In the localized regime to reach as the strength of scatterings become significant, electrons cannot move and $D(\epsilon, t)$ goes down to zero $D(\epsilon, t \to +\infty) \to 0$. These observations show that the transport properties in extended (diffusive) and localized regimes are determined from the evaluation of the time-dependent diffusion coefficient $D(\epsilon, t)$.

Instead of calculating the numerical diagonalization of the matrix $\hat{H}_e(t)$ of huge systems which requires $O(N^3)$ calculations, we evaluate $D(\epsilon, t)$ using wave-packets by the statistical average of the operator $\hat{A} = (\hat{x}(t) - \hat{x}(0))^2$ for several initial wave-packets

$|\psi_n\rangle$, which includes information of wavefunctions for various energies [7]. We can show that if we use sufficient numbers of initial wave-packets, the time-dependent diffusion coefficient $D(\epsilon, t)$ approaches statistically to the one that is represented by the complete sets of the eigenfunctions. Note that this method is suitable for the use of parallel computing to perform the time-evolution calculation of each initial condition. We compute $\langle\psi_n|\delta(\epsilon - \hat{H}_e)\hat{A}|\psi_n\rangle$ and $\langle\psi_n|\delta(\epsilon - \hat{H})|\psi_n\rangle$ separately using the Haydock's recursion method. The trajectories of wave-packets $\hat{x}(t)$ are calculated purely quantum-mechanically based on the time-evolution operator by the Chebyshev polynomial development by

$$\exp\left[-\frac{i}{\hbar}\hat{H}_e\Delta t\right] = \sum_{n=0}^{\infty} \exp\left(i\frac{a}{\hbar}\Delta t\right) h_n i^n J_n\left(\frac{b\Delta t}{\hbar}\right) T_n\left(\frac{\hat{H}_e - a}{b}\right) \tag{69}$$

where J_n is the Bessel functions and T_n is the Chebyshev polynomials of the first kind. The energy spectrum of $\hat{H}_e$ has the interval $[a - b, a + b]$, $h_0 = 1$ and $h_n = 2$ $(n \geq 1)$. Since the Chebyshev polynomials for a operator $\hat{x}$ obey the recursive relation

$$T_{n+1}(\hat{x}) = 2\hat{x}T_n(\hat{x}) - T_{n-1}(\hat{x}) \qquad \text{for} \quad n \geq 1 \tag{70}$$

with $T_0(\hat{x}) = 1$ and $T_1(\hat{x}) = \hat{x}$, we can reduce the number of expansion n for J_n by taking a small time step Δt in Eq. (69). This is because the higher order terms of J_n decreases to zero rapidly as n increases and the recursive procedure of T_n is much reduced.

Note that TD-WPD method becomes $O(N)$, while NEGF method needs $O(N^3)$ calculations for the matrix inverse. The memory usage of TD-WPD method needs also $O(N)$ due to the storage of non-zero components of Hamiltonian, compared with $O(N^2)$ for NEGF method. Thus the TD-WPD method requires low resources of computer facilities [8].

3.2. Van der Waals Interaction in DFT Calculations

We extract the parameters for the transport properties of organic semiconductors used here, such as electron-phonon coupling constants and elastic constants from the calculations based on the density-functional theory, including van der Waals interactions for the energy called as the DFT-D approach [15].

The recent development of introduction of van der Waals interaction energy due to the dipole-dipole interaction $\sim -1/R^6$ is well described by the formula as

$$E_{\text{disp}} = -s_6 \sum_{i=1}^{N_{at}-1} \sum_{j=i+1}^{N_{at}} \frac{C_6^{ij}}{R_{ij}^6} f_{\text{dmp}}(R_{ij}), \tag{71}$$

which are shown to be accurate for the molecular systems. Here N_{at} is the number of atoms in the systems, C_6^{ij} are the dispersion coefficients for atom pair ij, s_6 is the scaling factor, and $f_{\text{dmp}}(R) = 1/(1 + e^{-\alpha(R/R_0 - 1)})$ with R_0 is the sum of atomic van der Waals radii.

Fig. 12 (left) shows the schematic picture of the pentacene molecule coupled with neighboring molecules, which affect the translational and rotational motions. Fig. 12 (right) show

[7] TD-WPD method treats both the carrier wave-packet dynamics and the atomic-motion dynamics by MD.

[8] Note that the present TD-WPD method is not the full DFT calculations to obtain charge and potential self-consistently, but an analog of TB-DFT coupled with MD for the parameters to obtain from the DFT calculations.

the temperature dependence of the rotational motions L_{11}, L_{22}, L_{33}, which interacting with neighboring molecular systems. We see that the calculated results using the DFT-D approach are in good agreement with experimental observations by TLS analysis.

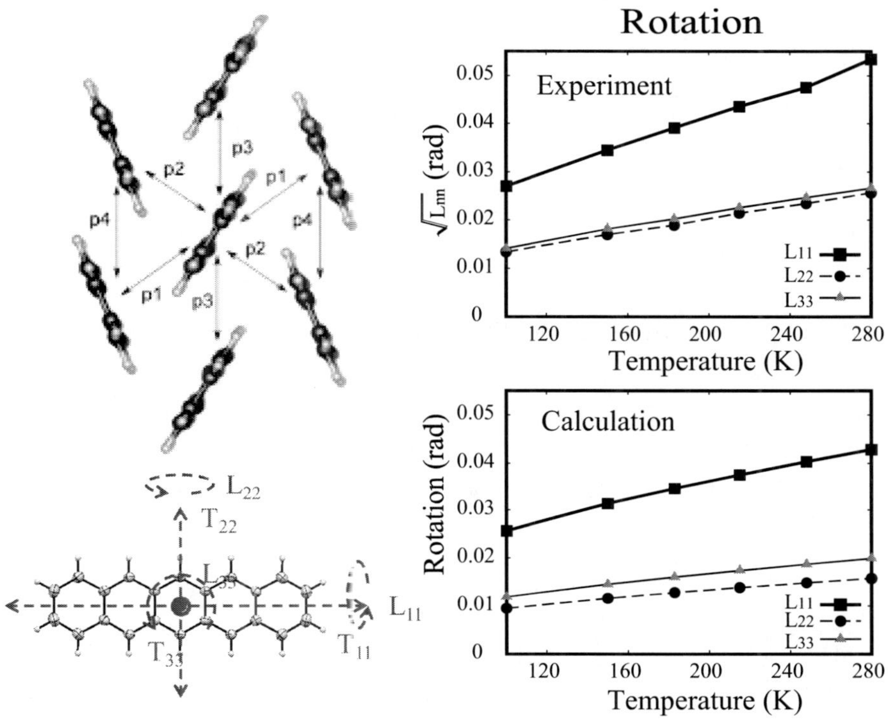

Figure 12. (Left) Schematic picture of the pentacene molecular systems. The pentacene molecule P1 interacts with neighboring molecules represented by P2-P7 with translations (T) and rotations (L). (Right) Temperature dependence of the rotational motions interacting with neighboring molecules by TLS experiments (upper) and DFT-D calculations (lower). Calculations are done with use of the GAMESS program at B3LYP-D3/6-31G(d) basis sets.

The effective transfer integrals γ^{HOMO} are extracted from the DFT-D calculations as follows [16]. When we have HOMO and HOMO-1 orbitals Ψ_i from the interaction of two molecules with a distance d and an angle θ including the effects from other environmental molecules, the effective orbital energies are described from the 2×2 secular equation of $\det|\bar{H} - \varepsilon\bar{S}| = 0$. Here $\bar{H}$ and $\bar{S}$ are expressed by

$$\bar{H} = \begin{pmatrix} e_1 & J_{12} \\ J_{12} & e_2 \end{pmatrix} \quad \bar{S} = \begin{pmatrix} 1 & S_{12} \\ S_{12} & 1 \end{pmatrix} \tag{72}$$

with the matrix elements of $e_i = \langle\Psi_i|\bar{H}|\Psi_i\rangle$, $J_{ij} = \langle\Psi_i|\bar{H}|\Psi_j\rangle$, and $S_{ij} = \langle\Psi_i|\Psi_j\rangle$. The matrix elements of e_i and J_{ij} have the same physical meaning of the parameters of the on-site energy and the transfer integral, although these two sets are not identical. We use Lödin's symmetric transformation to have an orthogonal basis sets which maintain the initial character of the orbitals as much as possible. Then the effective Hamiltonian becomes

$$\bar{H}^{\text{eff}} = \begin{pmatrix} e_1^{\text{eff}} & \gamma^{\text{HOMO}} \\ \gamma^{\text{HOMO}} & e_2^{\text{eff}} \end{pmatrix} \tag{73}$$

with

$$e_{1,2}^{\text{eff}} = \frac{1}{2}\frac{(e_1+e_2) - 2J_{12}S_{12} \pm (e_1-e_2)\sqrt{1-S_{12}^2}}{1-S_{12}^2},$$
$$\gamma^{\text{HOMO}} = \frac{J_{12} - \frac{1}{2}(e_1+e_2)S_{12}}{1-S_{12}^2}. \tag{74}$$

In this way, the secure equation becomes the standard form and the resulting energy splitting between HOMO and HOMO-1 levels becomes $\Delta\varepsilon_{12} = \sqrt{(e_1^{\text{eff}} - e_2^{\text{eff}})^2 + (2\gamma^{\text{HOMO}})^2}$. In Fig. 13, we show an example of the transfer integral γ^{HOMO} as a function of the angle θ. The data for the evolution of γ^{HOMO} would be useful for evaluating the charge transport of organic semiconductors, where the molecular structures are flexible and change variously.

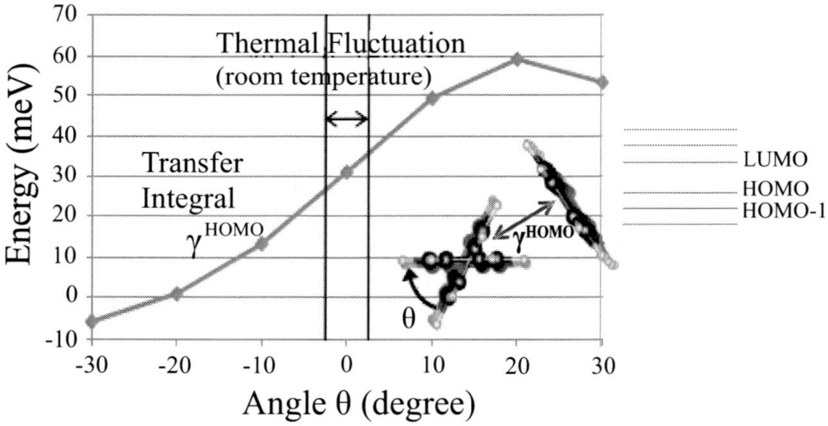

Figure 13. Evolution of the transfer integrals γ^{HOMO} as a function of the tilt angle θ with $P1$–L_{11} direction between pentacene molecules.

3.3. Band-Like vs. Hopping-Like Transport – Polaron Problem

Different from the inorganic materials which are formed by the strong covalent bond with neighboring "atoms" having typically $\sim$ eV bandwidth W_B, the organic semiconductors are formed by the weak van der Waals interactions with neighboring "molecules" having $10 \sim 100$ meV narrow bandwidth W_B, which make these materials much more flexible than inorganic materials. W_B is comparable to the room temperature k_BT and the energy level spacing $\Delta\epsilon$ due to the carrier traps, static structural disorders, impurities, and grain boundaries. Therefore the charge transport is determined by the competition of these energies [17]. The states are extended and charge transport becomes "band-like" for $\Delta\epsilon \ll W_B$, while the states are localized and charge transport becomes "hopping-like" by the thermal activation k_BT for $\Delta\epsilon \gg W_B$. This implies that the accurate evaluations of transfer energies and molecular orbital energies by the DFT calculations are essential to account for the transport properties of organic semiconductors.

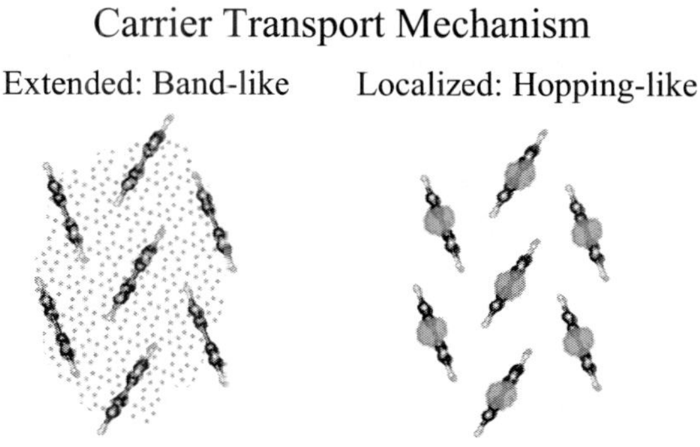

Figure 14. Charge transport in organic semiconductors. (left) Band-like transport via the extended states. (b) Hopping-like transport via the localized states.

It is important to note that the charge carriers are strongly scattered and are localized in a molecule, creating a quasi-particle state called polaron as a result of the strong interaction with electron and molecular vibrations, which makes the charge-carrier transport different from those for covalently bonded semiconductors. The large (Peierls) polaron is formed by the coupling between the inter-molecular vibration and electron, while the strong coupling of an electron with the intra-molecular vibartion forms the small (Holstein) polaron having a large binding energy, which describes the charge transport in the localized hopping regime due to static disorders.

To investigate the charge-carrier transport of organic semiconductors, we need to take both the Holstein and Peierls polarons into account for the carriers interacting tightly with molecular distortion. Here the Holstein model is based on the local electron–phonon coupling which acts purely intra-molecule, i.e., at the single molecule ionized by the charge of carrier. On the other hand, inter-molecular vibration influences the time-dependence of transfer integrals between adjacent molecules. The resulting nonlocal coupling leads to the Peierls model such as the Su-Schrieffer-Heeger model used for the polymers.

3.4. Coupling with Molecular Dynamics for Polaronic States

First, we describe the electron motion in Fig. 15 (a) where the Hamiltonian for a hole in the highest occupied molecular orbital (HOMO) band is written as

$$\hat{H}_e(t) = \sum_{n,m} \tilde{\gamma}_{nm}^{\mathrm{HOMO}}(\Delta R_{nm}(t)) \cdot \left(\hat{c}_n^\dagger \hat{c}_m + \hat{c}_m^\dagger \hat{c}_n\right) + \sum_n \left\{\tilde{\varepsilon}_n^{\mathrm{HOMO}}(\Delta u_n(t)) + W_n\right\} \hat{c}_n^\dagger \hat{c}_n, \quad (75)$$

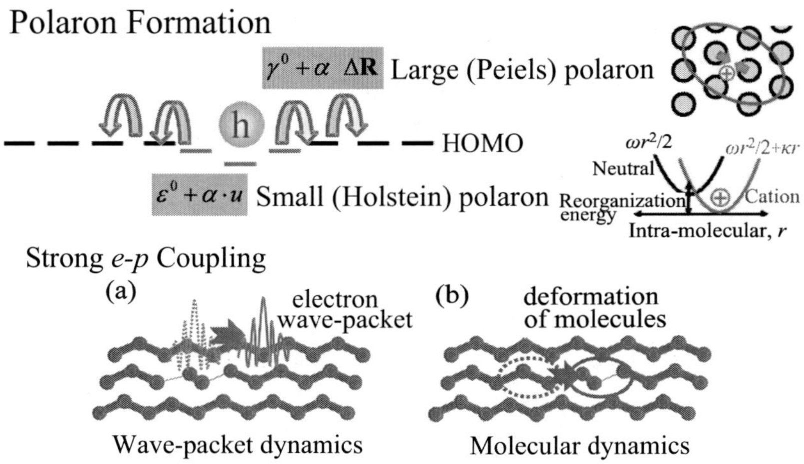

Figure 15. (Top) Large and small polaron formation in organic semiconductors. The intra-molecular deformation by the reorganization Δu changes the HOMO level to form the small (Holstein) polaron, while the inter-molecular deformation by the bond-length ΔR changes the transfer energy to form the large (Peierls) polaron. (Bottom) Schematic picture polaron formation due to strong electron–phonon coupling.

where the operators $\hat{c}_n$ and $\hat{c}_n^\dagger$ are annihilation and creation operator of a hole at the HOMO with the orbital energy $\varepsilon_n^{\rm HOMO}$. The transfer integral $\tilde{\gamma}_{nm}^{\rm HOMO}$ and the orbital energy $\tilde{\varepsilon}_n^{\rm HOMO}$ coupled with the inter- and intra- molecular lattice distortions are defined by

$$\begin{aligned}\tilde{\gamma}_{nm}^{\rm HOMO}(\Delta R_{nm}) &= \gamma_{nm}^{\rm HOMO} + \alpha_{\rm P}^{\rm HOMO}\cdot\Delta R_{nm},\\ \tilde{\varepsilon}_n^{\rm HOMO}(\Delta u_n) &= \varepsilon_n^{\rm HOMO} + \alpha_{\rm H}^{\rm HOMO}\cdot\Delta u_n,\end{aligned} \quad (76)$$

where $\gamma_{nm}^{\rm HOMO}$ is the bare transfer integral between nth and mth molecules. Here $\alpha_{\rm P}^{\rm HOMO}$ is the Peierls-type electron-phonon coupling constant. When we represent the displacement of nth molecule as ΔR_n, the change of inter-molecular distance is given by $\Delta R_{nm} = |(\mathbf{R}_m^e + \Delta\mathbf{R}_m) - (\mathbf{R}_n^e + \Delta\mathbf{R}_n)| - |\mathbf{R}_m^e - \mathbf{R}_n^e|$ where $\mathbf{R}_n^e$ is the equilibrium position of the molecule. On the other hand, $\alpha_{\rm H}^{\rm HOMO}$ is the Holstein-type electron–phonon coupling constant. When we replace the intra-molecular displacement $\Delta\mathbf{u}_n$ with the effective scalar coordinate Δu_n, we can express the shift of orbital energy by $\alpha_{\rm H}^{\rm HOMO}\cdot\Delta u_n$.

In addition to these polaron effects, we also consider the effects of static disorder, which inevitably exists in the molecular crystals. We introduce the disorder potentials W, which modulate the on-site orbital energies randomly within the width $[-W/2, +W/2]$.

Next, we describe the molecular dynamics (MD) simulations in Fig. 15 (b) for the dynamical lattice distortions, which include both the thermal fluctuation and reorganization upon ionization by charge carriers. The equation of motion for the nth site with the mass M is derived from the equation as $Md^2R_n/dt^2 = -\partial E_{tot}/\partial R_n$, which includes the elastic force and the lattice distortion due to the polaron formation. When elastic energies of both inter- and intra-molecular deformations are approximated by a harmonic form, we obtain the energies of molecular lattice as

$$E_{tot} = \sum_n \frac{1}{2} M \left\{ \frac{d\Delta R_n(t)}{dt} \right\}^2 + \sum_{n,m} \frac{1}{2} K_{\mathrm{P}} \left\{ \Delta R_{nm}(t) \right\}^2 + \sum_n \frac{1}{2} K_{\mathrm{H}} \left\{ \Delta u_n(t) \right\}^2, \tag{77}$$

where the inter- and intra-molecular elastic constants represent K_{P} and K_{H}. As for the lattice distortion, we determine Δu_n by the variational principles for the minimum total energy, $\partial E_{tot}/\partial \Delta u_n = 0$, resulting in $\Delta u_n(t) = (\alpha_{\mathrm{H}}/K_{\mathrm{H}}) \cdot \{\rho_{nn}(t)/\sum_m \rho_{mm}(t)\}$ with the density matrix ρ_{nm}. We note that the temperature T is fixed by normalizing the kinetic energy of lattice vibration at each time step on the condition of $\sum_{n=1}^{N} M\dot{R}_n^2/2 = dNk_BT/2$.

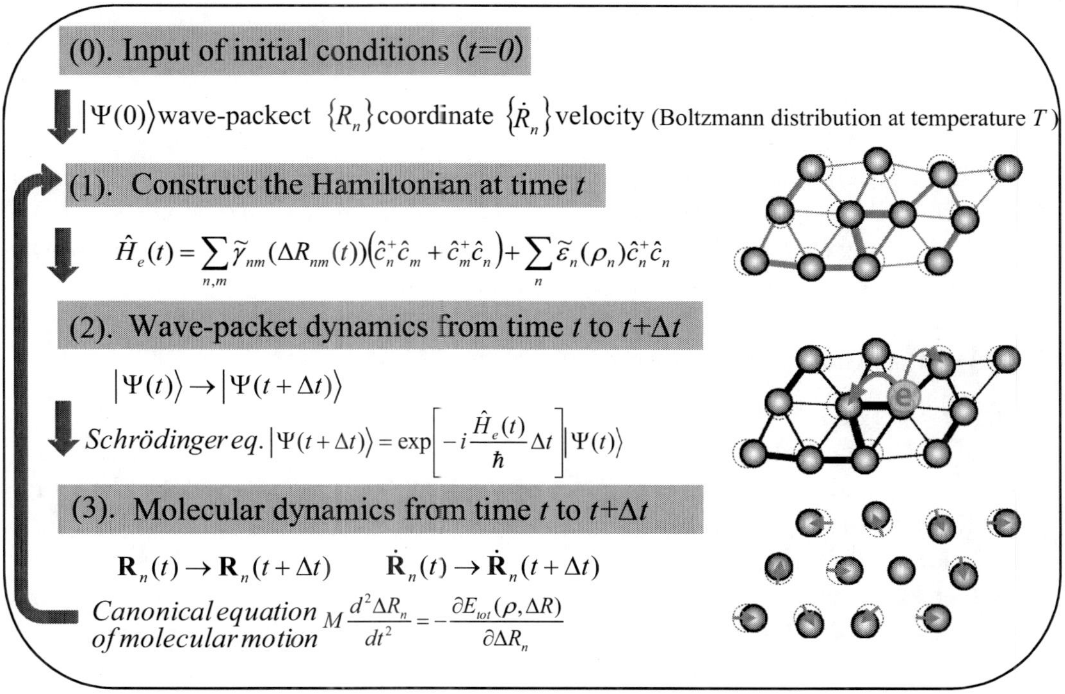

Figure 16. Flowchart of numerical computation according to the coupling of TD-WPD methodology with the MD simulations in case of the Holstein-Peierls model.

The flowchart of the calculations is shown in Fig. 16. Solutions of the coupled equations Eqs. (64) and the molecular dynamics allow us to describe the electron motion strongly coupled with both the thermal fluctuating lattice and the distortion by the polaron formation. Repeating the processes shown, we take into account the complicated trajectory of an electron motion in its induced time-dependent lattice distortion.

3.5. Formation of Polaron States

Let us see the electron–phonon coupling effect on the polaron formation of single-crystal pentacene molecules without both thermal fluctuations and static disorders [18].

Table 1. Polaron binding energies E_B by various models.

	Peierls	Holstein	Peierls-Holstein
Binding Energy	1.3 meV	7.5 meV	14.3 meV

3.5.1. Holstein-Type Coupling Effects

The Holstein-type electron–phonon coupling and the elastic constant are $\alpha_{\mathrm{H}}^{\mathrm{HOMO}} = -93$ meV and $K_{\mathrm{H}} = 92$ meV. First, we focus on the interaction between the electron and intra-molecular distortion inducing the energy shift of the HOMO levels. A discrete polaron state is observed in the energy spectrum of DOS located above the HOMO band whose binding energy E_B is evaluated as 14.3 meV, the same energetic order of transfer integrals γ and the thermal excitation energy k_BT. This implies that the Holstein-type coupling affects the charge transport properties through the polaron formation.

In Table 1, the calculated polaron binding energies are presented for three different models; the Holstein-Peierls model, the Peierls model with $\alpha_P \neq 0$ and $\alpha_H = 0$, and the Holstein model with $\alpha_P = 0$ and $\alpha_H \neq 0$. The Holstein-Peierls model gives rise to a much larger binding energy, which indicates that both the Holstein-type and the Peierls-type electron–phonon coupling should be taken into account simultaneously.

3.5.2. Peierls-Type Coupling Effects and Thermal Fluctuation

The Peierls-type elastic constant is estimated as $K_P = 2.19$ eV/Å^2 by fitting the computed bond-length dependent total energy $U(\Delta R)$ with the parabolic form $U = K_{\mathrm{P}}(\Delta R)^2/2$ around the equilibrium molecular position. Then the inter- molecular vibration has the continuous phonon-band structure as $\hbar\omega_{\mathrm{P}}(q) = \hbar\sqrt{2K_P(1-\cos qa)/M}$ with the bandwidth of 3.8 meV and the bond length $a = 5.16$ Å. The phonon dispersion is in good agreement with an acoustic phonon branch of oligo-acene crystals. By coupling the wave-packet dynamics with the classical molecular dynamics, we can take various phonon modes from $q = 0$ to $\pm\pi/a$ thermally excited at a finite temperature to study the realistic phonon-scattering effects on the charge transport properties.

As for the Peierls-type electron–phonon coupling effects on the transfer integrals, which describes the interaction between the electron transfer integral and the inter- molecular vibration called as the lattice phonon, we can observe the formation of the Peierls-type large polaron from the shrink of the molecular bond in the calculations. On the other hand, since the organic semiconductor devices are operated in the room temperature, it is essential to consider how the thermal fluctuations of molecular motions, as well as the polaron formation, affects the transfer integrals. The thermal lattice vibrations give rise to a large dynamic disorder in the inter-molecular transfer integrals. These results indicate that the charge transport properties of organic semiconductors are determined in the subtle balance of energies among the thermal-fluctuation of molecular motion, the thermal excitation, and polaron formation.

3.6. Experiments on Organic Semiconductor Devices

With recent rapid progress of the technology, we can construct flexible thin-film transistors (TFTs) from single-crystal organic semiconductors with carrier mobility up to $\mu \sim 40$ cm^2/Vs. Due to a decrease of structural disorders, it exceeds the mobility of amorphous silicon [19, 20]. In single-crystal TFTs, the mobility increases with a decrease of temperature with power-law dependence as $\mu \propto T^{-n}$, which shows a delocalized "band-like" transport with the carrier scattering from phonons, coupling with the inter-molecular vibrations. Also the recent Hall effect measurements show the nature of coherent transport.

When the disorder is strong, the mobility shows the exponential behavior as described by $\mu \propto \exp(-\Delta\epsilon/k_B T)$ with a decrease as the temperature decreases, which is a typical character of localized "hopping-like" transport due to the thermal excitation. Therefore temperature dependence of mobility changes remarkably with the strength of disorder.

3.7. Calculations of Organic Semiconductor Transport

In the TD-WPD approach coupled with MD simulations, we evaluate the carrier mobility μ, mean free path l_{mfp}, and diffusion coefficient D including both the large (Peierls-type) and the small (Holstein-type) polaronic states based on the linear-response Kubo formula with use of electron wave-packet dynamics.

3.7.1. Polaron Effects on Intrinsic Charge Transport

First, we study intrinsic (without static disorder) charge transport for single-crystal organic semiconductors using the time-evolution of wave-packets. The charge transport is determined by the coupling with inter- and intra- vibrations of large and small polaron characters. This calculations would provide us the possible maximum value of the mobility.

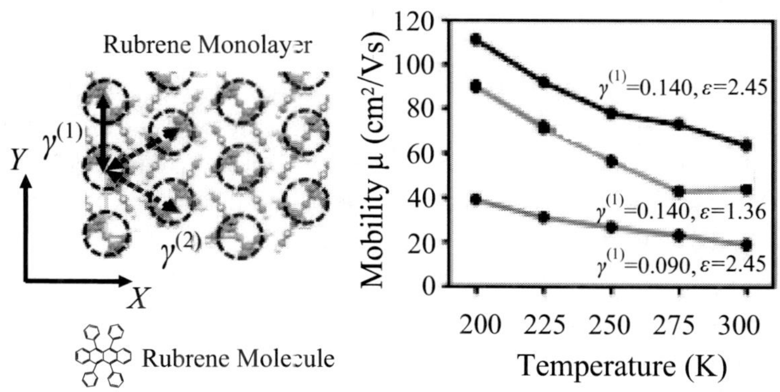

Figure 17. (Left) Schematic picture of rubrene monolayer. We assume $\gamma^{(2)} = 0.22\gamma^{(1)}$. (Right) Intrinsic carrier mobility μ for various conditions as a function of temperature T in case of no static disorders. The mobility μ decreases monotonically with increasing T.

We study rubrene monolayer which takes herringbone structure with the anisotropy of transfer integrals [21]. Here we approximate van der Waals interaction between molecules by the Lenard-Jones potential $E_{disp} = 4\epsilon \left\{ (\sigma/R_{ij})^{12} - (\sigma/R_{ij})^{6} \right\}$ with only two parameters. We use $\epsilon = 12.12$ bohr and $\epsilon = 1.36$ eV or 2.45 eV. The two kinds of transfer integrals are taken as $\gamma^{(1)} = 0.09$ eV or 0.14 eV and $\gamma^{(2)} = 0.22\gamma^{(1)}$. The other parameters are $K_P = M\Omega^2$ with $\Omega = 6.3$ meV and $M = 532M_0$, $K_H = m\omega^2$ with $\omega = 0.174$ eV for MD simulations, and $\alpha_P = 0.261\ eV$, $\alpha_H = 166$ meV for electron-phonon interaction. The on-site energy level ϵ_n for an electron is taken zero. The quantum wave-packet is propagated by the time-evolution with the classical MD simulations with $\Delta t = 0.5$ fs.

Figure 17 (right) shows the calculated carrier mobility μ as a function of temperature T for different interactions between rubrene molecules. The mobility increases monotonically with decreasing the temperature from 300 K to 200 K approximately by the power-law dependence due to the decrease of thermal fluctuation of lattice phonon, which shows an apparent evidence of the delocalized "band-like" transport.

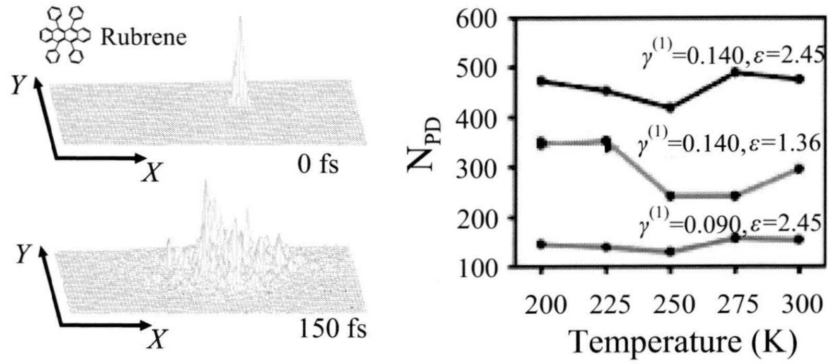

Figure 18. (Left) Examples of the wave-packet dynamics at $t = 0$ and $t = 150$ fs for rubrene monolayer at $T = 300$ K. (Right) Polaron delocalization number at the initial condition of $N_{PD} = \pi\langle R(0)^2\rangle / S_{\text{unit}}$ where S_{unit} is the area of the Wigner-Seitz cell of 160 bohr^2.

Let us see the propagation of wave-packet, which reflects the carrier motions through its time-dependent behavior. Fig. 18 (left) show the examples of wave-packet dynamics. To study the polaron behavior, we define the polaron delocalization number by $N_{PD} = \pi\langle R(0)^2\rangle / S_{\text{unit}}$, where $\langle R(0)^2\rangle$ is the mean square displacement of the spread of eigenvectors averaged with the Boltzmann factor and S_{unit} is the area of the Wigner-Seitz cell. We take 100 samples of lattice configurations at a given temperature for an average of $\langle R(0)^2\rangle$. Thus N_{PD} indicates the extent to which the polaron is delocalized over molecules. Fig. 18 (right) shows N_{PD} as a function of T. We see N_{PD} increases with increasing the transfer integral, similar to the trent of mobility. Under a large transfer integral, the polaron can be delocalized over few hundreds of molecules. The localization of polaron originates from the inhomogeneity of intermolecular coupling due to the thermal fluctuations.

3.7.2. Static Disorder Effects on Charge Transport

Here, we investigate how the transport properties are affected by the static disorders W [18], which inevitably exist in the molecular crystals in terms of static disorder such as crystal imperfections, the presence of impurities, and static dipole disorder caused by interaction of the charge with induced dipole moments in the gate dielectric. Several experimental evaluations show that the potential depths of these static disorders are estimated as about 50 meV. Thus we change the parameter W for the random distribution from 50 meV to 200 meV. We note that these static disorders are comparable to the HOMO band width of the present organic semiconductors. Therefore, the charge-transport properties are expected to be strongly disturbed by the static disorders in the various temperature regimes.

Figure 19 shows the normal linear behavior of the carrier mobility μ of single-crystal pentacene as a function of the temperature T for the strengths of static disorders W = 50 and 200 meV. The parameters used are those described in the previous sections. The existence of static disorders decreases the carrier mobility significantly from 100 cm^2/Vs (no static disorder) up to 2 cm^2/Vs (W = 200 meV) around 300 K, and more importantly their temperature dependences are changed completely.

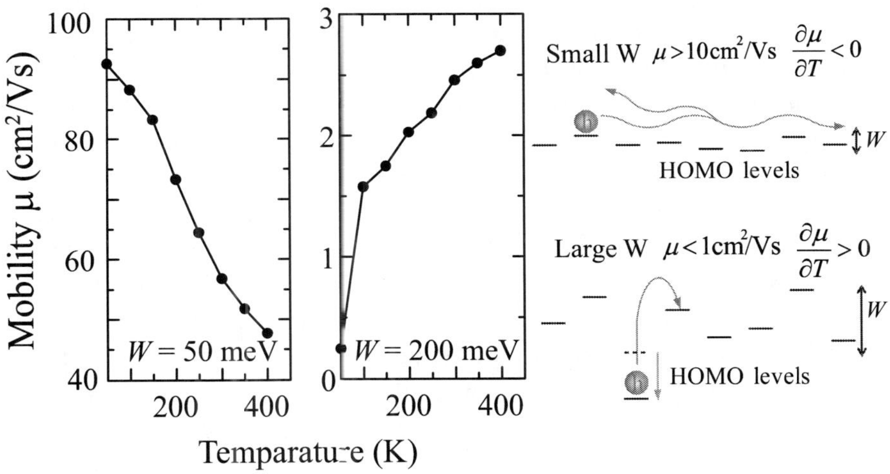

Figure 19. Plots of the carrier mobility μ of pentacene as a function of temperature T for the static disorders W = 50 meV (left) and 200 meV (center). As W increasing, T-dependence of μ around the room temperature changes from nearly power-law dependence to thermally activated behavior. (Right) Schematic pictures of the charge transport in the static disorder systems. For small W (W = 50), the carriers are extended over the crystals and are scattered from thermal fluctuations due to the dynamical disorder effects. For large W (W = 200), the carriers are trapped in the molecules and form the polaron state.

In case of W = 50 meV, the magnitude of μ becomes larger than 50 cm^2/Vs around the room temperature and the $d\mu/dT$ takes negative values in whole temperature regime, showing the monotonic increase of μ as a decrease of T. The T-dependence of μ higher than 200 K is close to the power-law dependence. On the other hand, the slope of an increase of μ becomes dull in the low T regime since the scatterings by static disorders with no temperature dependence becomes dominant for the carrier motion. The time dependences of diffusion constant $D(t)$ in Eq. (66) indicates the realization of the diffusive transport rather than the hopping transport of a localized carrier. The mean free paths $l_{\rm mfp}$ shows 4-6 times longer than the lattice constant a in case of W = 50 meV.

When the static disorders become significant to $W = 200$ meV, the magnitude of μ takes much lower value in comparison with that for $W = 50$ meV. The sign of $d\mu/dT$ becomes positive in whole temperature regime where the carrier is trapped tightly by disorders. The carrier is localized by the static disorders and induces the intra- molecular distortions, resulting in a low-mobility polaron formation. The calculated l_{mfp} is about 10 Å, less than the molecule-to-molecule distance. This renders the concept of band-like transport meaningless. The charge transport properties are close to typical thermally activated behaviors, as observed for the organic TFTs with low quality.

As T increases or W decreases, the carrier begins to move and its motion changes from the thermally activated hopping transport due to the localized carrier to the diffusive transport character where the polaronic localization state is destroyed because the thermal excitation energy beyond the polaron binding energy is given to the carrier. These behaviors are schematically described in the right of Fig. 19; the charge transport in the static disorder systems. For small W ($W = 50$), the carriers are extended over the crystals and are scattered from thermal fluctuations due to the dynamical disorder effects. We note that this behavior is mainly due to the transfer-integral modulation originated from the thermal fluctuation of molecular motion, whereas the polaron formation gives a subtle contribution to the transport properties. As the static disorders increase, the carrier becomes localized and couples with the intra- molecular distortions, forming the polaron state. For large W ($W = 200$), the carriers are trapped in the molecules and form the polaron state. Therefore, the magnitude of mobility is dramatically decreased and its temperature dependence changes gradually from the "band-like" transport behavior $d\mu/dT < 0$ to the thermal- activated "hopping transport" behavior $d\mu/dT > 0$. The present calculated results indicate that thermal fluctuation, polaron formation, and static disorders play important contributions to the understanding of transport mechanism of realistic organic semiconductors.

Let us estimate the disorder strength W due to the randomly oriented static dipole disorder, long-range electrostatic interaction of the charge with induced dipole distribution in the dielectric interface of the gate [22]. Here we estimate the dipole interaction by

$$\Phi(\mathbf{r}, \mathbf{p}) = -\frac{\sum_i p_i r_i}{4\pi\epsilon r^3}, \quad \langle p_i p_j \rangle = \frac{p^2}{3}\delta_{i,j}, \tag{78}$$

where $\mathbf{p}$ is an effective dipole strength in the dielectric with constants ϵ. Then the strength of static dipole disorder from the interface with the distance h is obtained from

$$W^2 = \int_{-\infty}^{h}\int_{-\infty}^{\infty}\int_{-\infty}^{\infty} \langle \Phi^2 \rangle n d\mathbf{r} = \frac{np^2}{48\pi\epsilon^2 h} \tag{79}$$

with a Gaussian distribution. This approximation assumes screening expression for the effective dipole strength to describe the microscopic interaction between the charge and its nearest dipole. For example, $p = 1.290$ debye, $\epsilon = 3.15\epsilon_0$, and $n = 5.60$ nm^{-3} for the polymethylmethacrylate (PMMA), which produces $W = 33$ meV for $h = 0.8$ nm.

3.7.3. ESR Measurements for Disorder Distribution

The electron spin resonance (ESR) experiments have recently been used to observe and investigate the localized electronic states of organic molecules due to static disorder [23].

This unique technique is expected to reveal the local distribution of disorder [24]. Indeed, the information on the rate of localization of electrons is provided from the stochastic optimization method (SOM) analysis of ESR measurements [25].

The principles of ESR are described in Fig. 20, where the microwave is provided to the organic semiconductor sample in a magnetic field to observe the resonance of unpaired electron, which are split by the Zeeman effects. Hyperfine interactions with nuclear spins due to the inhomogeneities of the sample broaden the ESR signal to spread in a Gaussian form. For example, the hyperfine coupling in pentacene molecules is due to 14 proton nuclear spins. The narrowing of linewidth and the random energy shift due to the disorder distribution would give us information on the spatial extension of wavefunctions.

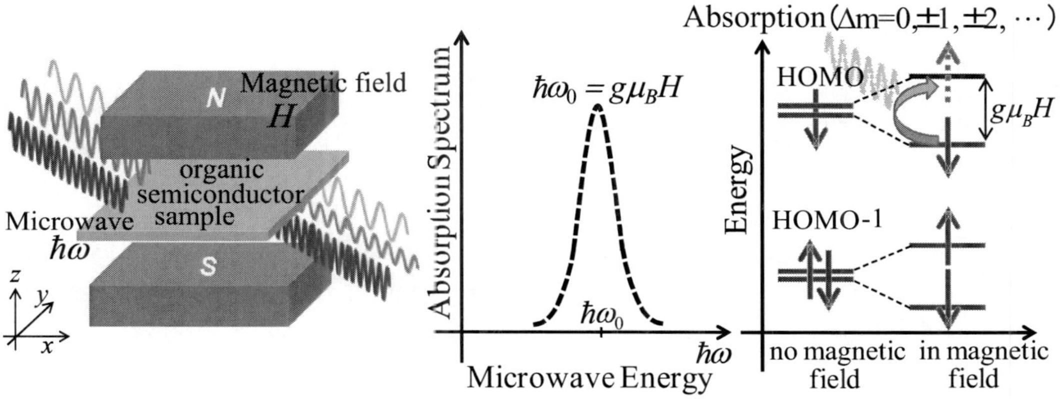

Figure 20. (Left) Schematic picture of the ESR measurement. Microwave with various frequencies $\hbar\omega$ is provided to the sample in magnetic fields of H. The microwave in resonance with the unpaired electronic states is absorbed with the condition of $\hbar\omega_0 = g\mu_B H$. (Center) Absorption spectrum as a function of microwave energy with a peak at resonance.(Right) Energy diagram of the molecular states without and with magnetic fields H.

Here let us derive the formula for ESR measurements. The Hamiltonian is given by

$$\hat{H}_S = g\mu_B H\hat{S}_z + \sum_{n=1}^{N_S} \sum_{\zeta=x,y,z} A^{(n)} \hat{I}_\zeta^{(n)} \cdot \hat{S}_\zeta \tag{80}$$

where the first term represents the coupling of unpaired electron spin with an applied magnetic field H and the second term represents the coupling with hyperfine nuclear spins $\hat{I^{(n)}}$. Here μ_B is the Bohr magnetron, g is the g-factor of an electron, and the coefficients $A^{(n)}$ take $A^n = 0.004\ \mu\text{eV} \sim 0.03\ \mu\text{eV}$ for the coupling with hyperfine nuclear spins of hydrogens in pentacene molecules. Then the absorption spectrum is obtained by the formula

$$I(\omega) = H_R^2\, \omega \cdot \chi''(\omega) \tag{81}$$

where H_R is the strength of microwave with frequency ω and $\chi''(\omega)$ is the imaginary part of the complex magnetic susceptibility $\chi(\omega) = \chi'(\omega) + i\chi''(\omega)$. Using the Kubo formula, the response function is calculated as [26]

$$\chi''(\omega) = \frac{1}{2}\left(1 - e^{-\frac{\hbar\omega}{k_B T}}\right) \mathrm{Re}\left[\lim_{\nu\to+0}\int_0^{+\infty} e^{-i(\omega-i\nu)t}\langle\hat{S}_x(t)\hat{S}_x(0)\rangle dt\right] \tag{82}$$

with the correlation function of

$$\langle\hat{S}_x(t)\hat{S}_x(0)\rangle = \frac{\mathrm{Tr}\left[e^{-\hat{H}_S/k_BT}\hat{S}_x(t)\hat{S}_x(0)\right]}{\mathrm{Tr}\left[e^{-\hat{H}_S/k_BT}\right]}. \tag{83}$$

Here $\hat{S}_x(t)$ is the Heisenberg picture as $\hat{S}_x(t) = e^{+i\hat{H}_S t/\hbar}\hat{S}_x e^{-i\hat{H}_S t/\hbar}$. Since it is written

$$\begin{aligned} &\mathrm{Tr}\left[e^{-\hat{H}_S/k_BT}\hat{S}_z(t)\hat{S}_z(0)\right] \\ &= \mathrm{Tr}\left[e^{-\hat{H}_S/2k_BT}e^{+i\hat{H}_S t/\hbar}\hat{S}_x e^{-i\hat{H}_S t/\hbar}\hat{S}_x e^{-\hat{H}_S/2k_BT}\right] \\ &= \langle\Psi_S|e^{-\hat{H}_S/2k_BT}e^{+i\hat{H}_S t/\hbar}\hat{S}_x e^{-i\hat{H}_S t/\hbar}\hat{S}_x e^{-\hat{H}_S/2k_BT}|\Psi_S\rangle, \end{aligned}$$

this shows we can use the previous technique for dynamics of wave-packet to calculate $e^{-i\hat{H}_S t/\hbar}\hat{S}_x e^{-\hat{H}_S/2k_BT}|\Psi_S\rangle$ for $\langle\hat{S}_x(t)\hat{S}_x(0)\rangle$, and to obtain $\chi''(\omega)$ and $I(\omega)$ [27]. Here we assign random phases $e^{i\theta_{1\sim N_{\max}}}$ to $N_{\max}$ elements of $|\Psi_S\rangle$ for statistical average.

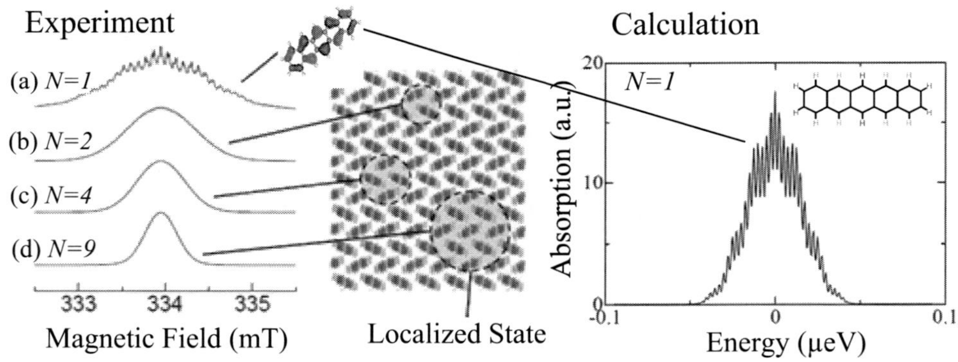

Figure 21. (Left) Experimental ESR measurements for various pentacene molecules, which observe the localized states of electrons to spread within N atoms. (Right) ESR signal calculation for $N = 1$ pentacene molecule from the wave-packet method. A number of spiky peaks are seen due to the resonances of electrons with nuclear spins of hydrogens.

Figure 21 (left) show the ESR measurements of pentacene molecules for $N = 1, 2, 4, 9$ and (right) the calculation for $N = 1$ case. The ESR signal shows the coupling of electrons with hyperfine nuclear spins of hydrogen atoms at the edges of pentacene molecules, thus it reveals how much the localized states of electrons are spread out and their rate of distribution. For $N = 1$, we see a number of (total 196) spiky peaks which show the resonances of spins, corresponding to the experimental observation. We note the relation of $g\mu_B H = 0.116$ μeV for $H = 1$ mT to estimate the width of resonance. We also note

$g\mu_B H = 38.7\ \mu$eV for the Zeeman splitting by $H = 334$ mT, sufficient to observe the hyperfine structures. This shows that from the comparison of the calculations for large N electronic states with static disorder within the width of $[-W/2, +W/2]$ to the EPS experiments, we will obtain information on the nature of disorder, the distribution of trap states, and the spatial extent of carrier wavefunctions.

3.7.4. Hall Effect Measurements for Extended Electronic States

Recently Hall effects are measured for the single-crystal organic semiconductors with high mobility. The Hall voltage is observed when the charge exhibits coherent electronic state, since the Hall effect appears due to the coupling of vector potential $\mathbf{A}$ with wave vector $\mathbf{k}$. Therefore Hall-effect measurements reveal the intrinsic electronic states and show that the charge transport of single-crystal organic semiconductors is "band-like" diffusive with extended states, not incoherent site-to-site "hopping-like" with spatially localized states.

Fig. 22 (left) shows the experiments of Hall conductivity $1/R_H$ of single-crystal rubrene as a function of gate voltage V_G. Here the Hall conductivity R_H due to charge transport in the presence of magnetic fields is defined and is described in the free-electron model by [9]

$$R_H = \frac{V_H}{I_D H}, \qquad \frac{1}{R_H} = qn. \tag{84}$$

The coincide of $1/R_H (= qn)$ with $Q = C|V_G - V_{th}|$ clearly shows that all the carriers are extended as free-electron-like and contribute to the charge transport. Thus in the rubrene organic semiconductors the charge carrier is "band-like" with extended electronic states.

Fig. 22 (right) shows the conductivity σ as a function of $1/R_H$ where the slope $\sigma/(1/R_H)$ corresponds to carrier mobility μ through the relation of $\sigma = qn\mu$. Since the charge transport is diffusive without influences of grain boundaries, high mobility $\mu = 25-30$ cm^2/Vs is achieved for the single-crystal rubrene organic semiconductors [28].

To investigate the Hall conductivity σ_{xy} of organic semiconductors based on the quantum mechanics from atomistic calculations, let us extend the TD-WPD method to treat charge transport in a magnetic field, which gives us the magnetoconductivity and Hall mobility [29]. The off-diagonal part σ_{xy} is obtained from an integral with respect to λ of the generalized Kubo formula with the two- velocities correlation function in the real-time domain

$$\sigma_{\zeta\xi} = e^2 \int_0^{1/k_B T} d\lambda \int_0^{+\infty} \mathrm{Tr}[\hat{\rho}_0 \hat{v}_\xi(0) \hat{v}_\zeta(t + i\hbar\lambda)] dt. \tag{85}$$

This yields the following expression

$$\sigma_{xy} = -e^2 \lim_{\eta \to +0} \int_0^{+\infty} dt$$

[9]In the Hall experiment, a magnetic field H_z is applied in addition to an electric field E_x. Then the carrier flowing between source-drain I_D feels the Lorenz force $q\mathbf{v} \times \mathbf{H}$ and there appears a Hall voltage V_H in the y-direction, because carriers are accumulated at the sample edges. Since V_H is proportional to H and I_D, the Hall coefficient $R_H = V_H/I_D H$ is related to carrier density n. From the equation of motion to balance an electric field qV_H/L_y with the Lorenz force. $V_H = (H/qn)I_D$ and the Hall coefficient becomes $1/R_H = qn$.

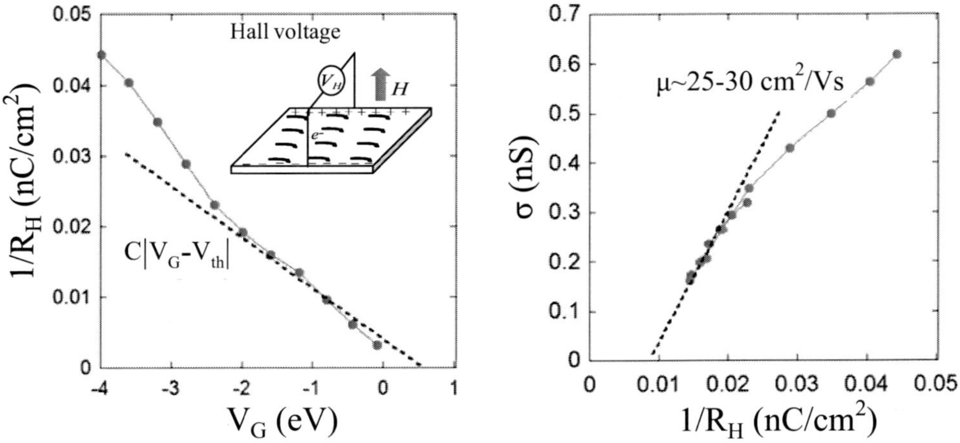

Figure 22. (Left) Hall coefficient $1/R_H = V_H/I_DH$ as a function of gate voltage V_G. Inset is the schematic picture of Hall effect measurements of organic semiconductor films. (Right) Conductivity σ (nS) as a function of $1/R_H$.

$$\sum_{n=1}^{N} f(\epsilon_n - \epsilon) 2\text{Re} \left(\langle n | \hat{v}_y e^{+i\frac{\hat{H}}{\hbar}t} \frac{1}{\epsilon_n - \hat{H} + i\eta} \hat{v}_x e^{-i\frac{\hat{H}}{\hbar}t} | n \rangle \right), \tag{86}$$

where $|n\rangle$ is the eigenvector of $\hat{H}$ with the eigenvalue ϵ_n. Here $\hat{\rho}_0$ is the density matrix in equilibrium and $\hat{v}_\xi$ is the electron velocity operator along the ξ- direction, which is obtained from the Heisenberg equation $i\hbar\hat{v}_\xi(t) = [\hat{\xi}(t), \hat{H}]$ where $\hat{\xi}$ and $\hat{H}$ represent the position operators and the Hamiltonian for electrons. Note that the diagonal part σ_{xx} represents the previous result of $\sigma_{xx} = e^2\nu(\epsilon) \lim_{t\to+\infty} D(\epsilon, t)$.

Eliminating eigenvalue dependence by an insertion of energy integral of the delta function and replacing the set of all eigenvectors with respect to any complete orthonormal basis set, we get eigenvector-free expression of the off-diagonal part of the Hall conductivity as

$$\sigma_{xy} = -e^2 \lim_{\eta\to+0} \int_0^{+\infty} dt \int_{-\infty}^{+\infty} d\epsilon\, f(\epsilon - \epsilon_F) \sum_{n,m} 2\text{Re} \Big(\langle \psi_n | \delta(E - \hat{H}) | \psi_m \rangle \times \langle \psi_m | \hat{v}_y e^{+i\frac{\hat{H}}{\hbar}t} \frac{1}{\epsilon - \hat{H} + i\eta} \hat{v}_x e^{-i\frac{\hat{H}}{\hbar}t} | \psi_n \rangle \Big), \tag{87}$$

where $|\psi_n\rangle$ is set as an initial wave packet. This shows that σ_{xy} is obtained without using the direct eigenvectors. We note that the present TD-WPD method is suitable for use of parallel computing, since we perform the time-evolution calculation of "each" initial wave packets independently. Using σ_{xx} and σ_{xy}, the Hall coefficient R_H is given by [10]

$$R_H = \frac{\sigma_{xy}}{(\sigma_{xx}^2 + \sigma_{xy}^2)H} \tag{88}$$

[10] In the 2D system, $j_x = \sigma E_x = \sigma_{xx}E_x + \sigma_{xy}E_y$ and $j_y = 0 = \sigma_{yx}E_x + \sigma_{yy}E_y$. Taking $\sigma_{xx} = \sigma_{yy}$ and $\sigma_{xy} = -\sigma_{yx}$ from the reciprocity relation, we have $R_H = V_H/I_DH = E_y/j_xH = \sigma_{xy}/(\sigma_{xx}^2 + \sigma_{xy}^2)H$.

and the diagonal conductivity defined by $j_x = \sigma E_x$ becomes $\sigma \equiv (\sigma_{xx}^2 + \sigma_{xy}^2)/\sigma_{xx}$.

Here, let us show how this formula for σ_{xy} works for the Hall effects and show the conductivities when a magnetic field is applied perpendicular to the systems. We consider the 2D "square" lattice of 30×30 sites with a lattice constant a, where the periodic boundary condition is employed. The Hamiltonian is simply written by the sum of transfer with energies of γ_{ij} to nearest-neighbor sites and the on-site energies W_i. The magnetic-field effect is introduced by multiplying γ^0 by a phase factor

$$\gamma_{ij} = \gamma^0 \exp\left\{\frac{ie}{\hbar}\int_{\mathbf{R}_i}^{\mathbf{R}_j} \mathbf{A}(\mathbf{r})\cdot d\mathbf{r}\right\}, \tag{89}$$

where $\mathbf{R}_i$ is the position vectors of the ith site and $\mathbf{A}(\mathbf{r})$ is the vector potential. W_i is the random potential energy at ith site in the energy width $[-W/2, +W/2]$. As an initial wave packet, we employ a state localized at one site and simulate the time evolution of the wave packet with time step of $\Delta t = 0.05 \times h/\gamma^0$ up to the total evolution time of 80 time steps.

We note that the energy-dependences of off-diagonal Hall conductivities σ_{xy} by the TD-WPD method show good agreements with those obtained from the Kubo formula by

$$\sigma_{xy} = -i\hbar e^2 \lim_{\eta\to+0}\sum_{n,m} f(\epsilon_n - \epsilon)\frac{\langle n|\hat{v}_x|m\rangle\langle m|\hat{v}_y|n\rangle - \langle n|\hat{v}_y|m\rangle\langle m|\hat{v}_x|n\rangle}{(\epsilon_n - \epsilon_m - i\eta)(\epsilon_n - \epsilon_m + i\eta)}, \tag{90}$$

where $|n\rangle$ represents the nth eigenvector of $\hat{H}$ with eigenvalue ϵ_n. Here eigenvectors and eigenvalues are obtained by the direct numerical diagonalization of $\hat{H}$. Indeed, in extended and weak magnetic field regime where we introduce static disorders $W/\gamma^0 = 3.2$, much larger than the cyclotron energy $\hbar\omega_c$ and $\omega_c\tau \simeq 0.1 \ll 1$ estimated from $1/\tau = (2\pi/\hbar)|W|^2\nu$, we can show that σ_{xy} as a function of energy by the TD-WPD method agree with those obtained from Eq. (90), showing anti-symmetrical behaviors which reflect that the square lattice with AB sub-lattice structure has a symmetrical band structure.

We calculate the Hall mobility μ_H and compare it with the field-effect carrier mobility $\mu = q\tau/m^*$. We evaluate the Hall coefficient by Eq. (88) using the conductivity tensors σ_{xx} and σ_{xy} by the TD-WPD method quantum mechanically, taking a phase factor into consideration. Fig. 23 (left) shows $1/R_H$ as a function of ϵ around the band bottom. In a parabolic band approximation, carrier density is given by $qn = qm^*\epsilon/2\pi\hbar^2$, which is extracted from two expressions of the diagonal conductivity, $\sigma_{xx} = q^2 n\tau/m^*$ and $\sigma_{xx} = q^2\nu v_x^2\tau$ with the DOS of $\nu = m^*/2\pi\hbar^2$. At the bottom of band dispersion for the "square" lattice, we obtain an effective mass $m^* = -\hbar^2/2\gamma^0 a^2$ and velocity $v_x^2 = \epsilon/m^*$. Then we show the linear dependence of a carrier density qn for ϵ by the solid line. We can see, around the band edge, the obtained $1/R_H$ is close to the carrier density.

From the Hall coefficient, Hall mobility defined by $\mu_H = \sigma R_H$ is obtained in experiments as a linear slope of conductivity σ for $1/R_H$. In Fig. 23 (center), we see that σ increases monotonically as a function of $|1/R_H|$ in the small $1/R_H$ regime, where the parabolic-band approximation is well defined. Therefore we can evaluate the Hall mobility μ_H from the slope of σ, which is shown as the solid line. We compare the carrier mobility $\mu = q\tau/m^*$ from the diagonal part of conductivity with no magnetic field, which is given by $\mu = eD(\epsilon)/(m^*(\epsilon)v(\epsilon)^2)$. Fig. 23 (right) shows two mobilities μ_H and μ as a function of the energy. We see that both of mobilities agree with each other quantitatively for most

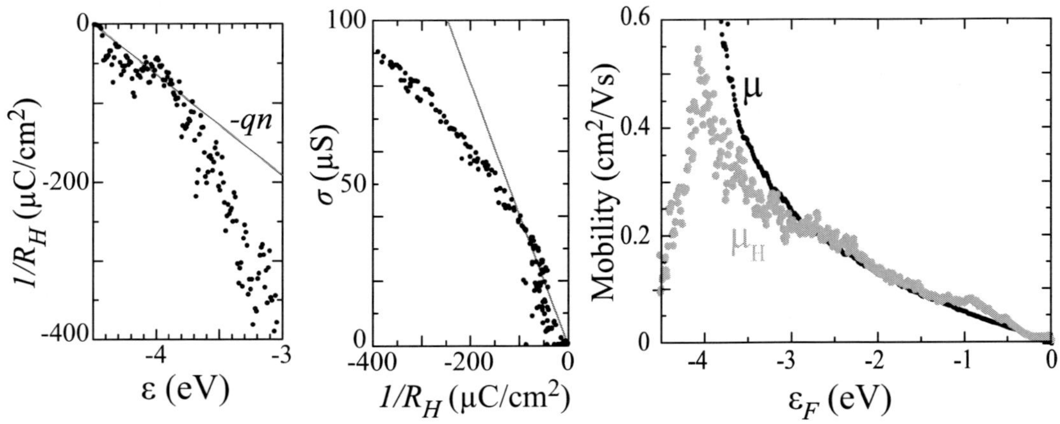

Figure 23. (Left) $1/R_H$ as a function of energy ϵ around the band edge. Solid line shows an asymptotic behavior in the parabolic band approximation. (Center) σ as a function of $1/R_H$. The solid line represents the relation of $\sigma = \mu_H \cdot R_H^{-1}$. (Right) Hall mobility μ_H and field-effect carrier mobility $\mu = e\tau/m^*$ as a function of ϵ.

Table 2. Fluctuation of transfer energy for various organic semiconductors.

	Tetracene	Pentacene	DNT-V	DNTT
$\Delta\gamma/\gamma$	0.27	0.17	0.14	0.08

of energy regime except for the band edge, where we cannot use m^* and v directly for the mobility. We note that Hall mobility μ_H is obtained in the TD-WPD calculations even near the band edge. These calculations show that Hall mobility in the extended electronic states regime under a magnetic field agrees well with the carrier mobility from the diagonal part of the conductivity tensor. Therefore the extended carriers observed in Hall conductivities work as conducting carriers for the transport, as we see in experiments.

To find the organic semiconductor materials which have high mobility is very important for the development of next-generation flexible electronics devices. For that purpose, the measurement of Hall conductance is essential, since it reveals the "band-like" nature with the extended nature of wavefunctions for the charge transport. Since organic semiconductors have much more complicated structures than the present "square" lattice and the behaviors of phases are strongly influenced by the structural configurations, it is indispensable to perform the calculations of transport properties for such structures including small and large polaron effects, with parameters taken from ab initio DFT-D calculations.

Finally, let us consider the dimensionless quantity $\Delta\gamma/\gamma$, the fluctuation of transfer energy, which would give us an estimate for the search of organic materials with high mobility. When $\Delta\gamma/\gamma$ is large, due to the large fluctuation of transfer energy, the electronic states would be localized. On the other hand, hight mobility is expected for small $\Delta\gamma/\gamma$ since the fluctuation of transfer energy is small. Table 2 give $\Delta\gamma/\gamma$ for four kinds of organic semiconductor materials from the calculations based on the DFT-D approach [11]. We see that DNTT molecules have the smallest value. It would be interesting that DNTT shows very high mobility for charge transport in experiments among these materials, much larger than pentacene. This shows the relevance of the fluctuation of transfer energy, which determines the extent of localization, with the charge transport for high mobility [12].

4. Conclusion

In this review article, we describe our recent works on transport calculations of organic materials, which include single molecules, molecular wires, and organic semiconductors.

First based on the nonequilibrium Green's function (NEGF) coupled with the density-functional theory (DFT), we investigate transport properties of single molecules coupled to electrodes and molecular wires. The problems such as n-type and p-type of conductance behavior and its eigenchannel analysis, possibility of switching devices, bias drop inside the molecules, and local heating at the contact to electrodes are explored and presented.

Next we describe the charge transport of organic semiconductors, which are molecular systems assembling weakly with van der Waals interactions. Recent advances of single-crystal organic transistors with high mobility require us to understand the fundamental transport mechanisms in these mechanically flexible organic materials. Using the time-dependent wave-packet diffusion (TD-WPD) method with parameters taken from the DFT-D calculations which include van der Waals energy, we treat the temperature dependence of charge transport with strong electron-phonon interaction in small and large polarons, anisotropic field effects, static disorder problems with ESR, and the crossover from the localized hopping-transport to delocalized band-transport in view of Hall measurements.

Acknowledgments

This work has been performed in collaborations with a number of researchers in theory and experiments. The author acknowledges T. Fukami, T. Hasegawa, M. Homma, H. Ishii, N. Kobayashi, H. Matsui, T. Okamoto, H. Takaki, J. Takeya, H. Tamura, M. Tsukada, and T. Uemura for long collaborations and for a number of valuable comments and suggestions.

References

[1] Hirose, K., & Kobayashi, N. *Quantum Transport Calculations for Nanosystems*, Pan-Stanford Publishing: Singapore, in printing.

[2] Kadanoff, L. P., & Baym, G. *Quantum Statistical Mechanics*, Benjamin: New York, 1962.

[11] We perform the calculations for various directions shown in Fig. 12 by changing the distances T_{11}, T_{22}, T_{33} and the angles L_{11}, L_{22}, L_{33} infinitesimally from equilibrium configurations and take RMS statistical average.

[12] This value might be related to the dimensionless parameter $\alpha = C/\left[\partial(1/R_H)/\partial V_G\right]$ by the ratio of $1/R_H$ and the carrier Q in experiments, which describes the extent of coupling between partially extended electronic states and the external electromagnetic wave. From the definition $0 \leq \alpha \leq 1$. $\alpha = 1$ corresponds to the situation that all the states are extended. Note $\alpha \approx 0.5$ for pentacene and $\alpha \approx 1$ for DNTT [30].

[3] Keldysh, L. V. *Sov. Phys. JETP* 1965, 20, 1018.

[4] Haug, H., & Jauho, A. P. *Quantum Kinetics in Transport and Optics of Semiconductors*, Springer-Verlag: Berlin, 1996.

[5] Fisher, D. S., & Lee, P. A. *Phys. Rev.* 1981, B23, 6851.

[6] Taylor, J., Guo, H., & Wang, J. *Phys. Rev.* 2001, B63, 245407.

[7] Xue, Y., Datta, S., & Ratner, M. A. *Chem. Phys.* 2002, 281, 151.

[8] Brandbyge, M., Mozos, J. -L., Ordejón, P., Taylor, J., & Stokbro, K. *Phys. Rev.* 2002, B65, 165401.

[9] Hirose, K., & Tsukada, M. *Phys. Rev.* 1995, B51, 5278.

[10] Hirose, K., Kobayashi, N., & Tsukada, M. *Phys. Rev.* 2004, B69, 245412.

[11] Kobayashi, N., Ozaki, T., Tagami, K., Tsukada, M., & Hirose, K. *Jpn. J. Appl. Phys.* 2006, 45, 2151.

[12] Hirose, K., Ishii, H., & Kobayashi, N. *Appl. Surf. Sci* 2012, 258, 2121.

[13] Frederiksen, T., Paulsson, M., Brandbyge, M., & Jauho, A. P. *Phys. Rev* 2007, B75, 205413.

[14] Ishii, H., Kobayashi, N., & Hirose, K. *Phys. Rev.* 2010, B82, 085435.

[15] Grimme, S. *J. Comp. Chem.* 2004, 25, 1463.

[16] Valeev, E. F., Coropceanu, V., da Silva Filho, V. D. A., Salman, S., & Brédas, J. -L. *J. Am. Chem. Soc.* 2006, 128, 9882.

[17] Troisi, A., & Orlandi, G. *Phys. Rev. Lett.* 2006, 96, 086601.

[18] Ishii, H., Honma, M., Kobayashi, N., & Hirose, K. *Phys. Rev.* 2012, B85, 245206.

[19] Gershenson, M. E., Podzorov, V., & Morpurgo, A. F. *Rev. Mod. Phys.* 2006, 78, 973.

[20] Hasegawa, T., & Takeya, J. *Sci. Technol. Adv. Mater.* 2009, 10, 024314.

[21] Tamura, H., Tsukada, M., Ishii, H., Kobayashi, N., & Hirose, K. *Phys. Rev.* 2012, B86, 035208.

[22] Richard, T., Bird, M., & Sirringhaus, H. *J. Chem. Phys.* 2008, 128, 234905.

[23] Marumoto, K., Kuroda, S., Takenobu, T., & Iwasa, Y. *Phys. Rev. Lett.* 2006, 97, 256603.

[24] Matsui, H., Hasegawa, T., Tokura, Y., Hiraoka, M., & Yamada, T. *Phys. Rev. Lett.* 2008, 100, 126601.

[25] Matsui, H.. Mishchenko, A. S., & Hasegawa, T. *Phys. Rev. Lett.* 2010, 104, 056602.

[26] Kubo, R., & Tomita, K. *J. Phys. Soc. Jpn.* 1954, 9, 888.

[27] Machida, M., Iitaka, T., & Miyashita, S. *J. Phys. Soc. Jpn.* 2005, 74, 107.

[28] Takeya, J., Kato, J., Hara, K., Yamaguchi, M., Hirahara, R., Yamada, K., Nakazawa, Y., IKehata, S., Tsukagoshi, K., Aoyagi, Y., Takenobu, T., & Iwasa, Y. *Phys. Rev. Lett.* 2007, 98, 196804.

[29] Ishii, H., Kobayashi, N., & Hirose, K. *Phys. Rev.* 2011, B83, 233403.

[30] Uemura, T., Yamagishi, M., Soeda, J., Takatsuki, Y., Okada, Y., Nakazawa, Y., & Takeya, J. *Phys. Rev.* 2012, B85, 035313.

In: Density Functional Theory
Editors: J. Morin and J.M. Pelletier, pp. 227-257
ISBN: 978-1-62417-954-9

Chapter 8

MODERN DENSITY FUNCTIONAL THEORY, A USEFUL TOOL FOR COMPUTATIONAL CHEMISTS

Reinaldo Pis Diez*
CEQUINOR, Center of Inorganic Chemistry, Department of Chemistry
Faculty of Exact Sciences, National University of La Plata, Argentina

Abstract

Since the 1980's, the Density Functional Theory has evolved to become nowadays the most used tool for the calculation of electron structure of molecules, clusters and solids at a first-principles level of theory. In this chapter, the formal evolution of the theory from ancient local models to the modern hybrid meta generalized gradient approximations and double hybrid generalized gradient approximations is discussed. Moreover, the success and limitations of current Density Functional Theory in the prediction of a variety of properties, such as geometries, energies and thermodynamic functions, as well as in the interpretation of spectroscopic data are commented from a critical point of view.

Keywords: Density Functional Theory, DFT, LDA, GGA, meta-GGA, hybrid meta-GGA

1. Introduction

The first attempts to develop a theory of the electronic structure of atoms and molecules based exclusively on the electron density can be traced back to the 1920's and 1930's. The statistical nature of such attempts, however, led to very poor results for systems containing a few electrons, specially when they were compared with results obtained within the framework of Hartree-Fock and post Hartree-Fock methodologies.

Despite some efforts to improve the methods mentioned above by including the gradient of the electron density, it was in the mid 1960's that the formal framework for a theory of the electronic structure of atoms and molecules, in which the electron density is the basic variable, was developed. The pioneer work of Hohenberg, Kohn and Sham [1,2] gave place to the birth of what it is known nowadays as Modern Density Functional Theory (DFT).

*E-mail address: pis_diez@quimica.unlp.edu.ar

However, and due to the lack of efficient and reliable correlation functionals, the routine use of DFT by theoretical chemists was delayed up to early 1980's. From those days, DFT has exhibited an enormous progress both from methodological and practical point of views. It is completely safe to state that DFT is currently the most used computational tool both for theoretical chemists and for experimental researchers that look for an elegant explanation for their empirical results.

In 2000, Perdew [3] used the concept of the Jacob's Ladder found in the Bible to make an analogy with the development of DFT from the earliest models, based only on the electron density, to the modern approaches making explicit use of orbitals. Thus, the most basic approach to the electronic structure problem within the DFT is equivalent to the first rung in the Ladder, and involves the electron density only. To ascend to the second rung, the gradient of the electron density must be added to the calculations. The inclusion of the kinetic energy density (instead of the Laplacian of the electron density) allows the Ladder's user to climb to the third rung. The approach that gives place to the fourth rung includes the occupied orbitals through a given amount of the exact exchange energy, that is, the one that is obtained from Hartree-Fock. Finally, the inclusion of unoccupied orbitals by means of perturbation theory allows the user to reach the Heaven of Chemical Accuracy after climbing the fifth rung.

In the present chapter, the Jacob's Ladder analogy is used as a guide to show the evolution of DFT in the last fifty years. Special emphasis is laid on the pros and cons of the theory in the prediction of those properties of interest for chemists. It is important to make clear at this point that the amount of existing density functionals is enormous. Thus, only a few functionals, representative of every rung in the Ladder, will be described in detail. The election of such functionals is rather arbitrary, though, but it is mainly based on the contribution of the most important researchers in the field according to the author's viewpoint.

2. The Early Days

In 1927, Thomas [4] and Fermi [5], independently and simultaneously, developed the mathematical expression for the kinetic energy of a set of independent particles in terms of their density and in form of a *functional* of the density

$$T_{TF}[\rho] = C_K \int \rho^{5/3}\, d\mathbf{r}, \tag{1}$$

where ρ is a shorthand notation for $\rho(\mathbf{r})$ and $C_K = 3(3\pi^2)^{2/3}/10$. The electrostatic terms adopt a classic form

$$E_{Ne}[\rho] = \sum_{\alpha} \int Z_{\alpha} \frac{\rho}{|\mathbf{R}_{\alpha} - \mathbf{r}|}\, d\mathbf{r}, \tag{2}$$

and

$$J[\rho] = \frac{1}{2} \int \frac{\rho\rho'}{|\mathbf{r} - \mathbf{r}'|} d\mathbf{r} d\mathbf{r}', \tag{3}$$

where Z_{α} and R_{α} are the atomic number and coordinates of nuclei.

The three equations given above form the first attempt to define a theory to calculate electronic structure: the Thomas-Fermi model

$$E_{TF}[\rho] = T_{TF}[\rho] + E_{Ne}[\rho] + J[\rho]. \tag{4}$$

It is easily recognized that no exchange nor correlation terms are included in the Thomas-Fermi model. A few years later, Dirac [6] proposed an exchange energy density functional

$$K_D[\rho] = C_X \int \rho^{4/3} d\mathbf{r}, \tag{5}$$

where $C_X = -3(3/\pi)^{1/3}/4$. In this way, the electronic energy as a functional of the electron density could be written as

$$E_{TFD}[\rho] = T_{TF}[\rho] + E_{Ne}[\rho] + J[\rho] + K_D[\rho], \tag{6}$$

giving place to the Thomas-Fermi-Dirac (TFD) model. The TFD model performs very poor for atoms with a few electrons, with errors of about 40% with respect to Hartree-Fock results. Most important are the facts that no negative ions nor molecules exist within the TFD model due to the crude description of the kinetic energy.

The first correction to the TFD model was introduced by von Weizsäcker [7], who improved slightly the kinetic energy functional with a term including the gradient of the electron density

$$T_{TFvW}[\rho] = C_K \int \rho^{5/3} d\mathbf{r} + \frac{1}{8} \int \frac{|\nabla\rho|^2}{\rho} d\mathbf{r}. \tag{7}$$

Another important shortfall of the original TFD model is the absence of correlation effects. In spite of some efforts to include a correlation energy functional into the TFD or the TFvWD models, they have nowadays a historical interest only.

3. The 1960's - The Birth of Modern Density Functional Theory

In 1964, Hohenberg and Kohn (HK) [1] established by means of two theorems the basis of what it is known as Modern Density Functional Theory nowadays. In the first theorem, HK deal with a system formed by an arbitrary number of electrons moving under the influence of an external potential, $v(\mathbf{r})$, and the Coulomb repulsion. The external potential could be due to the presence of a nucleus or of a set of nuclei in the case of atoms or molecules, respectively, although it is not restricted to Coulombic potentials. The keystone in the HK argument is that the electron density should be thought of as the independent variable in an electronic structure determination problem as it defines the external potential. Once the external potential is known, the Hamiltonian is fixed and all the properties of the system could be obtained after solving the Schrödinger equation. Thus, it is said that the external potential is a functional of the density.

In the second theorem, HK define an electronic energy density functional as

$$E_{HK}[\rho] = F_{HK}[\rho] + \int v\rho \, d\mathbf{r}, \tag{8}$$

where F_{HK} includes the kinetic energy and the electron-electron interaction energy, and could be considered as a universal functional as it is independent of the external potential. HK state that the functional in equation 8 assumes its minimum value, say E_0, for the ground state density, ρ_0, which fixes in turn the external potential, v. Then, any trial density $\rho \neq \rho_0$, will deliver an electronic energy, E, that is an upper limit to E_0, $E \geq E_0$.

Unfortunately, equation 8 could not be used in practical calculations due to the uncertainty in the mathematical form of the universal functional, F_{HK}, beyond the obvious relationship

$$F_{HK}[\rho] = T[\rho] + V_{ee}[\rho]. \tag{9}$$

One year later, Kohn and Sham (KS) [2] proposed a solution to the problem on the uncertainty in F_{HK}. Indeed, KS proposed a smart solution to solve the problem of the uncertainty in the kinetic energy functional, $T[\rho]$. They invoked a reference system formed by N non interacting electrons, described by an electron density ρ, under the influence of an external potential v. Thus, the HK electronic energy density functional could be written as

$$E_{HK,s}[\rho] = T_s[\rho] + \int v\rho\, d\mathbf{r}. \tag{10}$$

The key ingredient in equation 10 is that due to the non interacting nature of the system, the T_s functional could be written as

$$T_s[\rho] = -\frac{1}{2}\sum_i^N \int \psi_i^* \nabla^2 \psi_i, \tag{11}$$

where the set $\{\psi\}$ is a set of auxiliary functions that helps in the construction of the density ρ

$$\rho = \sum_i^N \psi_i \psi_i^*. \tag{12}$$

It should be stressed again that ψ is a shorthand notation for $\psi(\mathbf{r})$. The HK electronic energy density functional can now be written for the real system, formed by N interacting electrons, as

$$E_{HK}[\rho] = T[\rho] + \int v\rho\, d\mathbf{r} + V_{ee}[\rho]. \tag{13}$$

If it is accepted that the density of the interacting system is the *same* as the density of the non interacting system, then it is possible to re-write equation 13 as

$$E_{KS}[\rho] = T_s[\rho] + \int v\rho\, d\mathbf{r} + \frac{1}{2}\int \frac{\rho\rho'}{|\mathbf{r}-\mathbf{r}'|} d\mathbf{r}d\mathbf{r}' + E_{xc}[\rho], \tag{14}$$

where the subscript HK has been changed to KS to indicate that this new functional differs from the original one due to Hohenberg and Kohn. Furthermore, the electron-electron interaction term has been divided into the classical Coulomb term and the non classical term E_{xc}

that describes exchange and correlation effects and also accounts for the error introduced when T_s replaces T,

$$E_{xc}[\rho] = V_{ee}[\rho] - \frac{1}{2}\int \frac{\rho\rho'}{|\mathbf{r}-\mathbf{r}'|} d\mathbf{r}d\mathbf{r}' + T[\rho] - T_s[\rho]. \tag{15}$$

The set of auxiliary functions that minimizes the KS functional, equation 14, can be found by writing the new functional,

$$\Omega[\{\psi\}] = E_{KS}[\rho] + \sum_i^N \sum_j^N \varepsilon_{ij}\left(\int \psi_i^* \psi_j d\mathbf{r} - \delta_{ij}\right), \tag{16}$$

where the set of Lagrange multipliers, $\{\varepsilon\}$, assures the orthonormality of the auxiliary functions. The condition

$$\delta\Omega[\{\psi\}] = 0, \tag{17}$$

leads to a set of N one-particle equations in canonical form

$$\left[-\frac{1}{2}\nabla^2 + v_{ef}[\rho]\right]\psi_i = \varepsilon_i\psi_i,\ i = 1,\ldots,N \tag{18}$$

where the *effective* potential, v_{ef}, is

$$v_{ef}[\rho] = v + \int \frac{\rho(\mathbf{r}')}{|\mathbf{r}-\mathbf{r}'|} d\mathbf{r}' + v_{xc}[\rho], \tag{19}$$

and $v_{xc}[\rho]$ is the *exchange-correlation* potential

$$v_{xc}[\rho] = \frac{\delta E_{xc}[\rho]}{\delta\rho}. \tag{20}$$

It is evident from equations 18 and 19 that the KS scheme must be solved self-consistently.

In summary, the merits of the proposal made by Kohn and Sham is that they solve the uncertainty in the kinetic energy density functional by invoking a reference, non interacting system of N electrons with density ρ, which in turn is the same as the density of the real, interacting system of N electrons. Furthermore, the KS equations introduce both exchange and correlation effects at a computational cost similar to the Hartree or Hartree-Fock methods. However, KS did not propose any solution to the non classical part of the electron-electron interaction. Moreover, they introduced a set of auxiliary functions that helps in modeling the electron density and must be obtained from a self-consistent procedure, loosing in this way the original appealing of a theory of the electronic structure of atoms and molecules based solely on the electron density.

The next step to put equation 14 to work is to give $E_{xc}[\rho]$ a practical mathematical form.

4. The 1980's - The Local Density Approximation, the First Rung in the Jacob's Ladder

It is common practice to separate the exchange contribution from the correlation contribution to $E_{xc}[\rho]$

$$E_{xc}[\rho] = E_x[\rho] + E_c[\rho], \tag{21}$$

to facilitate the modeling of the functionals needed to accomplish practical calculations. It is also accepted that those functional can be written in terms of the energy per particle o energy density

$$E_{xc}[\rho] = E_x[\rho] + E_c[\rho] = \int \varepsilon_x(\rho)\rho\, d\mathbf{r} + \int \varepsilon_c(\rho)\rho\, d\mathbf{r}. \tag{22}$$

The simplest form for the exchange energy density functional is the expression derived by Dirac, equation 5, that can be written in a more general form to account for spin polarized cases

$$E_x[\rho] = 2^{1/3} C_X \int \{\rho_\alpha^{4/3} + \rho_\beta^{4/3}\}\, d\mathbf{r}. \tag{23}$$

The first effort to model $E_c[\rho]$ can be traced back to 1972, when von Barth and Hedin [8] proposed a simple form for the correlation energy, in which the correlation energy per particle is

$$\varepsilon_c^{vBH}(\rho, \xi) = \varepsilon^0(r_s) + [\varepsilon^1(r_s) - \varepsilon^0(r_s)] f(\xi), \tag{24}$$

where r_s is the Wigner-Seitz radius,

$$\frac{4}{3}\pi r_s^3 = \frac{1}{\rho}, \tag{25}$$

ξ is the spin polarization

$$\xi = \frac{\rho^\alpha(\mathbf{r}) - \rho^\beta(\mathbf{r})}{\rho(\mathbf{r})}, \tag{26}$$

and the function $f(\xi)$ adopts the form

$$f(\xi) = \frac{(1+\xi)^{4/3} + (1-\xi)^{4/3} - 2}{2^{4/3} - 2}. \tag{27}$$

In equation 24 the superscripts 0 and 1 indicates the spin compensated and ferromagnetic cases, respectively. von Barth and Hedin suggested the use of analytic expressions for $\varepsilon^0(r_s)$ and $\varepsilon^1(r_s)$ due to Hedin and Lundqvist [9].

In 1980, Ceperley and Alder [10] carried out a set of Quantum Monte Carlo simulations on the homogeneous electron gas to obtain the correlation energy for a wide range of densities. They presented their results in tabular form, however, and then, their use in real calculations was impractical due to the necessity of interpolation to obtain intermediate values of $E_c[\rho]$.

In the same year, Vosko, Wilk and Nusair (VWN) [11] proposed an expression for the correlation energy density obtained from a random-phase approximation analysis

$$\begin{aligned}\varepsilon_c^{VWN}(r_s,\xi) &= \varepsilon_c^{VWN}(r_s,0)+\alpha_c(r_s)\frac{f(\xi)}{f''(0)}(1-\xi^4)\\ &+[\varepsilon_c^{VWN}(r_s,1)-\varepsilon_c^{VWN}(r_s,0)]f(\xi)\xi^4,\end{aligned} \tag{28}$$

where $\alpha_c(r_s)$ is the spin stiffness. VWN offer analytic expressions for $\varepsilon_c^{VWN}(r_s,0)$, $\varepsilon_c^{VWN}(r_s,1)$ and $\alpha_c(r_s)$, which were obtained from accurate fits to the numerical results of Ceperley and Alder

$$\begin{aligned}\mathcal{G}_c(r_s) &= A\Bigg\{\ln\frac{x^2}{X(x)}+\frac{2b}{Q}\tan^{-1}\frac{Q}{2x+b}-\\ &\frac{bx_0}{X(x_0)}\left[\ln\frac{(x-x_0)^2}{X(x)}+\frac{2(b+2x_0)}{Q}\tan^{-1}\frac{Q}{2x+b}\right]\Bigg\},\end{aligned} \tag{29}$$

where $\mathcal{G}_c(r_s)$ is used to represent $\varepsilon_c^{VWN}(r_s,i)$ and $\alpha_c(r_s)$; $x=r_s^{1/2}$, $X(x)=x^2+bx+c$ and $Q=(4c-b^2)^{1/2}$ are functions, and A, x_0, b and c are parameters. The VWN correlation functional is accepted as one of the most accurate functional available for the homogeneous electron gas. The work of VWN presents various fits to the above equations. In particular, expressions III and V in reference [11] are the ones more often found in different implementations of the functional, leading obviously to slightly different results. The authors recommend the use of formula V.

The combination of the Dirac exchange functional and the VWN correlation functional conforms what it is labeled nowadays as the Local Density Approximation (LDA) or Local Spin Density Approximation

$$E_{xc}^{L(S)DA}=E_x^D+E_c^{VWN}. \tag{30}$$

The term *local* is used with a physical meaning in the sense that knowing the electron density at a given point then, it has the same value in any point around the reference one.

As expected, the L(S)DA fails when it is applied to inhomogeneous systems as atoms and molecules, providing energies larger than those obtained by post Hartree-Fock methods. Nevertheless, it is worth noting that equilibrium geometries are reasonably accurate, the bond lengths being slightly larger than experimental ones. Thus, L(S)DA geometries can be a good starting point for the study of large molecules.

As a rule of thumb, deviation of homogeneity can be considered non negligible when $|\nabla\rho|/\rho$ is comparable to $(3\pi\rho)^{1/3}$. As this comparison is routinely found in atomic and molecular systems, it is mandatory to go beyond the L(S)DA to consider DFT on a competitive basis with respect to post Hartree-Fock methods.

Before analyzing possible improvements to the L(S)DA, it is very important to mention that the exact Hohenberg-Kohn functional remains unknown. Thus, all the functionals proposed to carry out practical calculations are mere approximations to it. This fact indicates that the Variational Principle demonstrated by HK is no longer valid. In other words, the total electronic energy calculated within the KS scheme could approach the exact value well from above or from below. No systematic improvement can be achieved in modern DFT by the use of different functionals. The only possible enhancement concerns the improvement in the basis sets used in the calculations.

5. The Generalized Gradient Approximation, the Second Rung in the Jacob's Ladder

It seems reasonable to account for the L(S)DA shortfall in describing inhomogeneous systems by including some function of the electron density gradient. The so-called *reduced density gradient* or *dimensionless density gradient*

$$s = \frac{|\nabla\rho|}{(24\pi^2)^{1/3}\rho^{4/3}}, \tag{31}$$

is used to this end.

It was shown at the end of section 2 that the correction to the Thomas-Fermi kinetic energy density functional due to von Weiszäcker included the gradient of the electron density. It is then safe to say that von Wieszäcker was a pioneer in improving early local DFT by the inclusion of a density gradient.

The exchange energy density functional due to Dirac also possesses density gradient corrections

$$E_x[\rho] = C_X \int \rho^{4/3}\, d\mathbf{r} + \beta \int \frac{|\nabla\rho|^2}{\rho^{4/3}}\, d\mathbf{r}, \tag{32}$$

where β is a constant that accepts both pure theoretical and empirical values (see reference [12] for a comparison of the different values). Unfortunately, the potential associated to the above correction diverges when $r \longrightarrow \infty$.

In 1986, Perdew and Wang (PW) [13, 14] proposed for the first time the term Generalized Gradient Approximation (GGA) and presented a simple equation for the exchange energy density functional, in which the inhomogeneity is considered

$$E_x^{GGA}[\rho] = C_X \int \rho^{4/3} F^{GGA}(s)\, d\mathbf{r}, \tag{33}$$

where $F^{GGA}(s)$ is a function that accounts for some conditions that the exact exchange hole must satisfy and depends on the dimensionless density gradient. PW proposed a simple analytic form for $F^{GGA}(s)$

$$F^{GGA}(s) = (1 + 0.0864 s^2/m + bs^4 + cs^6)^m, \tag{34}$$

where m, and b and c are constants. This GGA exchange energy density functional is known as PW86.

In the same year, more precisely a few pages ahead in the same issue of the Physical Review B in which the PW86 exchange energy density functional was presented, Perdew [15, 16] proposed some modifications to the correlation energy density functional due to Langreth and Mehl [17–20]. This correlation density functional is known as P86 and has the form

$$E_c^{GGA}[\rho,\xi] = \int \varepsilon_c(\rho,\xi)\rho\, d\mathbf{r} + \int \frac{C(\rho)\exp(-\Phi)}{d}\frac{|\nabla\rho|^2}{\rho^{4/3}}\, d\mathbf{r}, \tag{35}$$

where $\varepsilon_c(\rho,\xi)$ is taken from another parameterization of Ceperley and Alder results for the correlation energy of the homogeneous electron gas [21]. The other functions present in equation 35 are written as

$$C(\rho) = 0.001667 + \frac{0.002568 + \alpha r_s + \beta r_s^2}{1 + \gamma r_s + \delta r_s^2 + 10^4 \beta r_s^3}, \tag{36}$$

$$\Phi = 0.19195 \frac{C(\infty)}{C(\rho)}\frac{|\nabla\rho|}{\rho^{7/6}}, \tag{37}$$

and

$$d = 2^{1/3}\left[\left(\frac{1+\xi}{2}\right)^{5/3} + \left(\frac{1-\xi}{2}\right)^{5/3}\right]^{1/2}, \tag{38}$$

where α, β, γ and δ are constants.

In 1988, Becke [22] concentrated his efforts in the development of an exchange energy density functional with the correct asymptotic behavior. His functional, known as B88, exhibits the following mathematical form

$$E_x^{B88}[\rho] = C_X \int \rho^{4/3} F^{B88}(s)\, d\mathbf{r}, \tag{39}$$

where the inhomogeneity function is

$$F^{B88}(s) = 1 - \beta \frac{s^2}{1 + 6\beta s \sinh^{-1}(s)}, \tag{40}$$

and β is a constant that best fits the Hartree-Fock exchange energy of the noble gases. The $\sinh^{-1}(x)$ function accepts a simple analytic expression, which facilitates its use in practical calculations

$$\sinh^{-1}(x) = \ln\left[x + \sqrt{1+x^2}\right]. \tag{41}$$

The extension of equations 39 and 40 to spin-polarized systems is straightforward.

In 1988, Lee, Yang and Parr (LYP) [23] used as starting point in their investigation an expression for the correlation energy due to Colle and Salvetti (CS) [24,25], which is based on the electron density and on the second-order reduced density matrix, both obtained from the Hartree-Fock wave function. LYP re-wrote the original CS equations in terms of the local Hartree-Fock kinetic energy and a density-dependent term, which is similar to the von Weiszäcker correction, see equation 7. By expanding the Hartree-Fock kinetic energy about the Thomas-Fermi kinetic energy functional, LYP obtained three different formulas depending on the degree of the expansion. The most general expression is

$$E_c^{LYP}[\rho] = -\int \frac{a}{1+d\rho^{-1/3}}\left(\rho + b\rho^{-2/3}\left[C_K\rho^{5/3} - 2t_{vW} + (\frac{1}{9}t_{vW} + \frac{1}{18}\nabla^2\rho)\right] \times \exp(-c\rho^{-1/3}\right) d\mathbf{r}, \qquad (42)$$

where a , b, c and d are constants and $t_{vW} = |\nabla\rho|^2/(8\rho) - \nabla^2\rho/8$. A similar equation was defined by LYP for open-shell systems.

Three possible formulas can be derived from equation 42. The first formula is called the second-order LYP correlation functional and agrees with equation 42. The zero-order formula can be easily obtained by discarding the two terms in parenthesis. Finally, a zero-order mean-path formula is obtained after replacing $(1/18)$ by $(-1/36)$ in equation 42.

In 1992, Perdew and Wang [26, 27] proposed a GGA version for both exchange and correlation energies, which is known as PW92 nowadays[1]. The exchange energy density functional is based on the B88 functional. Some modifications are introduced to fulfill bound and scaling conditions, giving place to the following inhomogeneity factor

$$F^{PW92}(s) = \frac{1 + as\sinh^{-1}(bs) + (c - d\exp(-100s^2))s^2}{1 + as\sinh^{-1}(bs) + es^4}. \qquad (43)$$

where a, b, c, d and e are constants. The correlation part of PW92 is far more involved as it defines a new inhomogeneity function, $H(t, r_s, \xi)$

$$E_c^{PW92}[\rho] = \int \rho\left(\varepsilon_c^{L(S)DA}(\rho, \xi) + H(t, r_s, \xi)\right) d\mathbf{r}, \qquad (44)$$

that depends on the Wigner-Seitz radius, the spin polarization and a new, scaled density gradient, t

$$t = \frac{|\nabla\rho|}{4g(\xi)(3/\pi)^{1/6}\rho^{7/6}}, \qquad (45)$$

where g is a function of the spin polarization

$$g(\xi) = \frac{(1+\xi)^{2/3} + (1-\xi)^{2/3}}{2}. \qquad (46)$$

The local part of the correlation energy per particle, $\varepsilon_c^{L(S)DA}(\rho, \xi)$, can be taken from any accurate fit to the Ceperley and Alder results for the homogeneous electron gas. Perdew and Wang proposed in fact an expression for $\varepsilon_c^{L(S)DA}(\rho, \xi)$ that becomes a simplification with respect to the VWN correlation functional. The *gradient-corrected* term is, in turn, divided into two contributions

$$H(t, r_s, \xi) = H_0(t, r_s, \xi) + H_1(t, r_s, \xi), \qquad (47)$$

where $H_0(t, r_s, \xi)$ is developed to simulate the behavior in the limit of high densities, that is, when $r_s \to 0$

[1]Sometimes, it is also referred to as PW91.

$$H_0(t,r_s,\xi) = g^3(\xi)\frac{\beta^2}{2\alpha}\ln\left[1+\frac{2\alpha}{\beta}t^2\frac{1+At^2}{1+At^2+A^2t^4}\right], \tag{48}$$

where α is a constant, $\beta = 16(3/\pi)^{1/3}C(0)$, $C(\rho)$ has been defined in equation 36, and

$$A = \frac{2\alpha}{\beta}\frac{1}{\exp(-2\alpha\varepsilon_c(\rho,\xi)/g^3(\xi)\beta^2)-1}. \tag{49}$$

On the other hand, $H_1(t,r_s,\xi)$ is included to recover the correct result when $s \to 0$, and it is negligible unless $s \ll 1$,

$$\begin{aligned} H_1(t,r_s,\xi) &= 16(\frac{3}{\pi})^{1/3}[C(\rho)-C(0)+0.0007144]g^3(\xi)t^2 \times \\ & \exp(-400g^4(\xi)t^2/\left((3\pi^5)^{1/3}\rho^{1/3}\right)). \end{aligned} \tag{50}$$

In 1996, Perdew, Burke and Ernzerhof (PBE) [28] recognize that the PW92 exchange-correlation functional has some problems, besides it incorporates inhomogeneity effects and obeys some conditions from the LDA. In particular, PW92 contains many parameters that are not joined smoothly, thus creating artifacts in the corresponding potentials in the regimes of small and large dimensionless density gradient. Then, instead of following the strategy of building up a density functional to fulfill as many exact conditions as possible, PBE designed an exchange-correlation functional that satisfy only those conditions that are *energetically* significant.

The PBE exchange energy density functional is extremely simple in its mathematical form, the inhomogeneity factor given by

$$F(s) = 1+\kappa - \frac{\kappa}{1+\mu s^2/\kappa}, \tag{51}$$

where κ and μ are constants.

The PBE correlation functional shows the same form as the PW92 functional

$$E_c^{PBE}[\rho] = \int \rho\left(\varepsilon_c^{L(S)DA}(\rho,\xi)+H(t,r_s,\xi)\right)d\mathbf{r}, \tag{52}$$

but the $H(t,r_s,\xi)$ function has only one term

$$H(t,r_s,\xi) = \gamma g^3(\xi)\ln\left[1+\frac{\beta}{\gamma}t^2\frac{1+At^2}{1+At^2+A^2t^4}\right], \tag{53}$$

where

$$A = \frac{\beta}{\gamma}\left[\frac{1}{\exp(-\varepsilon_c(\rho,\xi)/[\gamma g^3(\xi)])-1}\right], \tag{54}$$

where β and γ are constants.

The PW92 and PBE schemes are special cases within the GGA in the sense that can be invoked to model both the exchange energy and the correlation energy. The other functionals must be combined to get the desired DFT model as they are not complete schemes. Thus, GGA DFT models such as B88P86 (or simply BP86) or B88LYP (or simply BLYP) were usually used some years ago and still are in use when cheap models are needed.

6. The Meta Generalized Gradient Approximation, the Third Rung in the Jacob's Ladder

It has been mentioned before in this chapter that the consequence of the uncertainty in the form of the Hohenberg-Kohn functional is that no systematic improvement can be achieved as it is the case of post Hartree-Fock methodologies. Modern DFT is then based on the Kohn-Sham version of the DFT as it solves the problem of the uncertainty in the kinetic energy density functional. All the current efforts in methodological DFT are devoted to the modeling of accurate exchange-correlation energy density functionals.

The first two rungs in the Jacob's Ladder model the exchange-correlation energy density functional in terms of the electron density and its gradient

$$E_{xc}^{L(S)DA} = \int \varepsilon_{xc}^{L(S)DA}(\rho)\rho\, d\mathbf{r} \tag{55}$$

and

$$E_{xc}^{GGA} = \int \varepsilon_{xc}^{GGA}(\rho, \nabla\rho)\rho\, d\mathbf{r}, \tag{56}$$

where the explicit dependence of ε^{GGA} on both the electron density and its gradient in indicated.

It seems then reasonable to include higher-order derivatives of the electron density into the exchange-correlation energy per particle to go beyond the GGA, that is, to achieve a *meta* GGA (mGGA)

$$E_{xc}^{mGGA} = \int \varepsilon_{xc}^{mGGA}(\rho, \nabla\rho, \nabla^2\rho)\rho\, d\mathbf{r}. \tag{57}$$

However, Neumann and Handy [29] have demonstrated that the inclusion of derivatives of the electron density beyond the density gradient to a general form like equation 57, leads to no significant improvements in the accuracy of the results.

In 1996, Becke (B95)[2] [30] proposed a correlation energy functional that included the term

$$D_\sigma = \sum_i^{occ} |\nabla\psi_{i,\sigma}|^2 - \frac{1}{4}\frac{|\nabla\rho_\sigma|^2}{\rho_\sigma}, \tag{58}$$

in the parallel-spin part, σ being both α and β. D_σ derives from the second-order expansion of the exact non interacting $\sigma\sigma$ pair density within the Kohn-Sham version of the DFT. Moreover, by recalling that equation 15, the definition of E_{xc} within the KS context, also contains information about the kinetic energy, it seems natural to include the kinetic energy density of the occupied KS auxiliary functions in ε_{xc}

$$E_{xc}^{mGGA} = \int \varepsilon_{xc}^{mGGA}(\rho, \nabla\rho, \tau)\rho\, d\mathbf{r}, \tag{59}$$

where $\tau = \sum_i^{occc} |\nabla\psi_i|^2$.

[2]Although the paper was published in 1996, the functional is referred to as B95.

In 1998, Van Voorhis and Scuseria (VS98) [3] [31] developed an exchange energy density functional after expanding the reduced second-order density matrix, that enters the exact expression for the exchange energy, in terms of Bessel and Legendre polynomials. The VS98 exchange functional can be expressed as

$$E_x^{VS98} = C_X \int \rho^{4/3} F^{VS98}(s,z)\, d\mathbf{r}, \tag{60}$$

where the inhomogeneity factor is

$$F^{VS98}(s,z) = \frac{a}{\gamma(s,z)} + \frac{bs^2 + cz}{\gamma^2(s,z)} + \frac{ds^4 + es^2 z + fz^2}{\gamma^3(s,z)}, \tag{61}$$

where $z = \tau/\rho^{5/3} - C_K$ and $\gamma(s,z) = 1 + \alpha(s^2 + z)$. The constants a to f and α were adjusted to fit an extensive set of molecular properties. Besides the development of the VS98 functional was originally devoted to the exchange energy, Van Voorhis and Scuseria proposed that the correlation energy functional could have an expression similar to that of equation 61, but accepting different terms for same-spin and opposite-spin electrons

$$E_c^{VS98} = E_c^{VS98,\alpha\beta} + \sum_\sigma E_c^{VS98,\sigma\sigma}, \tag{62}$$

and

$$\begin{aligned} E_c^{VS98,\sigma\sigma} &= \int \rho \varepsilon_c^{L(S)DA} D_\sigma F^{VS98}(s,z)\, d\mathbf{r} \\ E_c^{VS98,\alpha\beta} &= \int \rho \varepsilon_c^{L(S)DA} F^{VS98}(s,z)\, d\mathbf{r}, \end{aligned} \tag{63}$$

where the function D_σ is similar to the one that appears in B95, see equation 58. In this way, the overall implementation of VS98 implies the evaluation of 21 parameters.

In 1999, Perdew, Kurth, Zupan and Blaha (PKZB) [32] felt unhappy with *semiempirical* density functionals containing about 20 parameters. To remedy this, PKZB proposed to construct a mGGA exchange-correlation functional based on the philosophy of the PBE functional [28], that is, retaining all its good properties and adding some others to enhance its predictive power. The exchange part of the PKZB density functional has the form

$$E_x^{PKZB} = C_X \int \rho^{4/3} F^{PKZB}(\rho,s,\tau)\, d\mathbf{r}, \tag{64}$$

where the inhomogeneity factor is

$$F^{PKZB}(\rho,s,\tau) = 1 + \kappa - \frac{\kappa}{1 + \frac{x}{\kappa}}, \tag{65}$$

and the function x is

$$x = \frac{10}{81}s^2 + \frac{146}{2025}q^2 - \frac{73}{405}qs^2 + \left[D + \frac{1}{\kappa}\left(\frac{10}{81}\right)^2\right]s^4, \tag{66}$$

[3] This functional is also known as VSXC.

where D is a constant set to 0.113 and the function q adopts the following form

$$q = \frac{3\tau}{2(3\pi^2)^{2/3}\rho^{5/3}} - \frac{9}{20} - \frac{s^2}{12}. \tag{67}$$

For the correlation part of the PKZB becomes

$$E_c^{PKZB} = \int \left\{ \rho\varepsilon_c^{PBE} \left[1 + C \left(\frac{\sum_\sigma \tau_\sigma^{vW}}{\sum_\sigma \tau_\sigma} \right)^2 \right] - (1+C) \sum_\sigma \rho_\sigma \varepsilon_{c,\sigma}^{PBE} \left(\frac{\tau_\sigma^{vW}}{\tau_\sigma} \right)^2 \right\} d\mathbf{r}, \tag{68}$$

where $\tau_\sigma^{vW} = \frac{1}{8}\frac{|\nabla\rho_\sigma|^2}{\rho_\sigma}$ and C is a constant set to 0.53. It can be seen that, despite the complexity of some formulas, the PKZB exchange-correlation density functional was formulated with only two adjustable parameters.

Interestingly, almost at the end of their work, PKZB states that although it is true that "there is no systematic way to construct density functional approximations... there are more or less systematic ways" to do so. The PBE and PKZB schemes demonstrate that.

In 2003, Tao, Perdew, Staroverov and Scuseria (TPSS) [33] proposed a revised version of the PKZB exchange-correlation density functional to overcome its failures in the prediction of some properties, notably bond lengths and hydrogen-bonded complexes. The philosophy to design the TPSS exchange-correlation energy density functional was to replace the empirical parameters D and C present in PKZB by other parameters chosen to enforce some exact conditions. Interestingly, the tests carried out by the authors on different sets of molecules indicated that the TPSS is indeed an improvement to the PKZB functional, with results similar to more sophisticated functionals containing a fraction of the exact exchange (see next section on hybrid GGA functionals). Also in their paper, the authors coined the term *nested functionals* to mean that the L(S)DA is inside PBE and PBE is nested into TPSS. Consistent with the Jacob's Ladder concept, the authors, too, anticipate the evolution of TPSS to include a certain percentage of the exact exchange, thus climbing naturally to the fourth rung of the Ladder.

In 2005, Truhlar and co-workers presented the first member of a prolific family of exchange-correlation energy density functionals. They called this functional M05 [34], but as it was designed to occupy the fourth rung in the Jacob's Ladder, the discussion is postponed to the next section. Now, the first mGGA member of that family is revised: the M06-L functional.

The exchange part of the M06-L functional [35] is based mainly on the VS98 functional

$$E_x^{M06-L} = \sum_\sigma \int \left\{ F_x^{PBE}(\rho_\sigma, \nabla\rho_\sigma) f(w_\sigma) + \varepsilon_x^{LSDA}(\rho_\sigma) h(s_\sigma, z_\sigma) \right\} d\mathbf{r}, \tag{69}$$

where $F_x^{PBE}(\rho_\sigma, \nabla\rho_\sigma)$ is given by equation 51 and $\varepsilon_x^{LSDA}(\rho_\sigma)$ is the Dirac exchange functional. The function $h(s_\sigma, z_\sigma)$ is identical to $F^{VS98}(s,z)$ given in equation 61 and

$$f(w_\sigma) = \sum_{i=0}^{m} a_i w_\sigma^i \tag{70}$$

$$w_\sigma = \frac{t_\sigma - 1}{t_\sigma + 1} \tag{71}$$

$$t_\sigma = \frac{2^{2/3} C_K \rho^{5/3}}{\tau}. \tag{72}$$

The correlation part of M06-L is divided in same-spin and opposite-spin contributions as in the VS98 case

$$E_c^{M06-L} = E_c^{M06-L,\alpha\beta} + \sum_\sigma E_c^{M06-L,\sigma\sigma}, \tag{73}$$

and

$$E_c^{M06-L,\alpha\beta} = \int e_{\alpha\beta} \left[g_{\alpha\beta}(s_\alpha, s_\beta) + h_{\alpha\beta}(s_{\alpha\beta}, z_{\alpha\beta}) \right] d\mathbf{r} \tag{74}$$

$$E_c^{M06-L,\sigma\sigma} = \int e_{\sigma\sigma} D_\sigma \left[g_{\sigma\sigma}(s_\sigma) + h_{\sigma\sigma}(s_\sigma, z_\sigma) \right] d\mathbf{r}, \tag{75}$$

where

$$g_{\alpha\beta}(s_\alpha, s_\beta) = \sum_{i=0}^{n} c_i \left(\frac{\gamma_{\alpha\beta}(s_\alpha^2 + s_\beta^2)}{1 + \gamma_{\alpha\beta}(s_\alpha^2 + s_\beta^2)} \right)^i \tag{76}$$

$$g_{\sigma\sigma}(s_\sigma) = \sum_{i=0}^{n} c_i' \left(\frac{\gamma_{\sigma\sigma} s_\sigma^2}{1 + \gamma_{\sigma\sigma} s_\sigma^2} \right)^i, \tag{77}$$

the $h_{\alpha\beta}$ and $h_{\sigma\sigma}$ functions are the same as in the exchange functional, and $s_{\alpha\beta}^2 = s_\alpha^2 + s_\beta^2$ and $z_{\alpha\beta} = z_\alpha + z_\beta$. The D_σ function is a self-correlation factor

$$D_\sigma = 1 - \frac{s_\sigma^2}{4(z_\sigma + C_K)}. \tag{78}$$

Finally, the functions $e_{\alpha\beta}$ and $e_{\sigma\sigma}$ are taken from reference [30] and follow the analysis made by Stoll, Pavlidou and Press [36]

$$e_{\alpha\beta} = \rho \varepsilon_c^{LSDA}(\rho_\alpha, \rho_\beta) - \rho_\alpha \varepsilon_c^{LSDA}(\rho_\alpha, 0) - \rho_\beta \varepsilon_c^{LSDA}(0, \rho_\beta) \tag{79}$$

$$e_{\sigma\sigma} = \rho_\sigma \varepsilon_c^{LSDA}(\rho_\sigma, 0). \tag{80}$$

The upper limit of the sum in equation 70 is set to 11, whereas the upper limit in the sums in equations 72 and 75 is set to 4. Thus, the M06-L exchange-correlation energy density functional needs 42 parameters to be used in practical calculations. Besides the broad applicability of the functional to the chemistry of main-group elements, it is worth noting its good performance to describe the energetic of systems containing transition metals.

Very recently, Peverati and Truhlar [37] proposed a new mGGA density functional, in which the exchange part is represented by a dual-range local functional. They called the functional M11-L.

The functional form of M11-L, in the spin-unpolarized version, can be written as

$$E_{xc}^{M11-L} = E_x^{SR-M11-L} + E_x^{LR-M11-L} + E_c^{M11-L} \tag{81}$$

where the *short-range* part of the exchange energy density functional is

$$E_x^{SR-M11-L} = \int \rho\varepsilon_x^{LDA}(\rho)G(\alpha)\left(f_1^{SR}(w)F_x^{PBE}(s) + f_2^{SR}(w)F_x^{RPBE}(s)\right) d\mathbf{r}, \tag{82}$$

and the inhomogeneity factors are the usual PBE exchange functional [28], and the revised version of the PBE functional, RPBE, proposed by Hammer, Hansen and Norskov [38] to improve the description of adsorption processes. The *range separation* or *attenuation* function is expressed as

$$G(\alpha) = 1 - \frac{8}{3}\alpha\left[\sqrt{\pi}\,\mathrm{erf}(\frac{1}{2\alpha}) - 3\alpha + 4\alpha^3 + (2\alpha - 4\alpha^3)\exp(-\frac{1}{4\alpha^2})\right], \tag{83}$$

where $\alpha = \omega/\left(2(6\pi^2\rho)^{1/3}\right)$ contains the *range-separation* parameter, ω [39]. The *long-range* part of the exchange functional is similar in form to the short-range one

$$E_x^{LR-M11-L} = \int \rho\varepsilon_x^{LDA}(\rho)\,(1 - G(\alpha))\left(f_1^{LR}(w)F_x^{PBE}(s) + f_2^{LR}(w)F_x^{RPBE}(s)\right) d\mathbf{r}. \tag{84}$$

The form of the correlation functional is quite similar to the PBE correlation functional, equation 52,

$$E_c^{M11-L} = \int \rho\left(f_3(w)\varepsilon_c^{L(S)DA}(\rho) + f_4(w)H(s,\rho)\right) d\mathbf{r}. \tag{85}$$

The six enhancement factors that depend on the kinetic energy density have the same functional form as the enhancement factors that appears in the development of M06-L, equation 70, and their argument, w, is given in equation 71. The upper limit in the sums that define the enhancement factors is set to 8. Considering the range-separation parameter as another adjustable parameter, the formulation of the M11-L contains 55 parameters.

7. The Hybrid (Meta) Generalized Gradient Approximation, the Fourth Rung in the Jacob's Ladder

The fourth rung in the Jacob's Ladder accounts for the occupied Kohn-Sham auxiliary functions by means of the inclusion of the exact exchange energy, the one that it is defined in the Hartree-Fock theory. Interestingly, the density functionals that constitute the fourth rung of the Jacob's Ladder were proposed a few years before than the mGGA functionals appeared in the DFT realm.

The first formal discussion on the possibility to include an explicit orbital dependence on the exchange-correlation functional was raised in 1993 and is due to Becke [40]. Considering the KS version of the DFT as starting point, the HK universal functional $F[\rho]$ can be written in terms of the constrained search formulation due to Levy [41]

$$F^{\lambda}[\rho] = \int \left(\Psi_{\rho}^{min,\lambda}\right)^{*} \left(\hat{T}_s + \lambda \hat{V}_{ee}\right) \Psi_{\rho}^{min,\lambda}\, d\mathbf{r} \tag{86}$$

where $\Psi_{\rho}^{min,\lambda}$ is the N-electron wave function that minimizes $F^{\lambda}[\rho]$ and leads to the density ρ, which, in turns, remains constant when λ goes from 0 to 1. $\hat{T}_s$ and $\hat{V}_{ee}$ are the operators that define the corresponding density functionals $T_s[\rho]$ and $V_{ee}[\rho]$, respectively, where $V_{ee}[\rho] = J[\rho] + E_{xc}[\rho]$ as can be deduced from equation 15. Moreover,

$$E_{xc}[\rho] = E_{xc}^{'}[\rho] + T[\rho] - T_s[\rho], \tag{87}$$

where $E_{xc}^{'}[\rho]$ represents *pure* exchange-correlation effects. From the above equations, it is clear that both the non interacting reference system and the interacting system of the KS scheme are closely related by the value of the parameter λ

$$F^{0}[\rho] = T_s[\rho] \tag{88}$$

$$F^{1}[\rho] = T_s[\rho] + J[\rho] + E_{xc}[\rho], \tag{89}$$

where the density, ρ, is constructed from the N-electron wave functions $\Psi_{\rho}^{min,0}$ and $\Psi_{\rho}^{min,1}$, respectively.

From the Hellmann-Feynman theorem it is easily seen that

$$\frac{\partial F^{\lambda}[\rho]}{\partial \lambda} = \int \left(\Psi_{\rho}^{min,\lambda}\right)^{*} \hat{V}_{ee} \Psi_{\rho}^{min,\lambda}\, d\mathbf{r}, \tag{90}$$

that can be written after integration as

$$F^{1}[\rho] - F^{0}[\rho] = \int_0^1 \int \left(\Psi_{\rho}^{min,\lambda}\right)^{*} \hat{V}_{ee} \Psi_{\rho}^{min,\lambda}\, d\mathbf{r}\, d\lambda. \tag{91}$$

Inserting equations 88 and 89 in equation 91 it is readily obtained

$$E_{xc}[\rho] = \int_0^1 \int \left(\Psi_{\rho}^{min,\lambda}\right)^{*} \hat{V}_{ee} \Psi_{\rho}^{min,\lambda}\, d\mathbf{r}\, d\lambda - J[\rho], \tag{92}$$

that can be written more concisely as

$$E_{xc}[\rho] = \int_0^1 E_{xc}^{\lambda}[\rho]\, d\lambda. \tag{93}$$

In his work, Becke proposed to use a sort of linear interpolation to solve the integral

$$E_{xc}[\rho] \approx \frac{1}{2}\left(E_{xc}^{0}[\rho] + E_{xc}^{1}[\rho]\right), \tag{94}$$

and, most important, related $E_{xc}^{0}[\rho]$ to the pure exchange energy that can be obtained from a single Slater determinant. Thus, Becke suggested to use the KS auxiliary functions for the calculation of the *exact* exchange, that is, using the Hartree-Fock expression. On the other hand, the term $E^1_{xc}[\rho]$ contains both exchange and correlation effects and can be estimated from the L(S)DA. Thus, the first *hybrid* density functional, known as *half-and-half*, has an exchange-correlation density functional of the form

$$E^{H\&H}_{xc}[\rho] = \frac{1}{2}\left(E^{HF}_x + E^{L(S)DA}_{xc}[\rho]\right). \tag{95}$$

In the same year, the same author, Becke (B3) [42], suggested to extend the half-and-half formula to include the electron density gradient. The new expression for the exchange-correlation energy functional contained three semiempirical coefficients, a_0, a_x and a_c

$$E^{B3PW92}_{xc} = a_0 E^{HF}_x + (1-a_0)\,E^{L(S)DA}_x + a_x \Delta E^{B88}_x + E^{L(S)DA}_c + a_c \Delta E^{PW92}_c, \tag{96}$$

where ΔE^{B88}_x is the GGA correction to the exchange energy proposed by Becke, see equations 39 and 40, and ΔE^{PW92}_x is the GGA correction to the correlation energy due to Perdew and Wang, see the discussion following equation 44.

In 1994, Stephens and co-workers [43] suggested the use of the LYP correlation energy density functional in the hybrid GGA of Becke instead of the PW92 functional. Moreover, due to the impossibility to get a local component from LYP, equation 96 is slightly modified to give

$$E^{B3LYP}_{xc} = a_0 E^{HF}_x + (1-a_0)\,E^{L(S)DA}_x + a_x \Delta E^{B88}_x + (1-a_c)\,E^{VWN}_c + a_c E^{LYP}_c, \tag{97}$$

where the values for the semiempirical coefficients are the same as in the B3PW92 case. B3LYP became one of the most popular density functionals adopted by the community of computational chemists. In the first years of the new century, B3LYP represented about 80% of the overall amount of citations involving any density functional [44]. In the last two years, the popularity poll of density functionals indicated that B3LYP occupies the second place in a list of forty functionals further separated in two groups [45].

In 1997, Becke (B97) [46] proposed a new hybrid GGA with only one parameter affecting the exact exchange

$$E^{B97}_{xc} = c_x E^{HF}_x + E^{B97}_x + E^{B97}_c. \tag{98}$$

Moreover, he also suggested a simplification in the form of the exchange and correlation functionals, inaugurating the philosophy of functionals design based on a *massive* parameterization

$$E^{B97}_x = \sum_\sigma \int \rho_\sigma \varepsilon^{LSDA}_{x,\sigma}(\rho_\sigma) g_{x,\sigma}(s^2_\sigma)\, d\mathbf{r}, \tag{99}$$

$$E^{B97,\alpha\beta}_c = \int e^{LSDA}_{\alpha\beta}(\rho_\alpha,\rho_\beta) g_{c,\alpha\beta}(s^2_\alpha, s^2_\beta)\, d\mathbf{r}, \tag{100}$$

$$E^{B97,\sigma\sigma}_c = \int e^{LSDA}_{\sigma\sigma}(\rho_\sigma) g_{c,\sigma\sigma}(s^2_\sigma)\, d\mathbf{r}, \tag{101}$$

and

$$E_c^{B97} = E_c^{B97,\alpha\beta} + \sum_{\sigma} E_c^{B97,\sigma\sigma}. \tag{102}$$

The functions $e_{\alpha\beta}^{LSDA}$ and $e_{\sigma\sigma}^{LSDA}$ adopt the form shown in equations 79 and 80, and the functions $g_{x,\sigma}$, $g_{c,\alpha\beta}$ and $g_{c,\sigma\sigma}$ share the same functional form

$$g(s^2) = \sum_{i=0}^{m} c_i u^i(s^2), \tag{103}$$

where $u(s^2) = \gamma s^2/(1+\gamma s^2)$. When g depends on both s_α^2 and s_β^2, a single average is formed, $s_{avg}^2 = (s_\alpha^2 + s_\alpha^2)/2$, and the γ values are fixed from atomic calculations. Becke found that a value of $m = 2$ provides reasonable results. Thus, the new hybrid GGA functional contains *only* 10 parameters.

In 1999, Adamo and Barone [47] proposed a hybrid GGA density functional based on the PBE GGA density functional. As the new functional does not contain any new additional parameter, except for the parameters already present in the PBE functional, it was known as PBE0 [4]

$$E_{xc}^{PBE0} = \frac{1}{4}E_x^{HF} + \frac{3}{4}E_x^{PBE} + E_c^{PBE}. \tag{104}$$

The coefficients for the exact exchange energy and the PBE exchange energy were obtained by Perdew, Ernzerhof and Burke [48] from the lowest order in perturbation theory that can yield reliable atomization energies for typical molecules.

Meta GGA density functionals also have their hybrid versions. As mentioned in section 6, Truhlar and co-workers presented in 2005 the first member of a family of exchange-correlation energy density functionals. They called this functional M05 [34], and it was classified as hybrid meta GGA. As usual, only a fraction of the exact exchange energy is mixed with the exchange energy density functional

$$E_x^{M05} = \left(1 - \frac{X}{100}\right) E_x^{mGGA} + \frac{X}{100} E_x^{HF}, \tag{105}$$

where E_x^{mGGA} is a simplified version of E_x^{M06-L}, equation 69[5],

$$E_x^{mGGA} = \sum_{\sigma} \int F_x^{PBE}(\rho_\sigma, \nabla\rho_\sigma) f(w_\sigma)\, d\mathbf{r}, \tag{106}$$

and $f(w_\sigma)$ is given by equations 70 to 72. The correlation energy density functional also takes a form very similar to that of the M06-L method, that is, there are different functionals for the opposite-spin and same-spin contributions. The self-correlation factor, D_α (see equation 78), takes a slightly different form.

Only a few months later, Truhlar *et al.* proposed another hybrid meta GGA functional, very similar to M05, but containing a larger amount of the exact exchange energy. In

[4] This hybrid GGA functional is also known as PBE1PBE.

[5] Although the M05 family of functionals was prior to the M06-L functional, they belong to different rungs in the Jacob's Ladder. This is why the M06-L functional, a meta GGA one, was presented in the previous section.

particular, the X variable in equation 106 increases from 0.28 to 0.56. The authors called the functional M05-2X [49]. The justification for the increase in the amount of exact exchange in the exchange functional was to provide a better functional with broad applications in systems containing main-group elements only. The original M05 functional, on the other hand, could still be used to calculate several properties of systems containing both main-group elements and transition metals with reasonable accuracy.

Also in 2006, the same group at the University of Minessota presented the M06-HF functional [50]

$$E_{xc}^{M06-HF} = E_{xc}^{M06-L} + E_x^{HF}, \tag{107}$$

where the original M06-L functional, but with a different set of parameters, was used. The objective of including the exact exchange at a full extent was not only to eliminate long-range self-interaction errors, but to achieve an overall performance as good as or even better that the hybrid GGA B3LYP functional.

In 2008, Zhao and Truhlar completed the M06 family of functionals with two new members that belong to the fourth rung of the Jacob's Ladder: M06 and M06-2X [51]. They return to the philosophy of the M05 and M05-2X functionals and give the new functionals the form

$$E_x^{M06,M06-2X} = \left(1 - \frac{X}{100}\right) E_x^{M06-L} + \frac{X}{100} E_x^{HF}, \tag{108}$$

where X is 27 and 54 for M06 and M06-2X, respectively. As in the case of the M05 functionals, M06 was parametrized to be used with systems containing both non metals and transition metals, whereas M06-2X was parametrized only for non metals.

Also in 2008, Zhao and Truhlar presented two new hybrid mGGA functionals, which were developed with the aim of improving self-consistent convergence and to fulfill some exact conditions up to second order in the reduced density matrix [52]. The functionals were called M08-HX and M08-SO. The exchange part of the functionals takes the form

$$E_x^{M08-HX,M08-SO} = \int \rho \varepsilon_x^{LDA}(\rho) \left(f_1(w) F_x^{PBE}(s) + f_2(w) F_x^{RPBE}(s)\right) d\mathbf{r}, \tag{109}$$

which is very similar to the exchange part of the mGGA M11-L functional, see equation 81. The correlation part of the M08 functionals is also similar to that of the M11-L one

$$E_c^{M08-HX,M08-SO} = \int \rho \left(f_3(w) \varepsilon_c^{L(S)DA}(\rho, \xi) + f_4(w) H(s, \rho, \xi)\right) d\mathbf{r}, \tag{110}$$

which, in turn, is taken from the correlation part of the PBE functional. The four enhancement factors, $f_{1-4}(w)$, are defined in such a way that the sums, see equation 70, goes from 0 to 11. Moreover, the hybrid scheme is defined for both functionals as

$$\begin{aligned} E_{xc}^{M08-HX,M08-SO} &= \left(1 - \frac{X}{100}\right) E_x^{M08-HX,M08-SO} + \frac{X}{100} E_x^{HF} + \\ & E_c^{M08-HX,M08-SO}. \end{aligned} \tag{111}$$

Every functional has 49 parameters to optimize, some of them are fixed due to the constraints imposed to the functionals, though. The exchange-mixing parameter, X, is 52.23 and 56.79 for M08-HX and M08-SO, respectively.

In 2011, Peverati and Truhlar presented a dual-range hybrid meta GGA functional called M11 [53]. The form of the functional is

$$E_{xc}^{M11} = \frac{X}{100}E_x^{HF} + \left(1 - \frac{X}{100}\right)\left(E_x^{SR-M11-L} + E_x^{LR-HF}\right) + E_c^{M11-L}, \tag{112}$$

where the short-range part of the exchange energy density functional is that of M11-L, equation 82 and the long-range functional is the exact Hartree-Fock exchange using the long-range of a range-separated Coulomb operator

$$E_x^{LR-HF} = \frac{1}{2}\sum_{\sigma}\sum_{i,j}^{occ}\int\int \psi_{i\sigma}^*(\mathbf{r_1})\psi_{j\sigma}^*(\mathbf{r_1})\frac{\mathrm{erf}(\omega r_{12})}{r_{12}}\psi_{i\sigma}(\mathbf{r_2})\psi_{j\sigma}(\mathbf{r_2})\,d\mathbf{r_1}d\mathbf{r_2}, \tag{113}$$

where ω is the usual range-separation parameter. The correlation part of the M11 functional is the same as that of the M11-L functional, see equation 85.

The dual-range hybrid meta GGA M11 functional is defined by 45 parameters as the four enhancement factors in $E_x^{SR-M11-L}$ and E_c^{M11-L} are polynomials of degree 10 and X is considered a parameter to optimize, too. The optimum value of X is 42.8

8. The Double-Hybrid (Meta) Generalized Gradient Approximation, the Fifth Rung in the Jacob's Ladder

The fifth rung in the Jacob's Ladder accounts for the inclusion of the unoccupied Kohn-Sham auxiliary functions by means of low-order Perturbation Theory.

In 2004, Zhao, Lynch and Truhlar presented two hybrid meta GGA density functionals containing a certain amount of correlation energy estimated through MP2 [54]. They called those functionals MC3 for Multi-Coefficient Three-Parameter. The form of the exchange-correlation energy proposed for the MC3 functionals is

$$E_{xc}^{MC3DFT} = c_1 E_x^{HF} + (1 - c_1)E_{xc}^{DFT} + c_2 E_c^{MP2}, \tag{114}$$

where E_{xc}^{DFT} is calculated, in turn, from a hybrid (meta) GGA scheme

$$E_{xc}^{DFT} = \frac{X}{100}E_x^{HF} + \left(1 - \frac{X}{100}\right)E_x^{DFT} + E_c^{DFT}. \tag{115}$$

The three parameters c_1, c_2 and X give the functionals their name. The first functional, MC3BB, uses the B88 and B95 functionals for exchange and correlation, respectively. The second functional, MC3mPW uses the PW92 functional for exchange and correlation, with the modification introduced by Adamo and Barone for the exchange [47].

In 2006, Grimme and co-workers [55, 56] suggested an expression for the exchange-correlation energy very similar to that in equation 114, but retaining the original philosophy of hybrid functionals, in which exchange and correlation bear different coefficients

$$E_{xc}^{B2-PLYP} = a_x E_x^{HF} + (1-a_x) E_x^{DFT} + (1-c) E_c^{LYP} + c E_c^{MP2}, \tag{116}$$

where pure GGA calculations could be performed with B88 or mPW92 for exchange, and the LYP functional is used for correlation. MP2 correlation energy is calculated according to

$$E_c^{MP2} = \frac{1}{4} \sum_{ia} \sum_{jb} \frac{[(ia|jb) - (ib|ja)]^2}{\varepsilon_i + \varepsilon_j - \varepsilon_a - \varepsilon_b}, \tag{117}$$

where i, j represent occupied KS functions and a, b are unoccupied KS functions, and the ε's are the eigenvalues of the corresponding functions. The functionals proposed by Grimme *et al.* were called B2-PLYP and mPW92-PLYP.

In 2008, Martin and co-workers [57, 58] carried out a series of benchmark studies to improve the original B2- and mPW92-PLYP double-hybrid functionals. The authors found that modifying both a_x and c much better results could be obtained for a series of properties. They called their modified functionals B2K-PLYP, mPW2K-PLYP, and B2GP-PLYP, respectively. The authors concluded that the B2GP-PLYP functional is a reliable general-purpose method, although B2K-PLYP should be used for nucleophilic substitution and hydrogen transfer reactions, in which its performance is much better.

In 2011, Brémond and Adamo proposed a double-hybrid functional free from adjustable parameters called PBE0-DH [59]. The authors used an adiabatic connection formula, which contains the Hartree exchange-correlation operator, to obtain the following form for the exchange-correlation energy

$$E_{xc}^{PBE0-DH} = \frac{1}{2}\left(E_x^{HF} + E_x^{PBE}\right) + \frac{1}{2}\left(\frac{7}{4}E_c^{PBE} + \frac{1}{4}E_c^{MP2}\right). \tag{118}$$

9. The Performance of Modern Density Functionals to Predict Properties of Interest to Chemists

In the last sections, the evolution of the Density Functional Theory was revised from the point of view of the increasing complexity in the exchange-correlation functionals needed in the Kohn-Sham version of the theory.

Now, it is time to see how well the functionals of different rungs perform in describing existing properties. As it was mentioned in section 1, the amount of existing functionals is extremely large and a thorough review of all of them is almost impossible. Thus, the present section will be devoted to those functionals shown and discussed in some extent in previous sections. Only in those cases in which a given functional, not mentioned in the present chapter, exhibits an outstanding performance for some specific property, it will be added to the discussion.

Moreover, benchmark studies are usually performed when a new functional is designed. Thus, comparative performance studies are biased to the performance of the new functional or to the performance of the family to which the new functional belongs. In other words, it is not easy to find benchmark studies including a large variety of functionals from every rung of the Jacob's Ladder.

Table 1. Performance of different exchange-correlation energy density functionals for various sub-databases, see text and reference [37]. The mean unsigned error, in kcal/mol is shown. The top performer for every sub-database is indicated in boldface.

Database	PBE	M06-L	M11-L	B3LYP	B97	M06	M06-2X	M06-HF	M08-HX	M08-SO	M11
SB1AE97	2.77	0.83	0.68	0.96	0.66	0.59	**0.42**	0.58	0.66	0.60	0.45
LB1AE12	12.96	2.25	2.94	**2.14**	2.22	2.28	2.18	5.67	2.43	2.36	2.58
SRMBE13	3.87	3.41	3.21	2.88	**2.64**	2.77	3.90	6.98	3.36	2.87	4.55
MRBE10	19.73	10.64	**6.14**	22.47	21.99	18.60	44.93	75.50	50.87	47.70	43.54
IsoL6	1.98	2.76	1.57	2.61	1.93	1.27	1.53	2.46	**0.59**	1.19	1.10
IP13/03	3.62	3.08	3.11	4.76	3.23	3.28	**2.56**	3.80	3.42	3.58	3.64
EA13/03	2.27	3.83	5.54	2.33	1.90	1.85	2.14	3.81	1.32	2.72	**0.89**
PA8/06	1.34	1.88	2.17	**1.02**	1.53	1.84	1.65	2.28	1.08	1.64	1.03
ABDE4/05	4.09	5.54	5.14	8.73	4.99	2.84	**2.12**	4.43	2.67	2.51	2.45
ABDEL8	7.16	8.85	6.98	10.40	7.71	4.72	**2.69**	4.56	2.87	3.88	3.48
HC7/11	3.97	3.35	2.42	16.80	8.46	2.78	**2.15**	2.29	4.89	4.60	3.74
πTC13	6.01	6.52	5.47	6.06	7.01	4.08	**1.51**	1.92	1.98	1.97	2.12
HTBH38/08	9.31	4.15	1.44	4.23	4.16	1.98	1.14	2.07	**0.72**	1.07	1.30
NHTBH38/08	8.42	3.81	2.86	4.55	3.31	2.33	**1.22**	2.53	**1.22**	1.23	1.28
NCCE31/08	1.24	0.58	0.56	0.96	0.70	0.41	0.29	0.41	0.35	0.37	**0.26**
DC10	39.22	19.76	10.38	20.66	15.34	9.45	10.46	15.63	9.60	10.87	**8.03**
AE17	54.44	3.86	6.42	13.6	9.25	6.85	6.18	**4.28**	6.54	10.54	5.15

The most relevant comparative studies are shown in reverse chronological order in the present chapter.

Peverati and Truhlar [37] performed an extensive study on the MXX family of functionals, where XX goes from 05 to 11, on the occasion of the development of the dual-range hybrid meta GGA M11 functional. Their authors used their own database, called BC338, where BC means *Broad Chemistry*, formed by 338 data taken from sub-databases containing atomization energies (SB1AE97, LB1AE12), single-reference metal bond energies (SRMBE13) and multireference bond energies (MRBE10), isomerization energies (IsoL6), ionization potentials, electron affinities and proton affinities (IP13/03, EA13/03, PA8), bond dissociation reaction energies (ABDE4/05, ABDEL8, HC7/11), hydrogen-transfer and non-hydrogen-transfer barrier heights (HTBH38/08, NHTBH38/08), thermochemistry of systems containing π electrons (πTC13), noncovalent complexation energies (NCCE31/05), atomic energies (AE17) and a dozen difficult cases for DFT (DC10). The performance of a series of functionals for every sub-database is shown in Table 1 and the overall performance for the BC338 database is depicted in Figure 1. It can be seen from the Table that the hybrid meta GGA M06-2X functional performs very well in almost all cases, except for bond energies in systems with appreciable multireference character (MRBE10) and for those cases described as difficult for DFT (DC10). Interestingly, the only functional with a mean unsigned error (MUE) well below 10 kcal/mol for the MRBE10 sub-database is the dual-range meta GGA M11-L. This fact is the main responsible for the lowest MUE exhibited by M11-L as shown in Figure 1. Only the M06, M11 and M11-L functionals present MUE's below 3 kcal/mol for the overall BC338 set, whereas B97, M06-L, M06-2X, M08-HX and M08-SO have MUE's between 3 kcal/mol and 4 kcal/mol. Surprisingly, the B3LYP functional is the second worse functional behind PBE.

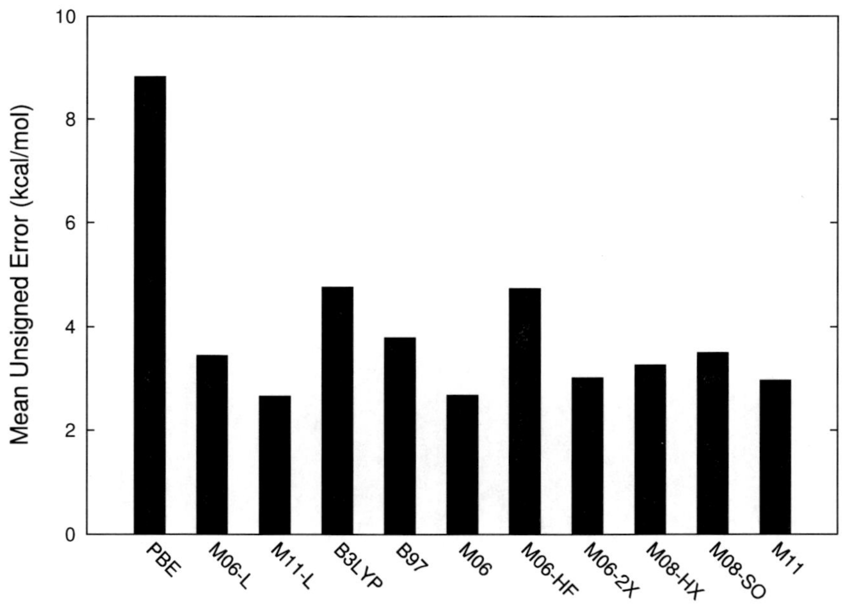

Figure 1. Overall performance of exchange-correlation energy density functionals for the BC338 database [37].

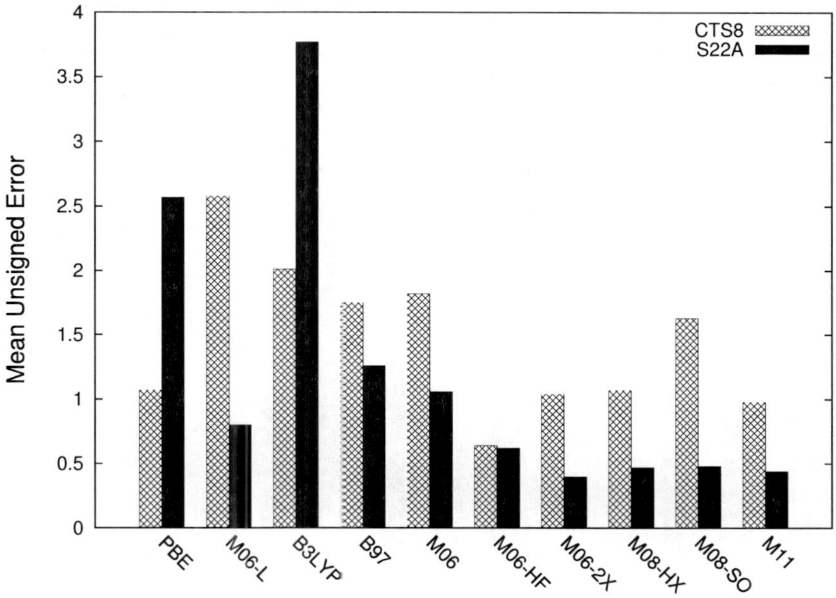

Figure 2. Performance of exchange-correlation energy density functionals for the CTS8 and S22A sub-databases, see text and reference [53]. MUE's are given in eV and kcal/mol for CTS8 and S22A, respectively.

Two other sub-databases were used to test the performance of the same functionals, except for M11-L [53]. The sub-databases contain charge-transfer electronic transitions (CTS8) and non covalent binding energies (S22A). Results are shown in Figure 2. It can be seen in the Figure that M06-HF and M06-2X are the top performers for CTS8 and S22A, respectively. Besides the good performance of some others functionals, it is interesting to see that B3LYP is the worst functional for S22A and the second worse behind M06-L.

Peverati, Zhao and Truhlar published another extensive study on the performance of a set of functionals on a smaller database, BC322 [60]. Besides the absence of hybrid GGA and hybrid meta GGA functionals, the interest of the benchmark study lies in the inclusion of various GGA and meta GGA functionals. Thus, within the GGA group formed by B88PW92, B88LYP, PW92, B88P86 and PBE, the best functional is B88PW92 with a MUE of 4.32 kcal/mol, whereas PBE becomes the worst with a MUE of 7.27 kcal/mol. The meta GGA TPSS functional has a MUE of 4.71 kcal/mol for the BC322 database.

In 2009, Zheng, Zhao and Truhlar performed a benchmark study on barrier heights of various reactions comprised in four sub-databases that, in turn, form the DBH24/08 database. The salient features of that work is that the computational cost, relative to the MP2/6-31+G(d,p) level of theory, is reported together with the partial and overall MUE's. The product of the overall MUE and the relative computational cost, both obtained with the 6-31+G(d,p) basis set, could be taken as a measure of the overall performance of a given method. Thus, the smaller the product, the better the performance index. Figure 3 summarizes the results. It can be seen in the Figure that M06-2X, M08-HX and M08-SO present the best performance indexes. The hybrid GGA functionals B3LYP, B3PW92 and PBE0 exhibit a performance index very similar to MP2. It is very interesting to note the very poor performance index shown by M06. As both M06 and M06-2X share the same relative computational cost, a rather large MUE should be the responsible for the bad performance of M06 when it is compared with M06-2X.

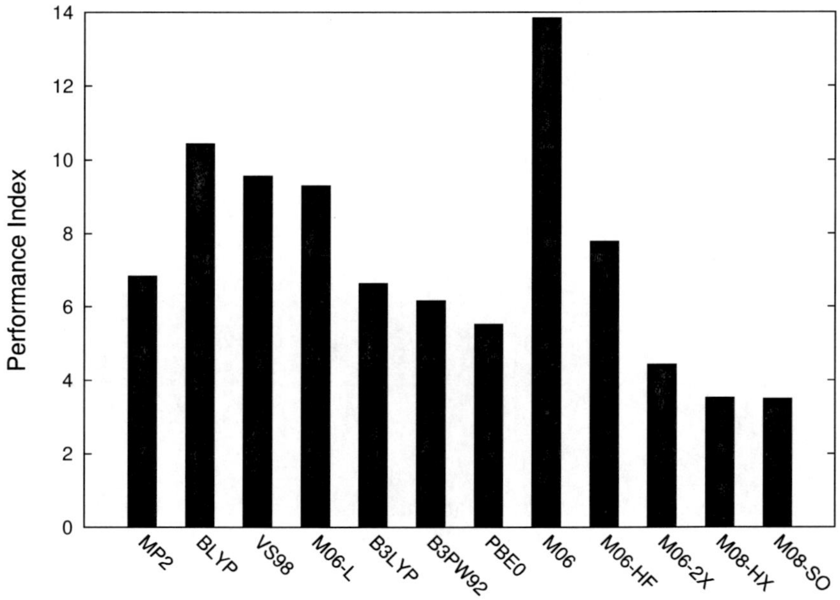

Figure 3. Performance index of exchange-correlation energy density functionals for the DBH24/08 database, see text and reference [61]. The performance index of the MP2 method is also shown for reference.

In 2008, Zhao and Truhlar carried out an extensive benchmark study of various functionals and several databases [51]. One of such databases, TMRE48, contains information on the thermochemistry of systems containing transition metals. In particular, atomization energies (TMAE9/05) and reaction energies (3*d*TMRE18/06) of systems containing transition metals, and metal-ligand bond energies (MLBE21/05) were calculated with a series of functionals. Results are summarized in Figure 4. It can be seen in the Figure that despite the good performance of BLYP for the TMAE9/05 database, and B3LYP and B97 for MLBE21/05, both M06 and M06-L show very good performances with overall MUE's of 5.6 and 5.7 kcal/mol, respectively. It is also very interesting to note the very bad performance of M06-2X for systems containing transition metals. This behavior should not be surprising, as it is a well known fact that the Hartree-Fock theory fails to describe open-shell systems, as those containing transition metals. The increasing contribution of the exact Hartree-Fock exchange energy should be the main responsible for that failure.

In the same work, Zhao and Truhlar show that bond lengths of systems containing main group elements are very well described by DFT, with an average MUE of 0.008 Å. When transition-metal systems are included in the benchmark, the average MUE notoriously increases up to about 0.055 Å, indicating the well known difficulty of dealing with transition metals. Surprisingly, the GGA PBE exchange-correlation functional is the top performer for bond lengths.

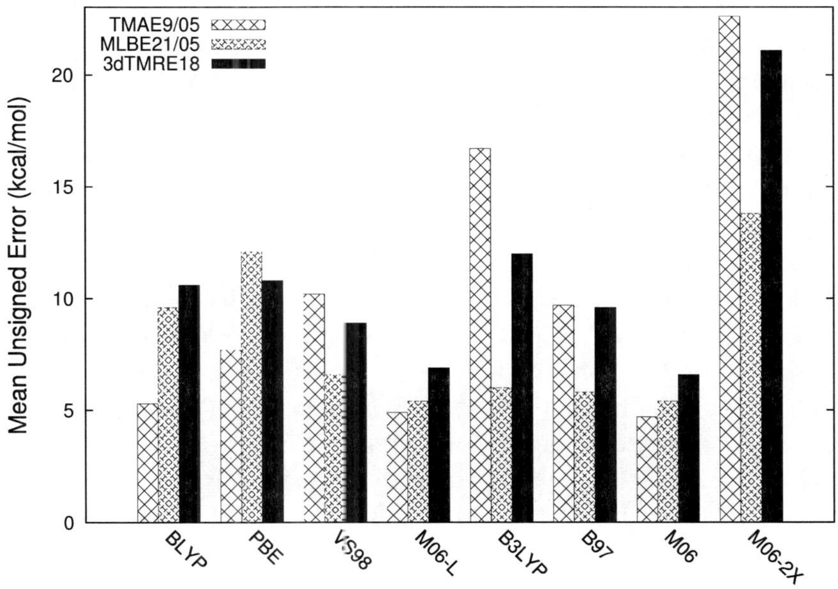

Figure 4. Performance of exchange-correlation energy density functionals for the TMAE9/05, MLBE21/05 and 3*d*TMRE18/06 databases, see text and reference [51].

10. Conclusion

The Density Functional Theory has greatly evolved from the 1920's, when the first attempts to devise a theory, in which the electron density is the basic variable were made.

Thanks to the work of Hohenberg, Kohn and Sham in mid 1960's, the Density Functional Theory was given a formal background, from which further evolution was possible.

The evolution of modern Density Functional Theory can be understood in terms of some basic ingredients. The first ingredient is the set of Kohn-Sham auxiliary functions, $\{\psi(\mathbf{r})\}$, which are needed to define the electron density, $\rho(\mathbf{r})$. The electron density is the basic ingredient of the Local (Spin) Density Approximation. The next step in the evolution is the inclusion of the gradient of the electron density through the dimensionless gradient $s = |\nabla\rho(\mathbf{r})|/\left((24\pi^2)^{1/3}\rho^{4/3}(\mathbf{r})\right)$. Functionals with both the electron density and the electron density gradient as arguments belong to the so-called Generalized Gradient Approximation. The kinetic energy density can be calculated from the Kohn-Sham auxiliary functions as $\tau = \sum_i^{occ} |\nabla\psi_i(\mathbf{r})|^2$. When τ is used as argument of a given functional, together with the electron density and the electron density gradient, the Meta Generalized Gradient Approximation is obtained. When the Kohn-Sham auxiliary functions are used to calculate the exact Hartree-Fock exchange energy and it is included into the exchange functional, the Hybrid Density Functional Theory is defined. Both Hybrid Generalized Gradient Approximations and Hybrid Meta Generalized Gradient Approximations can be used in practical calculations. Finally, when the *unoccupied* Kohn-Sham auxiliary functions are used to calculate the exact correlation energy up to second order and it is included into the correlation functional, the Double-Hybrid Density Functional Theory is obtained.

It is difficult to achieve a conclusive evidence about the reliability of different approximations to the DFT based only on benchmark studies. However, according to what it was shown in the last section, several useful tips can be pointed out. For systems containing main group elements, the hybrid meta GGA M06-2X functional performs very well for a variety of areas, such as thermochemistry, energy barriers and non covalent interactions. When transition metals are involved, the meta GGA M06-L functional is an excellent alternative as M06-2X is a poor choice due to the large amount of Hartree-Fock exchange. For more difficult cases, in which an appreciable multireference character is presented, such as bond dissociation or reactions involving different isomers, meta GGA, hybrid GGA and hybrid meta GGA are useless due to their single-range character. Thus, dual-range functionals, such as the meta GGA M11-L or the hybrid meta GGA M11 should be used instead. The more recent double-hybrid functionals seem to be an appealing alternative for practical calculations within the DFT. The relative computational cost of those functionals should be carefully considered, though, as the MP2 part of the calculation, if carried out self consistently, is rather demanding. In summary, a small subset of functionals formed by M06-L, M06-2X, M11 and M11-L could be well considered for routine calculations in various fields of Chemistry.

Finally, it is interesting to lay some emphasis on the mathematical form of modern functionals. Some researchers are devoted to design functionals with as few parameters as possible, whose values, in turn, are obtained after applying some well known boundary conditions that exchange and correlation functionals must fulfill. Other researchers prefer to develop functionals with the aim of chemical accuracy in mind. This philosophy involves mathematical forms with up to 50 adjustable parameters. Functionals designed following the former philosophy are unable to provide accurate results in certain fields of Chemistry. Functionals designed following the second philosophy, on the other hand, provide results accurate enough in several fields to consider modern DFT a reliable methodology for electronic structure calculations. This, however, seems to move modern DFT away from *ab-initio* or first principle methods and tend to *push* it to the side of semiempirical methods. Unfortunately, with this sort of *semiempirical* DFT the universal functional of Hohenberg and Kohn will surely remain unknown.

Acknowledgments

The author acknowledges CONICET, Argentina, for financial support under the PIP program. He is also member of the Scientific Researcher Career of CONICET.

References

[1] P. Hohenberg and W. Kohn. *Phys. Rev.*, 136:B864, 1964.

[2] W. Kohn and L.J. Sham. *Phys. Rev.*, 140:A1133, 1965.

[3] J. P. Perdew and K. Schmidt. In V. Van Doren, C. Van Alsenoy, and P. Geerlings, editors, *Density Functional Theory and Its Application to Materials*. AIP, Melville, NY, 2001.

[4] L. H. Thomas. *Proc. Camb. Phil. Soc.*, 23:542, 1927.

[5] E. Fermi. *Rend. Accad. Lincei*, 6:602, 1927.

[6] P. A. M. Dirac. *Proc. Camb. Phil. Soc.*, 26:376, 1930.

[7] E. von Weizsäcker. *Z. Physik*, 96:431, 1935.

[8] U. von Barth and L. Hedin. *J. Phys. C: Solid State Phys.*, 5:1629, 1972.

[9] L. Hedin and B. I. Lundqvist. *J. Phys. C: Solid State Phys.*, 4:2064, 1971.

[10] D. M. Ceperley and B. J. Alder. *Phys. Rev. Lett.*, 45:566, 1980.

[11] S. H. Vosko, L. Wilk, and M. Nusair. *Can. J. Phys.*, 58:1200, 1980.

[12] R. G. Parr and W. Yang. *Density Functional Theory of Atoms and Molecules*. Oxford University Press, New York, 1989.

[13] J. P. Perdew and Y. Wang. *Phys. Rev. B*, 33:8800, 1986.

[14] J. P. Perdew and Y. Wang. *Phys. Rev. B*, 40:3399, 1989.

[15] J. P. Perdew. *Phys. Rev. B*, 33:8822, 1986.

[16] J. P. Perdew. *Phys. Rev. B*, 34:7406, 1986.

[17] D. C. Langreth and J. P. Perdew. *Phys. Rev. B*, 21:5469, 1980.

[18] D. C. Langreth and M. J. Mehl. *Phys. Rev. B*, 28:1809, 1983.

[19] C. D. Hu and D. C. Langreth. *Phys. Scr.*, 32:391, 1985.

[20] C. D. Hu and D. C. Langreth. *Phys. Rev. B*, 33:943, 1986.

[21] J.P. Perdew and A. Zunger. *Phys. Rev. B*, 23:5048, 1981.

[22] A. D. Becke. *Phys. Rev. A*, 38:3098, 1988.

[23] C. Lee, W. Yang, and R. G. Parr. *Phys. Rev. B*, 37:785, 1988.

[24] R. Colle and D. Salvetti. *Theor. Chim. Acta*, 37:329, 1975.

[25] R. Colle and D. Salvetti. *J. Chem. Phys.*, 79:1404, 1983.

[26] J. P. Perdew, J. A. Chevary, S. H. Vosko, K. A. Jackson, M. R. Pederson, D. J. Singh, and C. Fiolhais. *Phys. Rev. B*, 46:6671, 1992.

[27] K. Burke, J. P. Perdew, and Y. Wang. In J. F. Dobson, G. Vignale, and M. P. Das, editors, *Electron Density Functional Theory: Recent Progress and New Directions*, page 81. Plenum, New York, 1997.

[28] J. P. Perdew, K. Burke, and M. Ernzerhof. *Phys. Rev. Lett.*, 77:3865, 1996.

[29] R. Neumann and N. C. Handy. *Chem. Phys. Lett.*, 266:16, 1997.

[30] A. D. Becke. *J. Chem. Phys.*, 104:1040, 1996.

[31] T. Van Voorhis and G. E. Scuseria. *J. Chem. Phys.*, 109:400, 1998.

[32] J. P. Perdew, S. Kurth, A. Zupan, and P. Blaha. *Phys. Rev. Lett.*, 82:2544, 1999.

[33] J. Tao, J. P. Perdew, V. N. Staroverov, and G. E. Scuseria. *Phys. Rev. Lett.*, 91:146401, 2003.

[34] Y. Zhao, N. E. Schultz, and D. G. Truhlar. *J. Chem. Phys.*, 123:161103, 2005.

[35] Y. Zhao and D. G. Truhlar. *J. Chem. Phys.*, 125:194101, 2006.

[36] H. Stoll, C. M. E. Pavlidou, and H. Preuss. *Theoret. Chim. Acta*, 49:143, 1978.

[37] R. Peverati and D. G. Truhlar. *J. Phys. Chem. Lett.*, 3:117, 2012.

[38] B. Hammer, L. Hansen, and J. Norskov. *Phys. Rev. B*, 59:7413, 1999.

[39] J.-D. Chai and M. Head-Gordon. *J. Chem. Phys.*, 128:084106, 2008.

[40] A. D. Becke. 98:1372, 1993.

[41] M. Levy. *Proc. Natl. Acad. Sci. USA*, 76:6062, 1979.

[42] A. D. Becke. *J. Chem. Phys.*, 98:5648, 1993.

[43] P. J. Stephens, F. J. Devlin, C. F. Chabalowski, and M. J. Frisch. *J. Phys. Chem.*, 98:11623, 1994.

[44] S. F. Souza, P. A. Fernandes, and M. J. Ramos. *J. Phys. Chem. A*, 111:10439, 2007.

[45] M. Swart, F. M. Bickelhaupt, and M. Duran. Density functionals poll, 2011.

[46] A. D. Becke. *J. Chem. Phys.*, 107:8554, 1997.

[47] C. Adamo and V. Barone. *J. Chem. Phys.*, 110:6158, 1999.

[48] J. P. Perdew, M. Ernzerhof, and K. Burke. *J. Chem. Phys.*, 105:9982, 1996.

[49] Y. Zhao, N. E. Schultz, and D. G. Truhlar. *J. Chem. Theory Comput.*, 2:364, 2006.

[50] Y. Zhao and D. G. Truhlar. *J. Phys. Chem. A*, 110:13126, 2006.

[51] Y. Zhao and D. G. Truhlar. *Theor. Chem. Account*, 120:215, 2008.

[52] Y. Zhao and D. G. Truhlar. *J. Chem. Theory Comput.*, 4:1849, 2008.

[53] R. Peverati and D. G. Truhlar. *J. Phys. Chem. Lett.*, 2:2810, 2011.

[54] Y. Zhao, B. J. Lynch, and D. G. Truhlar. *J. Phys. Chem. A*, 108:4786, 2004.

[55] S. Grimme. *J. Chem. Phys.*, 124:034108, 2006.

[56] T. Schwabe and S. Grimme. *Phys. Chem. Chem. Phys.*, 8:4398, 2006.

[57] A. Tarnopolsky, A. Karton, R. Sertchook, D. Vuzman, and J. M. L. Martin. *J. Phys. Chem. A*, 112:3, 2008.

[58] A. Karton, A. Tarnopolsky, J. F. Lamère, G. C. Schatz, and J. M. L. Martin. *J. Phys. Chem. A*, 112:12868, 2008.

[59] E. A. G. Brémond and C. Adamo. *J. Chem. Phys.*, 135:024106, 2011.

[60] R. Peverati and D. G. Truhlar. *J. Phys. Chem. Lett.*, 2:2810, 2011.

[61] J. Zheng, Y. Zhao, and D. G. Truhlar. *J. Chem. Theory Comput.*, 5:808, 2009.

In: Density Functional Theory
Editors: J. Morin and J.M. Pelletier, pp. 259-284
ISBN: 978-1-62417-954-9

Chapter 9

ALLOY-BASED DESIGN OF MATERIALS FROM FIRST PRINCIPLES: AN APPLICATION TO FUNCTIONAL HARD COATINGS

***David Holec*[1,*]*, Liangcai Zhou*[1,2]*, Richard Rachbauer*[1,3]**
***and Paul H. Mayrhofer*[1,2]**
[1]Department of Physical Metallurgy and Materials Testing,
Montanuniversität Leoben, Leoben, Austria
[2]Institute of Materials Science and Technology,
Vienna University of Technology, Vienna, Austria
[3]OC Oerlikon Balzers AG, Balzers, Liechtenstein

Abstract

Tailoring and improving material properties, as required in specific application fields, by alloying is a long-known and often used concept. The recent development of computational power has provided the opportunity to perform first principles calculations of systems with larger numbers of atoms ($\approx 10^2$), thus allowing for direct modelling of alloy-related phenomena. In this chapter we focus on applying the concept of *special quasi-random structures* (SQS) to quasi-binary and quasi-ternary nitride alloys, used as protective hard coatings.

In the first part we focus on concepts how to construct representative cells with 32-64 atoms. Subsequently, we use those to obtain ground state properties, such as lattice parameters, energies of formation, mixing enthalpies bulk modulus or phase stability, as functions of composition. The particular systems include quasi-binary $Y_{1-x}Al_xN$ and $X_{1-x}Y_xN$ and quasi-ternary $X_{1-x-y}Al_xY_yN$ systems (X and Y stand for transition-metal elements). It is shown that the lattice parameters do not strictly obey the linear Vegard's-like behaviour for the lattice parameters, but are "bowed out" towards larger values as a consequence of the continuously changing character of the bonds. Comparison of energies of formation for the same material system but various crystallographic variants, allows for discussion of the phase stability. Using the quasi-ternary models for $Cr_{1-x}Al_xN$ allows for modelling of its paramagnetic state. We show that inclusion of magnetism leads to non-negligible alternations in e.g., lattice

*E-mail address: david.holec@unileoben.ac.at

parameter or bulk modulus, as compared with non-magnetic case. Finally, we discuss the alloying trends in the elastic response of the single crystal and polycrystalline materials.

By comparing the calculated results with extended experimental work we demonstrate that the *ab initio* SQS calculations provide reliable trends, and hence can be used for the material selection in the alloy design process.

PACS 71.15.Mb, 61.66.Dk, 81.05.Je, 62.20.D-, 68.60.Dv, 79.20.Uv

Keywords: Density Functional Theory, alloys, nitrides, stability, elasticity

1. Introduction

Tailoring and improving material properties, as required by specific application fields, is a challenging task of modern materials research. To go beyond the performance of single elements or simple compounds, one can use multilayered or superlattice design of thin films, nanocomposites (altered microstructure by self-assembly even in bulk materials), or alloying.

An *alloy* is a mixture of two or more elements with the primary task to obtain properties not accessible by the single elements. This concept dates back to the prehistoric period now known as the bronze age. Bronze is an alloy of copper and tin; it is harder than pure copper and was originally used to make tools and weapons. Another long-time known example of an alloy is brass made from copper and zinc.

Alloying can be used to “tune” the lattice parameter for coherent interfaces of multilayers, to tailor the band gap for specific optoelectronic applications (so called “band gap engineering”, or to improve mechanical properties such as Young’s modulus or shear strength. These are consequences of either the altered chemistry of bonds (electronic structure), or purely the atomic size effect – larger/smaller atoms cause compressive/tensile forces on the neighbouring atoms, thus improving the resistance against external deformations. Sometimes, only a small amount of additional element is sufficient to change the material properties significantly. This is the case, for example, of steel, being made by alloying iron with small amounts of carbon, manganese, chromium, vanadium and/or tungsten.

1.1. Nitride Alloys

In order to demonstrate the capabilities of the Density Functional Theory (DFT) for materials design of alloys, in this chapter we focus on nitride pseudo-binary ($X_{1-y}Y_y\mathrm{N}$ or $(X\mathrm{N})_{1-y}(Y\mathrm{N})_y$) and pseudo-ternary ($X_{1-y-z}Y_yZ_z\mathrm{N}$ or $(X\mathrm{N})_{1-y-z}(Y\mathrm{N})_y(Z\mathrm{N})_z$) alloys. These are frequently used as *protective hard coatings*. The two widely spread material systems here are $\mathrm{Ti}_{1-x}\mathrm{Al}_x\mathrm{N}$ and $\mathrm{Cr}_{1-x}\mathrm{Al}_x\mathrm{N}$ [1, 2]. Their common features are that they are unstable, and at high temperatures decompose into the stable cubic B1 (NaCl prototype) TiN/CrN and hexagonal (wurtzite) B4 (ZnS prototype) AlN. Using a unique combination of the X-ray diffraction, high-resolution transmission electron microscopy and 3D atom-probe tomography, the $\mathrm{Ti}_{1-x}\mathrm{Al}_x\mathrm{N}$ alloy has been recently shown to decompose spinodally upon thermal load [3, 4]. The experimentally verified maximum aluminium content on the metallic sublattice that still preserves the cubic structure, is $x \approx 0.7$ for both systems, although values in the range from 0.4 to 0.91 have been reported experimentally depending

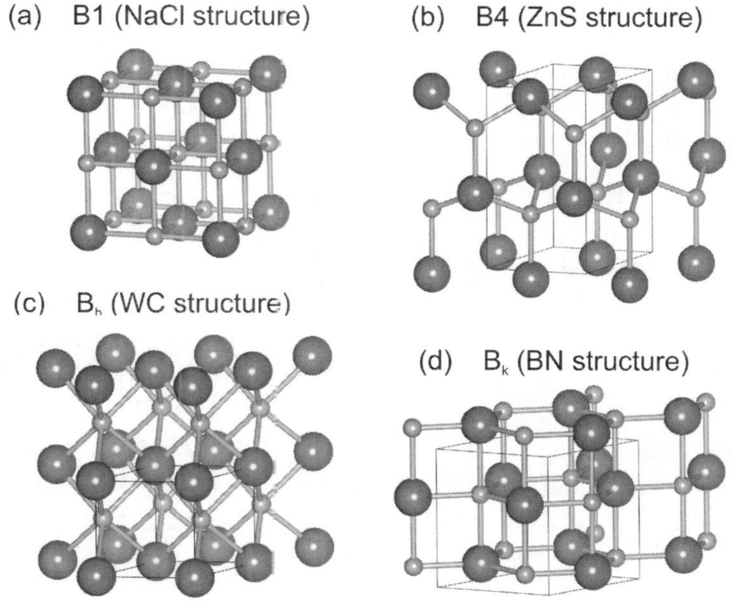

Figure 1. Crystallographic structures appearing in discussion of the results in this chapter: (a) B1 rock-salt cubic (NaCl prototype), (b) B4 hexagonal wurtzite (ZnS prototype), (c) B_h hexagonal (WC prototype), and (d) B_k hexagonal (BN prototype).

on the deposition conditions [5–10]. Higher AlN mole fractions result in wurtzite phase with largely deteriorated mechanical properties (hardness, Young's modulus) as compared with the cubic phase. These mechanical properties as well as the decomposition process have been many times demonstrated to depend strongly on the alloy composition [1, 3, 11]. Owing to their great application potential, a lot of attention has been paid to the structural properties, phase stability and mechanical properties of the pseudo-binary transition metal (TM) – aluminium – nitride systems [see, e.g., 3, 4, 9, 12–26], thus providing a vast amount of modelling as well as experimental results. The current tendency goes towards multi-component alloying (pseudo-ternary and multinary alloys) [27–35], hence the capabilities of the *ab initio* methods to reliably predict alloy-related trends also in these cases are briefly covered in this chapter.

Another important class of nitride alloys are *wide band gap semiconductors*, in particular alloys of GaN, AlN, InN (wurtzite B4 structures) and ScN (cubic B1 structure). These alloys, namely the $Al_xGa_{1-x}N$ system, are mentioned only briefly in discussion of the electron energy loss near edge structure (ELNES) as it possesses a different structure (wurtzite instead of cubic) and electronic character (semiconducting instead of metallic) when compared with $Ti_{1-x}Al_xN$ [36].

2. Modelling of Alloys

The crucial question when dealing with *ab initio* calculations of alloys is how to arrange atoms within the periodically repeated supercell. If the supercell contains N atom sites which can be occupied by atoms of either type A or type B (as is the case of a binary alloy), only the compositions $x = m/N$ are accessible, $0 \leq m \leq N$ and m is an integer. The problem is which m of the N sites to populate with atoms A.

2.1. Non-structural Theories

Early works approached the problem of alloys by developing the *virtual crystal approximation* (VCA) [37] and the *site-coherent potential approximation* (SCPA) [38, 39]. These are *non-structural theories* since they approach alloys by "average occupation" of atom sites. They have been applied to a wide variety of alloys and are able to capture effects with symmetry-preserving uniform volume changes. However, the principal drawback in these methods lies in the association of the average alloy properties with those of "effective atoms" on sites, bonds etc. Consequently, effects connected with locally broken symmetry (e.g, a tetrahedrally coordinated atom X in a cluster XA_4 has a point group T_d, while its local neighbourhood in a XA_3B cluster has C_{3v} symmetry), or with local relaxations (e.g., variations in lengths of A–A, A–B, and B–B bonds) cannot be described.

2.2. Structural Theories

On the contrary, *structural theories* consider the actual configuration of atoms in the structure. For a binary alloy with N sites, there are in total 2^N possible configurations (including those which are equivalent due to crystal symmetry). A measurable property of the alloy, for example the total energy E, is then the average over the ensemble of all these configurations

$$E = \langle E \rangle = \sum_{\sigma \in \Omega_m} \rho(\sigma) E(\sigma) \,. \tag{1}$$

Here, $\rho(\sigma)$ denotes the probability of finding the configuration σ in the ensemble and Ω_m is the subset of all possible configurations containing these with a fixed composition given by m. Although this is in principle a solution to the "alloy problem", it still constitutes a major obstacle: it is not possible to probe all 2^N configurations. For example, the ensemble size for $2 \times 2 \times 2$ supercell with 2 atoms in the unit cell is $65\,536$ which is far beyond any practical utilisation.

There are two practical approaches from here on: (i) the *cluster expansion method* and (ii) the *special quasi-random structures*. For a detailed overview of these techniques we refer the reader to Refs. [40] and [41].

2.2.1. Cluster Expansion Method

The concept of the cluster expansion method dates back to 1941 [42], and it is based on an observation that many physical properties depend strongly on the local environment. One starts with choosing a set of small *clusters* of atoms (k, m), where k defines by the number of atoms in the cluster ($k = 1$ for sites themselves, $k = 2$ for pairs of atoms, etc.), while m is the maximum neighbour distances separating them ($m = 1$ for the nearest neighbours, $m = 2$ for the second nearest neighbours, etc.). Sanchez et al. [43] proved that all clusters, $\{(k, m)\}$, posses a complete set for decomposition of any configuration σ. This allows expressing a property $E(\sigma)$ as

$$E(\sigma) = \sum_{(k,m)} \Pi_{(k,m)}(\sigma)\epsilon_{(k,m)} \tag{2}$$

where $\Pi_{(k,m)}(\sigma)$ are weights representing the statistical distribution of individual clusters (k, m) in a configuration σ, and $\epsilon_{(k,m)}$ is the cluster-related contribution to the macroscopic value of a property under question.

A desired property is calculated for several supercells with various arrangements of atoms (usually high-symmetry ones), which are subsequently used to fit $\epsilon_{(k,m)}$ values. The process is iterated (adding and removing clusters from the expansion set) until the fit reaches a prescribed accuracy. This is a somewhat tedious procedure; however, in the end of the day, one has an optimised set of cluster quantities $\epsilon_{(k,m)}$ which allow for extremely fast calculations of the desired property as a function of the alloy composition. Any composition that can be decomposed into the available clusters, can be thereafter modelled, hence yielding a dense mesh of accessible compositions. The main disadvantage is that nor the set of clusters neither the cluster quantities $\epsilon_{(k,m)}$ are transferable to other material systems. Additionally, any local phase transitions (e.g., a change in coordination) due to alloying are not captured unless they appeared during the optimisation of the set of clusters.

2.2.2. Special Quasi-Random Structures

In contrast to the cluster expansion method, the *special quasi-random structure* (SQS) approach aims for constructing a *single supercell* that can be used as a representative for a given fixed composition.

To describe the ordering in a structure, *Warren-Cowley short-range order parameters* (SROs) are defined as follows. Let N^j_{AB} be the number of $\{A, B\}$ pairs of atoms separated by the jth-neighbour distance in a given supercell. Let's assume that an overall concentration of atoms A and B in the alloys is x_A and $x_B = 1 - x_A$, respectively, and that every site has M^j neighbours in the distance j. A statistically random alloy with N atom sites in total contains $x_A N$ atoms of type A; every one of them has on average $x_B M^j$ neighbouring B atoms in the distance j. The Warren-Cowley SRO,

$$\alpha_j = 1 - \frac{N^j_{AB}}{x_A x_B N M^j}, \tag{3}$$

is a measure of randomness of the structure: for $\alpha_j = 0$, a statistically random supercell is obtained, while $\alpha_j < 0$ and $\alpha_j > 0$ correspond to the tendency for ordering and clustering, respectively.

In the case of the pseudo-binary alloys $X_{1-y}Y_y$N, where one sublattice is completely occupied with N, the Warren-Cowley SROs are evaluated only between the sites of the second sublattice with X and Y atoms. The arrangement of atoms is altered until α_j are 0 (or close to 0) for all j up to a chosen shell. The maximum meaningful j is given by the size of the supercell as one wants to avoid artificial ordering between the periodic images of atoms.

This procedure is relatively straightforward and the SQSs can be generated fast as compared with the cluster expansion method. On the other hand, one has to generate SQS for every composition of interest. In addition, the supercell size as well as the range of inter-atomic interaction should be carefully checked [17, 41].

It is straightforward to generalise the Warren-Cowley SROs also for the case multinary alloys (e.g., $A_{x_A}B_{x_B}C_{x_C}$, $x_A + x_B + x_c = 1$). In the case of three miscible elements, the possible mixed bonds are A–B, A–C, and B–C. Their numbers within a supercell having N atom sites are N_{AB}, N_{AC}, N_{BC}, respectively. The SROs for the jth neighbour distance (M^j coordinated) are

$$\alpha_j(AB) = 1 - \frac{N^j_{AB}}{x_A x_B N M^j}\,, \quad \alpha_j(BC) = 1 - \frac{N^j_{BC}}{x_B x_C N M^j}\,,$$
$$\alpha_j(AC) = 1 - \frac{N^j_{AC}}{x_A x_C N M^j}\,. \tag{4}$$

The ideal random configuration is characterised by all SROs being 0. In practise, one tries to find such arrangement of atoms which minimizes as many of the $\alpha_j(XY)$ as possible.

The SQS concept has been adopted in several cases to fit specific requirements. For example, the magnetic disorder in paramagnetic CrN (rock-salt structure) can be treated using SQS for $A_{0.5}B_{0.5}$ i.e., effectively working with an alloy $Cr^{\uparrow}_{0.5}Cr^{\downarrow}_{0.5}N$. The same approach has been applied also to $Cr_{1-x}Al_xN$ alloy being treated as a pseudo-ternary ($Cr^{\uparrow}$, $Cr^{\downarrow}$, and Al on one sublattice) system [44, 45]. The randomness of alloys becomes a contrary requirement to the overall e.g., cubic symmetry expected for cubic systems on the macroscale. For that reason, Holec et al. [46] have constructed SQS cells where the numbers of A–B bonds were summed separately in "cones" along x, y, and z axis, respectively, and their numbers were optimised separately. Such approach ensures that the three directions equivalent on the macroscale become similar also on the nanoscale within the model structure, resulting in elastic constants $C_{11} \approx C_{22} \approx C_{33}$, as expected for the cubic symmetry. Finally, the SQS approach can be combined with the cluster expansion in the way that not only the statistics of pairs distribution is optimised, but also the distribution of triplets, quadruplets, etc. are considered [22, 36, 47].

3. Ground State Properties

3.1. Equation of State

The equilibrium state of any particular phase corresponds to a state with minimum Gibbs free energy, G

$$G = E - TS + pV\,, \tag{5}$$

where E, S, and V are the internal energy, entropy, and volume (*intensive quantities*), and T and p are temperature and pressure (*extensive quantities*). At 0 K and 0 GPa external pressure, the minimum of G equals to the minimum of the internal energy, E, a quantity readily available from the *ab initio* calculations. $E(V)$ denotes a minimum energy cut through the phase space i.e., three lattice parameters, three lattice angles, and all the internal atom positions are optimised for a given specific volume V with respect to E.

The $E(V)$ relation is known as the *equation of state* (EOS). There exist several commonly used fitting equations, based on various assumptions for their derivation from thermodynamics. The most often employed one is *Birch–Murnaghan EOS* [48]

$$E(V) = E_0 + \frac{9V_0B}{16}\left(\left(\left(\frac{V_0}{V}\right)^{\frac{2}{3}} - 1\right)^3 B' + \left(\left(\frac{V_0}{V}\right)^{\frac{2}{3}} - 1\right)^2 \left(6 - 4\left(\frac{V_0}{V}\right)^{\frac{2}{3}}\right)\right). \quad (6)$$

Here, B is the bulk modulus, $B' = \partial B/\partial p$ its pressure derivative, E_0 and V_0 are the equilibrium energy and volume, respectively.

3.2. Equilibrium Volume

The cubic lattice parameter is $a = \sqrt[3]{V_0}$, where V_0 is the equilibrium volume of the cubic primitive cell. For non-cubic structures, the lattice parameters a, b, c, α, β, and γ can be estimated using fitted relations $a(V)$, $b(V)$, …, $\gamma(V)$ for $V = V_0$, where the data for fitting are obtained from the structures optimised during the $E(V)$ calculations. Fig. 2 shows the calculated results for several pseudo-binary and pseudo-ternary systems. In all cases, a quadratic polynomial yields a very good fit ($R^2 \approx 0.99$ or better) to the data. For an $X_{1-x}Y_x$N alloy, it is convenient to write this quadratic polynomial in a form of

$$a(x) = xa_{XN} + (1-x)a_{YN} + bx(1-x)\,, \quad (7)$$

where a_{XN} and a_{YN} are the lattice parameters of binary XN and YN, respectively. b is called a *bowing parameter*, and describes the deviation from the *Vegard's linear interpolation*. Its value is 0.072 Å, 0.460 Å, and 0.096 Å for $Ti_{1-x}Al_xN$, $Y_{1-x}Al_xN$, and

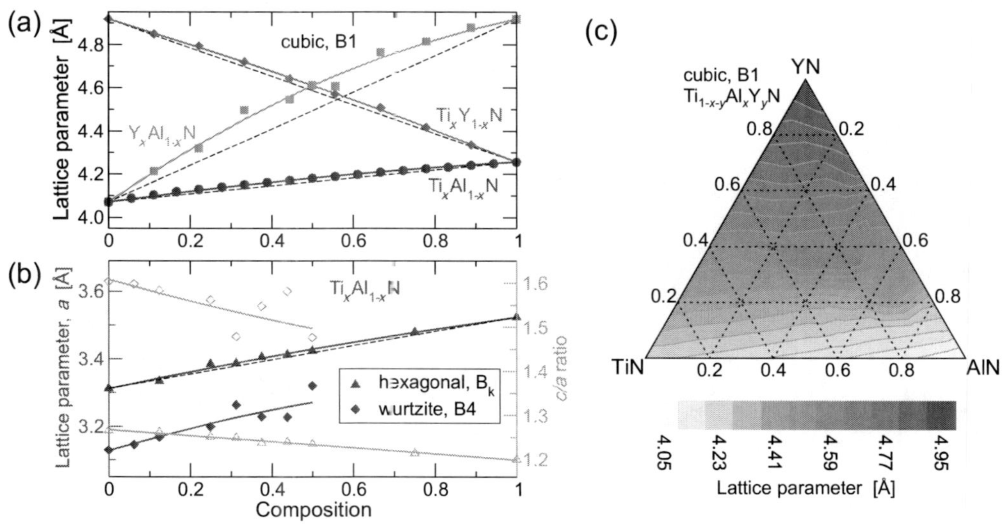

Figure 2. Lattice parameters of (a) cubic $Ti_{1-x}Al_xN$, $Y_{1-x}Al_xN$, and $Ti_{1-x}Y_xN$ alloys, (b) hexagonal B4 and B_k structures of $Ti_{1-x}Al_xN$, and (c) pseudo-ternary $Ti_{1-x-y}Al_xY_yN$ system.

$Ti_{1-x}Y_xN$, respectively, as shown in Fig. 2a. Consequently, the deviation from Vegard's estimation is larger than 2.5% for $Y_{0.5}Al_{0.5}N$. Fig. 2b presents the alloy related trends for the lattice parameters of hexagonal B_k and B4 phases (see Fig. 1) of $Ti_{1-x}Al_xN$. Similar to the cubic B1 phase, small deviations from the linear behaviour are predicted also here. More importantly, the calculated values are much more scattered, in particular for the wurtzite B4 phase at compositions close to $x \approx 0.5$. This is, however, close to the solubility limit of Al in the cubic phase. As it is discussed later in this chapter, such composition correspond to preferred cubic phase, and thus the scatter of the lattice constants is attributed partially to the structural instability (tendency for phase transition). Finally, Fig. 2c shows the contour plot of lattice parameter of the cubic phase of pseudo-ternary $Ti_{1-x-y}Al_xY_yN$ alloy. It varies smoothly with the composition, but once again exhibits non-linear behaviour which is demonstrated by the non-parallel contours.

Despite the fact that these *ab initio* calculations were obtained for 0 K, while experiments are typically performed at room temperature, the calculated trends agree well with the measured values. The agreement of the absolute values is in most cases below 1% for the nitride alloys (generalised gradient approximation (GGA) usually overestimates the lattice constant as compared with experiment) (see e.g., Ref. [49] for $Ti_{1-x}Al_xN$, and Ref. [31] for $Ti_{1-x-y}Al_xY_yN$).

3.3. Bulk Modulus

Another material property obtained from the Birch-Murnaghan EOS is the *bulk modulus*, B, a measure of compressibility under hydrostatic pressure. In general, it exhibits a non-linear dependence on the composition, similar to the lattice constants. Bulk modulus of cubic B1 $Ti_{1-x}Al_xN$ (Fig. 3a) shows a typical behaviour of many pseudo-ternary and quasi-quaternary systems: the B values lie between bulk moduli of its boundary systems.

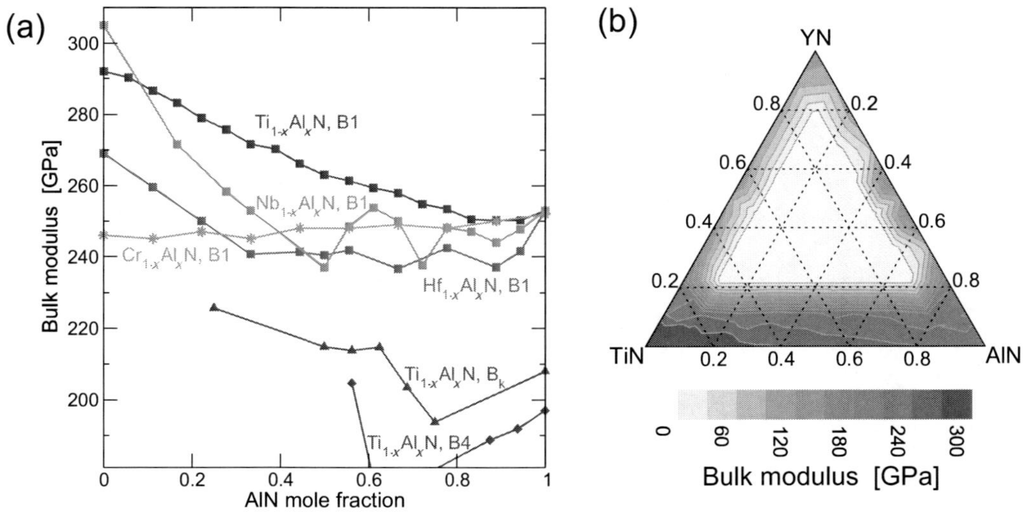

Figure 3. Bulk modulus of (a) quasi-binary and (b) quasi-ternary TM–Al–N alloys.

Cubic $Nb_{1-x}Al_xN$ and $Hf_{1-x}Al_xN$ systems exhibit softening with respect to the boundary systems (see Fig. 3a). The significant scatter of the $Nb_{1-x}Al_xN$ values is most likely a supercell-size related effect suggesting that the used 36-atom supercells do not describe this random alloy properly. On the other hand, scatter around 10–15 GPa is usually assumed to be acceptable for elastic constants. A more significant spread of bulk modulus data is obtained for the hexagonal B_k and wurtzite B4 phases of $Ti_{1-x}Al_xN$ (Fig. 3a). Similarly to the discussion of their lattice parameters, this scatter reflects the structural instability of the hexagonal phases for AlN mole fractions below ≈ 0.7. Paramagnetic $Cr_{1-x}Al_xN$ system represents a special case with respect of the bulk modulus, as B is practically constant over the whole compositional range $0 \leq x \leq 1$ [45]. Finally, Fig. 3b shows an extreme case of $Ti_{1-x-y}Al_xY_yN$, where the alloy bulk modulus lies far outside the range given by bulk moduli of the boundary systems, a fact related to the alloy instability. In conclusion, it is not possible to generalise the functional form of bulk modulus as this depends strongly on the material system.

3.4. Energy of Formation and Phase Stability

Energy of formation, E_f, defined as

$$E_f(X_{1-x}Y_x\text{N}) = E_0(X_{1-x}Y_x\text{N}) - \frac{1}{2}\left((1-x)E_0(X^{\xi}) + xE_0(Y^{\zeta}) + \frac{1}{2}E_0(\text{N}_2)\right), \quad (8)$$

where $E_0(X_{1-x}Y_x\text{N})$, $E_0(X^{\xi})$ and $E_0(Y^{\zeta})$ are the ground state energies of the alloy, and of the elements X and Y in their respective crystallographic structures ξ and ζ. $E_0(\text{N}_2)$ is the ground state energy of an isolated nitrogen molecule. E_f describes the energy gain when the alloy is formed as compared with individual elements in their crystalline form, hence it is a measure of the phase stability. In particular, when a competition of several phases takes place, the lowest E_f identifies the 0 K ground state.

Energy of formation of $Ti_{1-x}Al_xN$ is shown in Fig. 4a. Since the ground state structure of TiN is the rock-salt B1 cubic structure, while AlN crystallises in the wurtzite B4 hexagonal phase, there must be a cross-over of $E_f^{\text{B1}}(x)$ and $E_f^{\text{B4}}(x)$. The *maximum solubility* of Al in the cubic phase is calculated to be about ≈ 0.7 of the metallic sublattice sites, which agrees well with the experimentally estimated values $x_{\max} \approx 0.6 \div 0.7$ [22, 49]. The intermediate five-coordinated B_k hexagonal phase (see Fig. 1) is not predicted to be energetically favourable at any concentration.

A similar scenario is predicted and experimentally confirmed also for $Cr_{1-x}Al_xN$ alloy [45, 50], see Fig. 4b. There is a striking difference between the non-magnetic (NM) and paramagnetic (PM, where the SQS approach used to obtain both disorders, magnetic and chemical): the maximum solubility limit of AlN in the cubic phase increases from ≈ 0.61 for the NM to ≈ 0.73 for the PM case. It has been discussed several times in the literature that magnetism plays a crucial role for obtaining correct trends of the lattice parameters and bulk modulus of CrN-based alloys [44, 45, 51]. The upper boundary of the B1 phase stability range as shown in Fig. 4b, is yet another such example.

The predicted cross-over of E_f of different phases for $Ti_{1-x}Al_xN$ and $Cr_{1-x}Al_xN$ alloys depends sensitively on the applied stress state [52], a factor that should not be omitted

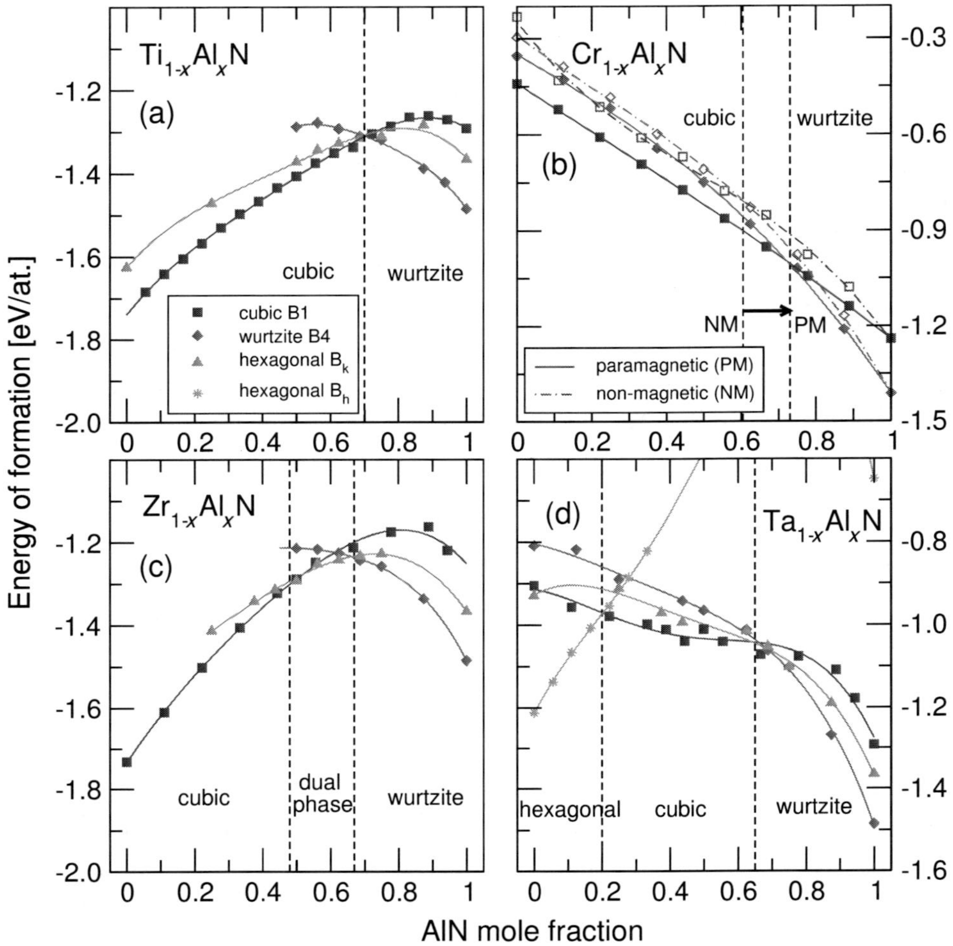

Figure 4. Energy of formation of (a) $Ti_{1-x}Al_xN$ (cubic, wurtzite, and hexagonal B_k phases), (b) $Cr_{1-x}Al_xN$ (cubic and wurtzite phases treated non-magnetically (NM) and with paramagnetic (PM) configuration), (c) $Zr_{1-x}Al_xN$ (cubic, wurtzite, and hexagonal B_k phases), and (d) $Ta_{1-x}Al_xN$ (cubic, wurtzite, and hexagonal B_k and B_h phases).

in the discussion of experimental results of thin films. In such case, enthalpy of formation, H_f, instead of the energy of formation, E_f is the correct thermodynamic potential. These quantities are related (cf. Eq. 5) as

$$H_f = E_f + pV. \tag{9}$$

The analysis in Ref. [52] shows that 4 GPa of external compressive pressure is enough to extend the maximum solubility x_{max} by ≈ 0.09 and ≈ 0.13 in the case of $Ti_{1-x}Al_xN$ and $Cr_{1-x}Al_xN$, respectively.

Somewhat different situation is predicted for the $Zr_{1-x}Al_xN$ alloy (isovalent with $Ti_{1-x}Al_xN$) [22]. Here, the energies of formation of the B1 and B_k phases overlap for AlN mole fractions between 0.4 and 0.7 (see Fig. 4c). Consequently, a dual phase region is predicted for these compositions. Indeed the experimental evidence suggests that single phase material can be deposited for $x \lessapprox 0.35$, while higher AlN mole fractions result in dual phase or amorphous material (experimental evidence from X-ray diffraction is given in e.g., Ref. [22]). Qualitatively the same behaviour is obtained also for the $Hf_{1-x}Al_xN$ alloy [22].

Yet a more complex picture is predicted for the case of $Ta_{1-x}Al_xN$ (Fig. 4d), where a fourth phase (hexagonal B_h) must be considered as a ground state of TaN. For low AlN mole fractions up to ≈ 0.20, the hexagonal B_h phase is predicted to be stable, followed by a region of preferred cubic B1 structure. The phase transition to wurtzite happens at composition ≈ 0.7. As a consequence of relatively small difference between the cubic B1, hexagonal B_k, and wurtzite B4 phases around $x = 0.7$, a multi-phase regime is predicted near to this composition. A similar situation (but with different hexagonal phase at $x \approx 0$) has been reported for the isovalent $Nb_{1-x}Al_xN$ alloy [53].

Using the structures for pseudo-ternary alloys allows for discussion of the alloying effects on the maximum stability. As reported in Ref. [54], the early transition metals from groups IVB (Zr and Hf) and VB (Nb and Ta) do not alter $x_{\max}$ within the calculation accuracy of either $Ti_{1-x}Al_xN$ or $Cr_{1-x}Al_xN$. A different trend is obtained for yttrium: 0.05 and 0.10 Y on the metallic sublattice of $Ti_{1-x}Al_xN$ is predicted to decrease the maximum solubility limit to ≈ 0.65 and ≈ 0.58, respectively'. The destabilisation of the dense cubic phase and favouring the less dense wurtzite phase, is predominantly connected with the large atomic radius of Y [54]. These trends have been largely confirmed by the extensive experimental work [28, 30–33, 55].

While the negative value of E_f means that a particular phase of an alloy $X_{1-x}Y_xN$ is stable with respect to individual elements X, Y, and N, the alloy itself could be unstable with respect to decomposition to XN and YN. A measure for the latter stability is *mixing enthalpy*, H_{mix}, defined as

$$
\begin{aligned}
H_{\text{mix}}(X_xY_{1-x}N) &= E_0(X_xY_{1-x}N) - xE_0(XN) - (1-x)E_0(YN) \\
&= E_f(X_xY_{1-x}N) - xE_f(XN) - (1-x)E_f(YN)\,. \qquad (10)
\end{aligned}
$$

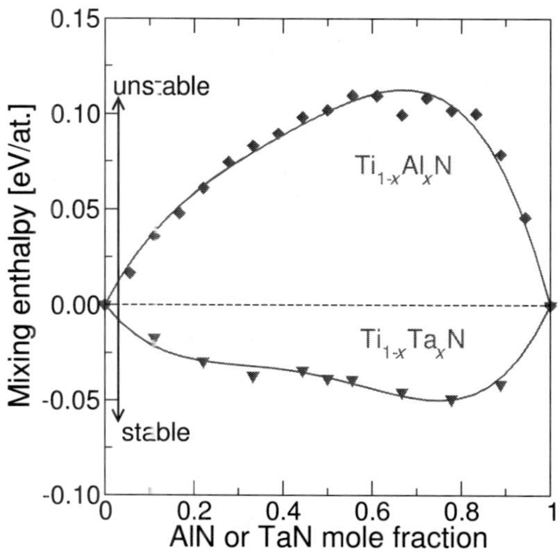

Figure 5. Mixing enthalpy of unstable $Ti_{1-x}Al_xN$ and stable $Ti_{1-x}Ta_xN$ alloys.

Mixing enthalpies of $Ti_{1-x}Al_xN$ and $Ti_{1-x}Ta_xN$ as examples of the unstable and stable alloys, respectively, are shown in Fig. 5. Recent 3D atom probe data complementing the previous transmission electron micrographs and X-ray diffractograms have confirmed, that the $Ti_{1-x}Al_xN$ alloy decomposes spinodally upon being thermally loaded [4]. This scenario is predicted to persist also under considerable hydrostatic loads [56]. $Ti_{1-x}Ta_xN$ alloy has been shown to crystallise in the cubic phase for the whole compositional range [57, 58]. Results in Fig. 5 in addition suggest that this pseudo-binary alloy is stable with respect to decomposition into cubic binary TiN and TaN. This is supported by the annealing experiments reported in Ref. [32].

Mixing enthalpies of pseudo-binary $X_{1-x}Al_xN$ and $Ti_{1-y}X_yN$ have been used to predict the decomposition products of pseudo-ternary alloys $Ti_{1-x-y}Al_xX_yN$ [30–33, 55]. When Y, Zr or Hf is added to $Ti_{1-x}Al_xN$, the resulting decomposition products are binary nitrides c-TiN, c-XN, and w-AlN. On the other hand, Nb and especially Ta containing coatings decompose into c-$Ti_{1-y}X_yN$ and w-AlN. Qualitatively the same trends for the onset of decomposition processes are obtained also for the pseudo-ternary $Cr_{1-x}Al_xN$+TM systems, though the decomposition route to the final products is more complex [59].

4. Elastic Properties

Density Functional Theory is a suitable tool for estimating elastic properties. There are two techniques for getting the *single crystal elastic constants*. The *strain–energy method* is based on calculating strain energy density, $\mathcal{U}$, as a function of applied strain ε_{ij}. $\mathcal{U}$ is obtained as the difference between total energies per unit volume of strained and equilibrium (strain-free) states. *Hooke's law* provides the relation between $\mathcal{U}$ and the strain tensor, ε_{ij}

$$\mathcal{U} = \frac{1}{V_0}(E(\varepsilon_{ij}) - E_0) = \sum_{ijkl} C_{ijkl}\varepsilon_{ij}\varepsilon_{kl} \ . \tag{11}$$

For a fixed deformation mode one can fit the strain energy as a function of the applied strain using the known analytical formula derived from Eq. 11 (a quadratic function within the linear elasticity theory). Doing so for several deformation modes provides the complete set of independent elastic constants. For example, cubic symmetry implies that there are only three independent elastic constants: $C_{11} = C_{xxxx}$, $C_{12} = C_{xxyy}$, and $C_{44} = C_{xyxy}$. As one deformation mode can be taken the hydrostatic volume expansion (EOS, Eq. 6) and employing a relation between C_{ijkl} and B: $B = \frac{1}{3}(C_{1111} + 2C_{1122})$. As the two additional deformation modes are often taken tetragonal and triclinic deformations or orthogonal and monoclinic deformations [60, 61]. Deformation modes suitable for general symmetries have been recently reviewed in Refs. [61–63], for practical aspects of such calculations the reader is referred to Refs. [60, 61].

The second approach called the *stress–strain method* is also based on the Hooke's law. It is applicable when the stress tensor, σ_{ij}, can be directly obtained from the *ab initio* calculations. The strain tensor, ε_{ij}, defines the geometry of the cell. Estimation of σ_{ij} for several different (and linearly independent) ε_{ij} allows for estimating the full tensor of elastic constants, C_{ijkl}

$$C_{\alpha\beta} = \sum_{\gamma=1}^{N} \Sigma_{\alpha\gamma} E^{-1}_{\gamma\beta} \tag{12}$$

where α and β are numbers from 1 to 6 representing the indices in Voigt's notation [64] ($xx \to 1$, $yy \to 2$, $zz \to 3$, $yz \to 4$, $xz \to 5$, and $xy \to 6$) converting the 3×3 matrix of strain and stress tensor to 6×1 vectors (note additional factors appearing for shear components of the strain tensor [64] e.g., $\varepsilon_{xy} = \frac{1}{2}\varepsilon_6$). Components of the strain and stress tensor of the N independent loading cases are arranged in $6 \times N$ matrices E and Σ, respectively. Finally, the elastic constants are calculated using a pseudo-inverse for matrix E. More details about this technique can be found in e.g., Ref. [65].

4.1. Supercell Size and Atom Distribution Effects

When dealing with random alloys represented by small supercells, the obvious problem is that the macroscopic symmetry (e.g., cubic) is not preserved at the atomic scale. The SQSs often have low or no symmetry, resulting in tensors of elastic constants having the most general form with 27 independent components. Tasnádi et al. [66] have recently shown that the statistically correct averaging of these 27 components to obtain elastic constants tensor with the macroscopic symmetry, $\bar{C}_{ij}$, is simple in the case of cubic systems:

$$\bar{C}_{11} = \frac{1}{3}(C_{11} + C_{22} + C_{33}) \,, \qquad \bar{C}_{12} = \frac{1}{3}(C_{12} + C_{13} + C_{23}) \,,$$
$$\bar{C}_{44} = \frac{1}{3}(C_{44} + C_{55} + C_{66}) \,. \tag{13}$$

Tasnádi et al. [66] have also performed analysis of the sensitivity of the calculated results on the SQS size and shape, and have concluded the best results are obtained for cells with around 100 atoms. Acceptable accuracy was obtained also in the case with 64 atoms, when the cell shape reflected the desired cubic symmetry (cubic $2 \times 2 \times 2$ supercell).

In order to improve the calculation efficiency, it has been recently proposed to use directionally resolved SQS [46] as described in Section 2.2.2, where the SROs are optimised independently along the x, y, and z directions, thus partly imposing the overall cubic symmetry on the SQS. The results for the cubic phase of the $Zr_{1-x}Al_xN$ system calculated with strain–energy method applied to 64-atom supercells, are plotted in Fig. 6. The cubic elastic constants were obtained using the symmetry projection technique as given by Eq. 13 [66], the error bars correspond to standard deviations of, for example, the $\{C_{11}, C_{22}, C_{33}\}$ data set. The calculated elastic constants vary smoothly with the composition, with some "wobbling" caused probably by the supercell size effects. The obtained C_{ij} lie between the respective values of the boundary systems, the only exception being C_{11} for $x \approx 0.8$. In all cases the compositional dependence is strongly non-linear, suggesting that to use Vegard's linear interpolation is not suitable in this case. A comparison with the "standard" SQS (i.e. without optimising x, y, and z directions independently), yielded some differences in particular for the shear C_{44} component, which are subject of discussion and experimental validation in Ref. [46].

A similar compositional behaviour of elastic constants has been reported for $Ti_{1-x}Al_xN$ [67] and $Cr_{1-x}Al_xN$ [45]. In the latter paper, both strain–energy and stress–strain methods were employed, clearly demonstrating the compatibility of the results.

4.2. Polycrystalline Materials

Experimentally obtained thin films are rarely single crystals, but instead exhibit textured polycrystalline nature. For such cases, an estimation of the polycrystalline elastic properties is of a special interest.

The simplest possible homogenisations are *Voigt's* (V) and *Reuss's* (R) averages corresponding to isotropic aggregates of randomly orientated grains (i.e., all grain orientations appear with the same probability). Voigt [68] assumed constant strains in all grains. The thus obtained polycrystalline shear and Young's moduli, G_V and E_V, respectively, provide the upper limit estimates. On the other hand, Reuss [69] proposed to apply constant stresses to all grains, which yields lower limits G_R and E_R. Taking $B_V = B_R = B$, where the bulk modulus, B, is obtained from the Birch-Murnaghan EOS [48], Eq. 6, one gets

$$G_V = \frac{C_{11} - C_{12} + 3C_{44}}{5}, \tag{14}$$

$$G_R = \frac{5}{4(S_{11} - S_{12}) + 3S_{44}}, \tag{15}$$

$$E_\alpha = \frac{9BG_\alpha}{3B + G_\alpha}, \quad \alpha = V,\, R\,. \tag{16}$$

The differences between Voigt and Reuss averages diminish for materials with isotropic

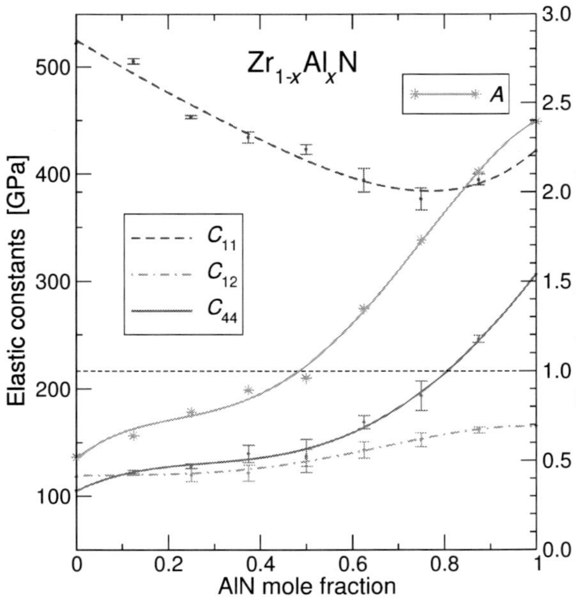

Figure 6. Single crystal elastic constants C_{11}, C_{12}, and C_{44} of cubic $Zr_{1-x}Al_xN$ alloy. A is the Zener's anisotropy ratio.

response. The *Zener's anisotropic ratio*, A,

$$A = \frac{2C_{44}}{C_{11} - C_{12}} \tag{17}$$

is a convenient measure of the degree of elastic isotropicity: $A = 1$, corresponds to a material with an ideal isotropic response. A of TiN, CrN, and ZrN is smaller than 1, while it is larger than 1 for AlN; consequently it is expected to reach $A = 1$ at a certain composition of these $TM_{1-x}Al_xN$ alloys. Indeed, this is the case of $Ti_{1-x}Al_xN$ for $x \approx 0.3$ [67], and $Zr_{1-x}Al_xN$ and $Cr_{1-x}Al_xN$ for $x \approx 0.5$ [45, 46]. For such compositions it is predicted that the microstructure of thin films (i.e., shape of grains) plays only a marginal role, especially when the effect of grain boundaries can be neglected.

Texture is described by *orientation distribution function* (ODF). Having such information at hand, the corresponding tensor of elastic constants is calculated as [70]

$$\widehat{C}_{ij} = \int_{\phi_1,\phi_2,\phi_3} \mathrm{ODF}(\phi_1, \phi_2, \phi_3) C_{ij}(\phi_1, \phi_2, \phi_3)\, d\phi_1\, d\phi_2\, d\phi_3 \,, \tag{18}$$

where $C_{ij}(\phi_1, \phi_2, \phi_3)$ is the single crystal elastic constants tensor corresponding to a grain rotated by the three Euler angles, ϕ_1, ϕ_2, and ϕ_3, with respect to a global coordination system. Fig. 7 shows an example of the texture effect on the directional Young's modulus of $Zr_{1-x}Al_xN$ as a function of the sharpness of the fibre texture (measured by the full width at half maximum (FWHT) of the ODF), for a texture composed of 50% of completely randomly oriented grains and 50% of fibres oriented according to the ODF. The effect of the texture diminishes for $x \approx 0.5$. This is indeed the composition previously predicted to behave isotropically, hence having the elastic response independent of the microstructure.

Consequently, texture is an important factor for correct evaluation and discussion of e.g., nanoindentation results of polycrystalline thin films with respect to the *ab initio* calculated data. The reader is referred to Ref. [46] for an in-depth discussion on this topic.

5. Electronic Properties

Electronic structure, such as charge distribution or density of states is the basic output of quantum mechanical *ab initio* calculations. Hence it is not surprising that a lot has been reported in the literature. For example, results summarised in Refs. [12, 22, 31, 35, 50, 53, 57, 58, 71] allowed to discuss alloying effects on bonding, strength, thermal stability and elastic properties in the $X_{1-x}Al_xN$ (X =Ti, Cr, Zr, Hf, Ta, Sc), $Ti_{1-x}Ta_xN$, $Ti_{1-x-y}Al_xY_yN$ and $Ti_{1-x-y}Al_xNb_yN$ systems. In this section we therefore focus on alloy-related trends in *energy electron loss spectroscopy* (EELS), another aspect of the electronic structure calculations.

EELS is a powerful analytical experimental technique in transmission electron microscopy, allowing for a direct probing of the electronic structure. The intensity, $I(E, \theta)$, of *electron energy loss near edge structure* (ELNES) is [72]

$$I(E, \theta) \propto \sum_{i,f} \langle i | \exp(i\vec{q} \cdot \vec{r}) | f \rangle |^2\, \mathrm{JDOS}(E_i, E_f) \tag{19}$$

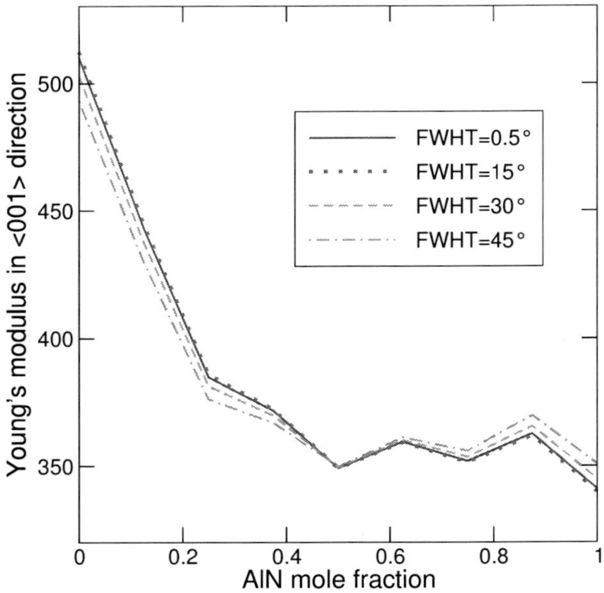

Figure 7. Young's modulus as a function of composition and full width at half maximum (FWHM) of the orientation distribution function.

where $|i\rangle$ and $|f\rangle$ are all allowed initial and final states, respectively, E_i and E_f are their respective energies, $E = E_f - E_i$ is the energy loss, θ is the scattering angle corresponding to the scattering vector $\vec{q}$, and JDOS is the joined density of states. Since the initial state $|i\rangle$ is well localised in energy, the JDOS can be replaced with density of states (DOS). The matrix element $|\langle i|\exp(i\vec{q}\cdot\vec{r})|f\rangle|^2$ gives the overall edge shape, while the fine structure of the edge comes from DOS, specifically from the (projected) density of unoccupied (final) states [72]. Although DFT is a ground state theory, using a concept of partially occupied final state (so called *core hole approach*) it is possible to obtain ELNES curves that are in excellent agreement with the measurements [73].

Figure 8 shows the compositional evolution of N K-edge (i.e., transitions from N $1s$ states to unoccupied $2p$ states) for the semiconducting $Al_{1-x}Ga_xN$ and metallic $Ti_{1-x}Al_xN$ alloys. The edge shape evolution of $Al_{1-x}Ga_xN$ seems to follow a smooth transition between the shapes of the two boundary system AlN and GaN, while an abrupt change in the shape takes place for the $Ti_{1-x}Al_xN$ alloy, as the phase transition from cubic to wurtzite phase happens at $x \approx 0.7$. The calculated curves (orange and blue) are confirmed by the experimental measurements (thick black lines). With this assurance one can use the calculated results to understand the origin of the individual peaks in the measured spectra. By comparing the N K-edge ELNES with projected density of states, Holec et al. [36] were, for example, able to show that the first peak in $Ti_{1-x}Al_xN$ N K-edge ELNES at about 4 eV above the edge onset comes from the interaction of the Ti d and N p states, while the peak at about 10 eV is a result of the interaction between Al p and N p states. In this particular case, the combination of modelling and experimental techniques provides a unique symbiosis to validate the calculations on one hand, and to interpret the experimental results on the other hand.

6. Conclusion

Atomistic modelling in general, and modelling of alloys in particular, are well established and indispensable parts of modern materials science. They allow for understanding and explaining experimental observations as well as for prediction of materials behaviour, thus enabling the computational design of materials. Density functional theory employs the quantum mechanical principles, and hence possesses an extremely accurate modelling technique. The drawback is that its hardware requirements limit its use to systems with maximum $\approx 10^2 - 10^3$ atoms. This becomes especially problematic for extended system and complex microstructures, an example of which are random solid solutions.

This chapter focused on special quasi-random structures, which is one of the possible approaches to deal with alloys. Using the example of pseudo-binary and pseudo-ternary transition metal–(aluminium)–nitrides, we have demonstrated the possibility for calculating structural, thermodynamic as well as electronic material trends. The lattice parameters always bow towards a larger value than a linear Vegard's-like interpolation. The phase stability is discussed in terms of the energy of formation, suggesting that the maximum solubility of Al on the metallic sublattice in cubic $Ti_{1-x-y}Al_xX_yN$ is not significantly affected by X =Zr, Nb, Hf, or Ta, while Y decreases it substantially. Mixing enthalpy is used to determine the driving force for the onset of the spinodal decomposition, as most of these systems are unstable. Single crystal elastic constants of $Zr_{1-x}Al_xN$ were calcu-

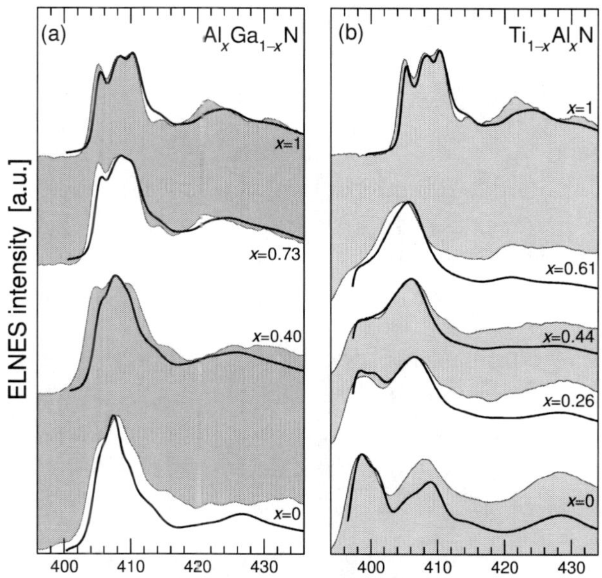

Figure 8. Compositional evaluation of N K-edge ELNES as predicted by *ab initio* calculations (thin colour lines) and confirmed by experiments (thick black lines) for (a) $Al_xGa_{1-x}N$ and (b) $Ti_{1-x}Al_xN$.

lated and implications for polycrystalline textured materials were discussed. The special quasi-random structures allowed us to identify the electronic origin of individual peaks in electron energy loss near edge structures. Last but not least, a comparison of the predicted trends with the extensive amount of experimental results available in the literature yields an excellent agreement.

In conclusion, we have demonstrated that alloying-related problems can be successfully addressed within the state-of-the-art density functional theory using the special quasi-random structures.

References

[1] S. PalDey and S. C. Deevi. Single layer and multilayer wear resistant coatings of (Ti,Al)N: a review. *Materials Science and Engineering A*, 342(1-2):58–79, February 2003. ISSN 09215093. doi: 10.1016/S0921-5093(02)00259-9. URL `http://linkinghub.elsevier.com/retrieve/pii/S0921509302002599`.

[2] Paul H. Mayrhofer, Christian Mitterer, Lars Hultman, and Helmut Clemens. Microstructural design of hard coatings. *Progress in Materials Science*, 51 (8):1032–1114, November 2006. ISSN 00796425. doi: 10.1016/j.pmatsci.2006.02.002. URL `http://linkinghub.elsevier.com/retrieve/pii/S0079642506000119`.

[3] Paul H. Mayrhofer, Anders Hörling, Lennart Karlsson, Jacob Sjölén, Tommy Larsson, Christian Mitterer, and Lars Hultman. Self-organized nanostructures in the TiAlN system. *Applied Physics Letters*, 83(10):2049–2051, 2003. ISSN 00036951. doi: 10.1063/1.1608464. URL `http://link.aip.org/link/APPLAB/v83/i10/p2049/s1\&Agg=doi`.

[4] Richard Rachbauer, S. Massl, E. Stergar, David Holec, Daniel Kiener, Jozef Keckes, J. Patscheider, M. Stiefel, Harald Leitner, and Paul H. Mayrhofer. Decomposition pathways in age hardening of Ti-Al-N films. *Journal of Applied Physics*, 110(2): 023515, 2011. ISSN 00218979. doi: 10.1063/1.3610451. URL `http://link.aip.org/link/JAPIAU/v110/i2/p023515/s1\&Agg=doi`.

[5] U. Wahlström, Lars Hultman, J.-E. Sundgren, F. Adibi, Ivan Petrov, and J.E. Greene. Crystal growth and microstructure of polycrystalline Ti1xAlxN alloy films deposited by ultra-high-vacuum dual-target magnetron sputtering. *Thin Solid Films*, 235(1-2): 62–70, November 1993. ISSN 00406090. doi: 10.1016/0040-6090(93)90244-J. URL `http://dx.doi.org/10.1016/0040-6090(93)90244-J`.

[6] J.Y. Rauch, C Rousselot, and N Martin. Structure and composition of TixAl1xN thin films sputter deposited using a composite metallic target. *Surface and Coatings Technology*, 157(2-3):138–143, August 2002. ISSN 02578972. doi: 10.1016/S0257-8972(02)00146-9. URL `http://linkinghub.elsevier.com/retrieve/pii/S0257897202001469`.

[7] Sang-Hyeob Lee, Byoung-June Kim, Hyung-Hwan Kim, and Jung-Joong Lee. Structural analysis of AlN and (Ti1XAlX)N coatings made by plasma enhanced chemical vapor deposition. *Journal of Applied Physics*, 80(3):1469, 1996. ISSN 00218979. doi: 10.1063/1.363015. URL http://link.aip.org/link/JAPIAU/v80/i3/p1469/s1\&Agg=doi.

[8] R Prange, R Cremer, and D Neuschütz. Plasma-enhanced CVD of (Ti,Al)N films from chloridic precursors in a DC glow discharge. *Surface and Coatings Technology*, 133-134:208–214, November 2000. ISSN 02578972. doi: 10.1016/S0257-8972(00)00941-5. URL http://linkinghub.elsevier.com/retrieve/pii/S0257897200009415.

[9] A.E. Reiter, V.H. Derflinger, B. Hanselmann, T. Bachmann, and B. Sartory. Investigation of the properties of Al1xCrxN coatings prepared by cathodic arc evaporation. *Surface and Coatings Technology*, 200(7):2114–2122, December 2005. ISSN 02578972. doi: 10.1016/j.surfcoat.2005.01.043. URL http://linkinghub.elsevier.com/retrieve/pii/S0257897205000976.

[10] Hiroyuki Hasegawa, Masahiro Kawate, and Tetsuya Suzuki. Effects of Al contents on microstructures of Cr1XAlXN and Zr1XAlXN films synthesized by cathodic arc method. *Surface and Coatings Technology*, 200(7):2409–2413, December 2005. ISSN 02578972. doi: 10.1016/j.surfcoat.2004.08.208. URL http://linkinghub.elsevier.com/retrieve/pii/S0257897204008655.

[11] Li Chen, Martin Moser, Yong Du, and Paul H. Mayrhofer. Compositional and structural evolution of sputtered Ti-Al-N. *Thin Solid Films*, 517(24):6635–6641, October 2009. ISSN 00406090. doi: 10.1016/j.tsf.2009.04.056. URL http://linkinghub.elsevier.com/retrieve/pii/S0040609009008633.

[12] Björn Alling, A. Karimi, and Igor A. Abrikosov. Electronic origin of the isostructural decomposition in cubic M1xAlxN (M=Ti, Cr, Sc, Hf): A first-principles study. *Surface and Coatings Technology*, 203(5-7):883–886, December 2008. ISSN 02578972. doi: 10.1016/j.surfcoat.2008.08.027. URL http://linkinghub.elsevier.com/retrieve/pii/S0257897208007330.

[13] Paul H. Mayrhofer, Franz D. Fischer, H.J. Böhm, Christian Mitterer, and J.M. Schneider. Energetic balance and kinetics for the decomposition of supersaturated Ti1xAlxN. *Acta Materialia*, 55(4):1441–1446, February 2007. ISSN 13596454. doi: 10.1016/j.actamat.2006.09.045. URL http://linkinghub.elsevier.com/retrieve/pii/S1359645406007312.

[14] S.H. Sheng, R.F. Zhang, and S. Veprek. Phase stabilities and thermal decomposition in the Zr1xAlxN system studied by ab initio calculation and thermodynamic modeling. *Acta Materialia*, 56(5):968–976, March 2008. ISSN 13596454. doi: 10.1016/j.actamat.2007.10.050. URL http://linkinghub.elsevier.com/retrieve/pii/S1359645407007483.

[15] Björn Alling, A. Karimi, Lars Hultman, and Igor A. Abrikosov. First-principles study of the effect of nitrogen vacancies on the decomposition pattern in cubic Ti[sub 1x]Al[sub x]N[sub 1y]. *Applied Physics Letters*, 92(7):071903, 2008. ISSN 00036951. doi: 10.1063/1.2838747. URL http://link.aip.org/link/APPLAB/v92/i7/p071903/s1\&Agg=doi.

[16] R. Sanjinés, C.S. Sandu, R. Lamni, and F. Lévy. Thermal decomposition of Zr1-xAlxN thin films deposited by magnetron sputtering. *Surface and Coatings Technology*, 200(22-23):6308–6312, June 2006. ISSN 02578972. doi: 10.1016/j.surfcoat.2005.11.113. URL http://linkinghub.elsevier.com/retrieve/pii/S0257897205013034.

[17] Björn Alling, A.V. Ruban, A. Karimi, O. Peil, S. Simak, Lars Hultman, and Igor A. Abrikosov. Mixing and decomposition thermodynamics of c-Ti1xAlxN from first-principles calculations. *Physical Review B*, 75(4):045123, January 2007. ISSN 1098-0121. doi: 10.1103/PhysRevB.75.045123. URL http://link.aps.org/doi/10.1103/PhysRevB.75.045123.

[18] Igor A. Abrikosov, Axel Knutsson, Björn Alling, Ferenc Tasnádi, Hans Lind, Lars Hultman, and Magnus Odén. Phase Stability and Elasticity of TiAlN. *Materials*, 4(9):1599–1618, September 2011. ISSN 1996-1944. doi: 10.3390/ma4091599. URL http://www.mdpi.com/1996-1944/4/9/1599/.

[19] Paul H. Mayrhofer, D. Music, and J. M. Schneider. Ab initio calculated binodal and spinodal of cubic Ti[sub 1x]Al[sub x]N. *Applied Physics Letters*, 88(7):071922, 2006. ISSN 00036951. doi: 10.1063/1.2177630. URL http://link.aip.org/link/APPLAB/v88/i7/p071922/s1\&Agg=doi.

[20] Richard Rachbauer, Jamie J. Gengler, Andrey A. Voevodin, Katharina Resch, and Paul H. Mayrhofer. Temperature driven evolution of thermal, electrical, and optical properties of TiAlN coatings. *Acta Materialia*, 60(5):2091–2096, March 2012. ISSN 13596454. doi: 10.1016/j.actamat.2012.01.005. URL http://linkinghub.elsevier.com/retrieve/pii/S1359645412000328.

[21] Florian Rovere, D. Music, S. Ershov, Moritz to Baben, H.-G. Fuss, Paul H. Mayrhofer, and J. M. Schneider. Experimental and computational study on the phase stability of Al-containing cubic transition metal nitrides. *Journal of Physics D: Applied Physics*, 43(3):035302, January 2010. ISSN 0022-3727. doi: 10.1088/0022-3727/43/3/035302. URL http://stacks.iop.org/0022-3727/43/i=3/a=035302?key=crossref.037a9860c0f6d3d9179eaf33ab7db121.

[22] David Holec, Richard Rachbauer, Li Chen, Lan Wang, Doris Luef, and Paul H. Mayrhofer. Phase stability and alloy-related trends in TiAlN, ZrAlN and HfAlN systems from first principles. *Surface and Coatings Technology*, 206(7):1698–1704, September 2011. ISSN 02578972. doi: 10.1016/j.surfcoat.2011.09.019. URL http://linkinghub.elsevier.com/retrieve/pii/S0257897211008942http://www.sciencedirect.com/science/article/pii/S0257897211008942.

[23] Li Chen, Jörg Paulitsch, Yong Du, and Paul H. Mayrhofer. Thermal stability and oxidation resistance of Ti-Al-N coatings. *Surface and Coatings Technology*, December 2011. ISSN 02578972. doi: 10.1016/j.surfcoat.2011.12.028. URL http://linkinghub.elsevier.com/retrieve/pii/S0257897211012382.

[24] Moritz to Baben, Leonard Raumann, Denis Music, and Jochen M Schneider. Origin of the nitrogen over- and understoichiometry in Ti(0.5)Al(0.5)N thin films. *Journal of Physics. Condensed matter*, 24(15):155401, April 2012. ISSN 1361-648X. doi: 10.1088/0953-8984/24/15/155401. URL http://www.ncbi.nlm.nih.gov/pubmed/22436621.

[25] Anders Hörling, Lars Hultman, Magnus Odén, Jacob Sjölén, and Lennart Karlsson. Thermal stability of arc evaporated high aluminum-content Ti[sub 1x]Al[sub x]N thin films. *Journal of Vacuum Science & Technology A: Vacuum, Surfaces, and Films*, 20 (5):1815, 2002. ISSN 07342101. doi: 10.1116/1.1503784. URL http://link.aip.org/link/JVTAD6/v20/i5/p1815/s1\&Agg=doi.

[26] K. Kutschej, Paul H. Mayrhofer, M. Kathrein, Peter Polcik, R. Tessadri, and Christian Mitterer. Structure, mechanical and tribological properties of sputtered Ti1xAlxN coatings with 0.5x0.75. *Surface and Coatings Technology*, 200(7):2358–2365, December 2005. ISSN 02578972. doi: 10.1016/j.surfcoat.2004.12.008. URL http://linkinghub.elsevier.com/retrieve/pii/S025789720401223X.

[27] H. Lind, R. Forsén, Björn Alling, N. Ghafoor, Ferenc Tasnádi, Mats P. Johansson, Igor A. Abrikosov, and Magnus Odén. Improving thermal stability of hard coating films via a concept of multicomponent alloying. *Applied Physics Letters*, 99(9):091903, 2011. ISSN 00036951. doi: 10.1063/1.3631672. URL http://link.aip.org/link/APPLAB/v99/i9/p091903/s1\&Agg=doi.

[28] Martin Moser and Paul H. Mayrhofer. Yttrium-induced structural changes in sputtered Ti1xAlxN thin films. *Scripta Materialia*, 57(4):357–360, August 2007. ISSN 13596462. doi: 10.1016/j.scriptamat.2007.04.019. URL http://linkinghub.elsevier.com/retrieve/pii/S1359646207002989.

[29] Martin Moser, Daniel Kiener, Christina Scheu, and Paul H. Mayrhofer. Influence of Yttrium on the Thermal Stability of Ti-Al-N Thin Films. *Materials*, 3(3):1573–1592, March 2010. ISSN 1996-1944. doi: 10.3390/ma3031573. URL http://www.mdpi.com/1996-1944/3/3/1573/.

[30] Richard Rachbauer, Andreas Blutmager, David Holec, and Paul H. Mayrhofer. Effect of Hf on structure and age hardening of Ti-Al-N thin films. *Surface and Coatings Technology*, 206(10):2667–2672, November 2011. ISSN 02578972. doi: 10.1016/j.surfcoat.2011.11.020. URL http://linkinghub.elsevier.com/retrieve/pii/S025789721101156X.

[31] Richard Rachbauer, David Holec, Martina Lattemann, Lars Hultman, and Paul H. Mayrhofer. Electronic origin of structure and mechanical properties in Y and

Nb alloyed Ti-Al-N thin films. *International Journal of Materials Research (formerly Zeitschrift für Metallkunde)*, 102(06):735–742, June 2011. ISSN 18625282. doi: 10.3139/146.110520. URL http://www.ijmr.de/directlink.asp?MK110520.

[32] Richard Rachbauer, David Holec, and Paul H. Mayrhofer. Phase stability and decomposition products of TiAlTaN thin films. *Applied Physics Letters*, 97(15):151901, 2010. ISSN 00036951. doi: 10.1063/1.3495783. URL http://link.aip.org/link/APPLAB/v97/i15/p151901/s1\&Agg=doi.

[33] Richard Rachbauer, David Holec, and Paul H. Mayrhofer. Increased thermal stability of TiAlN thin films by Ta alloying. *Surface and Coatings Technology*, pages 6–11, July 2011. ISSN 02578972. doi: 10.1016/j.surfcoat.2011.07.009. URL http://linkinghub.elsevier.com/retrieve/pii/S0257897211006931.

[34] M. Pfeiler, Christina Scheu, H. Hutter, J. Schnoller, C. Michotte, Christian Mitterer, and M. Kathrein. On the effect of Ta on improved oxidation resistance of TiAlTaN coatings. *Journal of Vacuum Science & Technology A: Vacuum, Surfaces, and Films*, 27(3):554, 2009. ISSN 07342101. doi: 10.1116/1.3119671. URL http://link.aip.org/link/JVTAD6/v27/i3/p554/s1\&Agg=doi.

[35] Florian Rovere, D. Music, J.M. Schneider, and Paul H. Mayrhofer. Experimental and computational study on the effect of yttrium on the phase stability of sputtered CrAlYN hard coatings. *Acta Materialia*, 58(7):2708–2715, April 2010. ISSN 13596454. doi: 10.1016/j.actamat.2010.01.005. URL http://linkinghub.elsevier.com/retrieve/pii/S135964541000008X.

[36] David Holec, Richard Rachbauer, Daniel Kiener, Peter D. Cherns, Pedro M. F. J. Costa, Clifford McAleese, Paul H. Mayrhofer, and Colin J. Humphreys. Towards predictive modeling of near-edge structures in electron energy-loss spectra of AlN-based ternary alloys. *Physical Review B*, 83(16):165122, April 2011. ISSN 1098-0121. doi: 10.1103/PhysRevB.83.165122. URL http://link.aps.org/doi/10.1103/PhysRevB.83.165122.

[37] L. Nordheim. The electron theory of metals. *Ann. Phys.*, 9:607–40, 641–78, 1931.

[38] B. Velický, S. Kirkpatrick, and H. Ehrenreich. Single-Site Approximations in the Electronic Theory of Simple Binary Alloys. *Physical Review*, 175(3):747–766, November 1968. ISSN 0031-899X. doi: 10.1103/PhysRev.175.747. URL http://link.aps.org/doi/10.1103/PhysRev.175.747.

[39] Paul Soven. Coherent-Potential Model of Substitutional Disordered Alloys. *Physical Review*, 156(3):809–813, April 1967. ISSN 0031-899X. doi: 10.1103/PhysRev.156.809. URL http://link.aps.org/doi/10.1103/PhysRev.156.809.

[40] S.-H. Wei, L. Ferreira, James Bernard, and Alex Zunger. Electronic properties of random alloys: Special quasirandom structures. *Physical Review B*, 42(15):9622–9649, November 1990. ISSN 0163-1829. doi: 10.1103/PhysRevB.42.9622. URL http://link.aps.org/doi/10.1103/PhysRevB.42.9622.

[41] A.V. Ruban and Igor A. Abrikosov. Configurational thermodynamics of alloys from first principles: effective cluster interactions. *Reports on Progress in Physics*, 71(4):046501, April 2008. ISSN 0034-4885. doi: 10.1088/0034-4885/71/4/046501. URL http://stacks.iop.org/0034-4885/71/i=4/a=046501?key=crossref.66d82c93d17391970d459423471d35ed.

[42] Joseph E. Mayer and Elliott Montroll. Molecular Distribution. *The Journal of Chemical Physics*, 9(1):2, 1941. ISSN 00219606. doi: 10.1063/1.1750822. URL http://link.aip.org/link/JCPSA6/v9/i1/p2/s1\&Agg=doi.

[43] J.M. Sanchez, F. Ducastelle, and D. Gratias. Generalized cluster description of multicomponent systems. *Physica A: Statistical Mechanics and its Applications*, 128(1-2): 334–350, November 1984. ISSN 03784371. doi: 10.1016/0378-4371(84)90096-7. URL http://www.sciencedirect.com/science/article/B6TVG-46DF7T0-N7/1/74b5907d6528c21f1108536439556baahttp://linkinghub.elsevier.com/retrieve/pii/0378437184900967.

[44] B. Alling, T. Marten, I. a. Abrikosov, and a. Karimi. Comparison of thermodynamic properties of cubic Cr[sub 1x]Al[sub x]N and Ti[sub 1x]Al[sub x]N from first-principles calculations. *Journal of Applied Physics*, 102(4):044314, 2007. ISSN 00218979. doi: 10.1063/1.2773625. URL http://link.aip.org/link/JAPIAU/v102/i4/p044314/s1\&Agg=doi.

[45] Liangcai Zhou, David Holec, and Paul H. Mayrhofer. First-principles study of elastic properties of Cr-Al-N. *submitted to JAP*, pages 1–24, 2012.

[46] David Holec, Peter Wagner, Ferenc Tasnádi, Martin Friák, and Paul H. Mayrhofer. Elastic constants of ZrAlN alloy. *in preparation*, 2013.

[47] David Holec. *Multi-scale modelling of III-nitrides: Selected topics from dislocations to the electronic structure*. VDM Verlag, Saarbrücken, Germany, 2010. ISBN 978-3-639-22941-7.

[48] F. Birch. Finite elastic strain of cubic crystals. *Physical Review*, 71(11):809–824, 1947. URL http://prola.aps.org/abstract/PR/v71/i11/p809_1.

[49] Paul H. Mayrhofer, D. Music, and J. M. Schneider. Influence of the Al distribution on the structure, elastic properties, and phase stability of supersaturated Ti[sub 1x]Al[sub x]N. *Journal of Applied Physics*, 100(9):094906, 2006. ISSN 00218979. doi: 10.1063/1.2360778. URL http://link.aip.org/link/JAPIAU/v100/i9/p094906/s1\&Agg=doi.

[50] Paul H. Mayrhofer, D. Music, Th. Reeswinkel, H.-G. Fuß, and J.M. Schneider. Structure, elastic properties and phase stability of Cr1xAlxN. *Acta Materialia*, 56(11):2469–2475, June 2008. ISSN 13596454. doi: 10.1016/j.actamat.2008.01.054. URL http://linkinghub.elsevier.com/retrieve/pii/S1359645408000736.

[51] Björn Alling, Tobias Marten, and Igor A. Abrikosov. Questionable collapse of the bulk modulus in CrN. *Nature materials*, 9(4):283–4; author reply 284, April 2010. ISSN 1476-1122. doi: 10.1038/nmat2722. URL http://www.ncbi.nlm.nih.gov/pubmed/20332781.

[52] David Holec, Florian Rovere, Paul H. Mayrhofer, and Péter B. Barna. Pressure-dependent stability of cubic and wurtzite phases within the TiN-AlN and CrN-AlN systems. *Scripta Materialia*, 62(6):349–352, March 2010. ISSN 13596462. doi: 10.1016/j.scriptamat.2009.10.040. URL http://linkinghub.elsevier.com/retrieve/pii/S1359646209006915.

[53] David Holec, Robert Franz, Paul H. Mayrhofer, and Christian Mitterer. Structure and stability of phases within the NbN-AlN system. *J. Phys. D: Appl. Phys.*, 43(14):145403, 2010. URL http://www.scopus.com/inward/record.url?eid=2-s2.0-77949888033\&partnerID=40\&md5=f66adeb2ab7564c8701b76522224f13d.

[54] David Holec, Liangcai Zhou, Richard Rachbauer, and Paul H Mayrhofer. Alloying-related trends from first principles : An application to the Ti–Al–X–N system. *submitted (2013)*, pages 1–9, 2013.

[55] Paul H. Mayrhofer, Richard Rachbauer, and David Holec. Influence of Nb on the phase stability of TiAlN. *Scripta Materialia*, 63(8):807–810, October 2010. ISSN 13596462. doi: 10.1016/j.scriptamat.2010.06.020. URL http://linkinghub.elsevier.com/retrieve/pii/S1359646210004057.

[56] Björn Alling, Magnus. Odén, Lars Hultman, and Igor A. Abrikosov. Pressure enhancement of the isostructural cubic decomposition in Ti[sub 1x]Al[sub x]N. *Applied Physics Letters*, 95(18):181906, 2009. ISSN 00036951. doi: 10.1063/1.3256196. URL http://link.aip.org/link/APPLAB/v95/i18/p181906/s1\&Agg=doi.

[57] L. E. Koutsokeras, G. Abadias, Ch. E. Lekka, G. M. Matenoglou, D. F. Anagnostopoulos, G. a. Evangelakis, and P. Patsalas. Conducting transition metal nitride thin films with tailored cell sizes: The case of δ-Ti[sub x]Ta[sub 1x]N. *Applied Physics Letters*, 93(1):011904, 2008. ISSN 00036951. doi: 10.1063/1.2955838. URL http://link.aip.org/link/APPLAB/v93/i1/p011904/s1\&Agg=doi.

[58] L. E. Koutsokeras, G. Abadias, and P. Patsalas. Texture and microstructure evolution in single-phase TixTa1xN alloys of rocksalt structure. *Journal of Applied Physics*, 110 (4):043535, 2011. ISSN 00218979. doi: 10.1063/1.3622585. URL http://link.aip.org/link/JAPIAU/v110/i4/p043535/s1\&Agg=doi.

[59] H. Willmann, P.H. Mayrhofer, P.O.Å. Persson, a.E. Reiter, L. Hultman, and C. Mitterer. Thermal stability of AlCrN hard coatings. *Scripta Materialia*, 54(11):1847–1851, June 2006. ISSN 13596462. doi: 10.1016/j.scriptamat.2006.02.023. URL http://linkinghub.elsevier.com/retrieve/pii/S1359646206001588.

[60] Martin Friák, Tilmann Hickel, Blazej Grabowski, Liverios Lymperakis, A. Udyansky, A. Dick, D. Ma, F. Roters, L. F. Zhu, A. Schlieter, U. Kühn, Z. Ebrahimi, R.A. Lebensohn, David Holec, J. Eckert, H. Emmerich, D. Raabe, and Jörg Neugebauer. Methodological challenges in combining quantum-mechanical and continuum approaches for materials science applications. *The European Physical Journal Plus*, 126(101):1–22, October 2011. ISSN 2190-5444. doi: 10.1140/epjp/i2011-11101-2. URL `http://www.springerlink.com/index/10.1140/epjp/i2011-11101-2`.

[61] David Holec, Martin Friák, Jörg Neugebauer, and Paul H. Mayrhofer. Trends in the elastic response of binary early transition metal nitrides. *Physical Review B*, 85(6): 064101, February 2012. ISSN 1098-0121. doi: 10.1103/PhysRevB.85.064101. URL `http://link.aps.org/doi/10.1103/PhysRevB.85.064101`.

[62] Michal Lopuszynski and Jacek A Majewski. Ab initio calculations of third-order elastic constants and related properties for selected semiconductors. *Physical Review B*, 76(4):045202, July 2007. ISSN 1098-0121. doi: 10.1103/PhysRevB.76.045202. URL `http://link.aps.org/doi/10.1103/PhysRevB.76.045202`.

[63] J. Zhao, J. M. Winey, and Y. M. Gupta. First-principles calculations of second- and third-order elastic constants for single crystals of arbitrary symmetry. *Physical Review B*, 75(9):094105, March 2007. ISSN 1098-0121. doi: 10.1103/PhysRevB.75.094105. URL `http://link.aps.org/doi/10.1103/PhysRevB.75.094105`.

[64] J.F. Nye. *Physical properties of crystals*. Clarendon Press, Oxford, first edition, 1957. URL `http://scholar.google.com/scholar?hl=en\&btnG=Search\&q=intitle:physical+properties+of+crystals\#0`.

[65] R. Yu, J. Zhu, and H.Q. Ye. Calculations of single-crystal elastic constants made simple. *Computer Physics Communications*, 181(3):671–675, March 2010. ISSN 00104655. doi: 10.1016/j.cpc.2009.11.017. URL `http://linkinghub.elsevier.com/retrieve/pii/S0010465509003932`.

[66] Ferenc Tasnádi, M. Odén, and Igor a. Abrikosov. Ab initio elastic tensor of cubic Ti_{0.5}Al_{0.5}N alloys: Dependence of elastic constants on size and shape of the supercell model and their convergence. *Physical Review B*, 85(14):1–9, April 2012. ISSN 1098-0121. doi: 10.1103/PhysRevB.85.144112. URL `http://link.aps.org/doi/10.1103/PhysRevB.85.144112`.

[67] Ferenc Tasnádi, Igor A. Abrikosov, Lina Rogström, Jonathan Almer, Mats P. Johansson, and Magnus Odén. Significant elastic anisotropy in Ti[sub 1x]Al[sub x]N alloys. *Applied Physics Letters*, 97(23):231902, 2010. ISSN 00036951. doi: 10.1063/1.3524502. URL `http://link.aip.org/link/APPLAB/v97/i23/p231902/s1\&Agg=doi`.

[68] W. Voigt. *Lehrbuch der Kristallphysik*. B. B. Teubner, Leipzig, 1928.

[69] A. Reuss. Berechnung der Fließgrenze von Mischkristallen auf Grund der Plastizitätsbedingung für Einkristalle . *Zeitschrift für Angewandte Mathematik und*

Mechanik, 9(1):49–58, 1929. ISSN 00442267. doi: 10.1002/zamm.19290090104. URL http://doi.wiley.com/10.1002/zamm.19290090104.

[70] K J Martinschitz, R Daniel, C Mitterer, and J Keckes. Elastic constants of fibre-textured thin films determined by X-ray diffraction. *Journal of applied crystallography*, 42(Pt 3):416–428, June 2009. ISSN 0021-8898. doi: 10.1107/S0021889809011807. URL http://www.pubmedcentral.nih.gov/articlerender.fcgi?artid=3246820\&tool=pmcentrez\&rendertype=abstract.

[71] D.G. Sangiovanni, L. Hultman, and V. Chirita. Supertoughening in B1 transition metal nitride alloys by increased valence electron concentration. *Acta Materialia*, 59(5):2121–2134, March 2011. ISSN 13596454. doi: 10.1016/j.actamat.2010.12.013. URL http://linkinghub.elsevier.com/retrieve/pii/S1359645410008347.

[72] V. J. Keast, A. J. Scott, R. Brydson, D. B. Williams, and J. Bruley. Electron energy-loss near-edge structure – a tool for the investigation of electronic structure on the nanometre scale. *Journal of microscopy*, 203(Pt 2):135–175, August 2001. ISSN 0022-2720. URL http://www.ncbi.nlm.nih.gov/pubmed/11489072.

[73] C. Hébert. Practical aspects of running the WIEN2k code for electron spectroscopy. *Micron*, 38(1):12–28, January 2007. ISSN 0968-4328. doi: 10.1016/j.micron.2006.03.010. URL http://www.ncbi.nlm.nih.gov/pubmed/16914318.

In: Density Functional Theory
Editors: J. Morin and J.M. Pelletier, pp. 285-310
ISBN: 978-1-62417-954-9

Chapter 10

ENERGY DENSITY FUNCTIONAL THEORY IN NUCLEAR PHYSICS

***Yoritaka Iwata*[1*] *and Joachim A. Maruhn*[2]**
[1]GSI Helmholtzzentrum für Schwerionenforschung, Germany
[2]Institut für Theoretische Physik, Universität Frankfurt, Germany

Abstract

The fundamentals with regard to energy density functional theory and nuclear physics are studied in this chapter. Our attention is given to the rigorous mathematical treatment involved in deriving the energy density functional theory. There are specific features that depict this density functional when studying many-nucleon systems, which are quite different from its use in many-electron systems. The intended audience for density functional theory research are physicists, chemists, and mathematicians. In particular, it is also intended for those eager to begin studying the density functional theory.

PACS 21.60.Jz, 21.30.-x, 13.75.Cs

1. Introduction

Among the several methods used in nuclear theory, the density functional method provides one of the most widely applicable treatments for many-nucleon systems. For example, the nucleus is a many-nucleon system, where more than 300 stable nuclei are already known. All atoms contain nuclei, and the chemical properties are identified by the number of protons included in the nucleus (cf. chemical elements: H, He, Li, $\cdots$). It follows that the nucleus is one of the most important ingredient discovered in our universe.

The theory based on the nucleonic degrees of freedom is presented, where the nucleon is not an elementary particle but consists of quarks and gluons. The validity of this theory is based on the fact that nucleons are quite stable quantum entities that can be regarded as effective physical units. Since nucleons are fermions, there are several common features relating this many-fermion system with other physical systems such as in many-electron

*E-mail address: y.iwata@gsi.de

systems and so on. However, there are several specific features in nuclear energy density functional theory. Indeed, nucleons, which have isospin and spin degrees of freedom, interact by two completely different forces called nuclear and Coulomb forces. In addition, it is worth noting that reactions between nuclei are not similar to reactions in the other physical systems to a large extent. Depending on the isospin degrees of freedom, there are two kinds of nucleons; i.e., protons and neutrons. While we understand the Coulomb force well, much is still not known about the nuclear force. Indeed, although the quantum chromodynamics (QCD) Lagrangian has already been established, its connection to the nuclear force is still developing. It makes many-nucleon system research quite difficult but also very fascinating.

From a scientific point of view, many-nucleon research is associated with clarifying the origin, existence, structure and reaction of chemical elements, where the origin of elements heavier than iron has not been understood well. It is an attempt to understand the time evolution of our universe with respect to the constituent chemical elements; "How and where were all the chemical elements created and why do they exist as they are ?" Chemical elements which do not naturally exist on earth are called superheavy elements. These elements are artificially synthesized in the laboratory. Such superheavy element research is related to clarifying the existence limit of chemical elements.

Finite-body quantum systems are often investigated in many-nucleon systems research. Although the spherical shape is expected to be energetically favored if the system is governed simply by surface tension; the nucleus, itself, has experimentally been shown to have several shapes: e.g., spherical, prolate, oblate shapes, and so on. This kind of research is associated with the structure of nuclei.

It is reasonable to have a unified theoretical framework describing both stationary and non-stationary states. The nuclear density functional theory is a possible candidate. The basic equation of the many-nucleon system is the Schrödinger equation containing the Hamiltonian. Based on the independent particle motion, a possible form of the Hamiltonian is generally provided by

$$H = -\sum_{i=1}^{A} \frac{\hbar^2}{2m} \triangle_i + \sum \sum_{i<j} v_{i,j} + \sum \sum \sum_{i<j<k} v_{ijk} \cdots .$$

The first term of the right-hand side arises from the kinetic energy, and the other terms given in the right-hand side are from the interaction energy. For the interaction part, the density functional theory in many-nucleon system is a theory that describes all the interaction by several densities (for many-electron systems, see Refs. [5, 6]). As already mentioned, the interaction part of the Hamiltonian is not perfectly known as far as the many-nucleon system is concerned. One important task in nuclear density functional theory is to find out the ultimate interaction, which can describe all the phenomena in the many-nucleon system. This point is indeed different from the study of many-electron systems. The process of finding the ultimate effective Hamiltonian mainly consists of two steps. In the first step, find an appropriate functional form of the effective Hamiltonian; and in the second step, find the best parameter sets.

In this chapter, our attention is to show the method that derives the effective interaction in many-nucleon systems. We present within this chapter a kind of modeling: the modeling of interacting femtometer-scale fermions. The nucleon-nucleon interaction is decomposed in Sec. 2. The effective Hamiltonian that arises from the nuclear force is discussed in Sec. 3. By starting from the zero-range nucleon-nucleon interaction (Sec. 3.1), we show a procedure to obtain the Hamiltonian density (Sec. 3.2). Then, the effective Hamiltonian is obtained by applying the variational principle (Sec.3.3). The effective Hamiltonian that arises from additional forces is briefly discussed in Sec. 4. In particular, a rigorous mathematical treatment is provided that applies the variational principle, where infinite-dimensional Hilbert/Banach spaces are considered. For the mathematics used in this chapter, refer to the textbooks of functional analysis such as Ref. [14].

2. Nucleon-Nucleon Interaction

There are two kinds of nucleons (isospin degree of freedom τ): these protons and neutrons are fermions. In addition, the nucleon also possesses a spin degree of freedom. Therefore, our interest will be in the many-fermion system will involve both spin and isospin degrees of freedom.

Nucleons interact by two kinds of force: nucleon-nucleon interaction V is represented by the sum of the nuclear potential V_K, the Coulomb potential V_C and the pairing potential V_{pair}:

$$V = V_K + V_C + V_{\text{pair}}. \tag{1}$$

Only protons interact by the Coulomb force. The two kinds of nucleons form a bound system that is called the atomic nucleus. The atomic nucleus is a finite quantum system containing Z protons and N neutrons, where $A = Z + N$ is called mass number. One goal in this theory is to determine for which combinations Z, N can exist.

3. Nuclear Force

Let us begin with the primitive picture of the nuclear force using the potential description. The potential of nuclear force depends on the positions, momenta, spins ($\boldsymbol{\sigma}_i = \pm\frac{1}{2}$) and isospins ($\boldsymbol{\tau}_i = \pm\frac{1}{2}$) of the two nucleons:

$$V_K = V_K(\boldsymbol{r}_i, \boldsymbol{r}_j, \boldsymbol{p}_i, \boldsymbol{p}_j, \boldsymbol{\sigma}_i, \boldsymbol{\sigma}_j, \boldsymbol{\tau}_i, \boldsymbol{\tau}_j),$$

where i and j are the indices that identify the two nucleons. It is rational to determine interactions obeying invariance, because the physical law must be invariant regardless of the observers. The following invariance are required for V_K: translational invariance, Galilean invariance, rotational invariance, isospin invariance, parity invariance, and time-reversal invariance (for each invariance, see Sec. 7.1.1 of Ref. [4]). It is worth to note here, that the radial dependence of the function V_K cannot be deduced from invariance principles. Among several attempts to determine the radial dependence, H. Yukawa proposed the Yukawa potential:

$$V_Y(r) = \frac{e^{-\mu\gamma}}{\mu\gamma}$$

based on meson field theory, where $1/\mu$ is the Compton wavelength of the pion.

Historically, components of the nuclear force are represented by using the identity operator, $(\boldsymbol{\sigma}_i \cdot \boldsymbol{\sigma}_j)$, $(\boldsymbol{\tau}_i \cdot \boldsymbol{\tau}_j)$ and $(\boldsymbol{\sigma}_i \cdot \boldsymbol{\sigma}_j)(\boldsymbol{\tau}_i \cdot \boldsymbol{\tau}_j)$ multiplied by the spin- and isospin- independent ingredient. In the traditional formulation, such a nucleon-nucleon interaction is represented by using several exchange operators:

$$\begin{aligned} V_K &= V_{\mathrm{W}}(\boldsymbol{r}) + V_{\mathrm{M}}(\boldsymbol{r})P_r + V_{\mathrm{B}}(\boldsymbol{r})P_\sigma - V_{\mathrm{H}}(\boldsymbol{r})P_\tau \\ &= V_{\mathrm{W}}(\boldsymbol{r}) + V_{\mathrm{M}}(\boldsymbol{r})P_r + V_{\mathrm{B}}(\boldsymbol{r})P_\sigma + V_{\mathrm{H}}(\boldsymbol{r})P_r P_\sigma, \end{aligned} \tag{2}$$

where P_r, $P_\sigma = \frac{1}{2}(1+\boldsymbol{\sigma}_i\boldsymbol{\sigma}_j)$ and $P_\tau = \frac{1}{2}(1+\boldsymbol{\tau}_i\boldsymbol{\tau}_j)$ are exchange operators for coordinates, spins, and isospins. In particular, the fermionic relation $P_r P_\sigma P_\tau = -1$ is used to replace the last term, and every term is represented without P_τ (for the property of fermions, refer to textbooks that cover quantum physics, such as Ref. [3]). The indices of the first, second, third, and fourth terms stand for Wigner, Majorana, Bartlett and Heisenberg, respectively. Another important ingredient is the tensor force represented by $(\boldsymbol{r}_i \cdot \boldsymbol{\sigma}_i)(\boldsymbol{r}_i \cdot \boldsymbol{\sigma}_j)$, where the tensor operator is

$$S_{ij} = \frac{3(\boldsymbol{r}_i \cdot \boldsymbol{\sigma}_i)(\boldsymbol{r}_i \cdot \boldsymbol{\sigma}_j)}{|\boldsymbol{r}_i|^2} - (\boldsymbol{\sigma}_i \cdot \boldsymbol{\sigma}_j).$$

Our aim is to have a density functional representation for the nuclear force (for the whole process, see Fig. 1), which includes the exchange properties represented by $(\boldsymbol{\sigma}_i \cdot \boldsymbol{\sigma}_j)$, $(\boldsymbol{\tau}_i \cdot \boldsymbol{\tau}_j)$, $(\boldsymbol{\sigma}_i \cdot \boldsymbol{\sigma}_j)(\boldsymbol{\tau}_i \cdot \boldsymbol{\tau}_j)$ and $(\boldsymbol{r}_i \cdot \boldsymbol{\sigma}_i)(\boldsymbol{r}_i \cdot \boldsymbol{\sigma}_j)$. The interaction described by spin and isospin degrees of freedom is specific to nuclear physics.

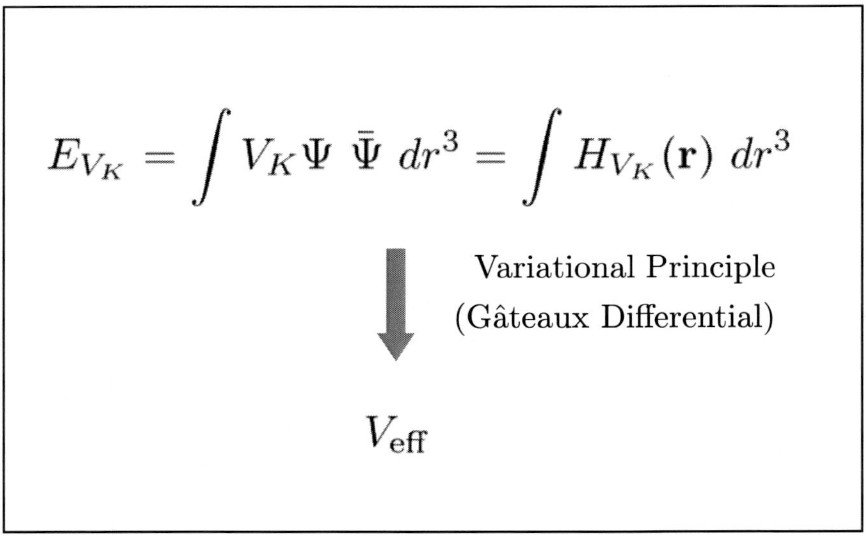

Figure 1. A process for obtaining the effective interaction. E_{V_K} denotes the interaction energy, V_K is the zero-range interaction, Ψ denotes a trial function, and $H_{V_K}(\mathrm{r})$ is the interaction part of the Hamiltonian density. V_{eff}, which is obtained by calculating the Gâteaux differential, is the effective interaction in the density functional formalism. The zero-range interaction V_K is presented in Sec. 3.1, the energy and the Hamiltonian density are calculated in Sec. 3.2, and the effective interaction is obtained in Sec.3.3.

3.1. The Zero-Range Force Formalism

For the purpose of obtaining the energy density functional theory in many-nucleon system, it is reasonable to begin with the nucleon-nucleon interaction of the following form:

$$V_K = \sum\sum_{i<j} v_{i,j} + \sum\sum\sum_{i<j<k} v_{i,j,k} \cdots , \tag{3}$$

where the first and second terms of the right-hand side are for the two-body (two-nucleon) and three-body (three-nucleon) interactions, respectively. In this formalism, all the interactions between each nucleon are summed up. With respect to the interaction used in density functional calculations, nucleon-nucleon interaction is usually considered up to a three-body term (cf. Skyrme interaction [10]).

In this context, special treatment (the short range approximation) is necessary to represent non-local potentials. First, the relative momentum between two nucleons, which is necessary to describe the interaction, is assumed and represented by

$$\boldsymbol{k} = \frac{1}{2i}(\nabla_i - \nabla_j), \quad \boldsymbol{k}' = -\frac{1}{2i}(\nabla'_i - \nabla'_j), \tag{4}$$

where, again, i and j are the indices that identify the two nucleons. $\boldsymbol{k}'$ acts on the wave function to the left. For this representation, it is useful to remember the Einstein-de Broglie formula: $\boldsymbol{p} = \hbar\boldsymbol{k}$ under the quantum mechanical correspondence: $\boldsymbol{p} \rightarrow (\hbar/i)\nabla$. This treatment is essential to obtain the zero-range interaction.

The two-body interaction, which contains momentum dependence as well as spin-exchange contributions, is derived by assuming the polynomial expansion in powers of $\boldsymbol{k}$ and $\boldsymbol{k}'$ with the low-momentum cut-off up to quadratic terms (linear terms in $\boldsymbol{k}$ are forbidden by time-reversal invariance):

$$\begin{aligned} v_{i,j}(\boldsymbol{k},\boldsymbol{k}') &= t_0(1+x_0P_\sigma)\delta(\boldsymbol{r}_i-\boldsymbol{r}_j) \\ &+\tfrac{t_1}{2}(1+x_1P_\sigma)\{\delta(\boldsymbol{r}_i-\boldsymbol{r}_j)\boldsymbol{k}^2+\boldsymbol{k}'^2\delta(\boldsymbol{r}_i-\boldsymbol{r}_j)\} \\ &+t_2(1+x_2P_\sigma)\boldsymbol{k}'\delta(\boldsymbol{r}_i-\boldsymbol{r}_j)\boldsymbol{k} \\ &+\tfrac{t_e}{2}[\{3(\boldsymbol{\sigma}_i\cdot\boldsymbol{k}')(\boldsymbol{\sigma}_j\cdot\boldsymbol{k}')-(\boldsymbol{\sigma}_i\cdot\boldsymbol{\sigma}_j)\boldsymbol{k}'^2\}\delta(\boldsymbol{r}_i-\boldsymbol{r}_j) \\ &+\delta(\boldsymbol{r}_i-\boldsymbol{r}_j)\{3(\boldsymbol{\sigma}_i\cdot\boldsymbol{k})(\boldsymbol{\sigma}_j\cdot\boldsymbol{k})-(\boldsymbol{\sigma}_i\cdot\boldsymbol{\sigma}_j)\boldsymbol{k}^2\}] \\ &+\tfrac{t_o}{2}[3(\boldsymbol{\sigma}_i\cdot\boldsymbol{k}')\delta(\boldsymbol{r}_i-\boldsymbol{r}_j)(\boldsymbol{\sigma}_j\cdot\boldsymbol{k})-(\boldsymbol{\sigma}_i\cdot\boldsymbol{\sigma}_j)\boldsymbol{k}'\delta(\boldsymbol{r}_i-\boldsymbol{r}_j)\boldsymbol{k}], \end{aligned} \tag{5}$$

where t_0, t_1, t_2, t_e and t_o are parameters (following the traditional manner of representation), $P_\sigma = \frac{1}{2}(1+\boldsymbol{\sigma}_i\cdot\boldsymbol{\sigma}_j)$ is the spin-exchange operator, and $\boldsymbol{\sigma}_i$ are the Pauli matrices. The quadratic cut-off implies that the interaction shown in Eq (5) is valid to low-energy situations (cf. the fermi energy). The t_0 term represents the central potential, the t_1 and t_2 terms stand for the non-local potential, and the t_o and t_e terms are derived from the tensor interaction. It is important to introduce the spin-orbit interaction, which is an additional effect necessary to reproduce the nuclear structure.

$$v_{i,j}^{LS}(\boldsymbol{k},\boldsymbol{k}') = iW_0\,\boldsymbol{k}'\,\delta(\boldsymbol{r}_i-\boldsymbol{r}_j)\,(\sigma_i+\sigma_j)\times\boldsymbol{k}. \tag{6}$$

Another important ingredient is a two-body interaction with density dependence:

$$v_{i,j}^{DD}(\boldsymbol{k},\boldsymbol{k}') = \tfrac{t_3}{6}(1+x_3P_\sigma)\rho(\tfrac{\boldsymbol{r}_i+\boldsymbol{r}_j}{2})^\alpha\delta(\boldsymbol{r}_i-\boldsymbol{r}_j). \tag{7}$$

This provides a treatment of medium effects that include three-body and many-body forces. Note that the introduction of this force is due to a phenomenological reason.

As a result, the nuclear interaction is represented by the zero-range Skyrme interaction:

$$V_{i,j}(\boldsymbol{k}, \boldsymbol{k}') = v_{i,j}(\boldsymbol{k}, \boldsymbol{k}') + v_{i,j}^{LS}(\boldsymbol{k}, \boldsymbol{k}') + v_{i,j}^{DD}(\boldsymbol{k}, \boldsymbol{k}'), \tag{8}$$

where it can be noted here, that the included momenta $\boldsymbol{k}$ and $\boldsymbol{k}'$ are represented by Eq. (4).

3.2. Hamiltonian Density

The Hamiltonian density is obtained by calculating the energy expectation value from the zero-range nuclear interaction (8). The following points are illustrated by a concrete discussion to provide this analysis:

- Fermionic relation is assumed in deriving the Hamiltonian density.
- The boundary condition is necessary.
- Differential operators appear in deriving non-local interaction part of the Hamiltonian density.

In this section (Sec. 3.2), wave functions are assumed to be sufficiently smooth. With respect to the many-nucleon system as a fermionic many-body system, the fermionic relation of ($P_r P_\sigma P_\tau = -1$), as already mentioned in Sec. 3.1, has to be taken into account:

$$V_{i,j}(\boldsymbol{k}, \boldsymbol{k}') = \frac{1}{2} V_{i,j}(\boldsymbol{k}, \boldsymbol{k}')(1 - P_r P_\sigma P_\tau). \tag{9}$$

Using Eqs. (8) and (9), the energy is obtained by calculating

$$\begin{aligned} E &= E_{t_0} + E_{t_1} + E_{t_2} + E_{t_3} + E_{t_e} + E_{t_o} + E_{W_0} \\ &= \tfrac{1}{2} \textstyle\sum_{l,m} \int \bar{\psi}_l(\boldsymbol{r}_i') \bar{\psi}_m(\boldsymbol{r}_j') V_{i,j}(\boldsymbol{k}, \boldsymbol{k}')(1 - P_r P_\sigma P_\tau) \psi_l(\boldsymbol{r}_i) \psi_m(\boldsymbol{r}_j) \\ & d\boldsymbol{r}_i d\boldsymbol{r}_j d\boldsymbol{r}_i' d\boldsymbol{r}_j', \end{aligned} \tag{10}$$

where E_{t_0}, $E_{t_1} \cdots$, E_{W_0} denote the energy arising from the terms with the coefficient t_0, t_1 $\cdots$, W_0, respectively. In Eq. (10), $\psi_l(\boldsymbol{r}_i)\psi_m(\boldsymbol{r}_j)$ plays a role of trial function that includes two single-particle wave functions. Furthermore, let us assume that there is no isospin mixing in the nuclear force. Since the expectation value E for the many-body Hamiltonian can also be written when using the Hamiltonian density $H(\boldsymbol{r})$:

$$E = \int dr^3 H(\boldsymbol{r}), \tag{11}$$

we obtain the representation for the Hamiltonian density. Among several terms, in order to see some principal mathematical features (e.g., the appearance of differential operators) for the density functional formalism in nuclear physics, we examine the following energies, which are choosen from Eq. (10).

$$E = E_{t_0} + E_{t_1} + E_{t_2} + E_{t_3},$$

these choices have pedagogical reasons.

Let ψ be a single wave function, and introduce the densities. Before calculating the Hamiltonian density, it is necessary to introduce the following two densities. The nonlocal particle density and the nonlocal spin density are defined by

$$\rho(\boldsymbol{r},\boldsymbol{r}') = \sum_{\sigma,q} \rho_q(\boldsymbol{r},\boldsymbol{r}',\sigma) = \sum_{l,\sigma,q} \bar{\psi}_l(\boldsymbol{r}',\sigma,q)\psi_l(\boldsymbol{r},\sigma,q),$$
$$\boldsymbol{s}(\boldsymbol{r},\boldsymbol{r}') = \sum_{\sigma,\sigma',q} \rho_q(\boldsymbol{r},\boldsymbol{r}',\sigma,\sigma') < \sigma'|\boldsymbol{\sigma}|\sigma >$$
$$= \sum_{l,\sigma,\sigma',q} \bar{\psi}_l(\boldsymbol{r}',\sigma',q)\ \boldsymbol{\sigma}\ \psi_l(\boldsymbol{r},\sigma,q)$$

using positions $\boldsymbol{r}$ and $\boldsymbol{r}'$, isospin q, and spins σ and σ'. According to Engel *et al.* [2], local densities are defined by

$$\begin{aligned}
\rho(\boldsymbol{r}) &= \rho(\boldsymbol{r},\boldsymbol{r}')|_{\boldsymbol{r}=\boldsymbol{r}'},\\
\tau(\boldsymbol{r}) &= \nabla\cdot\nabla'\rho(\boldsymbol{r},\boldsymbol{r}')|_{\boldsymbol{r}=\boldsymbol{r}'},\\
s_\mu(\boldsymbol{r}) &= s_\mu(\boldsymbol{r},\boldsymbol{r}')|_{\boldsymbol{r}=\boldsymbol{r}'},\\
j_\mu(\boldsymbol{r}) &= -\tfrac{i}{2}(\nabla_\mu - \nabla'_\mu)\rho(\boldsymbol{r},\boldsymbol{r}')|_{\boldsymbol{r}=\boldsymbol{r}'},\\
J_{\mu\nu}(\boldsymbol{r}) &= -\tfrac{i}{2}(\nabla_\mu - \nabla'_\mu)s_\nu(\boldsymbol{r},\boldsymbol{r}')|_{\boldsymbol{r}=\boldsymbol{r}'},\\
T_\mu(\boldsymbol{r}) &= \nabla\cdot\nabla' s_\mu(\boldsymbol{r},\boldsymbol{r}')|_{\boldsymbol{r}=\boldsymbol{r}'},
\end{aligned} \tag{12}$$

where the indices μ and ν indicate the Cartesian coordinate components x, y, and z. In particular, $\mu\nu$ in $J_{\mu\nu}(\boldsymbol{r})$ can take only unequal values to μ and ν (cf. exterior product). The first two densities, that is, the particle density $\rho(\boldsymbol{r})$ and the kinetic energy density $\tau(\boldsymbol{r})$ are scalar quantities; while the others, the spin density $s_\mu(\boldsymbol{r})$, the current density $j_\mu(\boldsymbol{r})$, the spin-current density $J_{\mu\nu}(\boldsymbol{r})$, and the spin kinetic energy density $T_\mu(\boldsymbol{r})$ are vector quantities. The index q for isospin is used here, instead of τ, because it was used for the kinetic energy density.

Galilean invariance restricts the form of the Hamiltonian density [2]. Galilean transformation replaces a single-particle wave function $\psi(\boldsymbol{r},\sigma,q)$ by $e^{i\boldsymbol{k}\boldsymbol{r}}\psi(\boldsymbol{r},\sigma,q)$. Thus, the densities transform as

$$\begin{aligned}
\rho(\boldsymbol{r},\boldsymbol{r}') &\rightarrow e^{i\boldsymbol{k}\cdot(\boldsymbol{r}-\boldsymbol{r}')}\rho(\boldsymbol{r},\boldsymbol{r}'),\\
\tau(\boldsymbol{r}) &\rightarrow \tau(\boldsymbol{r}) + 2\boldsymbol{k}\cdot\boldsymbol{j}(\boldsymbol{r}) + \boldsymbol{k}^2\rho(\boldsymbol{r}),\\
\boldsymbol{j}(\boldsymbol{r}) &\rightarrow \boldsymbol{j}(\boldsymbol{r}) + \boldsymbol{k}\rho(\boldsymbol{r}),
\end{aligned}$$

because

$$e^{-i\boldsymbol{k}\boldsymbol{r}'}\,\bar{\psi}(\boldsymbol{r}',\sigma,q)\,e^{i\boldsymbol{k}\boldsymbol{r}}\psi(\boldsymbol{r},\sigma,q) = e^{i\boldsymbol{k}(\boldsymbol{r}-\boldsymbol{r}')}\bar{\psi}(\boldsymbol{r}',\sigma,q)\,\psi(\boldsymbol{r},\sigma,q),$$

$$\begin{aligned}
&(\nabla e^{-i\boldsymbol{k}\boldsymbol{r}'}\,\bar{\psi}(\boldsymbol{r}',\sigma,q))\,(\nabla e^{i\boldsymbol{k}\boldsymbol{r}}\psi(\boldsymbol{r},\sigma,q))\\
&= (-i\boldsymbol{k}e^{-i\boldsymbol{k}\boldsymbol{r}'}\,\bar{\psi}(\boldsymbol{r}',\sigma,q) + e^{-i\boldsymbol{k}\boldsymbol{r}'}\,\nabla\bar{\psi}(\boldsymbol{r}',\sigma,q))\\
&(i\boldsymbol{k}e^{i\boldsymbol{k}\boldsymbol{r}}\psi(\boldsymbol{r},\sigma,q) + e^{i\boldsymbol{k}\boldsymbol{r}}\nabla\psi(\boldsymbol{r},\sigma,q))\\
&= \boldsymbol{k}^2 e^{i\boldsymbol{k}(\boldsymbol{r}-\boldsymbol{r}')}\bar{\psi}(\boldsymbol{r}',\sigma,q)\psi(\boldsymbol{r},\sigma,q) + e^{i\boldsymbol{k}(\boldsymbol{r}-\boldsymbol{r}')}\nabla\bar{\psi}(\boldsymbol{r}',\sigma,q)\nabla\psi(\boldsymbol{r},\sigma,q)\\
&-i\boldsymbol{k}e^{i\boldsymbol{k}(\boldsymbol{r}-\boldsymbol{r}')}\bar{\psi}(\boldsymbol{r}',\sigma,q)\,(\nabla\psi(\boldsymbol{r},\sigma,q))\\
&+i\boldsymbol{k}e^{i\boldsymbol{k}(\boldsymbol{r}-\boldsymbol{r}')}(\nabla\bar{\psi}(\boldsymbol{r}',\sigma,q))\,\psi(\boldsymbol{r},\sigma,q),
\end{aligned}$$

$$\begin{aligned}
&-(i/2)\{e^{-i\boldsymbol{k}\boldsymbol{r}'}\,\bar{\psi}(\boldsymbol{r}',\sigma,q)\,(\nabla e^{i\boldsymbol{k}\boldsymbol{r}}\psi(\boldsymbol{r},\sigma,q))\\
&-(\nabla e^{-i\boldsymbol{k}\boldsymbol{r}'}\,\bar{\psi}(\boldsymbol{r}',\sigma,q))\,e^{i\boldsymbol{k}\boldsymbol{r}}\psi(\boldsymbol{r},\sigma,q)\}\\
&= -(i/2)\{e^{-i\boldsymbol{k}\boldsymbol{r}'}\,\bar{\psi}(\boldsymbol{r}',\sigma,q)\,(i\boldsymbol{k}e^{i\boldsymbol{k}\boldsymbol{r}}\psi(\boldsymbol{r},\sigma,q) + e^{i\boldsymbol{k}\boldsymbol{r}}\nabla\psi(\boldsymbol{r},\sigma,q))\\
&-(-i\boldsymbol{k}e^{-i\boldsymbol{k}\boldsymbol{r}'}\,\bar{\psi}(\boldsymbol{r}',\sigma,q) + e^{-i\boldsymbol{k}\boldsymbol{r}'}\,\nabla\bar{\psi}(\boldsymbol{r}',\sigma,q))\,e^{i\boldsymbol{k}\boldsymbol{r}}\psi(\boldsymbol{r},\sigma,q)\}\\
&= -(i/2)\{e^{i\boldsymbol{k}(\boldsymbol{r}-\boldsymbol{r}')}\bar{\psi}(\boldsymbol{r}',\sigma,q)\nabla\psi(\boldsymbol{r},\sigma,q)\\
&-e^{i\boldsymbol{k}(\boldsymbol{r}-\boldsymbol{r}')}\nabla\bar{\psi}(\boldsymbol{r}',\sigma,q)\psi(\boldsymbol{r},\sigma,q) + 2i\boldsymbol{k}e^{i\boldsymbol{k}(\boldsymbol{r}-\boldsymbol{r}')}\bar{\psi}(\boldsymbol{r}',\sigma,q)\,\psi(\boldsymbol{r},\sigma,q)\}.
\end{aligned}$$

The transformation leaves $\rho\tau - \boldsymbol{j}^2$ invariant. Indeed, $e^{i\boldsymbol{k}\cdot(\boldsymbol{r}-\boldsymbol{r}')}\rho(\boldsymbol{r},\boldsymbol{r}')|_{\boldsymbol{r}=\boldsymbol{r}'} = \rho(\boldsymbol{r})$ gives to

$$\begin{aligned}
&\rho(\tau + 2\boldsymbol{k}\cdot\boldsymbol{j} + \boldsymbol{k}^2\rho) - (\boldsymbol{j} + \boldsymbol{k}\rho)^2\\
&= \rho(\tau + 2\boldsymbol{k}\cdot\boldsymbol{j} + \boldsymbol{k}^2\rho) - \boldsymbol{j}^2 - 2\boldsymbol{j}\cdot\boldsymbol{k}\cdot\rho - \boldsymbol{k}^2\rho^2 = \rho\tau - \boldsymbol{j}^2.
\end{aligned}$$

Consequently, $\boldsymbol{j}^2$ cannot appear alone and the corresponding ingredient of the Hamiltonian density is $\rho\tau - \boldsymbol{j}^2$ instead.

We use the following substitution through integration by parts:

$$(\nabla\rho(\boldsymbol{r}))^2 \;\rightarrow\; -\rho(\boldsymbol{r})\,(\nabla^2\rho(\boldsymbol{r})), \tag{13}$$

which can be valid under a suitable boundary condition. In this section (Sec. 3.2), we assume that the Dirichlet-zero boundary condition is imposed. Note that nonzero additional boundary integral remains, if we have no boundary condition. We have

$$\begin{aligned}
&\nabla\rho(\boldsymbol{r},\boldsymbol{r}')|_{\boldsymbol{r}=\boldsymbol{r}'} = \textstyle\sum_{l,\sigma,q}\bar{\psi}_l(\boldsymbol{r}',\sigma,q)\nabla\psi_l(\boldsymbol{r},\sigma,q)|_{\boldsymbol{r}=\boldsymbol{r}'}\\
&\quad= \textstyle\sum_{l,\sigma,q}\bar{\psi}_l(\boldsymbol{r},\sigma,q)\nabla\psi_l(\boldsymbol{r},\sigma,q)\\
&\quad= \textstyle\sum_{l,\sigma,q}\{\frac{1}{2}(\bar{\psi}_l(\boldsymbol{r},\sigma,q)\nabla\psi_l(\boldsymbol{r},\sigma,q)+\psi_l(\boldsymbol{r},\sigma,q)\nabla\bar{\psi}_l(\boldsymbol{r},\sigma,q))\\
&\quad+\tfrac{1}{2}(\bar{\psi}_l(\boldsymbol{r},\sigma,q)\nabla\psi_l(\boldsymbol{r},\sigma,q)-\psi_l(\boldsymbol{r},\sigma,q)\nabla\bar{\psi}_l(\boldsymbol{r},\sigma,q))\}\\
&\quad= \tfrac{1}{2}\nabla\rho(\boldsymbol{r})+i\boldsymbol{j}(\boldsymbol{r}),\\
&\nabla'\rho(\boldsymbol{r},\boldsymbol{r}')|_{\boldsymbol{r}=\boldsymbol{r}'} = \tfrac{1}{2}\nabla\rho(\boldsymbol{r})-i\boldsymbol{j}(\boldsymbol{r}),\\
&(\nabla^2+\nabla'^2)\rho(\boldsymbol{r},\boldsymbol{r}')|_{\boldsymbol{r}=\boldsymbol{r}'} = \textstyle\sum_{l,\sigma,q}\bar{\psi}_l(\boldsymbol{r}',\sigma,q)\nabla^2\psi_l(\boldsymbol{r},\sigma,q)+\\
&\quad\psi_l(\boldsymbol{r},\sigma,q)\nabla'^2\bar{\psi}_l(\boldsymbol{r}',\sigma,q)|_{\boldsymbol{r}=\boldsymbol{r}'}\\
&\quad= \textstyle\sum_{l,\sigma,q}\{(\bar{\psi}_l(\boldsymbol{r}',\sigma,q)\nabla^2\psi_l(\boldsymbol{r},\sigma,q)+\psi_l(\boldsymbol{r},\sigma,q)\nabla'^2\bar{\psi}_l(\boldsymbol{r}',\sigma,q)\\
&\quad+2\nabla'\bar{\psi}_l(\boldsymbol{r}',\sigma,q)\nabla\psi_l(\boldsymbol{r},\sigma,q))\\
&\quad-2\nabla'\bar{\psi}_l(\boldsymbol{r}',\sigma,q)\nabla\psi_l(\boldsymbol{r},\sigma,q)\})|_{\boldsymbol{r}=\boldsymbol{r}'}\\
&\quad= \nabla^2\rho(\boldsymbol{r})-2\tau(\boldsymbol{r}).
\end{aligned}\tag{14}$$

For exploring the historical milestones to obtain the Hamiltonian density in many-nucleon systems, see Refs. [2, 13].

3.2.1. Force Depending on the Parameter t_0

First, for the term with the coefficient t_0, we have

$$\begin{aligned}
&(1+x_0P_\sigma)(1-P_rP_\sigma P_\tau)\\
&\quad= (1+x_0P_\sigma)(1-P_\sigma P_\tau)\\
&\quad= 1+x_0P_\sigma-(x_0P_\sigma^2+P_\sigma)P_\tau\\
&\quad= 1+\tfrac{1}{2}x_0-\tfrac{1}{2}(1+2x_0)P_\tau+\tfrac{1}{2}x_0\boldsymbol{\sigma}_i\boldsymbol{\sigma}_j-\tfrac{1}{2}\boldsymbol{\sigma}_i\boldsymbol{\sigma}_jP_\tau,
\end{aligned}$$

where we can take $P_r = 1$ due to the fact that the zero-range force acts only in S-waves. It leads to the energy

$$\begin{aligned}
E_{t_0} &= \tfrac{t_0}{2}\textstyle\sum_{l,m}\int\bar{\psi}_l(\boldsymbol{r}_i')\bar{\psi}_m(\boldsymbol{r}_j')\left\{1+\tfrac{1}{2}x_0-\tfrac{1}{2}(1+2x_0)P_\tau\right.\\
&\quad\left.+\tfrac{1}{2}x_0\boldsymbol{\sigma}_i\boldsymbol{\sigma}_j-\tfrac{1}{2}\boldsymbol{\sigma}_i\boldsymbol{\sigma}_jP_\tau\right\}\psi_l(\boldsymbol{r}_i)\psi_m(\boldsymbol{r}_j)d\boldsymbol{r}_id\boldsymbol{r}_jd\boldsymbol{r}_i'd\boldsymbol{r}_j'|_{\boldsymbol{r}_i=\boldsymbol{r}_j=\boldsymbol{r}_i'=\boldsymbol{r}_j'}\\
&= \textstyle\int\{\frac{t_0}{2}(1+\frac{1}{2}x_0)\rho(\boldsymbol{r}_i,\boldsymbol{r}_i')\rho(\boldsymbol{r}_i,\boldsymbol{r}_i')-\frac{t_0}{2}(\frac{1}{2}+x_0)\delta_{q_1,q_2}\rho(\boldsymbol{r}_i,\boldsymbol{r}_i')\rho(\boldsymbol{r}_i,\boldsymbol{r}_i')\\
&\quad+\tfrac{t_0}{4}x_0\boldsymbol{s}(\boldsymbol{r}_i,\boldsymbol{r}_i')\boldsymbol{s}(\boldsymbol{r}_i,\boldsymbol{r}_i')-\tfrac{t_0}{4}\delta_{q_1,q_2}\boldsymbol{s}(\boldsymbol{r}_i,\boldsymbol{r}_i')\boldsymbol{s}(\boldsymbol{r}_i,\boldsymbol{r}_i')\}d\boldsymbol{r}_id\boldsymbol{r}_jd\boldsymbol{r}_i'd\boldsymbol{r}_j'|_{\boldsymbol{r}_i=\boldsymbol{r}_j=\boldsymbol{r}_i'=\boldsymbol{r}_j'}\\
&= \textstyle\int dr^3\left\{\frac{t_0}{2}(1+\frac{1}{2}x_0)\rho(\boldsymbol{r})^2-\frac{t_0}{2}(\frac{1}{2}+x_0)\sum_q\rho_q(\boldsymbol{r})^2+\frac{t_0}{4}x_0\boldsymbol{s}(\boldsymbol{r})^2-\frac{t_0}{4}\sum_q\boldsymbol{s}_q(\boldsymbol{r})^2\right\},
\end{aligned}\tag{15}$$

where P_τ is reduced to $\delta_{q1,q2}$ by the assumption that there is no isospin mixing in the nuclear force (q_i identifies the isospin of the single particle state i). The specific representation shown in the second line of Eq. (15), allows for the derivation of the Hamiltonian density, which has been used in physics, where the property of δ-function: $\int \delta(\boldsymbol{a}-\boldsymbol{r})f(\boldsymbol{r})d\boldsymbol{r} = f(\boldsymbol{a})$; and therefore,

$$\int\int \delta(\boldsymbol{r}-\boldsymbol{r}')f(\boldsymbol{r}')d\boldsymbol{r}'d\boldsymbol{r} = \int f(\boldsymbol{r})d\boldsymbol{r}$$

is taken into account.

3.2.2. Force Depending on the Parameter t_1

Second, for the term with the coefficient t_1, we have

$$\begin{aligned}
&(1+x_1P_\sigma)(\boldsymbol{k}^2+\boldsymbol{k}'^2)(1-P_rP_\sigma P_\tau)\\
&=\{(\tfrac{1}{2i}(\nabla_i-\nabla_j))^2+\left(-\tfrac{1}{2i}(\nabla_i'-\nabla_j')\right)^2\}(1+x_1P_\sigma)(1-P_\sigma P_\tau)\\
&=\{-\tfrac{1}{4}(\nabla_i^2-\nabla_i\nabla_j-\nabla_j\nabla_i+\nabla_j^2)-\tfrac{1}{4}(\nabla_i'^2-\nabla_i'\nabla_j'-\nabla_j'\nabla_i'+\nabla_j'^2)\}\\
&\quad(1-x_1P_\sigma^2P_\tau+x_1P_\sigma-P_\sigma P_\tau)\\
&=-\tfrac{1}{4}(\nabla_i^2+\nabla_j^2+\nabla_i'^2+\nabla_j'^2-2\nabla_i\nabla_j-2\nabla_i'\nabla_j')\\
&\quad\{1-x_1P_\tau+\tfrac{1}{2}(x_1-P_\tau)(1+\boldsymbol{\sigma}_i\boldsymbol{\sigma}_j)\}\\
&=-\tfrac{1}{4}(\nabla_i^2+\nabla_j^2+\nabla_i'^2+\nabla_j'^2-2\nabla_i\nabla_j-2\nabla_i'\nabla_j')\\
&\quad\{1+\tfrac{1}{2}x_1-(x_1+\tfrac{1}{2})P_\tau+\tfrac{1}{2}(x_1-P_\tau)\boldsymbol{\sigma}_i\boldsymbol{\sigma}_j\},
\end{aligned}$$

where we can take $P_r = 1$ due to the fact that the zero-range force acts only in S-waves. It accounts for the energy:

$$\begin{aligned}
E_{t_1} &= -\tfrac{1}{2}\tfrac{t_1}{8}\textstyle\sum_{l,m}\int\bar{\psi}_l(\boldsymbol{r}_i')\bar{\psi}_m(\boldsymbol{r}_j')\\
&(\nabla_i^2+\nabla_j^2+\nabla_i'^2+\nabla_j'^2-2\nabla_i\nabla_j-2\nabla_i'\nabla_j')\{1+\tfrac{1}{2}x_1-(x_1+\tfrac{1}{2})P_\tau\\
&\quad+\tfrac{1}{2}(x_1-P_\tau)\boldsymbol{\sigma}_i\boldsymbol{\sigma}_j\}\psi_l(\boldsymbol{r}_i)\psi_m(\boldsymbol{r}_j)d\boldsymbol{r}_id\boldsymbol{r}_jd\boldsymbol{r}_i'd\boldsymbol{r}_j'|_{\boldsymbol{r}_i=\boldsymbol{r}_j=\boldsymbol{r}_i'=\boldsymbol{r}_j'}\\
&\quad=-\tfrac{t_1}{16}\textstyle\int(\nabla_i^2+\nabla_j^2+\nabla_i'^2+\nabla_j'^2-2\nabla_i\nabla_j-2\nabla_i'\nabla_j')\\
&\quad\{(1+\tfrac{1}{2}x_1)\rho(\boldsymbol{r}_i,\boldsymbol{r}_i')\rho(\boldsymbol{r}_j,\boldsymbol{r}_j')-(x_1+\tfrac{1}{2})\delta_{q1,q2}\rho(\boldsymbol{r}_i,\boldsymbol{r}_i')\rho(\boldsymbol{r}_j,\boldsymbol{r}_j')\\
&\quad+\tfrac{1}{2}x_1\boldsymbol{s}(\boldsymbol{r}_i,\boldsymbol{r}_i')\boldsymbol{s}(\boldsymbol{r}_j,\boldsymbol{r}_j')\\
&\quad-\tfrac{1}{2}\delta_{q1,q2}\boldsymbol{s}(\boldsymbol{r}_i,\boldsymbol{r}_i')\boldsymbol{s}(\boldsymbol{r}_j,\boldsymbol{r}_j')\}d\boldsymbol{r}_id\boldsymbol{r}_jd\boldsymbol{r}_i'd\boldsymbol{r}_j'|_{\boldsymbol{r}_i=\boldsymbol{r}_j=\boldsymbol{r}_i'=\boldsymbol{r}_j'}
\end{aligned}$$

$$
\begin{aligned}
&= -\tfrac{t_1}{16}\int dr^3\{\ (1+\tfrac{1}{2}x_1)\{[(\nabla_i^2+\nabla_i'^2)\rho(\boldsymbol{r}_i,\boldsymbol{r}_i')\}\rho(\boldsymbol{r}_j,\boldsymbol{r}_j')\\
&+\rho(\boldsymbol{r}_i,\boldsymbol{r}_i')(\nabla_j^2+\nabla_j'^2)\rho(\boldsymbol{r}_j,\boldsymbol{r}_j')-2\nabla_i\rho(\boldsymbol{r}_i,\boldsymbol{r}_i')\nabla_j\rho(\boldsymbol{r}_j,\boldsymbol{r}_j')\\
&-2\nabla_i'\rho(\boldsymbol{r}_i,\boldsymbol{r}_i')\nabla_j'\rho(\boldsymbol{r}_j,\boldsymbol{r}_j')]\\
&-(x_1+\tfrac{1}{2})\textstyle\sum_q[(\nabla_i^2+\nabla_i'^2)\rho_q(\boldsymbol{r}_i,\boldsymbol{r}_i')\}\rho_q(\boldsymbol{r}_j,\boldsymbol{r}_j')\\
&+\rho_q(\boldsymbol{r}_i,\boldsymbol{r}_i')(\nabla_j^2+\nabla_j'^2)\rho_q(\boldsymbol{r}_j,\boldsymbol{r}_j')\\
&-2\nabla_i\rho_q(\boldsymbol{r}_i,\boldsymbol{r}_i')\nabla_j\rho_q(\boldsymbol{r}_j,\boldsymbol{r}_j')-2\nabla_i'\rho_q(\boldsymbol{r}_i,\boldsymbol{r}_i')\nabla_j'\rho_q(\boldsymbol{r}_j,\boldsymbol{r}_j')]\\
&+\tfrac{1}{2}x_1[(\nabla_i^2+\nabla_i'^2)\boldsymbol{s}(\boldsymbol{r}_i,\boldsymbol{r}_i')\}\boldsymbol{s}(\boldsymbol{r}_j,\boldsymbol{r}_j')+\boldsymbol{s}(\boldsymbol{r}_i,\boldsymbol{r}_i')(\nabla_j^2+\nabla_j'^2)\boldsymbol{s}(\boldsymbol{r}_j,\boldsymbol{r}_j')\\
&-2\nabla_i\boldsymbol{s}(\boldsymbol{r}_i,\boldsymbol{r}_i')\nabla_j\boldsymbol{s}(\boldsymbol{r}_j,\boldsymbol{r}_j')-2\nabla_i'\boldsymbol{s}(\boldsymbol{r}_i,\boldsymbol{r}_i')\nabla_j'\boldsymbol{s}(\boldsymbol{r}_j,\boldsymbol{r}_j')]\\
&-\tfrac{1}{2}\textstyle\sum_q[(\nabla_i^2+\nabla_i'^2)\boldsymbol{s}_q(\boldsymbol{r}_i,\boldsymbol{r}_i')\}\boldsymbol{s}_q(\boldsymbol{r}_j,\boldsymbol{r}_j')\\
&+\boldsymbol{s}_q(\boldsymbol{r}_i,\boldsymbol{r}_i')(\nabla_j^2+\nabla_j'^2)\boldsymbol{s}_q(\boldsymbol{r}_j,\boldsymbol{r}_j')\\
&= -\tfrac{t_1}{16}\int dr^3\{\ (1+\tfrac{1}{2}x_1)[2(\nabla^2\rho(\boldsymbol{r})-2\tau(\boldsymbol{r}))\rho(\boldsymbol{r})-2(\tfrac{1}{2}\nabla\rho(\boldsymbol{r})+i\boldsymbol{j}(\boldsymbol{r}))^2\\
&-2(\tfrac{1}{2}\nabla\rho(\boldsymbol{r})-i\boldsymbol{j}(\boldsymbol{r}))^2]-(x_1+\tfrac{1}{2})\textstyle\sum_q[2(\nabla^2\rho_q(\boldsymbol{r})-2\tau_q(\boldsymbol{r}))\rho_q(\boldsymbol{r})\\
&-2(\tfrac{1}{2}\nabla\rho_q(\boldsymbol{r})+i\boldsymbol{j}_q(\boldsymbol{r}))^2-2(\tfrac{1}{2}\nabla\rho_q(\boldsymbol{r})-i\boldsymbol{j}_q(\boldsymbol{r}))^2]+\tfrac{1}{2}x_1[2(\nabla^2\boldsymbol{s}(\boldsymbol{r})\\
&-2\boldsymbol{T}(\boldsymbol{r}))\boldsymbol{s}(\boldsymbol{r})-2(\tfrac{1}{2}\nabla\boldsymbol{s}(\boldsymbol{r})+i\boldsymbol{J}(\boldsymbol{r}))^2-2(\tfrac{1}{2}\nabla\boldsymbol{s}(\boldsymbol{r})-i\boldsymbol{J}(\boldsymbol{r}))^2]\\
&-\tfrac{1}{2}\textstyle\sum_q[2(\nabla^2\boldsymbol{s}_q(\boldsymbol{r})-2\boldsymbol{T}_q(\boldsymbol{r}))\boldsymbol{s}_q(\boldsymbol{r})-2(\tfrac{1}{2}\nabla\boldsymbol{s}_q(\boldsymbol{r})+i\boldsymbol{J}_q(\boldsymbol{r}))^2\\
&-2(\tfrac{1}{2}\nabla\boldsymbol{s}_q(\boldsymbol{r})-i\boldsymbol{J}_q(\boldsymbol{r}))^2]\}\\
&= \int dr^3\{\tfrac{3t_1}{16}[-(1+\tfrac{1}{2}x_1)\rho(\boldsymbol{r})\nabla^2\rho(\boldsymbol{r})+(x_1+\tfrac{1}{2})\textstyle\sum_q\rho_q(\boldsymbol{r})\nabla^2\rho_q(\boldsymbol{r})]\\
&+\tfrac{t_1}{4}[(1+\tfrac{1}{2}x_1)(\rho(\boldsymbol{r})\tau(\boldsymbol{r})-\boldsymbol{j}(\boldsymbol{r})^2)-(x_1+\tfrac{1}{2})\textstyle\sum_q(\rho_q(\boldsymbol{r})\tau_q(\boldsymbol{r})-\\
&\boldsymbol{j}_q(\boldsymbol{r})^2)]-\tfrac{3t_1}{32}[x_1\boldsymbol{s}(\boldsymbol{r})\nabla^2\boldsymbol{s}(\boldsymbol{r})-\textstyle\sum_q\boldsymbol{s}_q(\boldsymbol{r})\nabla^2\boldsymbol{s}_q(\boldsymbol{r})]\\
&+\tfrac{t_1}{8}[x_1(\boldsymbol{s}(\boldsymbol{r})\cdot\boldsymbol{T}(\boldsymbol{r})-\boldsymbol{J}^2(\boldsymbol{r}))-\textstyle\sum_q(\boldsymbol{s}_q(\boldsymbol{r})\cdot\boldsymbol{T}_q(\boldsymbol{r})-\boldsymbol{J}_q^2(\boldsymbol{r}))]\},
\end{aligned}
$$

where we use the relations shown in Eq. (14) and the reduction: $P_\tau \to \delta_{q_1,q_2}$.

3.2.3. Force Depending on the Parameter t_2

Third, for the term with the coefficient t_2, we have

$$
\begin{aligned}
&(1+x_2P_\sigma)(\boldsymbol{k}\cdot\boldsymbol{k}')(1-P_rP_\sigma P_\tau)\\
&= \{-\tfrac{1}{2i}(\nabla_i-\nabla_j)\times\tfrac{-1}{2i}(\nabla_i'-\nabla_j')\}(1+x_2P_\sigma)(1+P_\sigma P_\tau)\\
&= \tfrac{1}{4}(\nabla_i\nabla_i'-\nabla_i\nabla_j'-\nabla_j\nabla_i'+\nabla_j\nabla_j')(1+x_2P_\tau+x_2P_\sigma+P_\sigma P_\tau)\\
&= \tfrac{1}{4}(\nabla_i\nabla_i'-\nabla_i\nabla_j'-\nabla_j\nabla_i'+\nabla_j\nabla_j')(1+x_2P_\tau+\tfrac{1}{2}(x_2+P_\tau)(1+\boldsymbol{\sigma}_i\boldsymbol{\sigma}_j))\\
&= \tfrac{1}{4}(\nabla_i\nabla_i'-\nabla_i\nabla_j'-\nabla_j\nabla_i'+\nabla_j\nabla_j')(1+\tfrac{1}{2}x_2+(x_2+\tfrac{1}{2})P_\tau+\tfrac{1}{2}(x_2+P_\tau)\boldsymbol{\sigma}_i\boldsymbol{\sigma}_j),
\end{aligned}
$$

where we can take $P_r = -1$ since the odd powers of $\boldsymbol{k}$ are treated. It indicates the energy

$$
\begin{aligned}
E_{t_2} &= \tfrac{1}{2}\tfrac{t_2}{4}\textstyle\sum_{l,m}\int \bar{\psi}_l(\boldsymbol{r}'_i)\bar{\psi}_m(\boldsymbol{r}'_j)(\nabla_i\nabla'_i - \nabla_i\nabla'_j - \nabla_j\nabla'_i + \nabla_j\nabla'_j) \\
&(1+\tfrac{1}{2}x_2 + (x_2+\tfrac{1}{2})P_\tau + \tfrac{1}{2}(x_2+P_\tau)\boldsymbol{\sigma}_i\boldsymbol{\sigma}_j)\psi_l(\boldsymbol{r}_i)\psi_m(\boldsymbol{r}_j)d\boldsymbol{r}_i d\boldsymbol{r}_j|_{\boldsymbol{r}_i=\boldsymbol{r}_j=\boldsymbol{r}'_i=\boldsymbol{r}'_j} \\
&= \tfrac{t_2}{8}\int(\nabla_i\nabla'_i - \nabla_i\nabla'_j - \nabla_j\nabla'_i + \nabla_j\nabla'_j) \\
&\quad\{(1+\tfrac{1}{2}x_2)\rho(\boldsymbol{r}_i,\boldsymbol{r}'_i)\rho(\boldsymbol{r}_j,\boldsymbol{r}'_j) + (x_2+\tfrac{1}{2})\delta_{q1,q2}\rho(\boldsymbol{r}_i,\boldsymbol{r}'_i)\rho(\boldsymbol{r}_j,\boldsymbol{r}'_j) \\
&\quad+\tfrac{1}{2}x_2\boldsymbol{s}(\boldsymbol{r}_i,\boldsymbol{r}'_i)\boldsymbol{s}(\boldsymbol{r}_j,\boldsymbol{r}'_j) + \tfrac{1}{2}\delta_{q1,q2}\boldsymbol{s}(\boldsymbol{r}_i,\boldsymbol{r}'_i)\boldsymbol{s}(\boldsymbol{r}_j,\boldsymbol{r}'_j)\}\, d\boldsymbol{r}_i d\boldsymbol{r}_j|_{\boldsymbol{r}_i=\boldsymbol{r}_j=\boldsymbol{r}'_i=\boldsymbol{r}'_j} \\
&= \tfrac{t_2}{4}\int dr^3[(1+\tfrac{1}{2}x_2)\{\rho(\boldsymbol{r})\tau(\boldsymbol{r}) - (\tfrac{1}{2}\nabla\rho(\boldsymbol{r}) - \boldsymbol{j}(\boldsymbol{r}))(\tfrac{1}{2}\nabla\rho(\boldsymbol{r}) + \boldsymbol{j}(\boldsymbol{r}))\} \\
&\quad+(x_2+\tfrac{1}{2})\textstyle\sum_q\{\rho_q(\boldsymbol{r})\tau_q(\boldsymbol{r}) - (\tfrac{1}{2}\nabla\rho_q(\boldsymbol{r}) - \boldsymbol{j}_q(\boldsymbol{r}))(\tfrac{1}{2}\nabla\rho_q(\boldsymbol{r}) + \boldsymbol{j}_q(\boldsymbol{r}))\} \\
&\quad+\tfrac{1}{2}x_2\{\boldsymbol{s}(\boldsymbol{r})\cdot\boldsymbol{T}(\boldsymbol{r}) - (\tfrac{1}{2}\nabla\boldsymbol{s}(\boldsymbol{r}) - \boldsymbol{J}(\boldsymbol{r}))(\tfrac{1}{2}\nabla\cdot\boldsymbol{s}(\boldsymbol{r}) + \boldsymbol{J}(\boldsymbol{r}))\} \\
&\quad+\tfrac{1}{2}\textstyle\sum_q\{\boldsymbol{s}_q(\boldsymbol{r})\cdot\boldsymbol{T}_q(\boldsymbol{r}) - (\tfrac{1}{2}\nabla\boldsymbol{s}_q(\boldsymbol{r}) - \boldsymbol{J}_q(\boldsymbol{r}))(\tfrac{1}{2}\nabla\cdot\boldsymbol{s}_q(\boldsymbol{r}) + \boldsymbol{J}_q(\boldsymbol{r}))\}] \\
&= \tfrac{t_2}{16}\int dr^3[(1+\tfrac{1}{2}x_2)\{4\rho(\boldsymbol{r})\tau(\boldsymbol{r}) + \rho(\boldsymbol{r})\nabla^2\rho(\boldsymbol{r}) - 4\boldsymbol{j}(\boldsymbol{r})^2\} \\
&\quad+(x_2+\tfrac{1}{2})\textstyle\sum_q\{4\rho_q(\boldsymbol{r})\tau_q(\boldsymbol{r}) + \rho_q(\boldsymbol{r})\nabla^2\rho(\boldsymbol{r}) - 4\boldsymbol{j}_q(\boldsymbol{r})^2\} \\
&\quad+x_2\{2\boldsymbol{s}(\boldsymbol{r})\cdot\boldsymbol{T}(\boldsymbol{r}) + \tfrac{1}{2}\boldsymbol{s}(\boldsymbol{r})\cdot\nabla^2\boldsymbol{s}(\boldsymbol{r}) - 2\boldsymbol{J}(\boldsymbol{r})^2\} \\
&\quad+\textstyle\sum_q\{2\boldsymbol{s}_q(\boldsymbol{r})\cdot\boldsymbol{T}_q(\boldsymbol{r}) + \tfrac{1}{2}\boldsymbol{s}_q(\boldsymbol{r})\cdot\nabla^2\boldsymbol{s}_q(\boldsymbol{r}) - 2\boldsymbol{J}(\boldsymbol{r})^2\}],
\end{aligned}
$$

where we use the relations shown in Eq. (14) and the reduction: $P_\tau \rightarrow \delta_{q1,q2}$.

3.2.4. Force Depending on the Parameter t_3

Fourth, for the term with the coefficient t_3 (the density-dependent term), we have

$$
\begin{aligned}
&(1+x_3P_\sigma)\rho(\tfrac{\boldsymbol{r}_i+\boldsymbol{r}_j}{2})^\alpha(1-P_rP_\sigma P_\tau) \\
&\quad= (1+x_3P_\sigma)(1-P_\sigma P_\tau)\rho(\tfrac{\boldsymbol{r}_i+\boldsymbol{r}_j}{2})^\alpha \\
&\quad= \{1+x_3P_\sigma - P_\sigma P_\tau - x_3P_\sigma^2P_\tau\}\rho(\tfrac{\boldsymbol{r}_i+\boldsymbol{r}_j}{2})^\alpha \\
&\quad= \{1+\tfrac{1}{2}x_3 - \tfrac{1}{2}(1+2x_3)P_\tau + \tfrac{1}{2}x_3\boldsymbol{\sigma}_i\boldsymbol{\sigma}_j - \tfrac{1}{2}\boldsymbol{\sigma}_i\boldsymbol{\sigma}_jP_\tau\}\rho(\tfrac{\boldsymbol{r}_i+\boldsymbol{r}_j}{2})^\alpha,
\end{aligned}
$$

where we can take $P_r = 1$ due to the fact that the zero-range force acts only in S-waves. It gives the energy

$$
\begin{aligned}
E_{t_3} &= \tfrac{1}{2}\tfrac{t_3}{6}\textstyle\sum_{l,m}\int \bar{\psi}_l(\boldsymbol{r}'_i)\bar{\psi}_m(\boldsymbol{r}'_j)\,\{1+\tfrac{1}{2}x_3 - \tfrac{1}{2}(1+2x_3)P_\tau + \tfrac{1}{2}x_3\boldsymbol{\sigma}_i\boldsymbol{\sigma}_j - \tfrac{1}{2}\boldsymbol{\sigma}_i\boldsymbol{\sigma}_jP_\tau\} \\
&\quad\rho(\tfrac{\boldsymbol{r}_i+\boldsymbol{r}_j}{2})^\alpha\psi_l(\boldsymbol{r}_i)\psi_m(\boldsymbol{r}_j)d\boldsymbol{r}_i d\boldsymbol{r}_j|_{\boldsymbol{r}_i=\boldsymbol{r}_j=\boldsymbol{r}'_i=\boldsymbol{r}'_j} \\
&= \int\{\tfrac{t_3}{12}(1+\tfrac{1}{2}x_3)\rho(\boldsymbol{r}_i,\boldsymbol{r}'_i)\rho(\boldsymbol{r}_i,\boldsymbol{r}'_i) - \tfrac{t_3}{12}(\tfrac{1}{2}+x_3)\delta_{q1,q2}\rho(\boldsymbol{r}_i,\boldsymbol{r}'_i)\rho(\boldsymbol{r}_i,\boldsymbol{r}'_i) \\
&\quad+\tfrac{t_3}{24}x_3\boldsymbol{s}(\boldsymbol{r}_i,\boldsymbol{r}'_i)\boldsymbol{s}(\boldsymbol{r}_i,\boldsymbol{r}'_i) - \tfrac{t_3}{24}\delta_{q1,q2}\boldsymbol{s}(\boldsymbol{r}_i,\boldsymbol{r}'_i)\boldsymbol{s}(\boldsymbol{r}_i,\boldsymbol{r}'_i)\rho(\tfrac{\boldsymbol{r}_i+\boldsymbol{r}_j}{2})^\alpha\}d\boldsymbol{r}_i d\boldsymbol{r}_j|_{\boldsymbol{r}_i=\boldsymbol{r}_j=\boldsymbol{r}'_i=\boldsymbol{r}'_j} \\
&= \int\Big\{\tfrac{t_3}{12}(1+\tfrac{1}{2}x_3)\rho(\boldsymbol{r})^{2+\alpha} - \tfrac{t_3}{12}(\tfrac{1}{2}+x_3)\textstyle\sum_q\rho_q(\boldsymbol{r})^2\rho_q(\boldsymbol{r})^\alpha + \tfrac{t_3}{24}x_3\boldsymbol{s}(\boldsymbol{r})^2\rho(\boldsymbol{r})^\alpha \\
&\quad-\tfrac{t_3}{24}\textstyle\sum_q\boldsymbol{s}_q(\boldsymbol{r})^2\rho_q(\boldsymbol{r})^\alpha\Big\}\,dr^3,
\end{aligned}
$$

where we use the reduction: $P_\tau \rightarrow \delta_{q1,q2}$.

3.2.5. Hamiltonian Density of Interacting Many-Nucleon Systems

The expectation value E of the many-body Hamiltonian is written by an integral of the Hamiltonian density $H(\boldsymbol{r})$.

$$E = E_{t_0} + E_{t_1} + E_{t_2} + E_{t_3} = \int dr^3 H(\boldsymbol{r}).$$

The Hamiltonian density is obtained by collecting the results:

$$\begin{aligned}
H(\boldsymbol{r}) = \tfrac{1}{2}\big[& \tfrac{\hbar^2}{2m}\tau + \tfrac{t_0}{2}(1+\tfrac{1}{2}x_0)\rho^2 - \tfrac{t_0}{2}(\tfrac{1}{2}+x_0)\textstyle\sum_q \rho_q^2 + \tfrac{t_0}{4}x_0\boldsymbol{s}^2 - \tfrac{t_0}{4}\textstyle\sum_q \boldsymbol{s}_q^2 \\
& -\tfrac{3t_1}{16}(1+\tfrac{1}{2}x_1)\rho\triangle\rho + \tfrac{3t_1}{16}(x_1+\tfrac{1}{2})\textstyle\sum_q \rho_q\triangle\rho_q + \tfrac{t_1}{4}(1+\tfrac{1}{2}x_1)(\rho\tau - \boldsymbol{j}^2) \\
& -\tfrac{t_1}{4}(x_1+\tfrac{1}{2})\textstyle\sum_q(\rho_q\tau_q - \boldsymbol{j}_q^2) - \tfrac{3t_1}{32}x_1\boldsymbol{s}\triangle\boldsymbol{s} + \tfrac{3t_1}{32}\textstyle\sum_q \boldsymbol{s}_q\triangle\boldsymbol{s}_q \\
& +\tfrac{t_1}{8}x_1(\boldsymbol{s}\cdot\boldsymbol{T} - \boldsymbol{J}^2) - \tfrac{t_1}{8}\textstyle\sum_q(\boldsymbol{s}_q\cdot\boldsymbol{T}_q - \boldsymbol{J}_q^2)] + \tfrac{t_2}{16}(1+\tfrac{1}{2}x_2)\rho\triangle\rho \\
& +\tfrac{t_2}{4}(1+\tfrac{1}{2}x_2)(\rho\tau - \boldsymbol{j}^2) + \textstyle\sum_q\{\tfrac{t_2}{16}(x_2+\tfrac{1}{2})\rho_q\triangle\rho \\
& +\tfrac{t_2}{4}(x_2+\tfrac{1}{2})(\rho_q\tau_q - \boldsymbol{j}_q^2)\} + \tfrac{t_2}{32}x_2\boldsymbol{s}\cdot\triangle\boldsymbol{s} + \tfrac{t_2}{8}x_2(\boldsymbol{s}\cdot\boldsymbol{T} - \boldsymbol{J}^2) \\
& +\textstyle\sum_q\{\tfrac{t_2}{32}x_2\boldsymbol{s}_q\cdot\triangle\boldsymbol{s}_q + \tfrac{t_2}{8}x_2(\boldsymbol{s}_q\cdot\boldsymbol{T}_q - \boldsymbol{J}_q^2)\} + \tfrac{t_3}{12}(1+\tfrac{1}{2}x_3)\rho^{2+\alpha} \\
& -\tfrac{t_3}{12}(\tfrac{1}{2}+x_3)\textstyle\sum_q \rho_q^2\rho_q^\alpha + \tfrac{t_3}{24}x_3\boldsymbol{s}^2\rho^\alpha - \tfrac{t_3}{24}\textstyle\sum_q \boldsymbol{s}_q^2\rho^\alpha\big];
\end{aligned} \tag{16}$$

it is half of the standard Hamiltonian density [2, 13], which can be understood when applying the variational principle to the first term of the right-hand side (the contribution from the kinetic energy), where $1/2$ for the interaction part appears to avoid calculating the interaction twice. The justification of $1/2$ (in the term arising from the kinetic energy) is illustrated in the next section (Sec. 3.3). It is notable that differential operators appear due to the non-local interaction. Interacting many-nucleon systems cannot be described well if the differential operators are not utilized (cf. local density approximation).

3.3. Energy Density Functional Theory

The components in the effective interaction are obtained based on the variational principle. This corresponds to the procedure in obtaining the functional representation of an effective Hamiltonian. The following points are illustrated in a concrete discussion for this analysis:

- Once the energy is given, it is not necessary to have the fermionic relation or the Slater-determinant formalism to derive the effective interaction (the effective Hamiltonian).
- The boundary condition is indispensable.
- Regularity of wave functions is required; i.e., L^2-space is not necessarily sufficient.
- For mathematically strict treatment, the variational principle should be considered not only in the real Hilbert/Banach spaces (by varying "real" densities) but in the full complex Hilbert/Banach spaces (by varying "complex" wave functions).

Since the application of the variational principle includes differentiation, more careful treatment is required (differentiability, and so on). Furthermore, the application of variational principle means a process to obtain a functional representation in some complex functional spaces only from the real function (i.e., energy). In order to present the essential treatment of the variational principle, the discussion is developed for a spin-saturated and charge-conjugate nucleus (i.e., $\rho/2 = \rho_p = \rho_n$). In this situation Eq. (16) reduces to a model Hamiltonian density:

$$\mathcal{H}(\boldsymbol{r}) = \tilde{t}_1\tau + \tilde{t}_2\rho^2 + \tilde{t}_3\rho^{\tilde{\alpha}} + \tilde{t}_4(\rho\tau - \boldsymbol{j}^2) + \tilde{t}_5\rho\triangle\rho, \tag{17}$$

where $\tilde{\alpha}$ is a positive rational number, and $\{\tilde{t}_i\}$ is a parameter set related to the parameter sets $\{t_i\}$ and $\{x_i\}$ [2]. Based on this Hamiltonian model, the generalized higher order terms and terms with fractional powers are also treated. In this case, we need to discuss this within suitable functional spaces for a strict treatment of the variational principle. First, the term arising from the kinetic energy (the homogeneous term of Schrödinger equations) is treated in Ex. 3.3.1. Next, the derivation of terms arising from the interaction energy is demonstrated.

3.3.1. Linear and Nonlinear Laplacians

Let $L^p(\Omega)$ be the space of functions on Ω which are L^p for the Lebesgue measure, where $L^p(\Omega)$ are not (infinite-dimensional) Hilbert spaces but (infinite-dimensional) Banach spaces if p is not equal to 2. Let p be an even number satisfying $p \geq 2$ and ψ be included in $W^{1,p}(\Omega)$ with a open bounded set $\Omega \subset \mathbb{R}^3$, where $W^{1,p}(\Omega)$ is the space of functions in $L^p(\Omega)$ whose distribution derivatives of order ≤ 1 are in $L^p(\Omega)$ (for the "mathematical" distribution [9], referring to functional analysis). For our purposes it is necessary to introduce such a generalized differential in association with the spaces of integrable functions. Note here, that the δ-function, which cannot be defined using usual functions (C^k-function and so on), is successfully defined in the sense of a distribution. In this sense, the zero-range interaction formalism is based on the theory of distribution.

First of all, it is useful to consider a simple case when the energy is represented by

$$\mathcal{E}(\psi) = \frac{1}{p}\int_\Omega dr^3 |\nabla\psi|^p,$$

where the corresponding Hamiltonian energy density is $\mathcal{H}(\psi) = \frac{1}{p}|\nabla\psi|^p$. The differential operator is considered in the sense of distribution; this is true in other cases shown in this section (Sec 3.3). By minimizing this energy (i.e., the Gâteaux differential: $\mathcal{E}'(\psi) = \mathcal{A}$ is considered), we obtain

$$\mathcal{A}(\psi) = -\nabla\cdot(|\nabla\psi|^{p-2}\nabla\psi).$$

Indeed, for ψ and ϕ in $W^{1,p}(\Omega)$,

$$\begin{aligned}
|\nabla\psi + \lambda\phi|^p - |\nabla\psi|^p &= \textstyle\int_0^1 \frac{\partial}{\partial t}|\nabla\psi + t\lambda\nabla\phi|^p dt \\
&= \textstyle\int_0^1 \frac{\partial}{\partial t}(|\nabla\psi + t\lambda\nabla\phi|^2)^{p/2} dt \\
&= \textstyle\int_0^1 \frac{p}{2}(|\nabla\psi + t\lambda\nabla\phi|^2)^{(p-2)/2}\frac{\partial}{\partial t}|\nabla\psi + t\lambda\nabla\phi|^2 dt \\
&= \textstyle\int_0^1 \frac{p}{2}(|\nabla\psi + t\lambda\nabla\phi|^2)^{(p-2)/2}\cdot 2\mathrm{Re}\{(\nabla\psi + t\lambda\nabla\phi)\cdot\lambda\nabla\bar{\phi}\}dt \\
&= p\lambda\textstyle\int_0^1 |\nabla\psi + t\lambda\nabla\phi|^{p-2}\cdot\mathrm{Re}\{(\nabla\psi + t\lambda\nabla\phi)\cdot\nabla\bar{\phi}\}dt
\end{aligned} \tag{18}$$

is true. It follows that

$$\frac{\mathcal{E}(\psi+\lambda\phi)-\mathcal{E}(\psi)}{\lambda} = \int_\Omega dr^3 \int_0^1 |\nabla\psi + t\lambda\nabla\phi|^{p-2}\mathrm{Re}\{(\nabla\psi + t\lambda\nabla\phi)\cdot\nabla\bar{\phi}\}dt.$$

Due to $\lambda \to 0$ ($\lambda > 0$),

$$|\nabla\psi + t\lambda\nabla\phi|^{p-2}\mathrm{Re}\{(\nabla\psi + t\lambda\nabla\phi)\cdot\nabla\bar{\phi}\} \;\to\; |\nabla\psi|^{p-2}\mathrm{Re}(\nabla\psi\cdot\nabla\bar{\phi}) \quad \text{a.e.}$$

is valid ("a.e." means "almost everywhere", whose mathematical definition should be referred to functional analysis. Furthermore, due to $\lambda \to 0$, the following inequality is valid:

$$||\nabla\psi + t\lambda\nabla\phi|^{p-2}\mathrm{Re}\{(\nabla\psi + t\lambda\nabla\phi)\cdot\nabla\bar{\phi}\}| \;\le\; |\nabla\psi + t\lambda\nabla\phi|^{p-1}|\nabla\phi| \;\le\; (|\nabla\psi| + |\lambda||\nabla\phi|)^{p-1}|\nabla\phi|,$$

where it is worthy to note here, that the last term is integrable. We have

$$\begin{aligned}\frac{d}{d\lambda}\mathcal{E}(\psi+\lambda\phi)\big|_{\lambda=0} &= \lim_{\lambda\to 0}\frac{\mathcal{E}(\psi+\lambda\phi)-\mathcal{E}(\psi)}{\lambda} = \int_\Omega dr^3\, |\nabla\psi|^{p-2}\mathrm{Re}(\nabla\psi\cdot\nabla\bar{\phi}) \\ &= \mathrm{Re}\int_\Omega dr^3\, |\nabla\psi|^{p-2}\nabla\psi\cdot\nabla\bar{\phi}.\end{aligned}$$

If ψ and ϕ are sufficiently smooth,

$$\int_\Omega |\nabla\psi|^{p-2}\nabla\psi\cdot\nabla\bar{\phi}\, dr^3 = \int_{\partial\Omega} |\nabla\psi|^{p-2}\frac{\partial\psi}{\partial\mu}\bar{\phi}\, dS + \int_\Omega \mathcal{A}(\psi)\bar{\phi}\, dr^3$$

follows from the integration by parts, where $\partial/\partial\mu$ means the outward normal differential operator. Although we have discussed the limit-process in a roundabout sort of way, it is based on the fact that $\lim_{\lambda\to 0}\frac{\mathcal{E}(\psi+\lambda\phi)-\mathcal{E}(\psi)}{\lambda} = -\int_\Omega dr^3|\nabla\psi|^{p-2}\nabla\psi\nabla\bar{\phi}$ cannot follow from some simple treatments, because the left-hand side is real-valued and the right-hand side is complex-valued. It means that the differentiability of the energy cannot follow naively (for the differentiability, see Appendix (Sec. 6)). This fact is true for all the cases treated in Sec. 3.3.

The final treatment actually depends on the boundary condition. When the Dirichlet boundary condition is adopted (for the homogeneous part, Schrödinger operator is considered), $W^{1,p}(\Omega)$ is replaced by $W_0^{1,p}(\Omega)$, and we have

$$\int_\Omega |\nabla\psi|^{p-2}\nabla\psi\cdot\nabla\bar{\phi}\, dr^3 = \int_\Omega \mathcal{A}(\psi)\bar{\phi}\, dr^3 = (\mathcal{A}(\psi),\phi). \tag{19}$$

That is, for $f \in W^{-1,p'}(\Omega) = W_0^{1,p}(\Omega)^*$ and $\psi \in W_0^{1,p}(\Omega)$, the equality: $\int_\Omega |\nabla\psi|^{p-2}\nabla\psi\cdot\nabla\bar{\phi}\, dr^3 = (f,\phi)$ is valid to any $\psi \in W_0^{1,p}(\Omega)$, which is a generalized solution of

$$\begin{cases} \mathcal{A}(\psi) = f & \text{in } \Omega, \\ \psi = 0 & \text{on } \partial\Omega. \end{cases}$$

Meanwhile, when the Neumann boundary condition is adopted, we then have Eq. (19). However, the detail is different in the case with the Dirichlet boundary condition: for $f \in W^{1,p}(\Omega)^*$, the equality: $\int_\Omega |\nabla\psi|^{p-2}\nabla\psi\cdot\nabla\bar{\phi}\, dr^3 = (f,\phi)$ is valid to any $\psi \in W^{1,p}(\Omega)$. This is a generalized solution of

$$\begin{cases} \mathcal{A}(\psi) = f & \text{in } \Omega, \\ \frac{\partial\psi}{\partial\mu} = 0 & \text{on } \partial\Omega. \end{cases}$$

When the periodic boundary condition is adopted for cuboid Ω, $W^{1,p}(\Omega)$ and is replaced by $W^{1,p}_{per}(\Omega)$, there are space restrictions to Ω for periodic functions. We have Eq. (19) using this similar argument. As a result,

$$\tfrac{d}{d\lambda}\mathcal{E}(\psi+\lambda\phi)\big|_{\lambda=0} = \mathrm{Re}\int_\Omega dr^3\, |\nabla\psi|^{p-2}\nabla\psi\cdot\nabla\bar{\phi} = \mathrm{Re}\int_\Omega \mathcal{A}(\psi)\bar{\phi}\, dr^3 = \mathrm{Re}(\mathcal{A}(\psi),\phi) \tag{20}$$

is valid to the Dirichlet, Neumann, and periodic boundary conditions. The operator $\mathcal{A}$ where $p > 2$ is called a $p-$Laplacian or a nonlinear Laplacian. If we take $p = 2$, for the relation between $\mathcal{A}$, $\mathcal{H}$, and $\mathcal{E}$ that is between the Laplacian, we have the corresponding Hamiltonian density and the corresponding energy. The case with $p = 2$ has the direct connection to the standard kinetic energy term. Note that, for this discussion, it is sufficient that ψ is included in $W^{1,p}(\Omega)$, and it is not necessary for ψ to be a component of the Slater determinant.

In the case for many-particle situations, there is no need to introduce extraordinary treatment. Let the many-particle wave function be

$$\Psi = \sum_{l=1}^{N} \psi_l.$$

It is trivial that the energy becomes

$$\mathcal{E}_N(\Psi) = \frac{1}{p}\int_\Omega dr^3 \sum_{l=1}^{N} |\nabla_l\psi_l|^p,$$

where ∇_i acts on ψ_i, and the term in the effective Hamiltonian

$$\mathcal{A}_N(\Psi) = \sum_{l=1}^{N} -\nabla_l\cdot(|\nabla_l\psi_l|^{p-2}\nabla_l\psi_l)$$

follows. Here, the procedure to obtain a term in the effective Hamiltonian from a given energy consists only for the Gâteaux differential. So, the similar procedure is also valid if we assume many-particle situations. When we consider the linear situation $p = 2$, the term is equal to $\sum_{i=1}^{N}(-\Delta_i\psi_i)$, this corresponds to the homogeneous term of the typical type in the many-body Schrödinger equations. In the following, we do not discuss the many-particle situations explicitly. But, as previously shown, they can trivially follow; once the corresponding one-particle situations are well understood.

3.3.2. Nonlinear Interactions Depending on the Parameters $\tilde{t}_2$ and $\tilde{t}_3$

Again let $p \in \mathbb{Z}$ ($\mathbb{Z}$: a set of all integers), be an even number to satisfy $p > 2$. We consider the energy represented by

$$\mathcal{E}_{D,Z}(\psi) = \frac{1}{p}\int_\Omega dr^3 |\psi|^p, \tag{21}$$

where the corresponding Hamiltonian energy density is $\mathcal{H}_{D,Z}(\psi) = \frac{1}{p}|\psi|^p$. By minimizing this energy (i.e., the Gâteaux differential is considered), we obtain

$$\mathcal{F}_{D,Z}(\psi) = |\psi|^{p-2}\psi,$$

where this is one of the most typical ingredient for nonlinear Schrödinger equations. Indeed, for ψ and ϕ in $L^p(\Omega)$,

$$|\psi+\lambda\phi|^p - |\psi|^p = \int_0^1 \frac{\partial}{\partial t}|\psi+t\lambda\phi|^p dt = p\lambda\int_0^1 |\psi+t\lambda\phi|^{p-2}\cdot \mathrm{Re}\{(\psi+t\lambda\phi)\cdot\bar{\phi}\}dt \tag{22}$$

is true.

$$|\psi+t\lambda\phi|^{p-2}\mathrm{Re}\{(\psi+t\lambda\phi)\cdot\bar{\phi}\} \;\to\; |\psi|^{p-2}\mathrm{Re}(\psi\cdot\bar{\phi}) \quad \text{a.e.} \tag{23}$$

is valid due to $\lambda \to 0$ ($\lambda > 0$). In the same manner, we have

$$\begin{aligned}\tfrac{d}{d\lambda}\mathcal{E}_{D,Z}(\psi+\lambda\phi)\big|_{\lambda=0} &= \lim_{\lambda\to 0}\tfrac{\mathcal{E}_{D,Z}(\psi+\lambda\phi)-\mathcal{E}_{D,Z}(\psi)}{\lambda}\\ &= \int_\Omega dr^3\,|\psi|^{p-2}\mathrm{Re}(\psi\cdot\bar{\phi}) = \mathrm{Re}\int_\Omega dr^3\,|\psi|^{p-2}\psi\cdot\bar{\phi}.\end{aligned}$$

As a result, by taking into account the boundary condition: the boundary conditions shown in Sec. 3.3.1, we have

$$\tfrac{d}{d\lambda}\mathcal{E}_{D,Z}(\psi+\lambda\phi)\Big|_{\lambda=0} = \mathrm{Re}\int_\Omega dr^3\,|\psi|^{p-2}\psi\cdot\bar{\phi} = \mathrm{Re}\int_\Omega \mathcal{F}_{D,Z}(\psi)\bar{\phi}\,dr^3 = \mathrm{Re}(\mathcal{F}_{D,Z}(\psi),\phi). \tag{24}$$

Nonlinear Schrödinger equations that only include the interaction $\mathcal{F}_{D,Z}(\psi)$ with $p = 4$ are known as one of the most common type of nonlinear Schrödinger equations (cf. Ginzburg-Landau formalism).

Let us consider the energy $\mathcal{E}_{D,Q}$ when $p \in \mathbb{Z}$ in Eq. (21) is replaced by $q \in \mathbb{Q}$ ($\mathbb{Q}$: a set of all rational numbers). Nonlinear interaction with $q \in \mathbb{Q}$, is simply called a density-dependent force and is suggested to be indispensable in the explaination of the experimental results:

$$\mathcal{F}_{D,Q}(\psi) = |\psi|^{q-2}\psi,$$

where q satisfies $q > 2$. For example, if q satisfies "$(q-2)/2 = 1+1/6$" has been proposed as a possible candidate [1], as well able to satisfy $(q-2)/2 = 1+1/4$ [7]. Note here that the "density-dependent" force is a technical term for the force $\mathcal{F}_{D,Q}(\psi)$ satisfying $q \neq 4$. If we simply use the previous discussion shown in Eqs. (22) and (23), it is necessary to introduce the L^q-space of the fractional power. Here, it is not a problem to define L^q-space with noninteger q, and such L^q-space satisfying $1 \le q \le \infty$ holds the property of Banach spaces. Accordingly, the discussion shown in Eqs. (22) and (23) are valid even in this case, and we have

$$\tfrac{d}{d\lambda}\mathcal{E}_{D,Q}(\psi+\lambda\phi)\Big|_{\lambda=0} = \mathrm{Re}\int_\Omega dr^3\,|\psi|^{q-2}\psi\cdot\bar{\phi} = \mathrm{Re}\int_\Omega \mathcal{F}_{D,Q}(\psi)\bar{\phi}\,dr^3 = \mathrm{Re}(\mathcal{F}_{D,Q}(\psi),\phi). \tag{25}$$

3.3.3. Nonlinear Interaction Depending on the Parameter $\tilde{t}_4$

Again, let $p \in \mathbb{Z}$ ($\mathbb{Z}$: a set of all integers), be an even number satisfying $p > 2$. We consider the energy represented by

$$\mathcal{E}_{K,Z}(\psi) = \frac{1}{p}\int_\Omega dr^3|\nabla\psi|^{p-2}|\psi|^2,$$

where the corresponding Hamiltonian energy density is $\mathcal{H}_{K,Z}(\psi) = \frac{1}{p}|\nabla\psi|^{p-2}|\psi|^2$. By minimizing this energy (i.e., the Gâteaux differential is considered), we obtain

$$\mathcal{F}_{K,Z}(\psi) = \; -\tfrac{p-2}{p}\nabla\cdot(|\nabla\psi|^{p-4}|\psi|^2\nabla\psi) + \tfrac{2}{p}|\nabla\psi|^{p-2}\psi.$$

Indeed, for ψ and ϕ in $W^{1,p}(\Omega)$,

$$\begin{aligned}|\nabla(\psi+\lambda\phi)|^{p-2}|\psi+\lambda\phi|^2-|\nabla\psi|^{p-2}|\psi|^2 &= \int_0^1 \tfrac{\partial}{\partial t}\{|\nabla(\psi+t\lambda\phi)|^{p-2}|\psi+t\lambda\phi|^2\}\,dt\\ &= \lambda\int_0^1\big[(p-2)|\nabla(\psi+t\lambda\phi)|^{p-4}|\psi+t\lambda\phi|^2\cdot\mathrm{Re}\{\nabla(\psi+t\lambda\phi)\cdot\nabla\bar\phi\}\\ &\quad +2|\nabla(\psi+t\lambda\phi)|^{p-2}\cdot\mathrm{Re}\{(\psi+t\lambda\phi)\cdot\bar\phi\}\big]\,dt\end{aligned} \tag{26}$$

is true.

$$\begin{aligned}&(p-2)|\nabla(\psi+t\lambda\phi)|^{p-4}|\psi+t\lambda\phi|^2\cdot\mathrm{Re}\{\nabla(\psi+t\lambda\phi)\\ &\cdot\nabla\bar\phi\}\ \rightarrow\ (p-2)|\nabla\psi|^{p-4}|\psi|^2\mathrm{Re}(\nabla\psi\cdot\nabla\bar\phi)\quad \text{a.e.},\\ &2|\nabla(\psi+t\lambda\phi)|^{p-2}\cdot\mathrm{Re}\{(\psi+t\lambda\phi)\cdot\bar\phi\}\ \rightarrow\ 2|\nabla\psi|^{p-2}\mathrm{Re}(\psi\cdot\bar\phi)\quad \text{a.e.}\end{aligned} \tag{27}$$

are valid due to $\lambda\rightarrow 0$ ($\lambda>0$). We have

$$\tfrac{d}{d\lambda}\mathcal{E}_{K,Z}(\psi+\lambda\phi)\big|_{\lambda=0}=\mathrm{Re}\tfrac{1}{p}\int_\Omega dr^3\,\{(p-2)|\nabla\psi|^{p-4}|\psi|^2(\nabla\psi\cdot\nabla\bar\phi)+2|\nabla\psi|^{p-2}(\psi\cdot\bar\phi)\}.$$

If ψ and ϕ are sufficiently smooth,

$$\int_\Omega|\nabla\psi|^{p-4}|\psi|^2\nabla\psi\cdot\nabla\bar\phi\,dr^3=\int_{\partial\Omega}|\nabla\psi|^{p-4}|\psi|^2\frac{\partial\psi}{\partial\mu}\bar\phi\,dS-\mathrm{Re}\int_\Omega\nabla\cdot(|\nabla\psi|^{p-4}|\psi|^2\nabla\psi)\cdot\bar\phi\,dr^3.$$

As a result, by taking into account the boundary condition: the boundary conditions shown in Sec. 3.3.1, we have

$$\begin{aligned}\tfrac{d}{d\lambda}\mathcal{E}_{K,Z}(\psi+\lambda\phi)\big|_{\lambda=0}&=-\mathrm{Re}\int_\Omega dr^3\,\tfrac{p-2}{p}\nabla\cdot(|\nabla\psi|^{p-4}|\psi|^2\nabla\psi)\cdot\bar\phi+\mathrm{Re}\int_\Omega dr^3\,\tfrac{2}{p}|\nabla\psi|^{p-2}\psi\cdot\bar\phi\\ &=\mathrm{Re}\int_\Omega\mathcal{F}_{K,Z}(\psi)\bar\phi\,dr^3=\mathrm{Re}(\mathcal{F}_{K,Z}(\psi),\phi).\end{aligned} \tag{28}$$

Note that in this interaction which includes the differential operators cannot be bounded on $L^2(\Omega)$. If $p=4$, $\mathcal{F}_{K,Z}$ is reduced to

$$\mathcal{F}_{K,Z}(\psi)=\ -\tfrac{1}{2}\nabla\cdot(|\psi|^2\nabla\psi)+\tfrac{1}{2}|\nabla\psi|^2\psi.$$

Similar to Sec. 3.3.2, nonlinear interaction with $q\in\mathbb{Q}$ is derived from the energy $E_{K,Q}$, which is obtained by replacing $p\in\mathbb{Z}$ in $E_{K,Z}$ by $q\in\mathbb{Q}$.

$$\mathcal{F}_{K,Q}(\psi)=\ -\tfrac{q-2}{q}\nabla\cdot(|\nabla\psi|^{q-4}|\psi|^2\nabla\psi)+\tfrac{2}{q}|\nabla\psi|^{q-2}\psi.$$

where q satisfies $q>2$. If we simply use the previous discussion, it is necessary to introduce the $W^{1,q}$-space of the fractional power. Using the interpolation of Sobolev spaces, there is no problem to define $W^{1,q}$-space with noninteger q, and such $W^{1,q}$-space satisfying $1\le q\le\infty$ holds the property of Banach spaces. Accordingly, the discussion shown in Eqs. (26) and (27) are valid even in this case, and we have

$$\tfrac{d}{d\lambda}\mathcal{E}_{K,Q}(\psi+\lambda\phi)\big|_{\lambda=0}=\mathrm{Re}\int_\Omega dr^3\,|\nabla\psi|^{q-2}\psi\cdot\bar\phi=\mathrm{Re}\int_\Omega\mathcal{F}_{K,Q}(\psi)\bar\phi\,dr^3=\mathrm{Re}(\mathcal{F}_{K,Q}(\psi),\phi). \tag{29}$$

3.3.4. Nonlinear Interaction Depending on the Parameter $\tilde{t}_5$

The nonlinear term represented using the Laplacian operator is considered. We consider the energy represented by

$$\mathcal{E}_L(\psi)=\frac{1}{4}\int_\Omega dr^3|\psi|^2\triangle|\psi|^2,$$

where the corresponding Hamiltonian energy density is $\mathcal{H}_L(\psi) = \frac{1}{4}|\psi|^2\triangle|\psi|^2$. By minimizing this energy (i.e., the Gâteaux differential is considered), we obtain

$$\mathcal{F}_L(\psi) = (\triangle|\psi|^2)\psi.$$

Indeed, for ψ and ϕ in $W^{2,4}(\Omega)$,

$$|\psi + \lambda\phi|^2\triangle|\psi + \lambda\phi|^2 - |\psi|^2\triangle|\psi|^2 = \int_0^1 \frac{\partial}{\partial t}\{|\psi + t\lambda\phi|^2\triangle|\psi + t\lambda\phi|^2\}\, dt$$
$$= 2\lambda \int_0^1 \left[(\triangle|\psi + t\lambda\phi|^2)\, \mathrm{Re}((\psi + t\lambda\phi)\cdot\bar{\phi}) + \mathrm{Re}\triangle((\psi + t\lambda\phi)\cdot\bar{\phi})|\psi + t\lambda\phi|^2\right] dt$$

is true, where

$$(\triangle|\psi + t\lambda\phi|^2)\, \mathrm{Re}((\psi + t\lambda\phi)\cdot\bar{\phi}) \;\to\; (\triangle|\psi|^2)\, \mathrm{Re}(\psi\cdot\bar{\phi}) \quad \text{a.e.},$$
$$\triangle(|\psi + t\lambda\phi|\;\phi)|\psi + t\lambda\phi|^2 \;\to\; \triangle(|\psi|\phi)\;|\psi|^2 \quad \text{a.e.}$$

are valid due to $\lambda \to 0$ $(\lambda > 0)$. We have

$$\frac{d}{d\lambda}\mathcal{E}_L(\psi + \lambda\phi)\big|_{\lambda=0} = \frac{1}{2}\int_\Omega dr^3\; \{(\triangle|\psi|^2)\, \mathrm{Re}(\psi\cdot\bar{\phi}) + \mathrm{Re}\triangle(\psi\cdot\bar{\phi})\;|\psi|^2\}.$$

If ψ and ϕ are sufficiently smooth, by taking into account the boundary condition: the boundary conditions shown in Sec. 3.3.1, we have

$$\int_\Omega \triangle(|\psi|\;\phi)|\psi|^2\, dr^3 = -\int_\Omega \nabla(|\psi|\;\phi)\nabla|\psi|^2\, dr^3 = \int_\Omega \triangle|\psi|^2\, \mathrm{Re}(\psi\,\bar{\phi})\, dr^3.$$

Consequently,

$$\frac{d}{d\lambda}\mathcal{E}_L(\psi + \lambda\phi)\big|_{\lambda=0} = \mathrm{Re}\int_\Omega dr^3\; \{(\triangle|\psi|^2)\;\psi\cdot\bar{\phi}\} = \mathrm{Re}\int_\Omega \mathcal{F}_L(\psi)\bar{\phi}\, dr^3 = \mathrm{Re}(\mathcal{F}_L(\psi), \phi). \tag{30}$$

Note that in this interaction which includes the differential operators cannot be bounded on $L^2(\Omega)$.

3.3.5. Nonlinear Interaction Arising from Current (Depending on the Parameter $\tilde{t}_4$)

The momentum density $\boldsymbol{j}$, which is defined by $\boldsymbol{j} = -\frac{i}{2}(\bar{\psi}\nabla\psi - \psi\nabla\bar{\psi})$, plays a significant role more than one physical quantity to be utilized to describe the interaction. For a master equation: $i\partial_t\psi = (1/2)(-\Delta + V)\psi$ in a Hilbert space $L^2(\Omega)$,

$$\frac{\partial}{\partial t}(\bar{\psi}\psi) = (\partial_t\bar{\psi})\psi + \bar{\psi}(\partial_t\psi)$$
$$= \frac{1}{2}(i(-\Delta + V)\bar{\psi})\psi + \frac{1}{2}\bar{\psi}(-i(-\Delta + V)\psi$$
$$= \frac{i}{2}((-\Delta\bar{\psi})\psi - \bar{\psi}(-\Delta\psi)) + \frac{i}{2}((V\bar{\psi})\psi - \bar{\psi}(V\psi)).$$

If V is a self-adjoint operator in $L^2(\Omega)$, then $\frac{\partial}{\partial t}(\bar{\psi}\psi) = (i/2)((-\Delta\bar{\psi})\psi - \bar{\psi}(-\Delta\psi)) = (i/2)\nabla\cdot(\bar{\psi}\nabla\psi - \psi\nabla\bar{\psi})$ follows, and

$$\frac{\partial}{\partial t}(\bar{\psi}\psi) + \nabla\cdot\boldsymbol{j} = 0 \tag{31}$$

is obtained. This equation, which means the conservation of total particle density (represented by $\int dr^3|\psi|^2$), is known as the continuity equation. That is, the momentum density $\boldsymbol{j}$ plays a role of current.

We consider the energy represented by

$$\mathcal{E}_J(\psi) = \int_\Omega dr^3 \boldsymbol{j}^2 = \frac{(-i)^2}{4} \int_\Omega dr^3 (\bar{\psi}\nabla\psi - \psi\nabla\bar{\psi})^2,$$

where the corresponding Hamiltonian energy density is $\mathcal{H}_J(\psi) = -\frac{1}{4}(\bar{\psi}\nabla\psi - \psi\nabla\bar{\psi})^2$. By minimizing this energy (i.e., the Gâteaux differential is considered), we obtain

$$\mathcal{F}_J(\psi) = -2i\{2(\nabla\cdot\boldsymbol{j})\psi + \boldsymbol{j}\cdot\nabla\psi\}.$$

Indeed, for ψ and ϕ in $W^{1,4}(\Omega)$,

$$\begin{aligned}
&\{(\overline{\psi+\lambda\phi})\nabla(\psi+\lambda\phi) - (\psi+\lambda\phi)\nabla(\overline{\psi+\lambda\phi})\}^2 - (\bar{\psi}\nabla\psi - \psi\nabla\bar{\psi})^2 \\
&= \textstyle\int_0^1 \frac{\partial}{\partial t}\{(\overline{\psi+t\lambda\phi})\nabla(\psi+t\lambda\phi) - (\psi+t\lambda\phi)\nabla(\overline{\psi+t\lambda\phi})\}^2 dt \\
&= \textstyle\int_0^1 dt\ \big[2\{(\overline{\psi+t\lambda\phi})\nabla(\psi+t\lambda\phi) - (\psi+t\lambda\phi)\nabla(\overline{\psi+t\lambda\phi})\} \\
&\quad \tfrac{\partial}{\partial t}\{(\overline{\psi+t\lambda\phi})\nabla(\psi+t\lambda\phi) - (\psi+t\lambda\phi)\nabla(\overline{\psi+t\lambda\phi})\}\big] \\
&= 2\lambda \textstyle\int_0^1 dt\ \big[\{(\overline{\psi+t\lambda\phi})\nabla(\psi+t\lambda\phi) - (\psi+t\lambda\phi)\nabla(\overline{\psi+t\lambda\phi})\} \\
&\quad \{\bar{\phi}\nabla(\psi+t\lambda\phi) + (\overline{\psi+t\lambda\phi})\nabla\phi - \phi\nabla(\overline{\psi+t\lambda\phi}) - (\psi+t\lambda\phi)\nabla\bar{\phi}\}\big].
\end{aligned}$$

It follows that

$$\begin{aligned}
\tfrac{\mathcal{E}_J(\psi+\lambda\phi)-\mathcal{E}_J(\psi)}{\lambda} &= -\tfrac{1}{2}\textstyle\int_\Omega dr^3\ \int_0^1 dt\ \big[\{(\overline{\psi+t\lambda\phi})\nabla(\psi+t\lambda\phi) - (\psi+t\lambda\phi)\nabla(\overline{\psi+t\lambda\phi})\} \\
&\quad \{\bar{\phi}\nabla(\psi+t\lambda\phi) + (\overline{\psi+t\lambda\phi})\nabla\phi - \phi\nabla(\overline{\psi+t\lambda\phi}) - (\psi+t\lambda\phi)\nabla\bar{\phi}\}\big].
\end{aligned}$$

Due to $\lambda \to 0$ ($\lambda > 0$),

$$\begin{aligned}
&-\{(\overline{\psi+t\lambda\phi})\nabla(\psi+t\lambda\phi) - (\psi+t\lambda\phi)\nabla(\overline{\psi+t\lambda\phi})\} \ \to\ -\{\bar{\psi}\nabla\psi - \psi\nabla\bar{\psi}\} = 2i\boldsymbol{j} \quad \text{a.e.}, \\
&\{\bar{\phi}\nabla(\psi+t\lambda\phi) + (\overline{\psi+t\lambda\phi})\nabla\phi - \phi\nabla(\overline{\psi+t\lambda\phi}) - (\psi+t\lambda\phi)\nabla\bar{\phi}\} \\
&\ \to\ \bar{\phi}\nabla\psi + \bar{\psi}\nabla\phi - \phi\nabla\bar{\psi} - \psi\nabla\bar{\phi} \quad \text{a.e.}
\end{aligned}$$

are valid. If ψ and ϕ are sufficiently smooth, we have

$$\begin{aligned}
&\tfrac{d}{d\lambda}\mathcal{E}_J(\psi+\lambda\phi)\big|_{\lambda=0} = -\textstyle\int_\Omega dr^3\ (i\boldsymbol{j})\{(\nabla\psi)\bar{\phi} + \bar{\psi}\nabla\phi - (\nabla\bar{\psi})\phi - \psi\nabla\bar{\phi}\} \\
&= -\textstyle\int_{\partial\Omega} i(\boldsymbol{j}\cdot\boldsymbol{\nu})\bar{\psi}\phi\, dS + \int_{\partial\Omega} i(\boldsymbol{j}\cdot\boldsymbol{\nu})\psi\bar{\phi}\, dS \\
&\quad + \textstyle\int_\Omega dr^3\ \{-i\boldsymbol{j}(\nabla\psi)\bar{\phi} + i(\nabla\cdot\boldsymbol{j})\bar{\psi}\phi + i\boldsymbol{j}(\nabla\bar{\psi})\phi + i\boldsymbol{j}(\nabla\bar{\psi})\phi \\
&\quad - i(\nabla\cdot\boldsymbol{j})\psi\bar{\phi} - i\boldsymbol{j}(\nabla\psi)\bar{\phi}\} \\
&= -\textstyle\int_{\partial\Omega} i(\boldsymbol{j}\cdot\boldsymbol{\nu})\bar{\psi}\phi\, dS + \int_{\partial\Omega} i(\boldsymbol{j}\cdot\boldsymbol{\nu})\psi\bar{\phi}\, dS \\
&\quad + \textstyle\int_\Omega dr^3\ \{-2i\boldsymbol{j}(\nabla\psi)\bar{\phi} + 2i\boldsymbol{j}(\nabla\bar{\psi})\phi + i(\nabla\cdot\boldsymbol{j})\bar{\psi}\phi - i(\nabla\cdot\boldsymbol{j})\psi\bar{\phi}\} \\
&= -\textstyle\int_{\partial\Omega} i(\boldsymbol{j}\cdot\boldsymbol{\nu})\bar{\psi}\phi\, dS + \int_{\partial\Omega} i(\boldsymbol{j}\cdot\boldsymbol{\nu})\psi\bar{\phi}\, dS + \int_\Omega dr^3\ \{-4\mathrm{Re}(i\boldsymbol{j}(\nabla\psi)\bar{\phi}) \\
&\quad - 2\mathrm{Re}(i(\nabla\cdot\boldsymbol{j})\psi\bar{\phi})\},
\end{aligned} \tag{32}$$

where ν means the outward normal vector for the $\partial\Omega$. By taking into account the boundary condition: the boundary conditions shown in Sec. 3.3.1, the integral on the boundary surface cancels. Therefore

$$\tfrac{d}{d\lambda}\mathcal{E}_J(\psi+\lambda\phi)\big|_{\lambda=0} = \mathrm{Re}\textstyle\int_\Omega dr^3[-2i\{2\boldsymbol{j}(\nabla\psi)\bar{\phi} + (\nabla\cdot\boldsymbol{j})\psi\bar{\phi}\}] = (\mathcal{F}_J(\psi), \phi),$$

where it is very important to note that the second term on the right-hand side of the following equation:

$$(\mathcal{F}_J(\psi), \phi) = \mathrm{Re} \int_\Omega \frac{4}{i}\{\boldsymbol{j}(\nabla\psi)\bar{\phi} + (\nabla\cdot\boldsymbol{j})\psi\bar{\phi}\}\, dr^3 - \mathrm{Re}\int_\Omega \frac{2}{i}(\nabla\cdot\boldsymbol{j})\psi\bar{\phi}\, dr^3 \tag{33}$$

is missing in the standard formalism shown in the Appendix B of Ref. [2]. This term arises from the careful treatment of the integration by parts (see Eq.(32)). Note that in this interaction which includes the differential operators cannot be bounded on $L^2(\Omega)$.

3.3.6. Effective Hamiltonian

The corresponding effective Hamiltonian for the Hamiltonian density shown in Eq. (17) is obtained as

$$\begin{aligned}\mathcal{H}_{\rm eff}(\boldsymbol{r}) &= -\tilde{t}_1 2\triangle + \tilde{t}_2 4\rho + \tilde{t}_3 2(1+\tilde{\alpha})\rho^{\tilde{\alpha}}\\ &\quad +\tilde{t}_4\{-2\nabla\cdot(|\psi|^2\nabla\psi) + 2|\nabla\psi|^2\psi + 2i\{2\boldsymbol{j}(\nabla\psi) + (\nabla\cdot\boldsymbol{j})\psi\}\} + \tilde{t}_5 4\triangle\rho\\ &= 2[-\tilde{t}_1\triangle + 2\tilde{t}_2\rho + (2+\tilde{\alpha})\tilde{t}_3\rho^{1+\tilde{\alpha}}\\ &\quad +\tilde{t}_4\{-\nabla\cdot(|\psi|^2\nabla\psi) + |\nabla\psi|^2\psi + i\{2\boldsymbol{j}(\nabla\psi) + (\nabla\cdot\boldsymbol{j})\psi\}\} + 2\tilde{t}_5\triangle\rho],\end{aligned} \tag{34}$$

where we should pay attention to the coefficients with signs. Two points should be noticed: as shown in $\tilde{t}_1$, if we obtain the coefficient in the effective Hamiltonian as $-\hbar^2/2m$, the corresponding coefficient t_1 in the Hamiltonian density should be $\hbar^2/4m$ (it requires the modification of the standard choice of coefficients contained in the Hamiltonian density [13, 2]); it was already pointed out, it is necessary to subtract $2i(\nabla\cdot\boldsymbol{j})\psi$ (compared to the standard parametrization shown in the Appendix B of Ref. [2]) in order to have a complete set of terms arising from $\boldsymbol{j}^2$ in the Hamiltonian density. By this former point, it is useful to also remember that

$$\int_\Omega \tau dr^3 = \int_\Omega -(\triangle\psi)\bar{\psi}dr^3; \qquad \tau = |\nabla\psi|^2$$

does not follow from the variational principle, but from the integration by parts with a suitable boundary condition. Although the representation of the standard Hamiltonian density is not correct [13, 2], the standard effective Hamiltonian based on such a Hamiltonian density is exactly the same as the effective Hamiltonian obtained here (except for the terms arising from $\boldsymbol{j}^2$). On the latter point, the Galilean invariance has to be broken if the missing term is not included (for the Galilean invariance, see also the corresponding discussion in Sec. 3.2).

In this section the discussion has been developed in L^p spaces, where p is not necessarily equal to 2. Indeed, p ($\neq 2$) is necessary to consider nonlinear problems; e.g., for $\psi \in L^2(\Omega)$, $|\psi|^2$ is included in $L^1(\Omega)$ at least, and therefore it is not sure that $|\psi|^2\psi$ is included even in $L^1(\Omega)$. It is also important to note again, that $L^p(\Omega)$ are not Hilbert spaces, if p is not equal to 2. In such situations ($p \neq 2$), the inner product is not equipped, so that we cannot discuss the orthogonality of functions and so on. Here is the difficulty of considering nonlinear problems.

4. Remarks on the Additional Forces

4.1. The Coulomb Force

Only protons hold a charge, so that they interact by the Coulomb force. The charge is assumed to be equal to the probability distribution of the proton (cf. form factor). The Coulomb force consists of the direct and the exchange parts. The direct part of the Coulomb energy is represented by

$$E_C^{dir} = \frac{e^2}{2} \int \int dr_i^3 dr_j^3 \frac{\rho_p(\boldsymbol{r}_i)\rho_p(\boldsymbol{r}_j)}{|\boldsymbol{r}_i - \boldsymbol{r}_j|},$$

using the proton density, where the corresponding Hamiltonian density is equal to $\frac{e^2}{2} \int dr_j^3 \frac{\rho_p(\boldsymbol{r}_i)\rho_p(\boldsymbol{r}_j)}{|\boldsymbol{r}_i - \boldsymbol{r}_j|}$. On the other hand, the exchange part of the Coulomb energy is approximated by means of the Slater approximation [11] as

$$E_C^{ex} = -\frac{3e^2}{4}\left(\frac{3}{\pi}\right)^{1/3} \int dr_i^3 \rho_p(\boldsymbol{r}_i)^{4/3},$$

where the corresponding Hamiltonian density is equal to $-(3e^2/4)(3/\pi)^{1/3}\rho_p(\boldsymbol{r}_i)^{4/3}$. Note that the exchange part of the Coulomb force has the similar form as the term with the coefficient t_3, so that the treatment of obtaining the corresponding part of the effective Hamiltonian is similar to the cases with $E_{\tilde{t}_3}$.

Figure 2. An actual numerical calculation of the Coulomb interaction is presented in the coordinate space when the periodic boundary condition is imposed. A white-colored square means the computational cell. The solution is obtained by two fast Fourier transform operations in the enlarged region with periodic boundary condition. This treatment realizes the isolated charge distribution in the computational cell.

Let us have a short remark on the numerical calculation of the Coulomb force. Because of the long-range property of the Coulomb interaction, the periodic boundary condition is not necessarily appropriate, but the potential has to go to zero at infinity ("isolated charge distribution"). In practice this case is solved usually in one of two ways, either to obtain the boundary values from a multi-pole expansion or by Fourier techniques in embedding the computational cell in one of the large empty cells (Fig. 2). In the periodic case that is useful for astrophysical situations, the jellium approximation is used that corresponds to a constant background density of electrons that cancel the total charge.

4.2. Pairing Force

The pairing interaction is an important ingredient of quantum many-body systems. This interaction combines two fermions into one boson, so that condensation can take place as a new feature. For example, the superconductivity follows from the pairing interaction.

In order to introduce the pairing interaction to the density functional theory, it is necessary to have a functional representation of the pairing-interaction field. There have been two methods proposed to introduce the pairing interaction for many-nucleon systems [8]; one is the HFB (Hartree-Fock-Bogoliubov) approach, and the other is the BCS (Bardeen, Cooper, and Schrieffer) approach.

Following the review article [12], here we shall introduce the BCS type for the two pairing-field:

$$E_{\text{pair}} = \sum_q \frac{v_q}{4} \int dr^3 \, \chi_q(\boldsymbol{r})^2 \tag{35}$$

and

$$E_{\text{pair}}^{DD} = \sum_q \frac{v_{0,q}}{4} \int dr^3 \left[\left\{ 1 - \left(\frac{\rho(\boldsymbol{r})}{\rho_c} \right)^{\gamma} \right\} \chi_q(\boldsymbol{r})^2 \right], \tag{36}$$

where

$$\chi_q(\boldsymbol{r}) = \sum_{\hat{\alpha} \in q} w_{\hat{\alpha}} u_{\hat{\alpha}} v_{\hat{\alpha}} |\psi_{\hat{\alpha}}(\boldsymbol{r})|^2$$

denotes the pairing density with the phase-space weight $w_{\hat{\alpha}}$, occupation amplitude $v_{\hat{\alpha}}$, and non-occupation amplitude $u_{\hat{\alpha}} = \sqrt{1 - v_{\hat{\alpha}}^2}$. In Eqs. (35) and (36), v_q and $v_{0,q}$ are strength parameters, and ρ_0 is the nuclear saturation density, typically $\rho_0 = 0.16 \text{ fm}^3$. In contrast to Eq. (35), the Eq. (36) additionally includes the density dependence. The surface profile of the pairing interaction is controlled by the corresponding parameter γ, whose standard value is equal to 1. Note that the strong pairing takes place near the nuclear surface, so that Eq. (36) is expected to describe the pairing field better than Eq. (35). The corresponding Hamiltonian densities for Eq. (35) and Eq. (36) are equal to $\sum_q \frac{v_q}{4} \chi_q(\boldsymbol{r})^2$ and $\sum_q \frac{v_{0,q}}{4} [\{1 - (\frac{\rho(\boldsymbol{r})}{\rho_c})^{\gamma}\} \chi_q(\boldsymbol{r})^2]$, respectively. As readily shown, the treatment to obtain the corresponding parts of the effective Hamiltonian are similar to the cases with $E_{\tilde{t}_2}$ and $E_{\tilde{t}_3}$.

5. Summary

A whole process to derive an effective interaction in many-nucleon systems has been shown. This chapter presented a modeling for interacting many-nucleon systems. In particular, we have illustrated the appearance of both nonlinearity and differential operators in this formalism. In using the application of a variational principle, based on the functional analytic methods, we have presented a treatment of the Gâteaux differential in some generalized situations.

A phenomenological density-dependent force has been treated. The exchange part of the Coulomb force and the pairing force have fractional powers of density. Therefore, the density-dependent property is also true for both of these two forces. Such a fractional power dependence reasonably appears in physics when treating finite quantum systems. Indeed, as far as finite quantum systems are concerned, the existence of a ρ term in energy implies the emergence of a $\rho^{2/3}$ term in energy, due to the surface effects.

Two unknown features have been found. First, as shown in the first term on the right-hand side of Eq. (16), the coefficient of the kinetic energy part of the Hamiltonian density is equal to $\hbar^2/4m$. This fact requires the modification of the standard choice [13, 2] for coefficients contained in the Hamiltonian density. Second, as is shown in Eq. (33), we have pointed out that there is a missing term in the standard derivation of the effective interaction arising from $\boldsymbol{j}^2$ in the Hamiltonian density. This fact requires the modification of the standard choice [2] for coefficients contained in the effective interaction. The correct treatment of this term is necessary not only to hold the Galilean invariance, but also to clarify the time-odd contribution to the stationary and non-stationary states of many-nucleon system.

This work was supported by the Helmholtz Alliance HA216/EMMI. One of the authors (Y. I.) expresses his gratitude to Prof. Emeritus. Dr. Hiroki Tanabe (Department of Mathematics, Osaka University), who made many valuable comments with respect to the mathematical rigorous treatment of density functional.

6. Appendix -Differentiability

Let z be a notation for complex variable:

$$\frac{\partial}{\partial z} = \frac{1}{2}\left(\frac{\partial}{\partial x} - i\frac{\partial}{\partial y}\right), \quad \frac{\partial}{\partial \bar{z}} = \frac{1}{2}\left(\frac{\partial}{\partial x} + i\frac{\partial}{\partial y}\right).$$

We consider a function $f(z)$ that satisfies $f(z) = f(x,y)$ and $z = x + iy$. Here, let us assume that $f(x,y)$ is a differentiable function of two real variables (x,y). According to the Taylor's theorem,

$$f(z+\omega) - f(z) = f(x+\xi, y+\eta) - f(x,y) = \frac{\partial f(x,y)}{\partial x}\xi + \frac{\partial f(x,y)}{\partial y}\eta + \cdots,$$

where $\omega = \xi + \eta$ and

$$\frac{\partial}{\partial x} = \frac{\partial}{\partial z} + \frac{\partial}{\partial \bar{z}}, \quad \frac{\partial}{\partial y} = i\left(\frac{\partial}{\partial z} - \frac{\partial}{\partial \bar{z}}\right).$$

Therefore

$$\begin{aligned} f(z+\omega) - f(z) &= \left(\frac{\partial f(z)}{\partial z} + \frac{\partial f(z)}{\partial \bar{z}}\right)\xi + i\left(\frac{\partial f(z)}{\partial z} - \frac{\partial f(z)}{\partial \bar{z}}\right)\eta + \cdots \\ &= \frac{\partial f(z)}{\partial z}(\xi + i\eta) + \frac{\partial f(z)}{\partial \bar{z}}(\xi - i\eta) + \cdots \\ &= \frac{\partial f(z)}{\partial z}\omega + \frac{\partial f(z)}{\partial \bar{z}}\bar{\omega} + \cdots. \end{aligned} \tag{37}$$

In particular, f is Fréchet differentiable if $\partial f/\partial \bar{z} = 0$ (i.e., f is holomorphic). However, even though f is not holomorphic,

$$\lim_{\lambda \to 0} \frac{f(z+\lambda\omega) - f(z)}{\lambda} = \frac{\partial f(z)}{\partial z}\omega + \frac{\partial f(z)}{\partial \bar{z}}\bar{\omega}$$

is valid to real λ, if f is differentiable. In this situation it is possible to consider Gâteaux differential:

$$Df(z)(\omega) = \frac{\partial f(z)}{\partial z}\omega + \frac{\partial f(z)}{\partial \bar{z}}\bar{\omega}.$$

In particular, if f is real-valued, the Gâteaux differential is reduced to

$$Df(z)(\omega) = 2\mathrm{Re}\left(\frac{\partial f(z)}{\partial z}\omega\right), \tag{38}$$

where note that $\partial f(z)/\partial z$ and $\partial f(z)/\partial \bar{z}$ are complex-conjugate. In the case of Sec. (3.3.1), the right-hand side of Eq. (38) corresponds to

$$\frac{\lambda}{d\lambda}\mathcal{E}(\psi + \lambda\phi)|_{\lambda=0} = -\mathrm{Re}\int_{\Omega} dr^3 |\nabla\psi|^{p-2}\nabla\psi\nabla\bar{\phi},$$

but the limit ($\lambda \to 0$) cannot exist if λ is not real (on the other hand, λ should be complex with respect to finding the optimal condition in the complex Banach spaces). Indeed, according to Eq. (37), the limit of

$$\frac{\mathcal{E}(\psi+\lambda\phi)-\mathcal{E}(\psi)}{\lambda} = \frac{\partial\mathcal{E}(\psi)}{\partial\psi}\phi + \frac{\partial\mathcal{E}(\psi)}{\partial\bar{\psi}}\frac{\bar{\lambda}}{\lambda}\bar{\phi} + \cdots$$

due to $\lambda \to 0$ (λ is not real) exists only when $\mathcal{E}$ is holomorphic. Note that $\mathcal{E}$ shown in Sec. (3.3.1) is not holomorphic.

In summary, a mapping from a complex Banach space to a complex Banach space is a Fréchet differentiable, only if the function with complex variables is holomorphic. On the other hand, the real-valued functions are not holomorphic except for the constant. It provides a reason why we have lengthy treatments for the Gâteaux differential as demonstrated in Sec. 3.3.

References

[1] E. Chabanat, P. Bonche, P. Haensel, J. Meyer, and R. Schaeffer, *Nucl. Phys.* A635 (1998) 231; A643 (1998) 441(E).

[2] Y. M. Engel, D. M. Brink, K. Goeke, S. J. Krieger and D. Vautherin, *Nucl. Phys.* A249 (1975) 215.

[3] A. L. Fetter and J. D. Walecka, *Quantum Theory of Many-Particle Systems*, Dover Publications, 2003.

[4] W. Greiner and J. A. Maruhn, *Nuclear Models*, Springer-Verlag Berlin Heidelberg, 1996.

[5] P. Hohenberg and W. Kohn, *Phys. Rev.* B136 (1964) 864.

[6] W. Kohn and L. J. Sham, *Phys. Rev.* A140 (1965) 1133.

[7] P.-G. Reinhard, and H. Flocard, *Nucl. Phys.* A584 (1995) 467.

[8] P. Ring and P. Schuck, *The Nuclear Many-Body Problem*, Springer-Verlag Berlin Heidelberg, 2004.

[9] L. Schwartz, *Théorie des distributions 1 & 2*, Hermann, 1950 & 1951.

[10] T. H. R. Skyrme, *Nucl. Phys.* 9 (1959) 635.

[11] J. C. Slater, *Phys. Rev.* 81 (1951) 385.

[12] J. Stone and P.-G. Reinhard, *Rev. Mod. Phys.* 58 (2007) 587.

[13] D. Vautherin and D. M. Brink, *Phys. Rev.* C 5 (1972) 626.

[14] K. Yosida, *Functional Analysis*, Springer-Verlag Berlin Heidelberg New York, 1980.

INDEX

A

B

C

D

E

I

J

K

L

M

N

O

P

Q

R

T